International Conference on Mathematics and Natural Science (IConMNS 2017)

Journal of Physics: Conference Series Volume 1040

Bali, Indonesia
6 – 7 September 2017

ISBN: 978-1-5108-6411-5
ISSN: 1742-6588

Printed from e-media with permission by:

Curran Associates, Inc.
57 Morehouse Lane
Red Hook, NY 12571

Printed by Curran Associates, Inc. (2018)

For permission requests, please contact the Institute of Physics at the address below.

Institute of Physics
Dirac House, Temple Back
Bristol BS1 6BE UK

Phone: 44 1 17 929 7481
Fax: 44 1 17 920 0979

techtracking@iop.org

Additional copies of this publication are available from:

Curran Associates, Inc.
57 Morehouse Lane
Red Hook, NY 12571 USA
Phone: 845-758-0400
Fax: 845-758-2633
Email: curran@proceedings.com
Web: www.proceedings.com

Table of contents

Volume 1040

International Conference on Mathematics and Natural Sciences (IConMNS 2017)

6–7 September 2017, Bali, Indonesia

Accepted papers received: 15 May 2018
Published online: 4 June 2018

Preface

Papers

IOP Conf. Series: Journal of Physics: Conf. Series **1040** (2018) 011001 doi:10.1088/1742-6596/1040/1/011001

"Encountering Global Challenges through Innovation on Science, Technology, Engineering, and Mathematics (STEM), and Education"

Layout

Dr. I Wayan Puja Astawa, S.Pd, M.Stat.Sci

Putu Kartika Dewi, S.Pd, M.Sc

Ratih Ayu Apsari, S.Pd., M.Sc., M.Pd.

I Nyoman Budayana, S.Pd., M.Sc.

Cover Design

I Nyoman Laba Jayanta, M.Pd.

International Conference on Mathematics and Natural Sciences (IConMNS 2017) IOP Publishing
IOP Conf. Series: Journal of Physics: Conf. Series **1040** (2018) 011001 doi:10.1088/1742-6596/1040/1/011001

PROCEEDINGS

INTERNATIONAL CONFERENCE ON MATHEMATICS AND NATURAL SCIENCES (IConMNS) 2017

https://itembrowser.com/item.php?id=841

Denpasar, September 6th – 7th, 2017

IOP Conf. Series: Journal of Physics: Conf. Series **1040** (2018) 011001 doi:10.1088/1742-6596/1040/1/011001

COMMITTEE

Advisory Board

Prof. Dr. I Nengah Suparta, M.Si. (Ganesha University of Education, Indonesia)
Prof. Dr. Ida Bagus Jelantik Swasta, M.Si (Ganesha University of Education, Indonesia)
Dr. I Gusti Lanang Wiratma, M.Si. (Ganesha University of Education, Indonesia)
Dr. ir. W.T. van Horssen (Delft University of Technology, the Netherlands)
Prof. David James Young (Monash University, Australia)
Dr. Natthida Rakkapao (Prince of Songkla University, Thailand)
Dr. Komang Darmawan (Udayana University, Indonesia)
Prof. Dr. Ch. Rini Indrati (Gajah Mada University, Indonesia)
Prof. Dr. Don H. Faust (Northern Michigan University)
Prof. Hassan A. El-Morshedy (Damietta University, Egypt
Prof. Haydar Akca (Abu Dhabi University, United Arab Emirates)
Assoc. Prof. Murk Bottema (Flinders University, Australia)
Dr. Henk van der Kooij (Utrecht University, The Netherlands)
Dr. Agung Lukito, M.S. (The State University of Surabaya, Indonesia)
Edi Winarko, Ph.D. (Gadjah Mada University, Indonesia)
Dr. Wanty Widjaja (Deakin University, Australia)
Assoc. Prof. Dr. L. Michiel Doorman (Utrecht University, The Netherlands)
Prof. Dr. Sabine Fechner (Paderborn University, Germany)
Dr. Tran Ba Trinh (Hanoi National University Education, Vietnam)
Prof. Dr. Harrie M.C. Eijkelhof (Utrecht University, The Netherlands)

Scientific Committee

Drs. I Wayan Muderawan, M.S.,Ph.D.
Dr. rer. nat. I Wayan Karyasa, S.Pd., M.Sc.
Prof. Dr. Phil I Gusti Putu Sudiarta, M.Si.
Dr. rer. nat. I Gusti Ngurah Agung Suryaputra, S.T., M.Sc.
I Wayan Mudianta, S.Pd., M.Phil., Ph.D.
Dr. I Gede Aris Gunadi, S.Si., M.Kom.
Dr. I Nyoman Sukajaya, M.T.
Dr. I Wayan Sukrawarpala, M.Sc.
Prof. Dr. Ketut Suma, M.S.
Dr. Gede Iwan Setiabudi, S.Pd., M.Si.
Prof. Dr. Ni Putu Ristiati, M.Pd.

Organizing Committee

Prof. Drs. Sariyasa, M.Sc., Ph.D.
Ratih Ayu Apsari, S.Pd., M.Sc., M.Pd.
Dr. Dewa Ketut Sastrawidana, M.Si.
Dr. I Wayan Puja Astawa, S.Pd., M.Stat.Sci.
Dr. I Gede Ari Yudasmara, M.Si.
I Made Suarsana, S.Pd., M.Si.
I Made Pasek Anton Santiasa, S.Pd., M.Sc.
Prof. Dr. I Wayan Suastra, M.Pd.

IOP Conf. Series: Journal of Physics: Conf. Series **1040** (2018) 011001 doi:10.1088/1742-6596/1040/1/011001

PREFACE

Om Swastyastu, Assalamualaikum Wr. Wb., Namo Buddhaya, Shalom

First of all, let us praise and thank The God Almighty for His graces bestowed on us. Allow me to heartily welcome all participants to Bali. It is my pleasure and privilege to extend our warmest welcome to all participants of the International Conference on Mathematics and Natural Sciences (IConMNS). The theme of the conference is "Encountering Global Challenges through Innovation on Science, Technology, Engineering, Mathematics (STEM) and Education". The objective of the conference is to promote the exchange of ideas and presentation of the advancements of research in the area of mathematics, natural sciences, technology, engineering, and education.

The conference is expected to become a forum for researchers and academicians to discuss research and to establish collaboration between them. I hope that you enjoy the scientific atmosphere of the conference and that you can talk to each other and make friends to establish future collaborations.

The Faculty of Mathematics and Natural Sciences, Universitas Pendidikan Ganesha is proud to host this conference. The conference received 128 abstract submissions, of which 109 manuscripts are presented orally. Moreover, the conference is attended by 106 non-presenter participants. To all presenters and participants, thank for your enthusiastic responses to the conference. Your present in this conference is highly appreciated.

During the conference, there are four plenary talks and eight invited presentations by various experts. We are honoured to have distinguished keynote speakers: Prof. David James Young from University of the Sunshine Coast, Australia; Prof. Edward R T Tiekink from Sunway University, Malaysia; Dr. Wim T van Horssen from Delft University of Technology, the Netherlands; and Dr. Natthida Rakkapao from Prince of Songkla University, Thailand. The committee thanks all of you for having kindly and cordially accepted the invitation. The committee also thanks and appreciates all invited speakers from LPTK.

The committee appreciates with gratitude all parties for their support in various forms during the preparation for and the running of the conference. On behalf of the committee and the Faculty, I acknowledge the Rector of Universitas Pendidikan Ganesha for his endless commitment and financial support. To the Rector, who will officially open the conference, I would like to extend our gratitude and appreciation.

The committee has worked hard to organize this conference. This conference would not be possible without dedicated and coordinated efforts of the committee members. I thank all of you for making this a reality.

I wish you a fruitful conference and an exciting stay in Bali.

Om, Shanti, Shanti, Shanti, Om; Wassalamualaikum Wr. Wb.; Sadhu, Sadhu, Sadhu; Shalom

Denpasar, 6 September 2017
Chair,

Sariyasa

OPENING ADDRESS: RECTOR OF UNIVERSITAS PENDIDIKAN GANESHA

Honorable : Dean of Faculty of Mathematics and Natural Sciences
All Vice Rectors of UniversitasPendidikanGanesha
Keynote speakers, distinguished guests or participants of today's conference, ladies and gentlemen

Om Swastiastu
Assalamu'alaikumwarahmatullahiwabarakatuh
Good morning
May peace and God's blessing be upon us all

First of all, let us express our great gratitude to God the Almighty, because His Blessing has made us be in good condition or health, so that we all can be here today to attend *International Conference on Mathematics and Natural Sciences* (IConMNS). This conference is organized by the Faculty of Mathematics and Natural Sciences, UniversitasPendidikanGanesha, as an attempt to share information derived from the results of studies on sciences and technology conducted by researchers from various countries.

Distinguished Guests, Ladies and Gentlemen
It is realized that we nowadays exist in a period of time characterized by the development of sciences, especially the development of natural sciences, and technology, which happen exponentially. It should also be admitted that there is a close interrelation between sciences and technology. This can be seen from the fact that basic sciences are hadly required for the optimal development of technology. On the other hand, an effort of developing basic sciences greatly requires continuous support from technology. With regard to this point, wise people are always in agreement that **"Science without technology is fruitless and technology without science is rootless".**

An important thing that needs to be kept in mind is that in developing science and technology, mathematics plays very important role. Mathematics is considered as a tool and a communication language of science and technology. This quantitative nature of mathematics increases the predictive power and control of science. This can also be put in a different way that science provides much more exact responses, which enable to have more accurate and smarter problem solving. There is a possibility that mathematics enable science to undergo development from qualitative level of science to quantitative level.

For the sake of being able to follow the rapid development of science and technology, having adequate mastery of mathematics, science and technology becomes the need of every individual. It should also be noted that having a good knowledge in mathematics, science, and technology is getting more and more important in work places. This is due to the fact that there are quite a large number of job opportunities pre-requesting learning capacity, logical thinking, critical thinking, decision making, cooperation, and problem solving. Considering this phenomenon, all of the countries over the world, without exception Indonesia, pay tremendous attention to the development of mathematics, science, and technology, which can be conducted through researches and education in various levels.

Distinguished scientists, ladies and gentlemen,
Realizing the significances of the roles of mathematics, science, and technology, I as the rector of Ganesha University of Education would strongly support and highly appreciate the carrying out of International Conference on Mathematics and Natural Sciences (ICoMNS), presenting the theme:

IOP Conf. Series: Journal of Physics: Conf. Series **1040** (2018) 011001 doi:10.1088/1742-6596/1040/1/011001

"Encountering Global Challenge through Innovation on Science, Technology, Engineering, and Mathematics (STEM), and Education".

Through this conference, it is greatly expected that academicians and researches from different parts of the universe can mutually exchange experience, and expertise in the field of mathematics, sciences, technology, engineering and education. It is also hoped that there arises innovative ideas, the establishment of researches cooperation, and human resources sharing in terms of mathematics, sciences, technology, engineering, and education, both in national as well as international levels.

Honorable guest, ladies and gentlemen
It is obviously realized that today's conference would not be meaningful without the readiness and the presence of the key-note speaker coming from variety of countries and from distinctive discipline. My high appreciation is, there for, particularly directed to the key-note speakers for your tiring long journey to make this event successful. My great thanks and appreciation is also due to all presenters for the parallel sessions, who come from various countries, universities, and different scientific disciplines in mathematics, science, technology, engineering, and education, to present the result of your studies as well as your ideas which will enrich the outcome of today's conference. Thank you very much for your participation.

It is also expected that innovative ideas as to the result of the researches and mathematics, scientific, and technological education which will be presented in this international conference, will be able to give positive contribution to the improvement of the mastery of science and technology, and the improvement of the quality of mathematics, science, and technology education in various educational levels.

This international conference will not come into existence without the initiation of the Faculty of Mathematics and Natural Sciences, and without the hard work of the member of today's event committee. Consequently my thankfulness and appreciation are also devoted to the related faculty and the members of the committee. Thank you for your struggle and hard working to make this conference come true.

Ladies and Gentlemen.
That's all I could say in this special and exciting opportunity. I apologize if there is any inconvenient expression in my opening speech. I end my speech by saying
Om Santhi Santhi Santhi Om
Assalammualaikum Waramatulahiwabarakatuh
May God bless us all.

Denpasar, September 6th, 2017

Rector of Universitas Pendidikan Ganesha
I NYOMAN JAMPEL

As requested by the conference committee, I would like to declare that the International Conference on Mathematics and Natural Sciences, today Wednesday, 6 September 2017, is officially opened.

LIST OF PARTICIPANTS

The tolerance improvement of local soybean in waterlogging condition through the combination of irradiation and in vivo selection

TB Saputro, KT Purwani, VS Fatimah, EM Stevia and N Jadid

Identification of Microscopic Fungus in Gill of Skipjact Tuna (Katsuwonus pelamis L.) at Traditional Markets in Singaraja – Bali

I A P Suryanti, N L P M Widiyanti, S Tangguda, dan I A K Tridewi

Genotyping SNP Rs12255372 TCF7L2 Gene Using Three-Primer ARMS-PCR for Detection T2DM n Indonesian Batak Ethnic

Syamsurizal and H Kadri

Expression pattern of drought-responsive genes in burley tobacco under in vitro water deficit

N Jadid, E Estiasih, T B Saputro, K I Purwani, D Hidayati, E V Permatasari, W D Kurniawan

Characteristic of synthetic seeds from two medicinal plants (Moringa oleifera and Camellia sinensis)

W Muslihatin, N Jadid, T B Saputro, K I Purwani, C E S Himayani and A W Calandry

The effect of addition sucrose concentrations toward weight of Nata DE Lontar (*Borassus flabellifer*) Linn

N L P M Widiyanti, S Mulyadiharja, I N Sukarta, N W I Pradnyandari

Potential of yellowfin tuna catch in East Java-Indian Ocean based on length frequency and age distribution

D Hidayati, R Herlambang, N Jadid, N Nailis Sa'adah, N Maulidina and A P D Nurhayati

Cytotoxic activity of andrographolide in colon cancer through inhibition cox-2 by in silico study

N P L Laksmiani, K R Reynaldi, M I Widiastari, I P W Nugraha, I M K Suyadnya and R A I K Maharani

The inhibitory activity of peonidin purple sweet potato in human epidermal receptor-2 receptor (her-2) expression by in silico study

N P L Laksmiani, M I Widiastari and K R Reynaldi

The formation of mini illustrated dictionary of rare plants in the village forest of Penglipuran, Bangli, Bali

N Wijana and I G A N Setiawan

IOP Conf. Series: Journal of Physics: Conf. Series **1040** (2018) 011002 doi:10.1088/1742-6596/1040/1/011002

Peer review statement

All papers published in this volume of *Journal of Physics: Conference Series* have been peer reviewed through processes administered by the proceedings Editors. Reviews were conducted by expert referees to the professional and scientific standards expected of a proceedings journal published by IOP Publishing.

The tolerance improvement of local soybean in waterlogging condition through the combination of irradiation and in vivo selection

T B Saputro, K T Purwani, V S Fatimah, E M Stevia and N Jadid

Department of Biology, Faculty of Mathematics and Natural Sciences, Institut Teknologi Sepuluh Nopember. Jl. Arif Rahman Hakim, Keputih, Sukolilo, Surabaya, East Java 60111, Indonesia.

E-mail : trionobsaputro@bio.its.ac.id, trionobsaputro@gmail.com

Abstract. Soybean (*Glycine max* L.) is wellknown as a plant that posses an extra nutrition and many proteins and has a major role in improving people's nutrition status. The environmental factors are the main obstacle in soybean production. High rainfall frequency can lead the waterlogging phenomenon which potentially decrease the production rate of soybean. Crop improvement are the best solution to solve the problems. The purpose of this research was to determine the tolerance level and physiological response of irradiated soybean, Grobogan and Anjasmoro varieties in waterlogging condition. 60CGamma were used to irradiate the seeds with 25Gy, 50Gy, and 100Gy dosages. This research were conducted in vegetative stadia with 100%, 150%, 200% concentration of water that dispersed into medium and control as comparison. Based on the data, the irradiation has a complex responses in plant morphology. The irradiated plant height was not significantly different among all treatments. While the leaf's wide, root length, nitrogen in leaf, root nodules, chlorophyll content, fresh weight and dry weight are decline as the waterlogging concentration increased.

1. Introduction

Soybean (*Glycine max* L.) become the main food commodities that has long been cultivated in Indonesia after rice and corn. Soybean (Glycine max l.) currently not only positioned as a raw material for food industry, but also placed as raw material for non-food industry. Soybean (Glycine max l.) is one of the most important edible oil producers legumes plants because of its high nutritional value [1]. Soybean containing protein, oil, carbohydrates insoluble, carbohydrates soluble, the water level and various functional a material such as anthocyanin, isoflavon, saponin, and fibers food [1], [2] and [3].

The increased of demand were not followed by an increase in production rate. The average of soybean yearly needs as much as 2.2 million tons/year and will be increased every year. However, until 2016, soybean production in Indonesia were estimated to be difficult to improved from the range of 800-900 thousand tons. One of the main factors in decrement of soybean production is climate change including the prolonged periods of rainy season and distribution of precipitation that carries broad impact on various sectors of human life. The amount of water that exceeds the absorption average capacity of land could potentially causing waterlogging. The occurrence of waterlogging stress is defined when all the soils pores saturated with water and over capacity at least 20% of soils capacity [4].

Standing upon the conditions of water, the exceesive water can be classified into two, namely: (1) water-saturated condition (waterlogging) where only root crops is flooded, and (2) the condition of the whole part of the plant is flooded (complete submergence) [5]. The main problems in waterlogging stress is the lack of O2 on the submerged part of plants. This is a major factor that causes damage to the soybean crop, both physiological and physical.

The control in environmental condition of cultivation area can be simply conducted since has a complex factors and also cost consume [6; 7], so the tolerance characteristic of soybean againts waterlogging stress need to be improved. Recently, research on food crops especially on soybean tolerance to waterlogging stress is still limited. Furthermore, there is no source of tolerance genes to waterlogging stress in the soybean plant. Based on the background, we need to induce the variation by using a irradiation technique. Gamma irradiation has positive effect in improving growth of *Abelmoschus esculentus* L. Monech plant [8]. Irradiation will produce a random mutation, all the variants will be selected by using in vivo selection. The selected variant or surviving variants will be compared with its initial lines or original varieties. This information is important in development of resistant varieties of soybean to waterlogging stress.

2. Materials and Methods

This research were conducted in Laboratory of Plant Bioscience and Technology, Biology Department, Sepuluh Nopember Institute of Technology (ITS) Surabaya. Gamma rays was used and performed in BATAN.

Measurement of Field Capacity

Measurement of field capacity was aimed to determine the water volume as a linemark of the waterlogging treatment. Furthermore, planting medium weighed water saturated weight and dry weight. Water saturated weight was measured after no water dripped from polybags. Dry weight was measured after planting medium was dried in oven at 105° C until a constant weight was obtained.

Water requirement based on the field capacity is calculated by the following formula:

$$FC\ (\%) = \frac{Tb-Tk}{Tk} \times 100\%$$

FC: Field Capacity; Tb: Wet weight of soil; Tk: Dry weight of soil [9].

Preparation, irradiation and seedling of Seeds

The seeds of Grobogan and Anjasmoro varieties were obtained from Center of Bean and Tuber Crops Research Institute (BALITKABI), in Kendal Payak, Malang. Those two varieties were soaked for 6 hours using aquadest and then drained. The soybean seeds were then sterilized on the surface with 70% ethanol for 2 minutes and 1% NaOCl for a minute. Furthermore, seeds were placed on petridish and wrapped. The exposure of Gamma rays was performed at several levels of 0 Gy, 25 Gy, 50 Gy, and 75 Gy and 100 Gy at the Center of National Atomic Technology Agency (BATAN). Soybean seeds that have been exposed to Gamma rays are grown on prepared media, sowing until 2 leaves appear and then acclimatized for 7 days.

Waterlogging Selection

The soybean plants that have been acclimatized for 7 days are given the treatment of waterlogging stress according to the calculation obtained. The waterlogging was carried out for 14 days at all treatment levels by giving water into each polybag as much as the concentration determined based on the preliminary field capacity study results. Plants in each treatment were taken and cleaned from the rest of the soil using water and air drained. Furthermore, plants are inserted into a ziplock plastics and labeled. Several growth parameters were observed during and after the treatment.

Plant height (cm)

Measurement on plant height were carried out every 2 days for 14 days of waterlogging treatment. Soybean plants are measured from the base of the stem to the tip of the stem or the growth point.

Leaf area

Measurement of the leaf area was done after 14 days of waterlogging by gravimetric method, that is by drawing the leaf to the paper. The leaf area is calculated based on the weight ratio of the leaf replica to the total paper. The leaf samples used were taken from the same plant. The formula calculation is as follows:

$$LA = \frac{Wr\ (g)}{Wt\ (g)} \times PA$$

LA: leaf area; Wr: weight of paper replica (g); Wt: weight of total paper (g); PA:paper area [10].

Root parameters

Root length measurement were made after 14 days waterlogging from the base of the root until the tip of the longest root using a ruler. While, the observation of the root nodule was done after 14 days waterlogging stress by counting the number of root nodules on the root of each levels of waterlogging. The observation of adventitious root was performed after 14 days of waterlogging by counting the adventitious root that formed at the root of each level of waterlogging.

Wet weight and dry plant

The procedure of measuring the wet weight of the plant is cutting the plant based on each organ including leaves, stems, and roots, then weighted using analytical scales. The wet weight of the plants is the accumulation of the total number of all plant organs. While dry weight is done by drying the plants using the oven first until the constant weight and then weighed using an analytical scale. The dry weight obtained is the productivity of the plant.

Chlorophyll content

The measurement of chlorophyll content was conducted at each treatment of the waterlogging concentration. Soy leaves are weighed as much as 0,1 grams with analytical balance. Chlorophyll of soybean leaf extracted using 96% ethanol 10 ml and then macerated for 24 hours. The extract was filtered using whatman 42 filter paper and taken into cuvette and measured with UV-Vis spectrophotometre in 649 nm and 665 nm wavelength. The calculation formula as follow :

Total Chlorophyll Content = [20,0 x A649 + 6,10 A665]

3. Result and Discussion

Viability test needed after seeds completely irradiated in term to find the influence of irradiation of gamma rays to the ability of seed to germinate. The ability of seed germination is an important parameter related their resistance level to the environment condition. Seed viability is an ability of seeds to germinate demonstrated by the high rate of metabolic activity in germination and growth of seedling. Viability can be observed based on two categories i.e seedling ability and seedling rate. The results of testing the rate of germination can be seen in Table 1.

Table 1. Germination rate of soybean after irradiated with Gamma rays

Varieties	Irradiation Dosage (Gy)	Seedling percentage (%)	Seedling rate (day)
Anjasmoro	Kontrol	87,50	1,57
	25 Gy	87,50	1,33
	50 Gy	87,50	1,29
	75 Gy	87,50	1,95
	100 Gy	83,33	2,10
Grobogan	Kontrol	91,67	2,05
	25 Gy	91,67	2,14
	50 Gy	79,17	2,16
	75 Gy	79,17	2,16
	100 Gy	75,00	2,22

Seeds that have irradiated with gamma rays in 100 Gy doses showed the decrement in both Anjasmoro and Grobogan varieties. Anjasmoro were more stable in seed viability parameters eventhough indicate the decrease in the highest dosage. Seeds with treatment 100 doses Gy took more time to grow its radicle. The sensibility plant in the dose of radiation depending on plants type and even varieties in the same species. The maximum irradiated dose that can be given in soybean plant is not yet known. Hameed et al., 2008 reported that the germination percentage will be declined significantly on irradiated high doses ranged from 400 - 500 Gy and in 500 doses the germination percentage were declined significantly in chickpea plant [11]. High doses of irradiation will impede the entire germination by decreasing the seedling rate. Khawar et al., 2010 reported that the increase in irradiation dose of gamma rays on the several plant seed will negatively affect the germination [12]. But in low doses, irradiation will stimulate germination through a process for activation of RNA for protein synthesis [13].

Waterlogging stress can affect the plant growth including plant height, root length, root nodule, amount of adventif root, and chlorophyll. The elongation of plant stems caused by interactions between Giberellic acid (GA) and Ethylene. Rapid elongation process in internodal produced by Ethylene that induce the specific ratio between Giberelic acid (GA) and Absicic acid (ABA). Furthermore, in waterlogging there has been a decrease in the ABA indigenous and increase of GA [14]. The result show that in 25 and 50Gy plant able to elongate higher than the other dosages. Detail data about plant height can be shown in Figure 1.

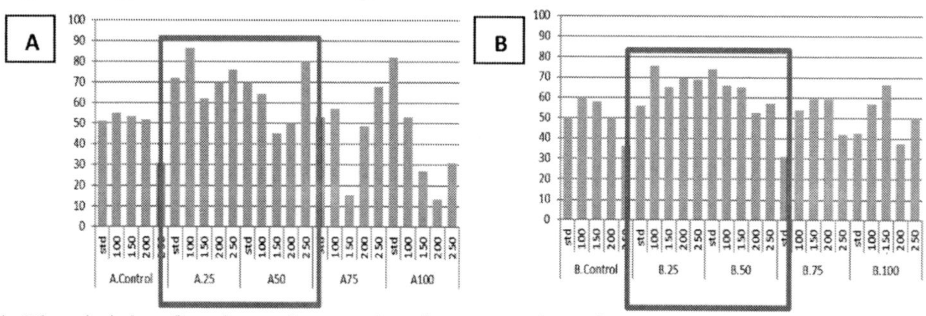

Figure 1. Plant height of soybeans in waterlogging stress A. Anjasmoro variety; B. Grobogan variety

Irradiation can increase the sensitivity of plants. According to Mudibu, et al., 2012, irradiation can improve the sensitivity of plant with reducing the number of Phytohormones, especially Cytokinins [15]. Irradiation can degrades mitochondria and the nucleus because the cell membrane and nucleus membrane is sensitive to the irradiation exposure, but Cytokinins that been accumulated in granules will not be degraded. The influence of water exceeds for vegetative stage were signed by the smaller leaves that can reduce the absorption of light, reduce chlorophyll and reduce the activity of some enzymes (e.g. nitrate reductase), the decline in plant productivity (biomass) due to the decline of primary metabolism, depreciation leaves wide and the photosynthesis.furthermore, waterlogging can affect the uptake of macro and several micronutrients N, P, K, Ca, Mg, S, Fe [16] . In addition, water also plays an important role in regulating cell turgidity that necessary for growth and cell enlargement. Those roles make a waterlogging stress causing decrement in metabolic processes directly or indirectly. The response of plants to waterlogging can be seen in metabolic activity, morphology, physiology and the growth level. The effect of waterlogging in fresh and dry weight of plant can be shown in Figure 2.

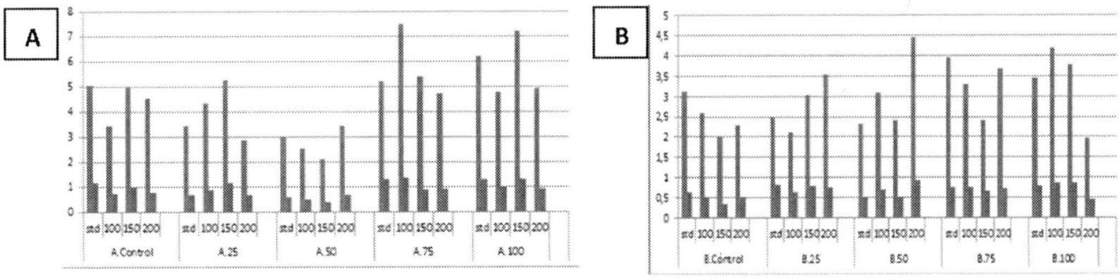

Figure 2. The fresh and dry weight of soybean in waterlogging stress 14 DAP, A. Anjasmoro variety; B. Grobogan variety; note : blue bars : fresh weight; red bars : dry weight.

Generally there are many physiological mechanism in plant that can improve its ability to stand or adapt with water exceeds condition. The concentration of several hormones also were decrease in responding the water exceeds condition [16]. In the otherhand, the ethylene production were increase. Ethylene can be the main mechanism in plant response to stress. Furthermore, ethylene can induce the formation of aerenchyma and also formation of adventif roots [17; 18]. Aeration stress can induce plant to held the plasticity mechanism. One of the main plasticity mode of plant can be shown in basal stem area, reforming that part to become unusual root (adventif root). The measurement of adventif root can be shown in Figure 3.

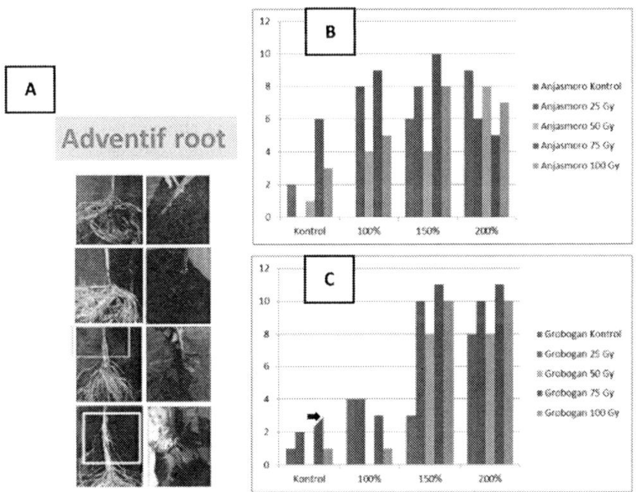

Figure 3. The adventif roots of plant in waterlogging conditions A. Adventif roots; B. Anjasmoro variety; C. Grobogan variety.

Waterlogging stress also affect the ratio of stomata that opens and closes and decrease of root physiological metabolism (roots permeability, root dry weight, the number of and effectiveness of roots nodules). In addition, the condition hypoxia or anoxia also affect in fixation and distribution of nitrogen and other minerals due to the dead roots and physiological drought that causes inhibition in root growth and nodulation. The Nitrogen content in soybean plant related to the formation of root nodes. Root nodules is a symbiotic mutualism between root cells to bacteria especially genus Rhizobium. This symbiotic relationship will produce nodules structure in roots that are mostly in plants legum.

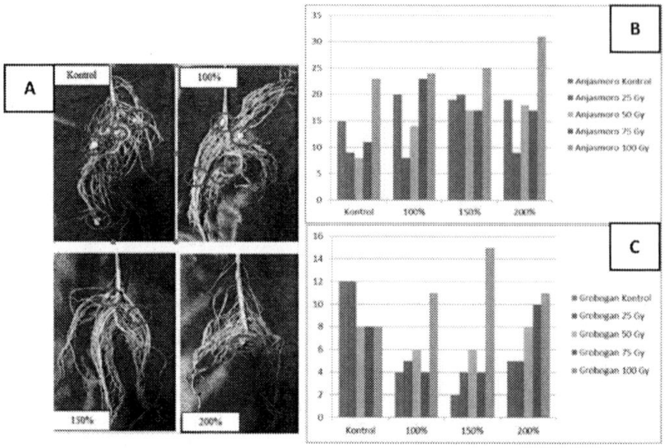

Figure 4. The amount of root nodules in waterlogging conditions A. Root nodules; B. Anjasmoro variety; C. Grobogan variety.

The result shows that the chlorophyll content were significantly decline along with the increase of water concentration in medium. Nitrogen serves in forming green pigmen substance in leaves. Hence, general visual symptoms of chlorophyll deficient is pale green or even pale yellow leaves. Nitrogen and Magnesium plays a vital role in determine the amount of chlorophyll in leaves. Figure 5 explain

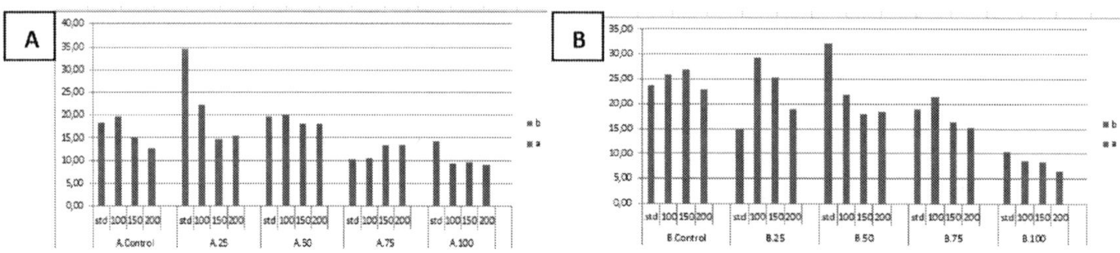

Figure 5. The chlorophyll content in waterlogging conditions A. Anjasmoro variety; B. Grobogan variety. Note : Red bars is chlorophyll A; Blue bars is Chlorophyll B.

4. Conclusion

Controlled experiments on vegetative stage of soybean plants conducted under greenhouse conditions confirmed results that the viability of seeds after irradiation were decrease, the significant decrease shown in 100 Gy dosage. The plant height and chlorophyl content shows that the optimum height were obtained in 25 and 50 Gy. while the amount of root nodules were decreased inline with the increasing of water concentration. The highest amount of adventif roots were obtained in 200% of water exceeds. All in all, Anjasmoro variety more tolerant compared to Grobogan variety in waterlogging. Further functional genomic in protein levels and transcriptomic analysis will be needed to evaluate the role of several genes involved in plant dynamic response to waterlogging stress.

Acknowledgements

We thank and grateful to Laboratory of Plant Bioscience and Technology, for supporting the equipments and greenhouse. Financial support from Sepuluh Nopember Institute of Technology (Grant No.771/PKS/ITS/2017) is gratefully acknowledged.

References

[1] Waqas M, Khan A L, Kang S M, Kim Y H and Lee I J 2014 Phytohormone-producing fungal endophytes and hardwood-derived biochar interact to ameliorate heavy metal stress in soybeans *Biol. Fertil. Soils.* **50** 1155–67

[2] Thomas J M G, Boote K J, Allen L H, Gallo-Meagher M and Davis J M 2003 Elevated temperature and carbon dioxide effects on soybean seed composition and transcript abundance *Crop Sci.* **43** 1548–57

[3] Bellaloui N, Hu Y, Mengistu A, Kassem M A and Abel C A 2013 Effects of foliar boron application on seed composition, cell wall boron, and seed δ15N and δ13C isotopes in water-stressed soybean plants *Front. Plant. Sci.* **4** 270

[4] Alam I, Lee D G, Kim K H, Park C H, Sharmin S A and Lee H 2010 Proteome analysis of soybean roots under waterlogging stress at an early vegetative stage *J. Biosci.* **35** 49–62

[5] Van Toai T T, Martin S K St, Chase K, Boru G, Schnipke V, Schmitthenner A F and Lark K G 2001 Identification of a QTL associated with tolerance of soybean to soil waterlogging *Crop Sci.* **41** 1247–52

[6] Chen X P, Cui Z L, Vitousek P M, Cassman K G, Matson P A and Bai J S 2011 Integrated soil–crop management system for food security *Proc. Natl. Acad. Sci. U.S.A.* **108** 6399–6404

[7] Ahmed F, Rafii M Y, Ismail M R, Juraimi A S, Rahim H A and Asfaliza R 2013 Waterlogging tolerance of crops: breeding, mechanism of tolerance, molecular approaches, and future prospects *Biomed. Res. Int.* **2013** 963525

[8] Hegazi A Z and Hamideldin N 2010 The effect of gamma irradiation on enchancement of growth and seed yield of okra (*Abelmoschus esculentus* L. Monech) and associated molecular changes *Journal of Horticulture and Forestry* 2 38-51

[9] Foth H D 1984 *Fundamentals of Soil Science, Sixth Edition* ed C E Millar and L M Turk (Canada: Jhon Willey and Sons, Inc.)

[10] Sitompul S M 1995 *Analisis Pertumbuhan Tanaman* ed Guritno B (Yogyakarta: Gadjah Mada University Press) 409-412

[11] Hameed A, Mahmud T S, Atta B M, Haq M A and Sayed H 2008 Gamma irradiation effects on seed germination and growth, protein content, peroxidase and protease activity, lipid peroxidation in desi and kabuli chickpea *Pak. J. Bot.* **40** 1033–41

[12] Khawar A, Bhatti I A, Khan Q M, Bhatti H N and Sheikh M A 2010 A germination test: an easy approach to know the irradiation history of seeds *Pak. J. Agri. Sci.* **47** 279-285

[13] Borzouei A, Kafi M, Khazaei H, Naseriyan B and Majdabadi A 2010 Effects of gamma radiation on germination and physiological aspects of wheat (*Triticum aestivum* L.) seedlings *Pakistan J Bot.* **42** 2281-90

[14] Kende H 1998 Deepwater rice: a model plant to study stem elongation *Plant Physiol.* **118** 1105–10

[15] Mudibu J, Nkongolo K K C, Kalonji-Mbuyi A and Roger V R 2012 Effect of gamma irradiation on morphoagronomic characteristics of soybean (*Glycine max* L.) *Am. J. Plant Sci.* **3** 331-337

[16] Akhtar I, Nazir N 2013 Effect of Waterlogging and Drought Stress in Plants International Journal of Water Resources and Environmental Sciences **2**(2): 34-40

[17] Rohmah E A and Saputro T B 2016 Analisis Pertumbuhan Tanaman Kedelai (Glycine max L.) Varietas Grobogan Pada Kondisi Cekaman Genangan. Jurnal Sains Dan Seni ITS **5**(2) 2337-3520

[18] Fatimah V S, Saputro T B 2016 Respon Karakter Fisiologis Kedelai (Glycine max L.) Verietas Grobogan terhadap Cekaman Genangan Jurnal Sains Dan Seni ITS **5**(2) 2337-3520

Identification of Microscopic Fungus in Gill of Skipjact Tuna (*Katsuwonus pelamis* L.) at Traditional Markets in Singaraja - Bali

I A P Suryanti,[1] N L P M Widiyanti[1], S Tangguda[1], dan I A K Tridewi[1]

[1] Universitas Pendidikan Ganesha, Singaraja-Bali, Indonesia

E-mail ayu.putu@undiksha.ac.id

Abstact. Skipjack tuna is a medium-sized fish from family of *Scombridae* (tuna) which is currently widely sold in traditional markets in Singaraja - Bali. The freshness of the fish in the markets is strongly influenced by microorganisms especially microscopic fungus that contaninated from market environment. The purposes of this research were determined the number of microscopic polyculture colonies on skipjack tuna that sold in traditional markets in Singaraja - Bali, and microscopic fungus found in the gill of skipjack tuna (*Katsuwonus pelamis* L.). This research is used a qualitative approach in the form of descriptive explorative by isolating and identifying microscopic fungi. The subjects of this research were skipjack fish from traders in three traditional markets spread in Singaraja – Bali, such as Anyar Market, Banyuasri Market and Banyuning Market. While the object of this research was gill of skipjack tuna. As much as one gram of skipjack tuna was introduced into the sterile distillation as the first dilution and was carried out until 10^{-6} dilution to calculate the number of microscopic policulture colonies per dilution. Furthermore, microscopic fungus suspected to have different morphological characteristics were isolated to find pure isolates. Identification of each fungus as macroscopically and microscopically was performed after the purification. The results of this research indicated that the average of total microscopic polyculture colony in Anyar market amounts to 5.3×10^5 cfu/ml; in the Banyuasri market of 2.2×10^5 cfu/ml and Banyuning market at 6.4×10^4 cfu / ml. There were 8 genera of microscopic fungus found in skipjack tuna in the three traditional markets in Singaraja city, namely Aspergillus, Penicillium, Rhizopus, Mucor, Trichoderma, Culvularia, Verticillium and Cladosporium. All the fungus that found as contaninant from the markets environment.

1. Introduction

Singaraja is one of the town that located in the northern part of Bali island. This town is very close to the beach. Singaraja's people are very interesting to eat food that comes from the sea, one of them is fish. It selected by people in singaraja because it contains high enough nutritional value. In some markets in singaraja city, the most popular fish for consumption is skipjack. One type of fish that is a marine biological resource in Bali is skipjack tuna that has a high economic value, both for local consumption and for export.

Based on the data of [5], The fish consumption for Bali province in 2010 to 2014 has increased 8.21%. The high interest of skipjack consumptions in Bali especially in Singaraja make the sellers in the market are doing various ways to extend the shelf life of the fish, ranging from preservation and so forth. But sometimes, bad handling and the condition of the markets cause skipjack's tuna

(*Katsuwonus pelamis* L.) decay very quickly. This is supported by the geographical location of Indonesia which includes tropical areas and has a high enough air humidity, making it suitable for the growth of microbes for food destruction. Process of decay in fish is generally caused by enzyme activity, chemical activity and microorganisms [7].

Body of the skipjack that susceptible to microorganisms is the gill. This section is a respiratory organ that binds oxygen and releases respirable carbon dioxide. In the gills of the skipjack is also have direct contact with the pollutants and the first organ directly related to the food consumed skipjack enters into the body, so this gill part becomes the main target of the growth of rotting microorganisms.

Microorganism is all small, invisible living things. One of the microorganisms such as microscopic fungus can be the cause of decay in skipjack. Most microscopic fungi are able to produce mycotoxins which are secondary metabolites that can damage the body of the fish so it becomes rotten (Sarjito, 2013). Based on the problems that have been described that the decay can occurs in skipjack is caused by microorganisms especially microscopic fungus from skipjack (*Katsuwonus pelamis* L.) on traditional markets in Singaraja, Bali.

2. Research Method

2.1 Tools and materials
Equipment used in this research is surgical board, autoclave, pipette, tweezers, incubator, analytical scales, glass object, needle ose, microscope, refrigerator, scissors, bunsen lamp, petri dish, test tube, measuring cup, laminary air flow and plastic bags, while the materials used in this research were gills of skipjack (*Katsuwonus pelanis* L.), Dextrose Agar (PDA), among others, potatoes, glucose (dextrose) and colorless agar, Media Malt Extra Agar (MEA), emersi oil, sterile distilled water, 70% and 96% alcohol.

2.2 Preparation of Skipjack Tuna Samples
Skipjack fish with an average body weight of 150-200 grams of which 9 samples were taken, sampling was conducted at three different locations representing the traditional market location in Singaraja town, Bali such as Anyar Market, Banyuasri Market and Banyuning Markets. Each of the market was taken for two fishes. The skipjack samples are taken randomly, and the test organ is gill. The sample was directly taken to Microbiology Laboratory, Department of Biology, F.MIPA, Ganesha University of Education (Undiksha) for examination of microscopic fungus to identification.

2.3 Isolation of Microscopic Fungus from Tuna Fish Samples
Isolation of fungi begins by taking gill samples and dissolved in sterile water as much as 90 ml to obtain a dilution rate of 10 times (10^{-1}). The sample was diluted further until the highest dilution rate was obtained by 10^{-6}. A 1 ml sample with a dilution rate of 10^{-1} to 10^{-6}. Afterthat it dispersed on the surface of the PDA medium, incubated at room temperature for 2-5 days. Then in the fifth day, count the colony and the suspected microscopic fungus colony was purified again on PDA and MEA media. Identification is done by looking at the macroscopic and microscopic forms of the Microscopic fungus and matched it by references [1].

2.4 Data Analysis
Data analysis of this research was done with qualitative approach. data on the number of microscopic fungus colonies in each dilution is made in tabular form, then the results were analyzed descriptively. the characteristics of each microscopic fungi isolate were also submitted in table form and described

3. Result and Discussion
The number of microscopic fungi colonies from skipjack (*Katsuwonus pelamis* L.) at traditional market in Singaraja, presented in table 1. form as follows.

Table 1. The Number of Microscopic Fungus Colonies from Gill of Cakalang (*Katsuwonus pelamis* L.) at Traditional Market in Singaraja Town, Bali

Sample	**Number of Polyculture Colonies of Microscopic Fungus (CFU/ml)**		
	Anyar Market (AM)	**Banyuasri Market (BM)**	**Banyuning Market (ByM)**
Skipjack 1	3.9×10^5	11×10^5	59×10^3
Skipjack 2	1.5×10^4	7.0×10^4	2.0×10^3
Skipjack 3	1.4×10^5	5.1×10^4	3.2×10^3
Total Number of Colonies	$\mathbf{5.3 \times 10^5}$	$\mathbf{2.2 \times 10^5}$	$\mathbf{6.4 \times 10^4}$

Based on table 1. it can be seen that the number of microscopic fungi colonies isolated from the skipjack tuna (*Katsuwonus pelamis* L.) per ml at traditional market in Singaraja Town. The highest number of microscopic fungi colonies in PA1 were 3.9×10^5 CFU / ml and the lowest number of microscopic fungi colonies were in PBy3 of 3.2×10^3 CFU / ml. The total number of microscopic fungi colonies from the three skipjacks (*Katsuwonus pelamis* L.) at Anyar Market is 5.3×10^5 CFU / ml, at Banyuasri market 2.2×10^5 CFU / ml and at Banyuning Market 6.4×10^4 CFU / ml.

Table 2. The Genus of Microscopic Fungus Isolated from the Gills of Skipjack (*Katsuwonus pelamis* L.) at Traditional Market in Singaraja Town.

Market	**Isolate**	**Result of Genus Identification**
1 (Anyar Market)	PA1	*Aspergillus*
	PA2	*Penicillium*
	PA3	*Penicillium*
	PA4	*Rhizopus*
	PA5	*Mucor*
	PA6	*Trichoderma*
	PA7	*Aspergillus*
	PA8	*Aspergillus*
2 (Banyuasri Market)	PBs1	*Penicillium*
	PBs2	*Curvularia*
	PBs3	*Verticillium*
	PBs4	*Cladosporium*
3 (Banyuning Market)	PBy1	*Aspergillus*
	PBy2	*Cladosporium*
	PBy3	*Penicillium*

The most common microscopic fungus genus is found in skipjack tuna in the three traditional markets namely the genus Aspergillus and Penicillium. The genus of Aspergillus can be found in Anyar Market and Banyuning Market, while Banyuasri Market is not found. The Penicillium genus can be found in all three markets. The genus Cladosporium can be found in the Banyuasri and Banyuning markets, whereas the Anyar Market is not found in this genus. The genera Rhizopus, Muchor and Trichoderma are found only in the Anyar Market, whereas in Banyuasri and Banyuning markets are not found these three genera. The genus Curvularia and Verticillium are not found in Anyar Market and Banyuning Market, but this genus is found in Pasar Banyuasri.

Table 3. Physical Parameters in Traditional Market in Singaraja Town

Market	Parameter		
	Air Humidity (%)	Light Intensity (Lux)	Temperature (^{0}C)
Anyar	71	240	25,4
Banyuasri	80	200	28,5
Banyuning	72	250	26,2

Based on table 3. it can be seen that each market where sampling has air humidity, light intensity and temperature are different, but the difference of this parameters were not significant. This condition could affect the condition of the skipjack samples, because the microbe that grew on the skipjack depends on the physical condition of each market and the condition of the skipjack itself.

Based on the research that has been done, it is known that the skipjack fish on Anyar Market has the highest number of microscopic fungus colony of 5.3 x 10^5 CFU / ml. This happens because it is influenced by several factors, including the state of the skipjack itself and the physical condition of the market when sampling is taken. According to the observations that have been made when sampling the skipjack fish, the atmosphere of Pasar Anyar is filled with many people who were buying and selling transactions and the location of selling fish traders close to landfills.

At Anyar Market, it was found that air humidity at the time of sampling that is 71% with light intensity 240 Lux and temperature at that time that is 25,4^{0}C. Such conditions may spur the growth of microscopic fungi, because low-grade fungus could grow optimally at room temperature and high humidity. It was in line with that proposed by [4], where micriscopic fungus generally required an environment with 70% -90% moisture to develop and optimum fungus temperature to be able to grow generally between 22^{0}C - 30^{0}C.

The second sampling location was at Pasar Banyuasri. The total number of polyculture microscopic fungi colonies in Banyuasri Market was less than in Anyar Market which is 2.2 x 10^5 CFU/ml. This occured because when sampling, the condition of the skipjack and market conditions were in normal condition. Based on the observations that have been made, that people who made buying and selling transactions at that time was not much. Handling sellers against skipjack was good. Skipjack is placed in a large box containing ice and closed, then the chances of microscopic fungus contamination from the environment against the skipjack tuna. So that, the number of polyculture of fungi and the number of pure isolates in skipjack fish was less than Anyar Market, It was in accordance with the proposed [7]. The amount of growth of a microbe was strongly influenced by human roles and activities. Through good and correct handling, it can suppress the population of harmful microbial growth. Handling is considered good by cold chain implementation, keeping fish at low temperatures, for example by using this ice is done to reduce the growth of microorganisms that can slow the process of decay in fish.

The third Sampling of skipjack (*Katsuwonus pelamis* L.) was in Banyuning Market. The total number of microscopic fungus colonies on the Banyuning market was at least in comparison with Anyar Market and Banyuasri Market at 6.4 x 10^4 CFU / ml. According to observations that have been done, this happens because when sampling, the condition of the skipjack and the condition of the market environment were in normal situation. Because at that time the market situation and skipjack buyers are not crowded. The fish sellers in Banyuning Market have done a good handling of skipjack.

Skipjack (*Katsuwonus pelamis* L.) at Anyar Market is indicated to be faster decayed by fungus, because the number of microscopic fungus colonies exceeds the maximum limit of microbial contamination in fresh fish that is 5 x 10^5 CFU / ml [11]. Meanwhile, skipjack fish at Pasar Banyuasri and Pasar Banyuning, the total number of microscopic fungus colonies were still below the maximum limit set by the Indonesian National Standard Board. So that the skipjack fish in these two markets have better quality than fish in Anyar Market.

The skipjack fishers mostly put the fish on the tray with the condition of the fish is not closed, so the fish are exposed to air that has been contaminated with microbes and cause the growth of microbes in large quantities and accelerate the occurrence of decay in fish, this is in accordance with the proposed [12], where the growth of microbial quantities and the process of fish quality decline is accelerated by poor handling or arrest, inadequate facilities and sanitation and limited distribution and marketing facilities.

The microscopic fungus that dominates growth in skipjack (*Katsuwonus pelamis* L.) on traditional markets in Singaraja City was the genus of Aspergillus and Penicillum. The genus Aspergillus is found in skipjack fish from Pasar Anyar and Pasar Banyuasri. According to [6], the Genus Aspergillus belongs to the phylum of Ascomycota, the order of Eurotiales, and the Trichocomaceae family. The genus Aspergillus generally has white, yellow or green base feathers with a thick brown to blackish conidiospora layer. This genus is found in nature as a saprophyte, grown in a tropical region with high moisture, its original habitat in the soil. The genus Aspergillus can also be found in cereals (maize, sorghum, rice, wheat) and various stored food items. Because in this market, fish's sellers located near from vegetable's sellers.

In this research also identified genus of Curvularia, Verticillium and Cladosporium fungi. These three genera are found in the Banyuasri Market and the Cladosporium genus is also found in Banyuning Market. The original habitat of these three fungi is generally on soil and grass. The Cladosporium genus can also be found in foodstuffs. .

Based on the exposure, it can be seen that of the 8 genera identified, the genus Aspergillus, Penicillium, Rhizopus and Mucor were fungal pathogen causes decay in skipjack (*Katsuwonus pelamis* L.) and mostly found in land. These four genera of fungi are capable of producing mycotoxins that cause rotting in skipjack. The genus Aspergillus capable of producing mycotoxins in the form of aflatoxins that attack the fish can cause fish to suffer physical damage until decay occurs. According Setiarto [11], Penicillium genus could produce mycotoxins in the form of okratoksin that can damage the fish body, causing decay in fish. According to [9] the genus Rhizopus and Mucor could cause zygomycosis and mucormycosis in fish, which can cause decay in fish because of physical, chemical, odor, color and texture damage to the fish. The genus of *Curvularia*, *Verticillium* and *Cladosporium* fungi did not produce mycotoxin compounds that endanger the skipjack, so this fungus plays a small role in the decay of fish [11].

4. Conclussion

Based on the results of the research, data analysis and discussion has been presented, it can be drawn conclusion as follows.

1) The total number of microscopic fungi colonies isolated from the skipjack tuna (*Katsuwonus pelamis* L.) on the Anyar Market market of 5.3 x 10^5 CFU / ml, at Banyuasri Market of 2.2 x 10^5 CFU / ml and on Banyuning Market 6.4 x 10^4 CFU / ml.
2) The total number of genus of microscopic mushroom isolated from gill of skipjack (*Katsuwonus pelamis* L.) is 8 genus of fungi. The eight genera of the fungus are, the genus Aspergillus, Penicillium, Rhizopus, Mucor, Trichoderma, Curvularia, Verticillium and Cladosporium. All the fungus that found as contaninant from the markets environment

References

[1] Cappucino J G and Sherman N 2013 *Manual Laboratorium Mikrobiologi Edisi 8* (Jakarta: EGC Medical Book Publishers)
[2] Dwidjoseputro D 1990 *Basics of Microbiology* (Jakarta: Djambatan)
[3] Gandjar I, Wellyzar S and Ariyanti O 2006 *Basic of Micology and Applied* (Jakarta: Yayasan Obor Indonesia)
[4] Gandjar I, Robert A S, Tweel-Vermeulen K V, Ariyanti O and Iman S 1999 *Introduction to General Tropical Fungi* (Jakarta: Yayasan Obor Indonesia)

[5] Marine and Fisheries Ministry 2014 *Fishing Cultivation Indonesia* Directorate General of Aquaculture Marine and Fisheries Ministry Available at http://djpb.kkp.go.id Retrieved on 6 April 2017
[6] McKane L 1996 *Microbiologi Applied & Practice* (New York: Mc-Graw. Hill Book Company)
[7] Murachman 2006 *Fish Handling* (Malang: Faculty of Fisheries and Marine Science Brawijaya University)
[8] Murniyati A S and Sunarman 2000 Pendinginan, Pembekuan dan Pengawetan Ikan (Yogyakarta: Kanisius)
[9] Noble E R and Noble G A 1989 *Parasitologi: Biologi Parasit Hewan Edisi kelima* (Yogyakarta: UGM Press)
[10] Sarjito, Prayitno S B and Haditomo A H C 2013 *Introduction to Parasites and Fish Disease* (Semarang: Diponegoro University)
[11] Setiarto R H B 2010 *Mycotoxins In Food* Biology Research Center Indonesian Institute of Sciences Available at http://www.biologi.lipi.go.id/bio_indonesia/ Retrieved on May 25, 2017
[12] SNI 7388-2009 *Maximum limit of microbial contamination in food* (Jakarta: Badan Standar Nasional Indonesia)
[13] Widiastuti I 2010 *Analysis of Fish Quality During Take Off Capture on Differences Preparation and Storage Time* (Bogor: IPB)

Genotyping SNP Rs12255372 TCF7L2 Gene Using Three-Primer ARMS-PCR for Detection T2DM n Indonesian Batak Ethnic

Syamsurizal[1] and H Kadri[2]

[1]Universitas Negeri Padang
[2]Universitas Andalas

E-mail: syam_unp@yahoo.co.id

Abstract. Genetic markers may indicate the increasing of individuals' susceptibility towards type 2 diabetes mellitus. The purpose of this research is to construct a primary SNP rs12255372 variant TGCCCAGGAATAAGGCAAGAATTCC (G/T) ATACCATATTCTGAATTACTCAGGC in the TCF7L2 gene and to determine the primary ability to detect polymorphisms of SNP rs12255372 variant TCF7L2 gene as a genetic marker. A case control study was used as the research method. Samples (62 people) were taken from the people of Batak ethnic with diabetes mellitus type 2 who came to the Metabolic Endocrinology clinic of Dr. M. Djamil Hospital. Controls were taken from healthy people without type 2 diabetes mellitus (62 people). EDTA blood was taken for DNA extraction. Polymorphisms of SNP rs12255372 in the TCF7L2 gene were detected by using three primer ARMS-PCR method and direct DNA sequencing method, and then were Analysed by bioinformatics. Based on the data analysis, it can be concluded that it has successfully constructed three primers ie RS12F forward primer, reverse primer and forward primer RS12R RS12C. All constructed primers are able to recognize SNP rs12255372 in TCF7L2 gene with three primer ARMS-PCR method, however, the SNP rs12255372 is genetic markers of type 2 diabetes mellitus for the Indonesian Batak ethnic.

1. Introduction

Worldwide, diabetes kills more humans than HIV/AIDS. Estimated number of people died of diabetes mellitus in 2000 to reach 6% (3.2 million people). One in ten people in the world died at the age of 35-64 years is a history of diabetes mellitus [1]. Every 10 seconds a person dies from complications of diabetes mellitus and at the same time discovered two new persons with diabetes mellitus [2].

Patients with diabetes mellitus in the world each year has increased, including in Indonesia and West Sumatra. The prevalence of diabetes mellitus in the world in 2000 amounted to 2.8% (171 million) and projected in 2030 by 4.4% (366 million people). Estimated number of people with diabetes mellitus in Indonesia in 2000 amounted to 4.1% (8.4 million Indonesian population of 205 132 000 votes). Projected in 2030 the number of cases of diabetes mellitus in Indonesia will increase by 7.8% (21.3 million out of 273 219 200 people). Estimated number of people with diabetes mellitus in Indonesia ranked fourth after India, China and America [2].

In clinical diabetes mellitus can be divided into four types, types I, II, gestational and other types. Diabetes mellitus type 2 is the most common type found 95% [3]. Diabetes mellitus type 2 occurs because the hormone insulin in the blood does not work effectively, although the amount of insulin produced by the beta cells of normal pancreatic islets of Langerhans. Glucose entry into the cell so that the cell is reduced so that the shortage of energy resources increased blood glucose [4]. Diabetes mellitus type 2 is influenced by several factors as follows: family history of diabetes, obese, risky *lifestyles,* lack of rest, and stress [4, 5].

Diabetes mellitus type-2 will appear in a person with a genetic defect after a genetic change in a long time. Acceleration and deceleration process of genetic change is highly dependent on environmental factors that influence it. If the genetic factors do not develop towards deterioration due to environmental factors, then theoretically diabetes mellitus type 2 will not come to the surface. Abnormalities or genetic abnormalities at an early stage without any symptoms that are clinically difficult to recognize [6]. Developing genetic markers toward deterioration but has not led to impaired sugar tolerance (IGT) can be detected through DNA analysis. DNA analysis is required to perform genetic data in the form of genes associated with type-2 diabetes mellitus. Some nations in the world already has a gene bank for diabetes mellitus type 2 such as the Caucasus, Denmark, USA, Britain, France and India [7].

Among the genes associated with type 2 diabetes mellitus are genes' transcription factor 7 like 2 (TCF7L2) "on chromosome 10q. TCF7L2 gene strongly associated with type 2 diabetes mellitus in ethnic Danish, Caucasian, Indian, and ethnicity of nations in Asia [7, 8]. TCF7L2 gene variant may be candidates for genetic markers in Indonesian Batak ethnic people with diabetes mellitus type 2.

One new surprises found in the *Human Genome Project* is the *single nucleotide polymorphisms* (SNPs). SNPs are minor elel the existence of more than 1%. If SNPs occur at the *gene coding regions* can lead *Synonymous* (does not cause amino acid changes) or *non Synonymous.* But in the last few years of research SNPs *Synonymous* led to the evolution that encourage the occurrence of a disease [9]. SNPs *Synonymous* can alter the structure, function, expression of the protein. Polymorphisms *Synonymous* can cause RNA splicing, stability and structure of proteins can be damaged. These changes may cause significant effects on protein function, changes in cellular response. *Single nucleotide polymorphisms* (SNPs) are DNA sequence variations that may be associated with susceptibility to a disease just as diabetes mellitus type 2. Most SNPs are non-coding region which is the basis of genetic variation in humans and refers to a single base differences between individuals [10].

The marker or haplotype is right will provide an indication of individual susceptibility to type 2 diabetes mellitus. Embodiment increased susceptibility is characterized by a relative risk of at least 1.2-1.4. TCF7L2 gene variant most strongly suspected association with diabetes mellitus type 2 is a variant rs12255372. Sequent rs12255372 is TGCCCAGGAATAAGGCAAGAATTCC (G/T) ATACCATATTCTGAATTACTCAGGC. The presence of the T allele in rs12255372 is indicative of increased susceptibility to diabetes mellitus type 2 [11].

The absence of genetic data on association the rs12255372 variant of the gene TCF7L2 with ethnic Batak people with diabetes mellitus type 2, it becomes its own interest to be investigated. In this research, a method of tracking the SNP rs12255372 with three primer ARMS-PCR method and DNA Direct Sequencing to identify TCF7L2 gene polymorphism in patients with type 2 diabetes mellitus ethnic Batak. *The purpose of research is to develop an early warning system with type 2 DM molecular fast, accurate so that it can assist in the prevention or treatment of type 2 DM in ethnic Batak. Specific target research are: primary construct for variants rs12255372. in genes TCF7L2 knowing the primary ability to detect polymorphisms TCF7L2 gene variant rs12255372.*

2. Method

This study is a descriptive study, in which researchers describe the results of the primary construction and confirmation of the primary ability to amplify the desired region. The research was conducted in the laboratory of Biomedical Universitas Andalas and the laboratory of Biotechnology Universitas Negeri Padang, West Sumatera, Indonesia.

The tools used in this study is water bath, engine thermo cycler *(Polymerase Chain* Reaction), micro centrifuge, micropipette, a tube eppendorf, microtube, vortex, chamber, tube rack micro, power supply, magnetic stirrer, tip, loop, LAFC, PCR tube, camera , Materials used are dNTP's, taq polymerase mL, 10X buffer, $MgCl_2 H_2O$, agarose, TAE IX, red gel, aquabidest, tris-base, EDTA, glacial acetic acid, 100bp DNA ladder.

In this study the DNA is derived from human peripheral blood. These isolates are needed to examine whether primary, constructed to work amplifying the DNA fragment of interest. Data were analysed qualitatively, the analysed data is the result of the primary construction and the primary ability to amplify the desired region.

Construction Primer. Primers used to detect SNP rs12255372 on gene TCF7L2 with ARMS-PCR method is constructed using computer software "primary designer". Will produce three results of primary construction RS12F forward primer, reverse primer and forward primer RS12C RS12R. Primary RS12F, RS12R reverse primer used to amplify DNA that includes the region ± 838 bp (hereinafter referred to as the external primer). Primary RS12C and RS12R used to fragment size and ± 384 bp, an area that includes the SNP rs12255372 (called internal primer). TCF7L2 gene sequences which will be used for the primary construction was obtained from *gene* NCBI bank. Possibility *mis prime* then the next primary construction results ready to be synthesis.

Confirmation is done using *software* to look at the possibility of *misprime* a primary with other regions in the gene TCF7l2 other than the area to be amplified. If not found zed into the oligonucleotide primer. The primary ability to amplify the desired region. Do the sequence of events as follows: Isolation of DNA using a kit from Invitrogen. Furthermore, the isolated DNA electrophoresis.

Amplification by ARMS-PCR method. DNA obtained from the insulation, then amplified using primers constructed with PCR mix RTG / Go Green Tag. To determine the results of amplification, electrophoresis on a 1.5% agarose gel. The steps performed during the study: 1) Isolation of DNA from the sample. 2) Design of primers for genes TCF7L2 primary use software designer. 3) Optimization of PCR reactions using the primer design results. 4) TCF7L2 gene amplification by PCR. 5) ARMS-PCR and sequencing for analysis of polymorphic sites. 6) bioinformatics analysis

3. Result and Discussion

One of the things that is very important in the PCR reaction is DNA constructs or appropriate primary election. Primary responsibility to recognize and mark theDNA segment *template* to be amplified. In this study generated three primary pieces that RS12F forward primer, reverse primer and forward primer RS12C RS12R. Primary RS12F, RS12R reverse primer used to amplify DNA that includes the region ± 838 bp (hereinafter referred to as the external primer). Primary RS12C and RS12R used to fragment size and ± 384 bp, for more details see Table 1.

Table 1. Results of the primary reconstruction RS12C

Sequence:	5'- **GGAATAGCCAGGCAAGAATG**-3'		
Criteria	Criteria setting	Result	
% GC	Min 50, Max 60	50	YES
Tm C	Min 55, Max 80	70	YES
No Hairpins	Energy cut off 0.0 kcal	-	YES
No 3' Dimers	Reject >= 3 matches at the end 3'	2	YES
No Dimers No	Reject >= 7 homologous base	3	YES
Runs	Reject >= 3 base runs	2	No
No 3'GC runs	Reject >= 3 G or C at the end 3'	1	YES

Table 2. The results of the primary construction and RS12RRS12F

	Sequence:	5'- TGTCTAATTGCCACAGCAGC -3'		
	Criteria	Criteria setting	Result	
Primer RS12F	% GC	Min 50, Max 60	50	YES
	Tm C	Min 55, Max 80	70	YES
	No Hairpins	Energy cut off 0.0 kcal	-	YES
	No 3' Dimers	Reject >= 3 matches at the end 3'	2	YES
	No Dimers No	Reject >= 7 homologous base	4	YES
	Runs	Reject >= 3 base runs	2	YES
	No 3'GC runs	Reject >= 3 G or C at the end 3'	2	YES
	Sequence:	5'- CAGAGGTGGTGATAAGCGGT -3' (Complementary strand)		
	Criteria	Criteria Setting	Result	
Primer RS12R	% GC	Min 50, Max 60	55	YES
	Tm C	Min 55, Max 80	70	YES
	No Hairpins	Energy cut off 0.0 kcal	-	YES
	No 3' Dimers	Reject >= 3 matches at the end 3'	1	YES
	No Dimers No	Reject >= 7 homologous base	2	YES
	Runs	Reject >= 3 base runs	2	YES
	No 3'GC runs	Reject >= 3 G or C at the end 3'	0	YES

Specificity primary construction made subsequently confirmed by the soft ware. This is done to avoid the possibility of mis prime a primary with other regions in the gene TCF7L2 other than the region to be amplified [12]. It can be seen that the position of the annealing rs12C are DNA sequences of TCF7L2 103 894. Annealing at that position as predicted earlier that the internal primary rs12C will identify areas that experienced the SNP. Theoretically, *annealing* primers 12C will start from position 103 894 and there were no possibility mis prime. The position and size of the tape primer / DNA fragments formed relatively can be seen in Figure 1.

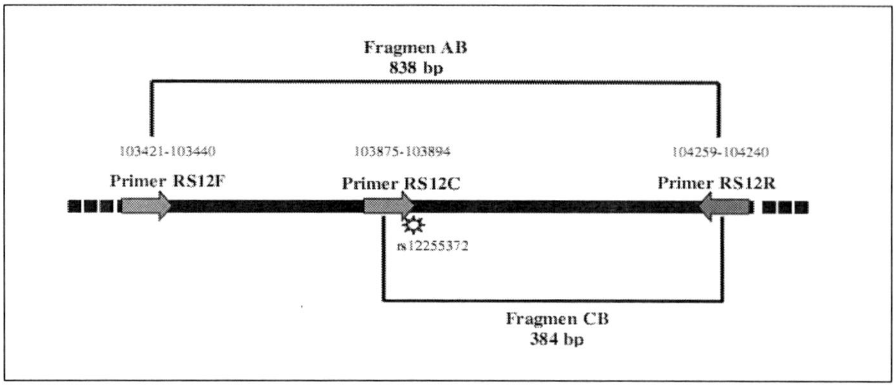

Figure 1. The position of primer and the amount of ribbon/DNA fragments formed

Determine the ability of primary constructed in detecting SNPs in the gene TCF7L2 particular SNP rs12255372, then was examined by PCR. The principle of PCR is multiplying exponentially a specific nucleotide sequence in vitro. In order to identify sequences that will be multiplied needed a special and specific primer. The primary area known is what will be multiplied by thousands or even millions of copies, about 10 6-10^7 times so as to be visible band after electrophoresis of the amplified DNA.

The initial phase of the primary confirmation is carried out separately in accordance with the conditions of each primer pair. Factors that must be considered in obtaining optimum results in the PCR is the amount / concentration mix used. Each of these components has a very important role in a PCR reaction. The composition of the enzyme, *template,* dNTPs, $MgCl_2$ buffer and appropriate primers determine the success of a PCR reaction. DNA obtained from the insulation, then electrophoresed on a 1.5% agarose gel to confirm the successful isolation of DNA from the sample. For more details, see Figure 2.

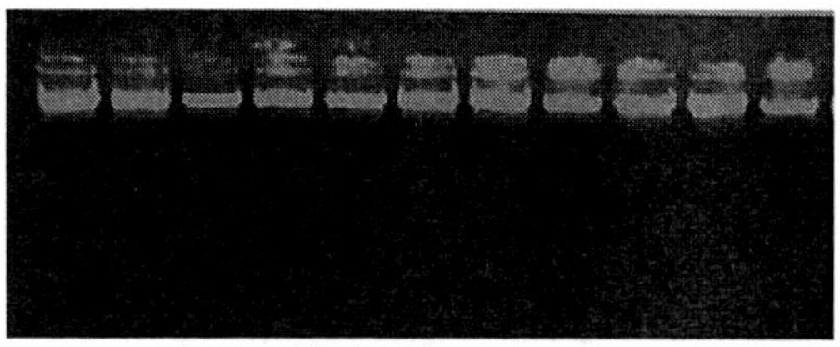

Figure 2. The results of DNA isolation from blood after the electrophoresis

DNA obtained from the insulation, then amplified using primers constructed with PCR mix RTG / Go Green Tag. The results of PCR amplification are analyzed using electrophoresis on agarose. Electrophoresis is the movement of charged molecules in an electric field. Speed moving molecules in an electric field depends on the charge, shape and size of the molecule. Agarose and poly acrylamide is a buffer matrix that is widely used for the separation of proteins and nucleic acids. In this experiment, agarose 1.5%. According to Sambrook and Russell, agarose 1.5% is suitable for separating DNA fragment size of 200-300 bases. The location of the DNA contained in the gel can be observed by staining using red gel, so that later can be seen as a gel placed over Gel Doc.

From these electrophoresis it can be seen that the ARMS-PCR reactions are performed can be used to detect SNPs in the gene TCF7L2 particular SNP rs12255372. But this method has its limitations include: the reaction may not be able to detect 100% of SNP in the gene TCF7L2. Nevertheless, the high specificity and sensitivity in detecting SNPs can be used as one factor why this method can be used. In addition, if compared with other SNP detection method, ARMS-PCR reaction has several advantages such as cheaper and easy to apply. Process/ shorter implementation time, starting from the preparation of reagents, equipment, including the addition of genomic DNA (template), three-ARMS-PCR amplification and electrophoresis on agarose can be completed in one day. Its application is quick and easy method to detect SNP rs12255372 is a very important value for the prevention of type 2 diabetes [13].

Sequencing. Besides using three-ARMS-PCR method to know SNP rs12255372 TCF7L2 used sequencing method [14]. Six samples were sequenced to accuracy of three-ARMS-PCR method. Based on the sequencing results are conformity with three-ARMS-PCR method. While the results showed polymorphisms can be seen in Figure 3. While the results showed not happen polymorphisms can be seen in Figure 4. Sequencing can be seen in the following figure:

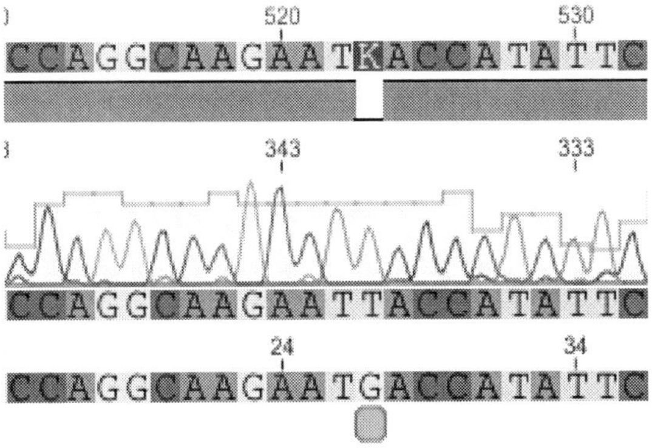

Figure 3. The results of sequencing the samples undergo polymorphisms at rs12255372 when the change in the base G into T.

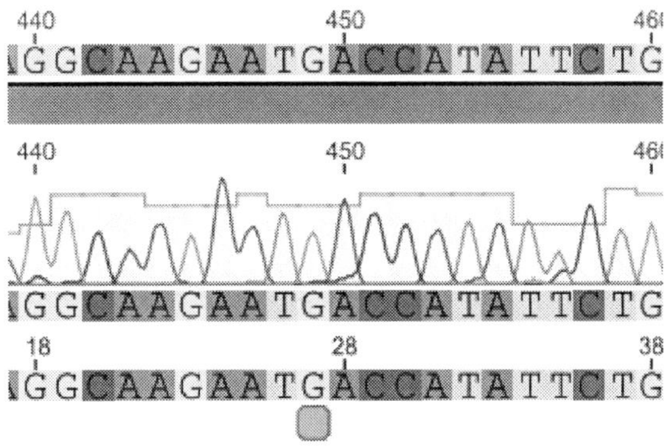

Figure 4. The results of sequencing the samples did not experience polymorphisms at rs12255372 where no change of base G.

4. Conclusion

Based on the results of data analysis can be concluded that it has successfully constructed three primary pieces that RS12F forward primer, RS12C reverse primer and RS12R forward primer. The third primer capable of recognizing constructed the SNP rs12255372 TCF7L2 gene with ARMS-PCR method.

Acknowledgments

On this occasion, we would like to thank the director of DRPM Kemristek and Dikti Republik Indonesia, as research sposors. So also all staff at Biomedical Laborotorium, Faculty of Medicine, Andalas University and laboratory of Genetics and Biotechnology, Faculty of Mathematics and Natural Sciences, Universitas Negeri Padang that helped the success of this research.

References

[1] IDF, *Diabetes Atlas*, IDF, Editor 2017, IDF: Brussels Belgium.

[2] IDF, *Diabetes Atlas*, 2015, IDF: Brussels.

[3] Syamsurizal S 2017 *Bioscience* **1** (1) 1-7

[4] Grant S F, et al. 2006 *Nature genetics* **38** (3) 320-323.

[5] ADA 2010 *Standards of Medical Care in Diabetes* CareDiab **33** (1) 1-5

[6] Chen R, et al. 2012 *Type 2 Diabetes Risk Alleles Demonstrate Extreme Directional Differentiation among Human Populations, Compared to Other Diseases.* PLoS Genetics, **8** (4)

[7] Mohan R V V 2007 *Indian J Med Res* **12** (5) 259-274

[8] Nordman S 2008 *Identification of susceptibility gene type-2 diabetes* (Stockhlom: Karolinski Institutet)

[9] Komar A 2009 *Single Nucleotide Polymorphisms: Methods and Protocols* (Cleveland, OH, USA: Humana Pres)

[10] Kwok P 2003 *Single nucleotide polymorphisms: Methods and Protocols* (Totowa, NJ: Humana Press)

[11] Lyssenko V, et al. 2007 *The Journal of clinical investigation* **117** (8) 2155-2163

[12] Alsmadi O, et al. 2009 *BMC Research Notes* **2** (48) 1-7.

[13] Syamsurizal and Sardi A. 2015 Transcription factor 7-like 2 as type-2 diabetes mellitus diagnostic marker in ethnic Minangkabau *Univ Med.* **33** (1) 206-13.

[14] Abdullah N, et al. 2015 *Characterizing the genetic risk for Type 2 diabetes in a Malaysian multi-ethnic cohort.* Diabet. Med. **1**(1)

Expression pattern of drought-responsive genes in burley tobacco under in vitro water deficit

N Jadid, E Estiasih, T B Saputro, K I Purwani, D Hidayati, E V Permatasari and W D Kurniawan

Department of Biology, Institut Teknologi Sepuluh Nopember, Surabaya-Indonesia

E-mail : nuruljadid@bio.its.ac.id

Abstract. Understanding molecular mechanism in tobacco adaptation to drought stress become indispensable strategy for the development of high-yield tolerant varieties. Profiling the pattern of drought-related gene expression is a promising strategy to preliminary evaluates the plant response to drought stress. This study aims to investigate the expression pattern of drought-responsive genes in burley tobacco under *in* vitro water deficit. In this study, we used three burley varieties including Marakot, Jepon Mawar and MKY. Drought stress treatment was applied *in vitro* using MS medium supplemented with several concentration of PEG 6000 (Polyethylen Glycol) (0%, 10%, 20% and 30%). The expression profile of drought-related genes including *NtLTP1*, *NtABA2* and *NtERD10B* were investigated using quantitative Reverse Transcriptase-PCR (qRT-PCR). The results showed that *NtABA2* and *NtERD10B* genes were parallely overexpressed following an increase concentrations of PEG in var. Jepon Mawar and var. MKY. However, the highest mRNA relative expression of these genes was occured in var. MKY treated with 30% of PEG, which increased 1,7 and 1,64-fold compared to control, respectively. *NtLTP1* was down-regulated in var. Jepon Mawar following an increase of PEG concentrations. In contrast, *NtLTP1* was up-regulated in var. MKY. Meanwhile, var. Marakot demonstrated similar pattern of all gene expressions, where treatment of 20% PEG induced highest expression compared to the other treatments. The overall result suggests that var. MKY is likely more tolerant to drought stress followed by var. Jepon Mawar and var. Marakot, respectively.

1. Introduction

Plant often perceives many environmental conditions including both biotic (pathogen, herbivores, etc) and abiotic (high temperature, low level of available nutrients, salinity, extreme pH condition, high concentration of heavy metals, cold and drought stress, etc) [1, 2]. Among them, drought stress is considered as the most influencing environmental stress that affect plant productivity. The incidence of low plant productivity due to the inability of the plant to escape is increasing. It also coincides with global climate change, which results in an unpredicted seasonal phenomenon [3].

Plants develop many interesting strategies to cope with extreme environmental stress. These include morphological, physiological and molecular adaptations. Plants are responding the environmental cues by re-designing their root architectures, inducing the stomatal closure, producing high level of cuticular wax, reducing the chlorophyll, inducing the formation of antioxidant to scavenge the reactive oxygen species (ROS), which is usually over-produced during the environmental stress [1]. Plant gene regulations have been extensively studied using both proteomic and transcriptomic approaches to better understand the plant molecular responses during environmental stresses. Transcriptomic approaches include measuring the expression of some transcript species such as mRNAs, non-coding RNAs and small RNAs under particular environmental stresses [4,5]. Previous studies have demonstrated that some genes might be used as drought stress markers [6]. Some of these genes include *NtLTP1* (Lipid Transfer Protein), which is involved in epicuticular wax deposition [7, 8], *NtABA2*, which encodes an important enzymes involved in the biosynthesis of abscisic acid hormone [9] and the *NtERD10B* that encodes the protein, which acts as important osmolites [10].

Solanaceae family is considered as one of important plant family that has been long-term cultivated. It comprises around 3,000 species, where tobacco is the most economically important species among the non-edible Solanaceae family. Not only serves as the main source for cigarette industry, tobacco is being utilized in the pharmaceutical industry and bio-energy sources [11]. Some interesting tobacco varieties have been identified and cultivated in Indonesia, such as Prancak 95, Dixie Bright, Marakot, Jepon Mawar, and MKY. Each variety possesses unique characteristics, both morphological and physiological characters. The genetic diversity that is responsible for diverse morpho-physiological characters also plays a central role in regulating the plant adaptation to environmental conditions. Therefore, each variety might possess distinct responses against environmental stress [2]. To date, many research have been conducted to study plant adaptation [5,6] but little is known about the molecular responses of Indonesian tobacco varieties against environmental adverse condition, including drought stress.

As previously reported, some genes are responsive to drought stress at the transcriptional level. The expression of these responsive genes might be decreased or increased when the plants are exposed to drought stress. It indicates that the dynamic expression of the genes is resulted as a consequence of the environmental stress. In addition, this molecular regulation certainly controls plant adaptation against drought stress [12]. Based on the above information, investigating the gene expression pattern could be utilized to preliminary evaluate plant response to drought stress. Even though many studies on the tobacco response to drought stress has been widely reported, until now, there is no research have been reported about the expression pattern of drought responsive genes of some burley tobacco, especially var. MKY, Marakot and Jepon Mawar. Therefore, this present study aims to investigate the expression pattern of drought-responsive genes in burley tobacco under *in* vitro water deficit.

2. Methods

2.1. In vitro culture of tobacco and drought stress treatment

Tobacco seeds used in this study include *Nicotiana tabacum* var. Marakot, Jepon Mawar and MKY. The seeds were obtained from PT. Sadhana. The seeds were sterilized by mixing in 1% NaOCl solution for 5 min and were subsequently washed with distilled water for 3 times. The seeds were then grown in the Murashige and Skoog (MS) medium (100 seeds per medium) [13] and kept in culture chamber with 12/12 h of light/dark photoperiod at 22°C. Seedlings aged 18-20 days after germination (dag) were moved to the MS medium containing Polyethylene glycol (PEG) 6000 at different concentrations (0%, 10%, 20% and 30%, w/v). After 8 days, the seedlings were collected and subjected to further analysis.

2.2. Total RNA extraction and quantification

Seedlings were treated as routine preparation prior to total rna extraction [14]. Briefly, samples were freezed using liquid nitrogen and then ground directly into powder. Subsequently, total RNA extraction was carried out using Total RNA Mini Kit (Plant) (Geneaid) according to the protocols provided by the manufacturer. Quantification of total RNA was done using Nano Drop (Themo Scientific™ NanoDrop 2000) and the concentration was expressed as ng/µL.

2.3. Gene expression analysis

Gene expression was analyzed using quantitative real time PCR (qRT-PCR). qRT-PCR was carried out using KAPA SYBR® FAST Universal One-Step qRT-PCR Kit (Kapa Biosystem) according to the manufacturer's protocols. The concentration of total RNA used in this analysis was 20 ng/μL. Three drought responsive genes were analyzed including *NtABA2* (accession no: EU123520), *NtERD10B* (accession no: AB049336), and *NtLTP1* (accession no: AAT45202). Meanwhile, *NtEF-1α* (accession no: AF120093) was used as gene of reference. The primers were designed using PrimerQuest Tool (Integrated DNA Technologies), *NtABA2* {5'- GTGGGATTGGAGGGTTAGTT -3' (forward) and 5'-GGTCCTCTATATTGCCCTTCTC -3' (reverse)}; *NtERD10B* {5'- ACGGACGAATACGGCAATC-3' (forward) and 5'- TCTCCTTAATCTTCTCCTTCATCC-3' (reverse)}; *NtLTP1* {5'-CATTGTTGGTGGTGGTGTG-3' (forward) and 5'-TGGAAGGGCTAATCTTGTAGG-3' (reverse)}; *NtEF-1α* {5'- TGAGATGCACCACGAAGCTC-3' (forward) and 5'-CCAACATTGTCACCAGGAAGTG-3' (reverse)}. The PCR condition were reverse transcription at 42 °C for 5 min, enzyme inactivation at 95 °C for 3 min followed by 39 cycles of 95 °C for 3 s, 61.3 °C for 20 s and extension step at 61.3 °C for 20 s. The gene expression was calculated using $\Delta\Delta C_t$ method [15]. All the relative mRNA levels were normalized to the *NtEF-1α* level in the same samples. All gene expression analysis was done in triplicate.

3. Results and discussion

3.1. Effect of in vitro water deficit on the expression of NtABA2 gene

NtABA2 encodes the ZEP (zeaxanthin epoxidase) enzyme that serves to catalyze the epoxidation of zeaxanthin to violaxanthin in the biosynthesis of ABA in plants. Furthermore, violaxanthin is converted to xanthoxin *via* neoxanthin intermediates. This reaction is catalyzed by NCED (9-cis-epoxicarotenoid dioxigenase). Subsequently, two enzymatic reactions complete the final stage of ABA synthesis. Based on these pathways, ZEP is a pivotal enzyme that functions in the early stages of ABA biosynthesis [16]. The results showed that each tobacco variety had a different response to PEG-induced drought stress. At 10% and 30% (w/v) of PEG, the expression of *NtABA2* in var. Jepon Mawar and Marakot were decreasing compared to control. However, the expression of *NtABA2* in var. Jepon Mawar is higher than that observed in var. Marakot on any given treatment. Whereas, var. MKY showed an increase of *NtABA2* expression following increased of PEG concentrations, compared to control (Figure 1). The highest expression of *NtABA2* is observed in var. MKY treated with 30% (w/v) PEG, which was 1.7-fold compared to untreated tobacco (control).

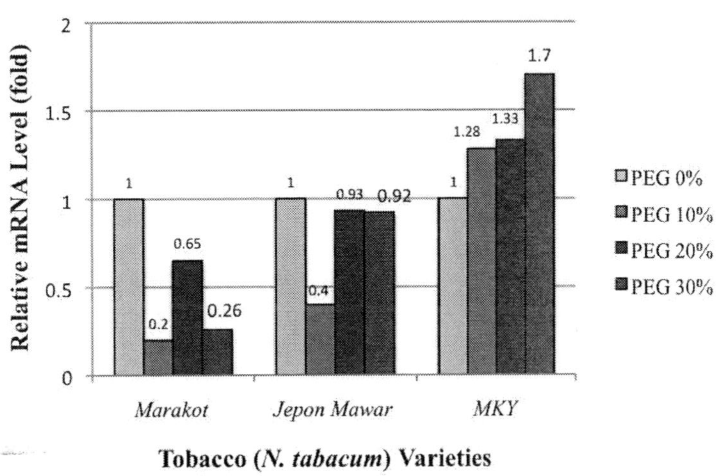

Figure 1. Tobacco *NtABA2* gene expression during *in vitro* water deficit.

The dynamic gene expression patterns is possibly due to overall changes in physiological responses to stress, which might be depend on the genotype of the tobacco. ABA biosynthesis generally increases following the severity of stress [17]. Previous study has showed that ABA augmentation is associated with an increase of *ABA2* expression and consequently result in better resistance of a plant to drought stress [18]. In addition, *ABA2* mutant study has also demonstrated that the plant hormone ABA plays a significant role in plant adaptation during severe drought condition [19]. ABA also serves to activate several other drought responsive genes including *ERD10* [20]. In addition, ABA functions in the stomatal closure regulation to prevent water loss through transpiration during water deficit. Based on our results, var. MKY is thought to have the best response to drought stress compared to other varieties, followed by var. Jepon Mawar and var. Marakot. This is indicated by a regular pattern of *NtABA2* expression in response to elevated level of PEG concentrations. Nevertheless, other study showed that there will be other genes involved in the plant adaptation mechanism against drought stress [21].

3.2. Effect of in vitro water deficit on the expression of NtERD10B gene

NtERD10B encodes the Late Embryogenesis Abundant (LEA) protein family (Group II LEA). The *ERD* gene is defined as an inducible gene and is rapidly activated under drought stress conditions [19]. LEA proteins are expressed in all plant organs and accumulated in high amounts when the plants are facing the drought stress [22]. Our results demonstrated that the *NtERD10B* gene expression profile showed different patterns in each variety against different level of PEG-induced drought stress concentration. Var. Marakot showed similar *NtERD10B* pattern of expression with those observed in *NtABA2*. While, the expression of *NtERD10B* in var. Jepon Mawar and var. MKY under water deficit also showed the same profile as observed in *NtABA2*. The expression of *NtERD10B* increased in a dose-dependent manner. However, the highest expression of *NtERD10B* was found in var. MKY, treated with 30% (w/v) of PEG (1.64-fold) compared to control (Figure 2).

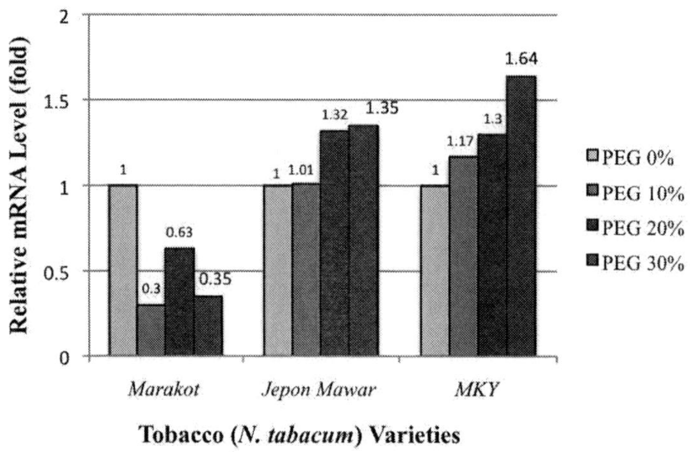

Figure 2. Tobacco *NtERD10B* gene expression during *in vitro* water deficit.

Similar pattern of expression between *NtERD10B* and *NtABA2* in this study, support the previous data, which stated that *NtABA2* induce the expression of other drought-responsive genes, including *NtERD10B* [19]. Our results are in accordance with previous study that also reported an increase of *ERD10B* expression. Furthermore, upregulation of this gene could reach 10-fold compared to untreated tobacco plant. Indeed, dramatic increase of *ERD10B* expression could also take place when plant is undergoing combined environmental stresses [23]. Enhance expression of this gene is followed by increase of transpiration rate and it is consequently reflected with an elevation of the ABA production in the plants. The latter might then induce stomatal closure [24]. It is also worth noting that overexpression of plasma membrane intrinsic proteins in *Medicago falcata* resulted in upregulation of *ERD10B* and thus enhanced the plant tolerance against drought stress [25].

The function of ERD10B in plant defence mechanism against drought stress include membrane protector, antioxidant, metal ion binding, protein stabilizer and water-binding molecule [26]. As protein stabilizer, LEA proteins have the capacity to protect proteins from inactivation and aggregation during drought stress. Its role as a stabilizer is supported by its ability to retain enzyme activity after partial dehydration or drought. During drought condition, membrane protection is important to maintain cellular and organelles integrity. The LEA protein has a charged amino acid residue, which allows it to absorb the ions. LEA proteins scavenge the ROS (reactive oxygen species) or ROS-producing metal ions. The production of ROS is increasing when the plants are suffering from environmental stress. Therefore, plants produce more antioxidants for minimizing the negative effect of an increase of ROS [2]. The function of LEA protein, including dehydrin, as an antioxidant is facilitated by the interaction of the amino acid residues of the protein with both the ROS and metal ions. This interaction could further results in oxidated amino acid residues and the formation of covalent bond between the metal ions and the LEA proteins [26]. Based on the relative mRNA expression of the *NtERD10B* gene on all three tobacco varieties, var. MKY is thought to be more responsive to drought stress followed by var. Jepon Mawar dan var. Marakot.

3.3. *Effect of in vitro water deficit on the expression of NtLTP1 gene*

NtLTP encodes Lipid Transfer Protein (LTP). LTP is encoded by a large multigenous family and these genes have been characterized in several crops including tobacco [27], *Oryza sativa* (52 members), *Triticum aestivum* (156 members) [28] and in *Arabidopsis thaliana* (110 members), of which 15 are named LTP1 to LTP15 [29]. The results of the qRT-PCR showed that the three tobacco varieties exhibited different expression pattern of the *NtLTP1* under *in vitro* water deficit. The pattern of *NtLTP1* expression on var. Marakot and var. MKY is similar to the expression of *NtABA2* and *NtERD10B* of both varieties. The expression of *NtLTP1* seems to be dose-dependent manner in var MKY, where the highest expression was showed in 30% (w/v) of PEG (1.52-fold compared to control). Meanwhile, we observed dynamic and irregular pattern of expression in var Marakot. All var. Marakot treated with different concentration of PEG exhibited low expression of *NtLTP1* (figure 3). Interestingly, in contrast to var. MKY, the *N. tabacum* var. Jepon Mawar exhibited different pattern of expression. The *NtLTP1* was down-regulated following an increase of PEG concentration. Highest level of expression was showed in this variety under 10% (w/v) PEG treatment (1.52-fold compared to control). This value was similar to that observed in var. MKY under 30% (w/v) of PEG treatment (Figure 3).

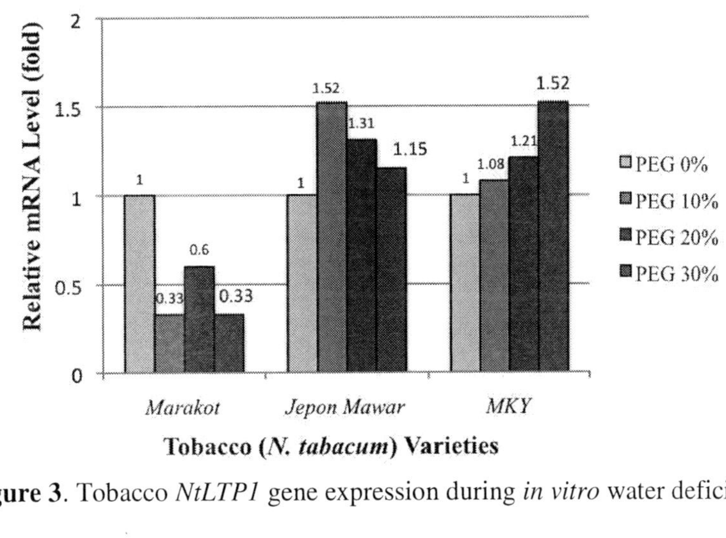

Figure 3. Tobacco *NtLTP1* gene expression during *in vitro* water deficit.

Decrease and increase of relative mRNA expression values is related to the stress resistance mechanism developed by different plants [30]. Upregulation of *NtLTP1* has been reported previously to be an example of plant response to drought stress. This is tightly associated with an enhance plant tolerance to drought stress [31]. In addition, variation in gene expression levels in response to drought stress also indicate that each plant possess their own limitation to this type of stress [32]. The LTP1 protein is involved in the intercellular lipid transfer trafficking, accumulation of wax and cutin deposition by bringing the epicuticular components to extracellular matrix. Furthermore, this protein is also involved in the embryogenesis, defence mechanism against pathogens, and crop adaptation to various environmental conditions [31]. In addition, another study also showed that LTP1 is involved in the secretion of lipid from glandular trichomes in *Nicotiana tabacum*. The latter could increase plant resistance to aphids [8]. Furthermore, LTP is considered as a pathogenesis-related protein family (PR-14) and is believed to be involved in several physiological processes including resistance to plant diseases [33].

The LTP has been reported to be induced by drought stress [34] and it will be upregulated in response to both osmotic and pathogenic stress [31]. The expression of *NtLTP1* gene in another species of Nicotiana (*Nicotiana glauca* L. Graham) increases when exposed to drought stress. Consequently, it increases the deposition of the cuticular wax. This elevated LTP1 might be due to plant strategy to prevent water loss through transpiration [7]. Based on the relative mRNA expression of the *NtLTP1* gene on all three treated varieties, var. MKY is thought to be responsive to drought stress followed by var. Jepon Mawar dan var. Marakot.

4. Conclusion

The present study demonstrated that molecular genetic factor plays important role in plant defense mechanism against drought stress. It was reflected with the distinct pattern of drought responsive genes in three treated tobacco varieties (*N. tabacum* var. MKY, Jepon Mawar and Marakot). Interestingly, *N. tabacum* var. MKY is found to be the most responsive variety compared to other treated varieties. However, evaluating other molecular genetic aspects is still required in order to complete our understanding on plant defence mechanism against water deficit condition.

Acknowledgments

We thank to all member of the Laboratory of Plant Biosciences and Technology, Institut Teknologi Sepuluh Nopember, Indonesia for the technical support. This work was supported by the Ministry of Research, Technology and Higher Education of the Republic of Indonesia.

References

[1] Wang W, Vinocur B and Altman A 2003 *Planta* **218** 1-14
[2] Jadid N, Maziyah R, Nurcahyani DD and Mubarokah NR 2017 *AIP Conf. Proc.* **1854** 020018
[3] Zhang S, Kang H and Yang W 2017 *PLOS ONE* **12** e0182012
[4] Wang Z, Gerstein M and Snyder M 2009 *Nat. Rev. Genet.* **10** 57-63
[5] Wehner G, Balko C, Humbeck K, Zyprian E and Ordon F 2016 *BMC Plant Biol.* **16** 3
[6] Jiang Q, Zhang J-Y, Guo X, Monteros MJ and Wang Z-Y 2009 *Int. J. Plant Sci.* **170** 969-78
[7] Cameron KD, Teece MA and Smart LB 2006 *Plant Physiol.* **140** 176-83
[8] Choi YE, Lim S, Kim H-J, Han JY, Lee M-H, Yang Y, Kim J-A and Kim Y-S 2012 *Plant J.* **70** 480-91
[9] Audran C, Borel C, Frey A, Sotta B, Meyer C, Simonneau T and Marion-Poll A 1998 *Plant Physiol.* **118** 1021-8
[10] Rai A, Suprasanna P, D'Souza S F and Kumar V 2012 *PLOS ONE* **7** e32658
[11] Apollonio D and Glantz SA 2017 *Am. J. Public Health* **107** 1636-42
[12] Rejeb IB, Pastor V and Mauch-Mani B 2014 *Plants* **3** 458-75
[13] Murashige T and Skoog F 1962 *Physiol. Plant.* **15** 473-97
[14] Jadid N, Mardika R K, Nurhidayati T and Irawan M I 2016 *AIP Conf. Proc.* **1744** 020042

[15] Livak K J and Schmittgen T D 2001 *Methods* **25** 402-8
[16] Xiong L, Lee H, Ishitani M and Zhu J-K 2002 *J. Biol. Chem.* **277** 8588-96
[17] Zhou S, Medlyn B, Sabate S, Sperlich D, Prentice IC and Whitehead D 2014 *Tree Physiol.* **34** 1035-46
[18] Yue Y, Zhang M, Zhang J, Tian X, Duan L and Li Z 2012 *J. Exp. Bot.* **63** 3741-8
[19] Xiong L, Lee B-H, Ishitani M, Lee H, Zhang C and Zhu J-K 2001 *Genes Dev.* **15** 1971-84
[20] Alves MS, Reis PAB, Dadalto SP, Faria JAQA, Fontes EPB and Fietto LG 2011 *J. Biol. Chem.* **286** 20020-30
[21] Bartels D and Sunkar R 2005 *CRC Crit. Rev. Plant Sci.* **24** 23-58
[22] Battaglia M, Olvera-Carrillo Y, Garciarrubio A, Campos F and Covarrubias AA 2008 *Plant Physiol.* **148** 6-24
[23] Mackova H, Hronkova M, Dobra J, Tureckova V, Novak O, Lubovska Z, et al. *J. Exp. Bot.* **64** 2805-15
[24] Vysotskaya LB, Veselov SY and Kudoyarova GR 2010 *J. Exp. Bot.* **61** 3709-17
[25] Zhuo C, Wang T, Guo Z and Lu S 2016 *BMC Plant Biol.* **16** 138
[26] Hanin M, Brini F, Ebel C, Toda Y, Takeda S and Masmoudi K 2011 *Plant Signal. Behav.* **6** 1503-9
[27] Masuta C, Furuno M, Tanaka H, Yamada M and Koiwai A 1992 *FEBS Lett.* **311** 119-23
[28] Boutrot F, Chantret N and Gautier MF 2008 *BMC Genomics* **9** 86
[29] Arondel V, Vergnolle C, Cantrel C and Kader J-C 2000 *Plant Sci.* **157** 1-12
[30] Zhou R, Yu X, Ottosen C-O, Rosenqvist E, Zhao L, Wang Y, et al. 2017 *BMC Plant Biol.* **17** 24
[31] Jung H W, Kim W and Hwang B K 2003 *Plant Cell Environ.* **26** 915-28
[32] Ramakrishna A and Ravishankar G A 2011 *Plant Signal. Behav.* **6** 1720-31
[33] Buhot N, Gomes E, Milat M-L, Ponchet M, Marion D, Lequeu J, et al. 2004 *Mol. Biol. Cell* **15** 5047-52
[34] Colmenero-Flores JM, Campos F, Garciarrubio A and Covarrubias A A 1997 *Plant Mol. Biol.* **35** 393-405

Characteristic of synthetic seeds from two medicinal plants (*Moringa oleifera and Camellia sinensis*)

W Muslihatin, N Jadid, T B Saputro, K I Purwani, C E S Himayani and A W Calandry

Departement of Biologi, Faculty of Science, Institut Teknologi Sepuluh Nopember Surabaya Indonesia 60111

E-mail: w_muslih@bio.its.ac.id

Abstract. *Moringa oleifera* and *Camellia sinensis* are medicinal plants and usefull for health. Generative propagation of *M. oleifera* plant is quite difficult because the ability of seed germination is low in its survival is low, the growth takes a long time and the lack of vegetative propagation method. *C. sinensis* has difficulty in rooting plantation. One alternative to overcome these obstacles is by planting vegetative propagation through biotechnology approach that is in vitro culture techniques and encapsulation techniques that can produce synthetic seeds. The purpose of this research is to know characteristics synthetic seeds of *Moringa oleifera* and *Camelia sinensis*. Stages done in this research is induction of embryogenic callus in MS medium which has been modified. The somatic embryo of *M. oleifera* and *C. sinensis* obtained will be encapsulated and placed at $CaCl_2.2H_2O$, when the seed has been hardened rinsed with aquades. Synthetic seed C. sinensis yellowish green and 5.5 to 5.7 mm in diameter and fresh weight of synthetic seed produced ranges from 260-290 mg. Characteristics of synthetic M. oleifera produced seeds have clear and yellowish-looking embryo, and are 5.8-6.0 mm in diameter and have a fresh weight of 250-300 mg.

1. Introduction

Medicinal plants are an important element of indigenous medical systems that has persisted in developing countries [1] . Traditional medicine has a long history of serving peoples all over the world [2]. Recently, the use of traditional medicine based on plants has received considerable interest. Different plants produce diverse products like Moringa oleifera and Camellia sinensis.

Moringa (*Moringa oleifera* Lam) belongs to the Moringaceae family. *Moringa oleifera* is one of thirteen species most widely cultivated. The leaves contain more vitamin A compared to carrots, more calcium than milk, more iron than spinach, more vitamin C than oranges, and more potassium than bananas, and that the protein quality of *Moringa* leaves rivals that of milk and eggs [3] The extracts of *Moringa* leaves shows antimicrobial and antifungal activity in addition to cancer preventive effect [4][5]. Some of the compounds isolated from *Moringa* preparations which are reported to have hypotensive, anticancer and antibacterial activity include 4-(4'-*O*-acetyl-a-L-rhamnopyranosyloxy), Benzyl isothiocyanate, 4-(a-L-rhamnopyranosyloxy) benzyl isothiocyanate, niazimicin, pterygospermin, benzyl isothiocyanate, and 4-(a-L-rhamnopyranosyloxy) benzyl glucosinolate [6]. According to ethanobotanical studies its roots are bitter, acrid, thermogenic, digestive, carminative, anthelmintic, constipating, anti inflammatory, emmenagogue, diuretic, ophthalmic, expectorant and stimulant. They are useful in dyspepsia, anorexia, verminosis, diarrhea, colic, flatulence, paralysis, inflammations, amenorrhea, dysmenorrheal fever, strangury, vesicle and renal calculi. It is used in cough, asthma, bronchitis, pectoral diseases, splenomegaly, epilepsy and cardiopathy [7]. Due to these medicinal values, this plant have to be conserved and multiplied to reach commercial requirement.

Moringa is now found in all tropical regions around the world ranging from parts of South Asia through West Africa, including Indonesia. Although the spread of Moringa is vast, but in Indonesia these plants become very rare nowadays. The spread of Moringa is also rare for seed germination and viability or survival are low, and there is the lack of vegetative propagation method [5]. Moringa distribution limited is the main reason that drives the need to make efforts so that the necessary conservation plant propagation techniques to produce many seeds of Moringa in a short time.

Tea (Camellia sinensis L.) is one of the most important plantation crops in the world. The medicinal effects of tea have a history dating back almost 5000 years. Tea is the oldest non-alcoholic caffeine-containing beverage crop in the world and health benefits attributed to tea consumption are well proven. Tea is the chemical components of green tea chiefly include polyphenols, caffeine and amino acids. Tea also contains flavonoids, compounds reported to have anti-oxidant properties having many beneficial effects. Tea flavonoids reduce inflammation, have antimicrobial effects and prevent tooth decay. The pleasing astringent taste and refreshing boost it provides is so deep-pervasive that its potential health benefits and medicinal properties are often overlooked. Ongoing scientific exploration points that the certain potential health benefits derived from tea have important implications on human health. The American Medical Association shows that green tea can lower cholesterol levels, high blood pressure, and reduce the risk of strokes (especially in men). The National Cancer Institute reports that because of the highly effective anti-oxidants in green tea, it can ward off various types of cancer. There are many therapeutic values in green tea, including, aiding digestion, blood purification, ensuring regularity, lowering body temperature, strengthening teeth and bones, boost immune system, enhance heart function, suppress aging, deter food poisoning, fights virus, and lowers blood sugar levels [8].

The chemical components of tea leaves include polyphenols (catechins and flavonoides), alkaloids (caffeine, theobromine, theophylline, etc.), volatile oils, polysaccharides, amino acids, lipids, vitamins (e.g., vitamin C), inorganic elements (e.g., aluminium, fluorine and manganese), etc. However, the polyphenols are primarily responsible for the beneficial healthful properties of tea. The flavonoides have antioxidant, anti-inflammatory, antiallergic and anti microbial effects. Green tea contains six primary catechin compounds namely catechin, gallocatechin, epicatechin, epigallocatechin, epicatechin gallate and epigallocatechin gallate (EGCG), the later being the most active component [8]. Single leaf-node cuttings conventionally propagate cultivated tea. This method of production becomes limiting when large numbers are required from new clones of which only very few stock plants are available. Moreover, the process in establishing tea plants from cutting is lengthy and labour-intensive, rendering it ineffective fore rapid dissemination of new clones for large-scale commercial plantings [9]. Conventional tea breeding is well stablished, though time-consuming and labors intensive due to its perennial nature and long gestation period (4-5 years). additionally,tea breeding has been slowed by lack of reliable selection criteria. Vegetative propagation is standard, yet limited by slow multiplication rate, poor survivability of some clones, and need for copious initial planting material [9].

One alternative techniques to get the plants in large quantities and quickly as conservation effort is using biotechnological approaches such as synthetic seed technology through in vitro culture. Encapsulation technique is a prospective technology developed for seed multiplication and conservation. Encapsulation technique is a technique of wrapping explants (callus, somatic embryos or meristem or shoot tips) with a wrapper specially made explant is not easily damaged and can grow. Packaging plays a role as the endosperm-containing carbon source, nutrients, and plant growth regulator (PGR) synthetic seed The technology can provide advantages such as long-lasting storage, simplify distribution or dissemination, as well as to protect the seed from disease [10].

In recent years, alginate encapsulation technique used in the in vitro alternative to synthetic seed production using somatic embryos. The definition of the seed synthetic that somatic embryos, seedlings or meristematic tissue other encapsulated artificially and can develop into a plant in conditions in vitro or in vivo encapsulation provide physical protection in somatic embryos, carrying nutrients, growth regulators, antibiotics, fungicides needed during germination and protector of the danger of pathogens [11][12]. Synthetic seed technology can provide benefits, among others, can be stored in the long term, facilitate the distribution or dissemination, as well as to protect the seed from disease [10]. Synthetic seed is a plant biotechnology applications in vitro which offers many benefits, especially *M. oleifera* and *C. sinensis*. The purpose of this research is to know characteristics synthetic seeds from somaticembryos of *Moringa oleifera* and *Camillia sinensis.*

2. Material and methods

2.1. Plant material

Leaf of *Moringa oleifera and C. sinensis* were used as explants for callus induction and its plant regeneration. These were collected from Surabaya and Malang East Java, Indonesia.

2.2. Sterilization and inoculation of explant

Sterilization was carried out by washing the leaves with distilled water for 10 minutes, and then soaked in surfactant solution for 10 minutes, followed by three times of rinses with distillated water for 5 minutes. The explant were soaked in antifungal for 10 minutes, followed in NaOCl solution for 2 minutes. The explant were further surface sterilized using 96% alcohol for 3 minutes, then rinsed with sterile distilled water. The explants were surface dried on sterile filter paper and cultured on MS medium supplemented with different levels of plant growth regulators for callus induction and somatic embryogenesis

2.3. Medium composition and culture condition

MS (Murasige and Skoog) medium consisting of MS basal medium with vitamins, and 30% of sucrose. The pH was adjusted to 5.6-5.8 with NaOH atau HCl before autoclaving at 121°C for 20 minutes. The growth regulators were added to the culture media before autoclaving. Cultures were incubated in a growth chamber at 24°C, 70% of relative humidity and 16/8 h our (light/dark) photoperiod with light supplied by cool white fluorescent lamps at an intensity 60 μmol $m^{-2} s^{-1}$

2.4. Medium of induction and maturation of embryonic calluses

The *M. oleifera* explants were placed on MS medium supplemented with 1 mg L^{-1} NAA and 1 mg L^{-1} kinetin for the induction of embryonic callus formation. Embryonic calluses were induced for 4 weeks. The *C. sinensis* explant was cultured in MS medium supplemented BAP medium 1mg / l and NAA 0.1 mg / l.

Somatic embryos were induced within 4 weeks on MS media supplemented with 3 mg L^{-1} BAP. The C. sinensis explant were placed on MS + BAP 1mg/l + NAA 0,1 mg/l for maturation of embryonic embryos. The greenish- yellow, globular pro- embryoids were observed and these were sub-cultured on the same medium for maturation. Matured somatic embryos (heart- and torpedo shaped) were sub-culturing every 2 weeks.

2.5. Encapsulation technique

The embryosomatic of *M. oleifera* and *C. sinensis* ware encapsulated by the same method. 4% sodium alginate was added with 0.3 mg L^{-1} NAA and 3 mg L^{-1} BAP in a calcium free liquid MS medium containing 30% sucrose. Encapsulation was done by mixing embryo on Na- alginate gel and MS. The somatic embryos was introduced into a mixture of sodium alginate and liquid MS and then taken with a disposable sterile plastic pipette with the position of all parts of the embryo covered by sodium alginate. The explant was release drop by drop into 75 mM $CaCl_2$ solution for 15 to 20 minutes, then the synthetic seeds were washing in aquadest to remove residue. The observed parameters of synthetic seed characteristics were diameter of synthetic seed, color and texture.

IOP Conf. Series: Journal of Physics: Conf. Series **1040** (2018) 012005 doi:10.1088/1742-6596/1040/1/012005

3. Result and discussion

The embryogenic callus formed is transferred to the maturation medium. Maturation is the maturation stage of embryogenic callus (Oetami, 2015). The maturation stage of the embryogenic callus is the developmental stage of the globular structure to form cotyledons. The first stage of embryosomatic is the formation of globular phase. The embryoid growth begins with a globular phase with the formation of a rounded bulge, usually from the callus. The next stage is the heart phase, marked by the formation of the gap so that there is a hollow between two bulges and looks like a heart shape, and greenish-white. The heart phase embryo is characterized by a gap in which SAM (Shoot Apical Meristem) is formed. At the torpedo stage, it looks that the cell is growing longer, and is whiteish-green.

3.1. Synthetic seeds of Moringa oleifera

Production and application of synthetic seeds has been widely practiced in the field of plant biotechnology such as massive clonal propagation and in large quantities, germplasm conservation, facilitate storage and transport of seeds [13]. Seed technology is a highly perspective technique to be developed in seed propagation and conservation.

The maturation medium used was MS with 3 ppm BAP, according to the [5], the right hormone for maturation was BAP concentration of 3 ppm. BAP is a cytokinin that is often used in tissue culture techniques. Cytokinins are derivatives of adenine. Cytokines are very important in the regulation of cell division and morphogenesis. The effect of cytokinin in plant tissue culture is related to the process of cell division, root growth inhibition, and micro root induction.

The resulting somatic embryo will be encapsulated using alginate with a concentration of 4%. Alginate is one of the most appropriate hydrogels in the manufacture of synthetic seeds because it is enriched with nutrients, growth regulators, has low toxicity, low cost, not too sticky, rapidly agglomerates and has biocompatibility properties [14]. Alginate with a concentration of 4% is the most appropriate concentration for the germination of synthetic seeds. Alginate that has too high concentration will cause the seeds produced too hard, thus inhibiting the ability of the embryo to germinate. In accordance with the statement [15] that the conditions of very dense seeds affect the life of the embryo in the seed because the condition does not support the growth and development of the next embryo. High hardness in the seeds is expected to cause the environment to become anaerobic, will further inhibit respiration, and delay the rate of respiration will inhibit the process of seed germination. The process of making synthetic seed begins with 4% alginate solution using dropper drops, then bulb on the pipette is pressed slowly. When the alginte solution comes out of the tip of the pipette immediately the somatic embryo callus portion to be encapsulated is inserted into the droplets of alginate at the tip of the pipette using tweezers rapidly and precisely until all the explants are wrapped. Then the eyeball bullet is pressed back slowly until the alginate on the tip of the pipette trickles and falls on a solution of 75mM CaCl2. Furthermore, the synthetic seeds that have been subjected to hardening are removed from the 75 mM CaCl 2 solution and soaked into the aquades. The hardening process is caused by the exchange of Na + ions from Na-alginate with Ca + when dropped into CaCl2 solution to form Ca alginate. The amount of Na + ions exchanging with Ca + ions determines the level of hardness of the capsule. The capsule is expected to be solid enough to prevent leakage [16].

$$2\text{Na Alg}_{(aq)} + \text{Ca}^{2+}_{(aq)} \longrightarrow \text{Ca (Alg)2}_{(s)} + 2\text{Na}^{+}_{(aq)}$$

Characteristics of the resulting synthetic seeds are seeds have a diameter ranging from 5.2 to 5.7 mm with a clear color and yellowish white embryo. he synthetic seeds that have been measured in diameter are then weighed to determine the initial fresh weight of the seed and then the seeds are grown on germination medium with ZPT composition and different concentrations of sucrose. The initial fresh weight of the seed ranges from 250 to 300 mg.

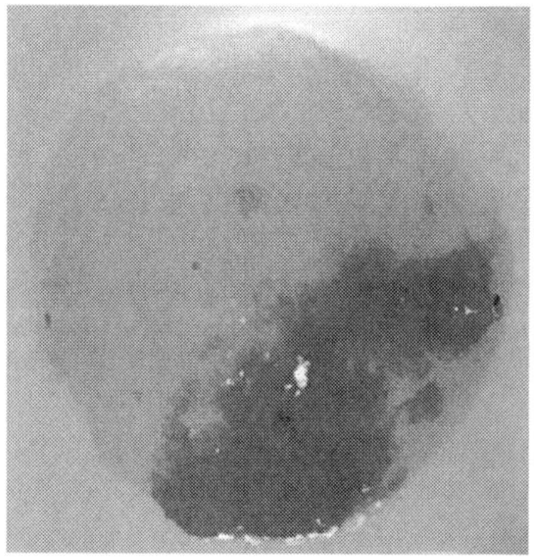

Figure 1. Synthetic seed of Moringa oleifera

3.2. Synthetic seeds of Camillia sinensis

Synthetic seed incubated for 30 days. After observation, the texture of synthetic seeds solid but not too hard, still vulnerable to damage if not careful when moving. All the synthetic seeds that are formed have the same texture, and the embryo appears to have developed into a cotyledon phase and penetrates the alginate. Synthetic seeds are yellowish green with a diameter of 5.5 to 5.7 mm and a wet weight of 260-290 mg. The synthetic seed forming process is carried out using alginate because it is not toxic to the embryo or the material to be encapsulated, its density can protect the fragile embryo, and serves as a reservoir of nutrients that the embryo uses to survive and even accelerate its growth [17].

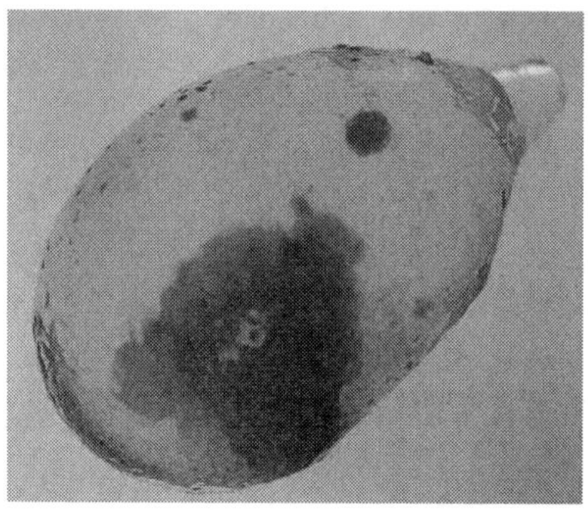

Figure 2. Synthetic seed of tea

Synthetic seeds are formed in greenish white, but the embryo has a compact structure. The color changes that occur in the embryo due to pigment and are influenced by nutrients and environmental factors such as light. Greenish embryo caused by chlorophyll content, due to interaction of NAA and BAP, especially BAP (cytokines) that play a role in the formation of chlorophyll in callus and environmental factors ie exposure to light. The change of color embryo to white-green, because embryo cells have started to form chloroplasts so that embryos become more complex and further differentiate into specific organs.

4. Conclusion

Characteristics of synthetic seeds produced is a soft texture and clear. All the synthetic seeds that are formed have the same texture, Synthetic seed C. sinensis yellowish green and 5.5 to 5.7 mm in diameter and fresh weight of synthetic seed produced ranges from 260-290 mg. Characteristics of synthetic M. oleifera produced seeds have clear and yellowish-looking embryo, and are 5.8-6.0 mm in diameter and have a fresh weight of 250-300 mg.

Acknowledgement
The authors would like to thank *Direktorat inovasi kerjasama dan kealumnian Institut Teknologi Sepuluh Nopember (ITS)* foundation for financial support of this research through industry-scale innovation Bach II 2017.

References

[1] Khalafalla M M, Dafallal H M, Nassrallah A, Aboul-Enein K M, El-Shemy H A, Abdellatef E 2011 *African Journal of Biotechnology* **10** 2746-2750
[2] Cheng J T 2000 *J. Clin. Pharmacol.* **40** 445-450
[3] Fuglie L J 2000 The Miracle Tree: *Moringa oleifera*: Natural Nutrition for the Tropics. *Church World Service*, Dakar; revised in 2001 and published as The Miracle Tree: The Multiple Attributes of Moringa, pp. 172.
[4] Siddhuraju P and Becker K 2003 *Journal of Agricultural and Food Chemistry* **51** 2144-2155
[5] Devendra B N, Prasad Talluri V S S L and Srinivas N 2012 *Journal of Agricultural Technology* **8** 1953-1963
[6] Bennett R N, Mellon F A, Foidl N, Pratt J H, DuPont M S, Perkins L and Kroon P A 2003 *Journal of Agricultural and Food Chemistry* **51** 3546-3553.
[7] Nepolean P, Anitha J and Emilin R R 2009 *Current Biotica* **3** 33-39
[8] Sharagi A B 2009 *Food Research International* **42** 529–535
[9] Tahardi J S, Imron R and Dodd W A 2003 *Journal Bioteknologi Pertanian* **8** 1-7.
[10] Lambardi M, Benelli C, Ozudogru EA and Tokatli Y O 2006 *Floriculture, Ornamental and Plant Biotechnology* **2** 347-354
[11] Rai M K, Asthana P, Singh S K, Jaiswal V S and Jaiswal U 2009 *A review Biotech Ad.m.* **27** 671-679
[12] Singh A K and Chand S 2010 *International Journal of Biotechnology* **9** 319-324
[13] Ravi D and Anand P 2012 *International Research Journal of Biological Sciences* **1** 74 – 78
[14] Reddy M C, Murthy K S R and Pullaiah T 2012 *Biotechnology* **11** 14254-14275
[15] Warnita I and Suliansyah 2008 *Jerami* **1** 139-143
[16] Sumaryono and Saptarini R T 2015 *Menara Perkebunan* 2015 **83** 54-59
[17] Ara H, Jaiswal U and Jaiswal V S 2000 *Current Science* **78** 1438-1444

The effect of addition sucrose concentrations toward weight of Nata DE Lontar (*Borassus flabellifer*) Linn

N L P M Widiyanti[1], S Mulyadiharja[1], I N Sukarta[2], N W I Pradnyandari[1]

[1]Biology Department of Education Faculty of Mathematics and Natural Sciences Ganesha University of Education Indonesia
[2]Chemical Analyst Department Faculty of Mathematics and Natural Sciences Ganesha University of Education Indonesia

E-mail : manikwidiyanti@gmail.com/putu.manik@undiksha.ac.id, sanusimulyadiharja@yahoo.com, sukartainyoman@gmail.com, pradnyandariintan@yahoo.com

Abstract. Making nata made from older *Borrasus flabellifer* wine (namely Nata de Lontar), it needs addition sucrose to produce nata with the optimum weight. The research aimed to know (1) The effect of addition sucrose with different concentrations on older palm wine toward the weight of "nata de lontar" (*Borrasus flabellifer*) (2) The concentration of sucrose is the most influential part to produce optimum weight of "nata de lontar". This study belonged to an true experimental research with posttest only control group design. The independent variable in this research was the addition of sucrose with different concentrations of 6% , 9 % , and 12 %. The dependent variable in this study was the weight of nata. The results were analyzed using one-way anova and then followed by a test of Least Significant Difference at the 5 % significance level. The results of the experiment show (1) There have been a weight difference "nata de lontar" from bacterial fermentation in older palm wine *Borassus flabellifer* Linn medium with the addition of different concentrations of sucrose with p-value less than < 0.05. (2) Addition 12% sucrose concentration to the medium has influenced the formation of nata optimum results with average 46.13 grams.

1. Introduction

Nata comes from the Spanish word meaning cream. Nata is solid, white and transparent of colors and chewy [1]. Nata is health food product, using cereals, legumes, sweet corn, unpolished rice, mung bean and nata de coco as sources of dietary fiber [2]. Nata fermentation by acetic acid bacteria *Acetobacter xylinum* to produce cellulose [3].

Nata is a fermented product made of high sugar solution using *Acetobacter xylinum*. Enzyme from these bacteria converted sugar in fermentation media into a million of cellulose fibril (fiber) which will become white or transparent and compact. Fiber as supplement was high in insoluble fiber, the results seem to imply that the insoluble fiber in the supplement was responsible for the reduction in serum triglyceride of the hyperlipidemic subjects and blood cholesterol (TC) [2]. Cellulose produced by *Acetobacter xylinum* is named bacterial cellulose (BC) or microbial cellulose [4]. A cellulose derivative using the bacterial cellulose nata de coco has also been synthesized [5].

Carboxymethyl cellulose (CMC) has been widely used in edible coatings for fresh fruits and vegetables [6]; [7] The derivative is herein referred to as carboxymethylnata (CMN). The CMN

IOP Conf. Series: Journal of Physics: Conf. Series **1040** (2018) 012006 doi:10.1088/1742-6596/1040/1/012006

coatings reduced the rate of ripening of bell pepper fruits. [8] and [9] investigated the use of edible natural biodegradable coatings to replace commercial synthetic waxes as coatings for citrus fruits.

Under a microscope, the nata looks like irregular fibril mass resembling threads or cotton. Generally, nata is made by using coconut water known as nata de coco. Nata can also be made from other materials containing sugar such as pineapple, tomato, cocoa [1] and palm wine from *Borrasus flabellifer* Linn.

Palm plant (*Borrasus flabellifer* Linn.) is used by human, whether used for food, building materials, household furniture and handicrafts. *Borrasus flabellifer* is one type of palm that grows in coastal areas in dry climates, such as in Buleleng and Karangasem Bali Indonesia. Papyrus plant (*Borrasus flabellifer* Linn) cultivated by Bali's farmers as a shade plant whose leaves are used as craft materials such as fans, hats, mats and to write media of traditional words. Whereas wood of palm plant can be used as building materials, tools and materials hand-crafted. Palm fruit can be used as a mixture of iced drinks or eaten fresh. Palm flower containing sap which can be consumed as a fresh beverage or through processed into brown sugar.

In Buleleng regency the province of Bali of Indonesia, precisely in the village of Abasan, district Tejakula. Palm (*Borrasus flabellifer* Linn.) sap is sold to manufacturers of brown sugar and palm wine. Best quality sap used as raw material for making palm wine and brown sugar are freshly harvested sap. Palm wine that is old with age more than one day (older palm wine) result palm wine value decrease. This happens because the sugar in palm wine decreases every day and the level of acidity increases. Circumstances sap acid can not generally be used as a basic ingredient of making palm wine and brown sugar. In an effort to improve the utilization of older palm wine, research must be done to produce a older palm wine processing technology to produce new products that have higher economic value. One of them is nata.

Making nata derived from wine raw materials, has not received much attention. This is caused by the lack of public knowledge in the use of wine into nata. Nata is a product of the bacterial *Acetobacter aceti* subspecies *xylinum* fermentation of a mixture material and glacial acetic acid were added to adjust pH to pH 3-4. A decrease in pH of the medium that passes through the pH optimum fermentation process causes disruption and disintegration of re nata cellulose into glucose that can be oxidized again to acetic acid. Sucrose passing optimum levels, causing a lot of sugar that is converted into acid. Nata sheet was harvested after 25 days [10].

Based on that explanation, then do research "The Effect of addition sucrose concentrations on older palm wine toward weight nata de lontar (*Borassus flabellifer* Linn)"

2. Methods

2.1 Research design
The Research design was used is a true experimental research, with the posttest only control group design has been using a Completly Randomized Design have been using simple random technique. The number of samples that have been used followed the formula $t(r-1)\geq 20$ [11] where t is the treatment, r is the replication. The treatments that have been used different concentrations of sucrose 6%, concentration sucrose 9% and concentration sucrose 12%.

2.2 Population and sample
Population that has been used is older palm wine with volume 1500 ml is derived from 3 strands mature in 1 palm plant of *Borassus flabellifer* Linn. The study sample was using older palm wine with a volume of 1200 ml and the sample was divided into 24 samples. Each sample containing 50 ml volume of older palm wine.

2.3 Materials and Equipment
Older palm wine, sucrose, *Diammonium phosphate* (DAP), *Zwavelzure Ammoniac* (ZA), Alcohol 70% concentration, glacial acetic acid, *starter (mother liquor) Acetobacter xylinum*, aquadest, cotton, and

aluminium foil. And equipment were : petri dish, filter, bekker glasses, trays, paper, erlenmeyer, pippet, spatula, analytical balance, autoclave, hot plate, latex for closed of bottles, bunsen, and glasses volume.

2.4 Data Collection technique

2.4.1 "Nata de Lontar" Production

Older palm wine that has been aged for about 3 days of tapping filtered and put into erlenmeyer respectively of the volume of 50 ml and then boiled in hot plate. Each older palm wine has been added ZA fertilizer as much as 0.2 g/50 ml of older palm wine, DAP 0.3 g/50 ml of older palm wine, 0.1 ml of glacial acetic acid/ 50 ml and sucrose according to treatment that is 0 g/ 50 ml, 6%= 3 g/50 ml, 9%= 4.5g/50 ml and 12%= 6 g/50 ml then boil. Hot older palm wine solution is removed and put into each petri dish. The petri dish is placed on trays and then was done closing the tray using paper. Cool solution approximately 1 hour, paper cover is opened and each solution was added to a solution of 8 ml starter. Trays that had been given starter has been closed with the paper. The next solution is incubated at room temperature (about 28^0C).

2.4.2 Data Analysis technique

The data that is heavy nata fermented taken on day 5 were measured using analytical balance. Data analysis technique using one-way anova for hypothesis testing. If in the analysis there is a meaningful difference, then do a further LSD (Least Significance Difference) at 5% significance level.

3. Results and Discussion

3.1 Results

Weight nata fermentation result are shown in **Table 1**. The optimal weight of nata that has been produced is at a concentration of 12%. The lowest nata weight was produced at a concentration of 6%. Increased nata weight has occurred, after administration of increased sucrose to the solution.

Table 1. Average effect of addition sucrose concentrations on older palm wine toward weight of "nata de lontar" (*Borrasus flabellifer*) Linn.

Treatment	Weight of nata per 50 ml older palm wine (gr)		
	6%	9%	12%
1	41.95	45.20	46.92
2	38.34	43.26	44.03
3	42.13	45.45	48.46
4	36.75	43.90	45.56
5	40,13	44.23	44.16
6	40,65	43.46	45.28
7	38.20	44.19	47.94
8	41.23	42.05	46.65
Total	319.38	351.74	369.00
Average	39.92	43.98	46.13

IOP Conf. Series: Journal of Physics: Conf. Series **1040** (2018) 012006 doi:10.1088/1742-6596/1040/1/012006

Differences in treatment as result of fermented older palm wine illustrated in Figure 1.

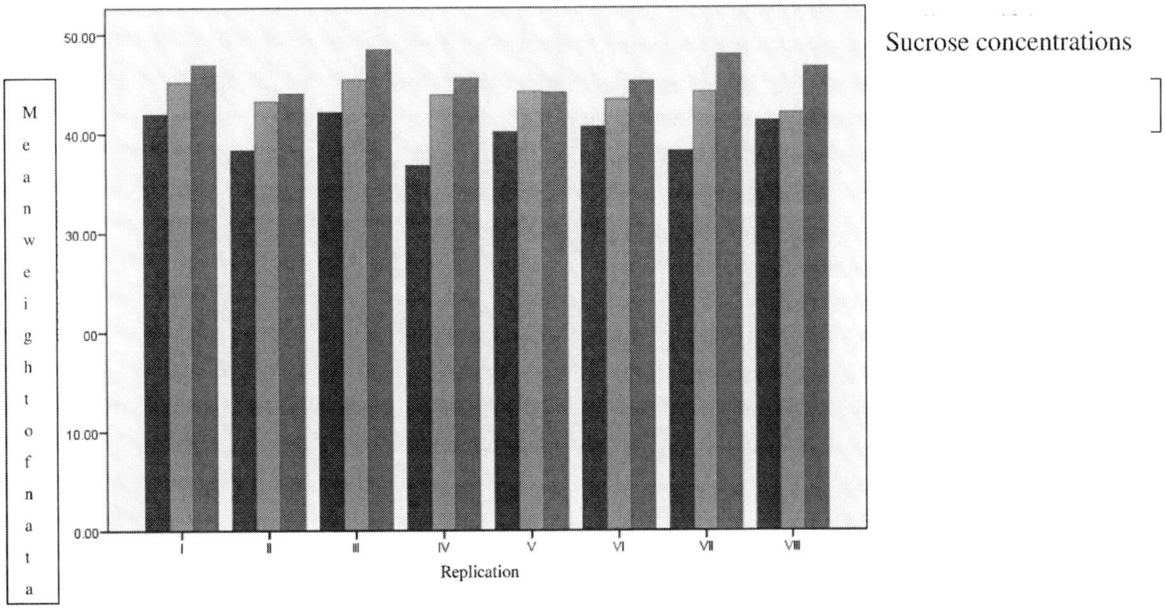

Figure 1. Grafic Heavy "nata de lontar" *Borrasus flabellifer* Linn

The observation results of nata color has shown the color is muddy brown as Figure 2 bellow.

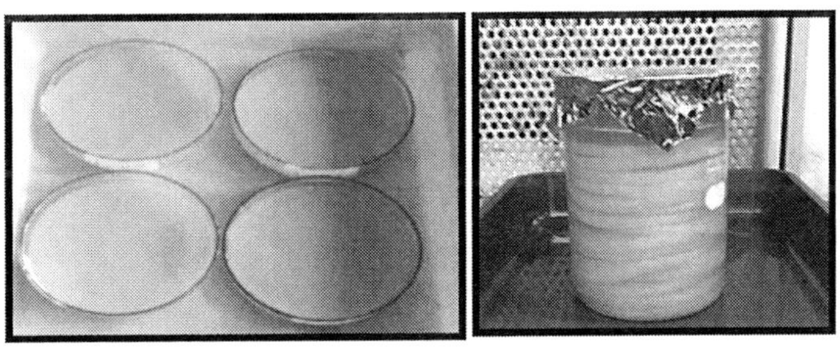

Figure 2. Nata production.

3.2 Data analysis

Data were analyzed using one-way anova with the stages of normality test using Kolmogorov-Smirnov, homogeneity data using Levene test and hypothesis test using one-way anova shown in **Table 2.**

Table 2. Hypothesis testing using one-way anova at 5% significance levels

Group	Average	Deviation Standard	F-value	p-value
6%	39.923	1.957	30.837	0.000
9%	43.968	1.087		
12%	46.125	1.645		

Based on Kolmogorov-Smirnov test results have shown the value of p (probability) is more than 0.05 shows the distribution of the data follow a normal curve distribution. Homogeneity test showed p-value greater than 0.05, the data shows homogenous distribution of the data.

Results of hypothesis testing showed p-value less than 0.05 have demonstrated an nata weight difference between treatments. Hypothesis testing continued with LSD test to analyze treatment significantly and the level of significance shown in Table 3.

Table 3. Results LSD testing using Post Hoc Test at 5% significance levels

Group	*Different*	Value *p(Sig.)*
6% with 9%	-4.04500*	0.000
6% with 12%	-6.20250*	0.000
9% with 6%	4.04500*	0.000
9% with 12%	-2.15750*	0.014
12% with 6%	6.20250*	0.000
12% with 9%	2.15750*	0.014

3.3 Discussion

Nata be made from various types of materials that contain sugar, protein, minerals [12]. One nata besides the manufacture of coconut water is older palm wine *Borrasus flabellifer* Linn. *Borrasus flabellifer* Linn reinforced by [13] stating that nata fermentation by using raw material produces nata is still low, ranging from 27.36 to 60.32% or average 34.31%. Low nata is due to the fermentation medium vulnerable to contamination by fungi. Nata fermenting bacteria is *Acetobacter xylinum* as the growing medium contains water, protein, fat, carbohydrate, ash and some minerals to nourish the growth of bacteria. Nutrients are added to the growth medium of bacteria are sucrose as a carbon source, ZA source of nitrogen, DAP is source of phosphate.

Under a microscope, the nata looks like irregular fibril mass resembling threads or cotton [1].

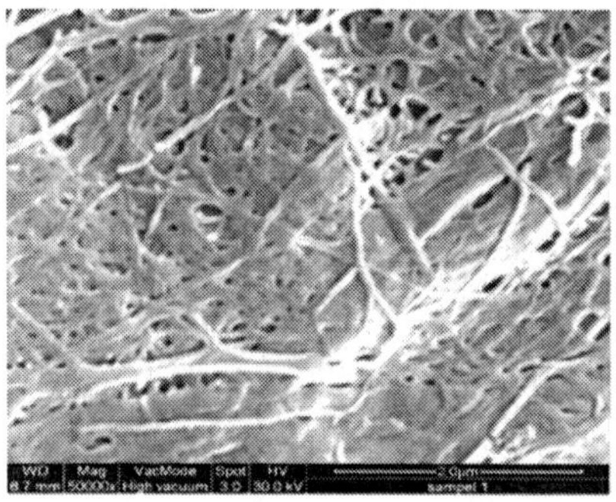

Figure 3. Irregular fibrill mass of nata under microscope

Source : [14]

Acetobacter xylinum bacteria use sucrose as a source of energy and carbon sources into extracellular cellulose so that the cells of bacteria trapped in a cellulose fiber layer.

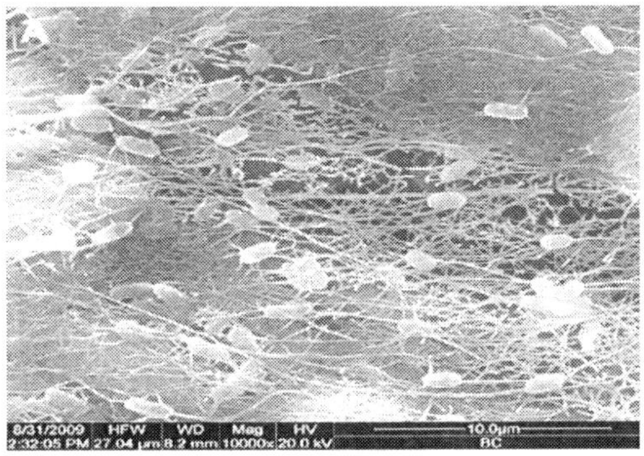

Figure 4. Bacteria trapped in cellulose fiber

Source : [15]

The longer incubation supplies nutrients begin to decrease so that it will increase competition between bacterial cells to sustain life [16]. Based on statistical analysis that shows the influence of the addition of different concentrations of sucrose by weight ratio nata de lontar with statistical F-value = 30.837 and significance of 0.000. Figures significance less than < 0.05 shows there is significant contribution to the weight variation of sucrose concentration nata of de lontar. This was confirmed by [17] that has proven different concentrations of sucrose in making nata of squash has given significantly different effect on weight nata at best is 0.76 cm thick. Sucrose has a heavy influence on nata also made by [18] who have proved that the addition of sucrose was significantly affected nata weight of liquid tofu waste with the highest nata weight 23.79 grams with addition of 10 grams per 200 ml of tofu whey.

In nata fermentation, the bacteria *Acetobacter xylinum* can break down sugars into cellulose components. Cellulose formation begins the growth of bacteria marked by the turbidity in growth medium after 24 hours of incubation. After 36 – 48 hours of incubation are formed in the surface layer of translucent medium and gradually thickens to form a new layer over the conditions allow. Extra long variations in the treatment that fermentation also has had a heavy influence on nata as research conducted by [19] to have proven that different concentrations of sucrose and fermentation time nata de mango have significantly different effect on the weight of nata de mango. The treatment with addition 10% sucrose and fermentation 16 days have provided a high nata weight is 1.325 cm. Whereas treatment with the addition of 6% sucrose concentration and fermentation time of 10 days give nata a low weight is 0.675 cm. [20] has got the addition of mass of sugar 7 grams up to 13 grams in 300 ml of boiling corn water for 2 weeks fermentation in making nata de corn obtained nata de corn with thickness that has been produced not significantly different. [10] has got different shape of cellulose production were used coconut water, coconut milk and skim used a spectrophotometer (Perkin Elmer), X-Ray Diffraction (X'Pert PRO PAN analytical), Thermogravimetric analyzer (Leco TGA 701), SEM-XRD (Hitachi S-3400 N) in nata de coco production.

The results the effect additional nutrition in making nata have also been supported by [21] who have proven that sugar and fermentation time had significant effect on the thickness of the producing nata de soya. From her research showed that the average thickness of nata was observed at the highest of nata giving treatment with the addition of sugar 100 grams/l which has been combined with long fermentation for 7 days, with the resulting thickness of 3.390 mm nata. While the thickness of the thinnest nata has been observed in the treatment with the addition of sugar 150 grams/l were combined with the length of fermentation 14 days with a thickness 2.403 mm nata. [22] has been done using pineapple in the manufacture of nata de pinna has got results the highest weight was reached in treatment B (additional sugar + ZA) has got results weight of nata 899 grams/5 L, followed by treatment A (without addition sugar and ZA) has got weight of nata 616.4 grams/5L, and C (stored for six months with the addition of nutrients both additional sugar +ZA) has got weight of nata 477.8

grams/5 L. The thickness of NP of the height and low as in treatment B (1.58 cm) followed by treatment A (1.24 cm) and C (0.88 cm), respectively.

The addition of organic nitrogen source derived from soybean germs has resulted in better nata weight compared with the addition of urea [1]. The addition of 20 ml greenbean germs extract/2000 ml solution was also done on nata de cassava research and has got results average 5.4 grams nata/50 ml; 9.9 grams nata/ 50 ml; 15.9 grams nata/ 50 ml and 19.7 grams nata/50 ml in treatments 0%; 0.75%; 1.5%; 2.25%, respectively of sugar [23].

In the process of fermentation, *Acetobacter xylinum* bacteria cells will grow rapidly from the initial cell number by experiencing exponential growth until it reaches the maximum amount with a generation time of approximately 2 hours [24]. When the number of cells reached a certain amount, aggregate form began to appear and the number of cells in solution are not rising as long as nata aggregate formation, some bacteria enter the cell cavities between the aggregate so that the bacterial cells are not free in solution.

On culture growth, the availability of oxygen in the surface of the medium and in the presence of glucose will stimulate an increase in cell mass and increase the amount of cellulase is enzyme forming cellulose. During the formation of cellulose, the cavities are filled by water nata causing nata become thick and contains water as much as ±95-98% and 2-5% cellulose [24].

Based on these results research by measuring the weight of "nata de lontar" has obtained an average yield of successive 39.92 grams, 43.98 grams, and 46.13 grams with the addition of sucrose concentrations in a row 6%, 9% and 12% respectively. Data have shown that the addition of sugar concentration of 12% by weight nata optimum yield.

4. Conclusion

The results of research that has been done statistically obtained there have been a weight difference nata from bacterial fermentation in older palm wine *Borassus flabellifer* Linn medium with the addition of different concentrations of sucrose with p-value less than < 0.05. Addition 12% sucrose concentration to the medium has influenced the formation of nata optimum results with average 46.13 grams.

References

[1] Fifendy M, Putri D H, Maria S 2011 *J Saintek* **3** (2) 165-170.
[2] Mesomya W, Pakpeankitvatana V, Komindr S, Leelahakul P, Cuptapun Y, Hengsawadi D, Tammarate P and Tangkanakul P 2006 *J Sci Technol* **28** (1) 23-28.
[3] Swissa M, Aloni Y, Weinhouse H, Benziman M 1980 *J Bacteriol* **143** 1142-1150.
[4] Panesar P S, et al. 2012 *Carbohydrate Polymers* **87** (1) 939-934
[5] Sabularse V C, Montalbo M N D, Hernandez H P and Serrano E P 2009 *Int J of Food Sci and Nut* **60** (7) 206–218
[6] Niazmand A, Ghodusi H B, Shahidi F and Niazmand R 2009 *J Agri&Environ Sci* **5** 740–744
[7] Bahri M H and Rashid M 2009 *Int J of Agr and Biol* **11** 443–447
[8] Arnon H, Zaitseva Y, Portat R and Poverenova E 2014 *J Postharvest Biol and Tech* **87** 21–26
[9] Arno H, Granit R, Porat R and Poverenov E 2015 *J Food Chem* **166** 465–472
[10] Andasuryani, Kasim A, Rahmi I D and Derosya V 2017 *J of Pharm Biol and Che Sci* **8** (2) 658-664
[11] Gazpersz 1995 *Teknik Analisis Dalam Penelitian Percobaan* (Bandung: Penerbit Tarsito)
[12] Wowor L Y, Muis M, Arinong A R 2007 *J Agrisyst* **3** (2) 77-86
[13] Lempang M 2007 *J.Penel Hasil Hutan* **25** (2) 147 – 157
[14] Halib N, Amin M C I M and Ahmad I 2012 *J Sains Malaysiana* **41** (2) 205–211
[15] Jagannath A, Raju P S, Bawa A S 2010 *LWT Food Sci Tech* **43** 1197-1203
[16] Nisa F C 2002 Penurunan Tingkat Polusi Dari Limbah Cair Tahu (Whey) Pada Produksi Nata de Soya (Review Waktu Inkubasi) *J of Agri Tech* **3** (2)

[17] Rona D P 2003 Pengaruh Konsentrasi Sukrosa Pada Karakteristik Nata dari Labu (*Sechium edule*). *Undergraduate Thesis Unpublished* (Program Studi Hasil Tekhnologi) (Padang : Universitas Andalas (In Indonesia)
[18] Nisa F C, Rani R H, Wastono H, Baskoro B, Moestijanto 2001 Produksi Nata dari Limbah Cair Tahu (Whey) : (Studi Penambahan Sukrosa dan Ekstrak Kacang Hijau) *J Agricul Tech* **2** (2) 74-78.
[19] Djajati S, Sarofa U, A Syamsul 2009 Pembuatan Nata de Manggo (Studi Pengaruh Sukrosa dan Lama Fermentasi) *Skripsi Tidak dipublikasi* (East Java : UPN Veteran)
[20] Rizal H M, Pandiangan D M, Saleh A 2013 Pengaruh Penambahn Gula, Asam Asetat dan Waktu Fermentasi Terhadap Kualitas Nata de Corn *J Tekh Kim* **19** (1)
[21] Nurhayati S *J Mat SainTek* **7** (1) 40-47
[22] Sutanto A 2012 *Makara Teknol* **16** (1) 63-67
[23] Widiyanti N L P M, Ristiati N P, Mulyadiharja S 2015 *Proceed Semnas MIPA* Peningkatan Konsentrasi Sukrosa Terhadap Perbedaan Berat *Nata de Cassava* Dari Limbah Cair Pembuatan Tepung Tapioka (Singaraja-Bali : Universitas Pendidikan Ganesha)
[24] Ratnawati D 2007 *J Gradien* **3** (2) 257 – 261

IOP Conf. Series: Journal of Physics: Conf. Series **1040** (2018) 012007 doi:10.1088/1742-6596/1040/1/012007

Potential of yellowfin tuna catch in East Java-Indian Ocean based on length frequency and age distribution

D Hidayati, R Herlambang, N Jadid, N N Sa'adah, N Maulidina and A P D Nurhayati

Biology Department, Institut Teknologi Sepuluh Nopember (ITS) Surabaya-Indonesia

E-mail: dewi_hidayati@ymail.com

Abstract. Tuna's catch in Indonesia is the largest in the world, which contributing approximately 16 percent of the world's tuna supply. Investigation of tuna length frequency distribution that caught in Indian ocean has conducted to know the Indonesian's fisheries potential included yellowfin tuna (YFT). The result study provide useful data for estimating fish growth as a fundamental for YFT stock assessments and managing exploitation. The 203 YFT samples were collected from fish landing area at Sendang Biru Malang, East Java Indonesia during April and May 2017. The research found that YFT in East Java- Indian ocean was dominated with length group of 151-180 cm (53.2 %) and followed by with length group of 121 – 150 cm (33.5%) with estimated age of 4.1 – 6.0 years and 3 ,1- 4,0 years respectively. There were 17 individual YFT at juvenile size in group length of 31-60 cm was found 8.4%, while the maximum group size (181-120 cm) was reported at level of 4.4%. The variation of YFT length and age indicates the variation of trophic level in East Java -Indian ocean water which very important for sustainable fisheries.

1. Introduction

Indonesia is the archipelago country that dominated with the sea area (76.94 %) or 6,653,341,439 km^2 [1], therefore opens great opportunities for the development of fisheries sector, including tuna. Indonesia's tuna catch is the largest in the world which contributes about 16 % of the world's total tuna supply [2].

As a top predators in the ocean, tuna plays an important role in marine ecosystems and provides the worldwide protein requirements [3]. The one of abundant and high economically value of tuna catch species in Indonesia is YFT (*Thunnus albacares*) which in 2014 reached 65.686 metric tons [2] and [4] . The geographic position of Indonesia included the East Java-Indian ocean is fits for YFT to breed hence supply high quality YFT across the global market. [5]. Moreover, YFT catch have increased in recent years Due to its potential, the regulation of yellowfin exploitation has been issuing by Marine and Fisheries Ministerial to regulate capture and the Sustainable Fisheries Partnership (SFP) of Indian Ocean Longline Tuna [2]. Accordingly, the stock assessment for exploitation management of YFT in East Java-Indian Ocean is required. The study of length frequency and age distributions of catch may provide directly and useful information the condition of fish [6] and importantly used data for estimating fish growth as a fundamental for fish stock assessments and managing exploited species [7] and [8], as well as to predict the future yields, the sustainability of biomass levels and value of the catch [9] Furthermore, this study provides fisheries data that has importantly contribution for the sustainable YFT catch in Indonesia.

2. Methods

2.1. Time, Place and Research Sample

Site sampling determination were conducted based on current information of forecast map of fishing area of Java, Bali and Nusa Tenggara Region that produced by the Ministry of Marine Affairs and Fisheries (KKP) in 2017. The site sampling determined the location of the study covering the South Coast of East Java and technically considering the basis of accessibility or accessibility aspects. In this study this criteria was represented by Sendang Biru coastal area, Malang Regency, East Java Indonesia (Figure 1). Sendang Biru is well-known fisheries area that produces the best handline tuna in Indonesia [10] Samples of YFT were collected from fish landing area at Sendang Biru during April and May 2017.

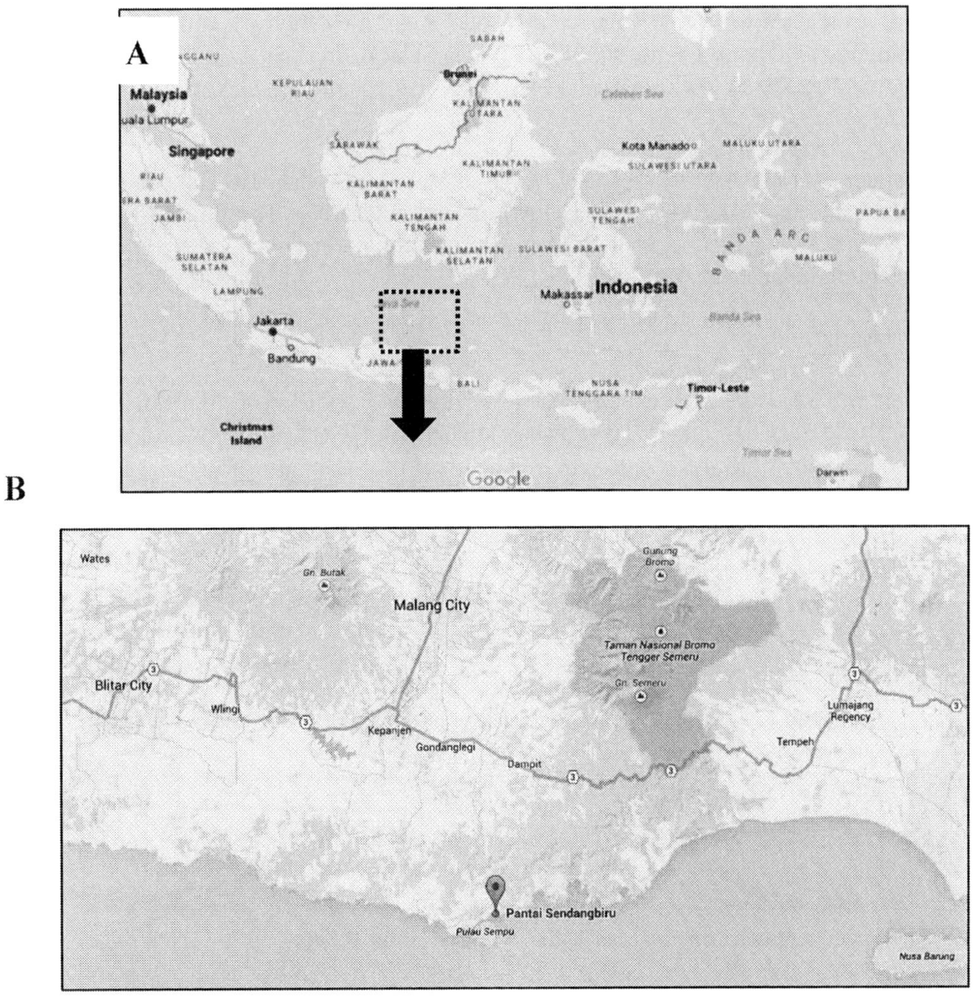

Figure 1. Sampling site of YFT: A. Map of Indonesia, the dashed square is region of East Java; B: YFT sampling site at coastal area of Sendang Biru, East Java (balloon mark). Source with modification :https://www.google.co.id/maps/place/Indonesia [11] https://afebrianas.files.wordpress.com/2015/02/map-sendang-biru-jpg-e1423730834224 [12]

2.2. Analysis of YFT body length and age frequency distribution

The body length measurement was conducted based on morphometric method adopts the usual measurement method used in fisheries research [13]. The measurement of body length, and weight of YFT were obtained on site of fish landing area at Sendang Biru. The correlation among biometric data were analyzed using scatter plot in excel. Meanwhile, the incomplete data of YFT length/weight were estimated using the length-weight relation analysis according to Costa et al. [14], with the following formula:

$$\text{Weight} = 2\times10^{-5} \times \text{Length}^{2.932}$$

The observed and estimated of length-weight data were then analyzed using Frequency function within Excel in Microsoft Office. Frequency of data that determined within a certain range, then used for created a graph which shows the frequency distribution of YFT body length. Moreover, the frequency distribution of YFT body length were used for age estimation that calculated according to YFT growth equation that made by to Costa et al. [14]

2.3. Additional data of Tuna catch fisheries

Monthly and annual data of tuna catch were analyzed based on the secondary fisheries data from local institution i.e. Pondok dadap Office of Maritime Affairs and Fisheries. The data of YFT fishing equipment that used in Sendang Biru as well as tuna catch location were obtained by interview with the fishermen that landing at Sendang Biru..

3. Result and Discussion

3.1. The trend of YFT catch at Sendang Biru

The tuna catch report at East Java-Indian ocean refer to Pondok Dadap Office of Maritime Affairs and Fisheries from year of 2013, 2015 and 2016 showed fluctuated total weight production i.e. 811.176 ton; 745.420 ton and 918.393 ton respectively.

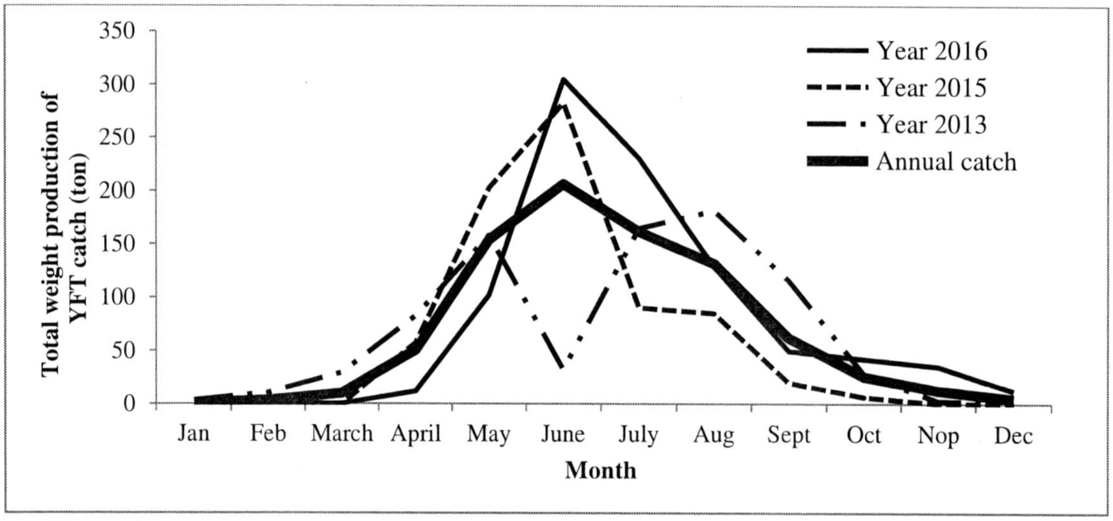

Figure 2. Monthly and annual catch of YFT production at Sendang Biru, Malang East Java during 2013-2016

Refer to Figure 2, there similarity trend showed between year of 2015 and 2016. During April to August the catch was showed increasing trend with the peak catch found at June (at 2015=. 304.84 ton; at 2016=282.742 ton). January found as the lowest month of YFT tuna catch in 2015 as well as at 2016 (0 ton), Another study by Isfandiary [15] also reported that in 2015 there was no capture of

YFT (0 ton / trip). Meanwhile, data of 2013 indicated two peaks of YFT catch i.e. on May (158.469 ton) and August (181.462 ton).

Indonesia has two seasons, wet and dry season. In Java, the wet season occurs between September to March and the dry season is from March to September. The average annual rainfall for Indonesia approximately in level of 3,175 mm, however in the eastern tip of Java tend to be dry until the rainfall decrease less than 1,000 mm [16]. These trend differences might be affected by disruption of the physical environment due to the climate patterns, which then impact natural stock behaviors such as migration, egg spawning hence decreasing the fish juveniles each year. For example, the annual fish stock in Pacific hake were different more than 100 times from year to year [17]

3.2 The correlation between body length and weight of YFT catch

The analysis of length-weight correlation that calculated using completely data of length-weight from 80 individual YFT samples presented at Figure 3. It is demonstrated the strong positive correlation between body length and weight ($R^2 = 0.92$). According to ICCAT was known that there is no significant size-weight difference between sexes, hence the length-weight applied for both sex. Length-weight relationships have been devised to obtain better estimates of catches in round weight from landed and processed catches [18]). Meanwhile, 123 individual YFT samples that were not completely yet measured (only consist of one data of length or weight) were estimated using length-weight correlation according to Costa et. al [14]

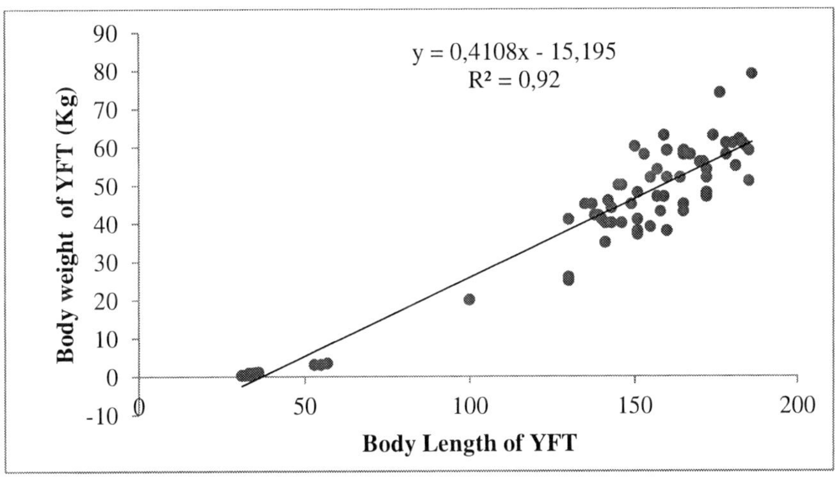

Figure 3. The correlation between body length and weight of YFT catch

3.3 *YFT body length and age frequency distribution*

Bard et al. and ICCAT reported there are three age categories: juveniles (50-65 cm), pre-adults (65-110 cm) and adults (110-170 cm). The YFT fork length of 50 cm is remain in the coastal areas, and show moderate migratory habits (30 miles). Pre-adult YFT migrates with similar patterns to that of juvenile, while adults make trophic migrations towards higher latitudes during the summer and migrations across the ocean [19].

The percentage of body length and age frequency distribution from 203 individual YFT samples showed at Figure 4 and table 1.

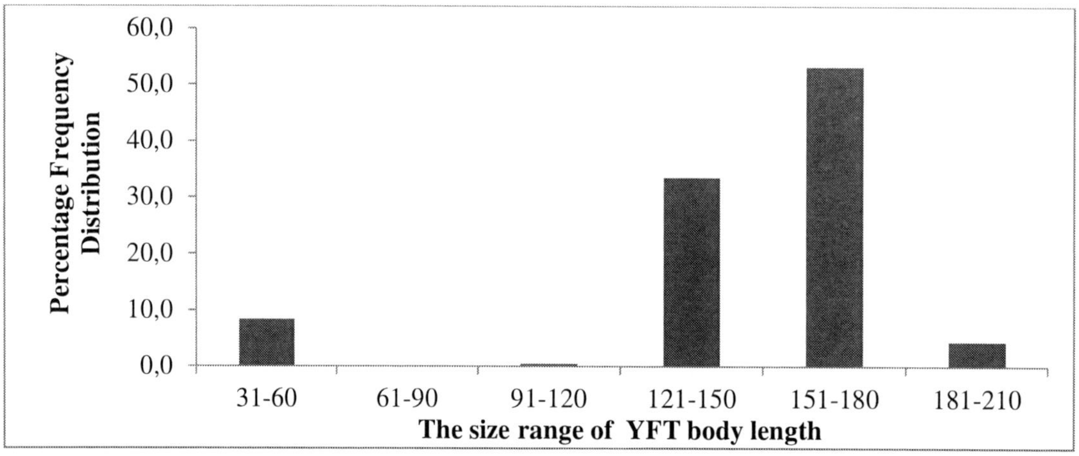

The research found that YFT in East Java- Indian ocean was dominated with length group of 151-180 cm (53.2 %) and followed by with length group of 121 – 150 cm (33.5%) with estimated age of 4.1 – 6.0 years and 3 ,1- 4,0 years. The both groups were categorized at adult size (table 1). There were 17 individual YFT at juvenile size in group length of 31-60 cm was found 8.4%, while the adult maximum group size (181-120 cm) was found at level of 4.4%.

Table 1. Frequency distribution of Body Length and Age of YFT catch

Range size	No of ind.	Percentage	Estimated age (years)	Age category (Bard et al. 1991)
31-60	17	8.4	0.5 - 1.5	Juvenile
61-90	0	0.0	1.6 - 2.5	Pre adult
91-120	1	0.5	2.6 - 3.0	Pre adult/adult
121-150	68	33.5	3.1 - 4.0	Adult
151-180	108	53.2	4.1 - 6.0	Adult
181-210	9	4.4	6.1-7.0	Adult

As previously described, that each age category have trophic migration at different area and possibly intersection area. According to the interview data with local tuna fishermen was known that they do fishing generally at approximately 30 miles, which available to find mostly adult YFT. The variation of YFT length and age indicates the variation of trophic level in East Java -Indian ocean water which very important for sustainable fisheries. The length frequencies data may use to estimate catch-at-age which is to be the one of important component in fisheries assessment [20]. Moreover, the age distribution of the fish is an important aspect of stock predictions included the sexual maturity, the numbers of parent and young fish stock as well as estimates and manages the condition of fish populations in future years. Fish stock assessments providing the information that necessary to make sound decisions, therefore support sustainable fisheries [16] and [21].

4. Conclusion

The YFT in East Java- Indian ocean was dominated with adult sizes (91.1%) included the varied adult size groups: 151-180 cm (53.2 %); 121 – 150 cm (33.5%) and (181-120 cm). There rest of YFT catch (8.9%) were composed of varied size of juveniles and pre-adult. The diversity of size and age, the high frequency of adult YFT and small frequency of YFT juvenile/ pre adult reflects the YFT catch sustainability at economically and ecology prospective.

References

[1] Ramdhan M and Arifin T 2013 Aplikasi sistem informasi geografis dalam penilaian proporsi luas laut Indonesia. (Application of Geographic Information System for Assessment of Indonesia Marine Proportion) *Jurnal Geomatika* **19** 141-146

[2] CEA (California Environmental Associates) 2016 Indonesia Fisheries: 2015 Review: A report on trends in coastal marine resources and fisheries management in Indonesia. Prepared for The David and Lucile Packard Foundation

[3] FAO (Food and Agriculture Organization of the United Nations) 2011 Tuna and tuna-like species, in: Review of the state of world marine fishery resources. *FAO Fisheries and Aquaculture Technical Paper* **569** 227-243.

[4] Albaina A, Iriondo M, Velado I, Laconcha U, Zarraonaindia I, Arrizabalaga H, Pardo M A, Lutcavage M, Grant WS and Estonba A 2013 Single nucleotide polymorphism discovery in albacore and Atlantic bluefin tuna provides insights into worldwide population structure. *Animal Genet* **44** 678-692

[5] Ikhwan F B 2015 Export New Indonesia: Yellowfin Tuna. Directorate General of National Export Development Ministry of Trade of The Republic of Indonesia Ditjen PEN/MJL/XXV/11/2015.

[6] Erismana B E, Ashley M, Apel , Alec MacCall D, Románd M J and Fujita R 2014 The influence of gear selectivity and spawning behaviour on a data-poor assessment of a spawning aggregation fishery . *Fisheries Research* **159** 75–87

[7] Francis R I C C 1988 Are growth parameters estimated from tagging and age-length data comparable? *Can. J. Fish. Aquat.Sci* **45** 936–942

[8] Gulland, J. A. 1988. Fish population dynamics: the implications for management, 422 p. John Wiley and Sons Ltd., New York.

[9] Sparre P and Venema S. C. 1998. Introduction to Tropical Fish Stock Assessment - Part 1: Manual. FAO fisheries technical paper 306/1 Rev. 2

[10] Arifin Z 2016 Capture Fisheries Assistant For Tuna Fisherman Community Engagement WWF-Indonesia).

[11] https://www.google.co.id/maps/place/Indonesia

[12] https://afebrianas.files.wordpress.com/2015/02/map-sendang-biru-jpg-e1423730834224 [8

[13] Gayanilo F C and Pauly D 1997 The FAO ICLARM Stock Assessment Tools (FISAT) reference manual. FAO Computer Information Series

[14] Costa F E S, Braga F M S, Amorim A F and Arfelli C A 2005 Fishery biology of the YFT, *Thunnus albacares* in southern Brazil *Col. Vol. Sci. Pap. ICCAT* **58** (1) 309-349

[15] Isfandiary X.I 2016 Pola Produksi Penangkapan Ikan Tuna Sirip Kuning (*Thunnus albacares*) dan Albakora (*Thunnus alalunga*) di Perairan Selatan Jawa Timur Berdasarkan Hasil Tangkapan Yang Didaratkan di Instalasi Pelabuhan P. Sarjana thesis, Universitas Brawijaya.

[16] Hays J 2015 Weather and climate in Indonesia http://factsanddetails.com/indonesia/Nature_Science_Animals/sub6_8a/entry-4079.html

[17] NOAA 2012 Science Advisor for Stock Assessments: Fish Stock Assessment 101 Series: Part 1—Data Required for Assessing U.S. Fish Stocks http://www.nmfs.noaa.gov/stories/2012

[18] ICCAT 2006 ICCAT Manual CHAPTER 2.1.1: Yellowfin tuna. author: IEO https://www.iccat.int/Documents/SCRS/Manual/CH2/2_1_1_YFT_ENG.pdf

[19] BARD, F.X. and Scott. E.D 1991. Sept traverses transatlantiques d'albacores marques thons migrateurs ou sédentaires? *Collect. Vol. Sci. Pap, ICCAT* **36** (1) 205-222.

[20] Andrade H A and Kinas Paul G 2004 Estimation of birthdates and catch-at-age using length frequency analysis (LFA), with application for skipjack tuna (*Katsuwonus pelamis*) caught in the Southwest Atlantic. *ICES Journal of Marine Science* **61** 798-811. doi: 10.1016/j.icesjms.03.002

[21] WOR2 (World ocean review) 2013 Living with the oceans. Published by maribus in cooperation with 2 The Future of Fish – The Fisheries of the Future

IOP Conf. Series: Journal of Physics: Conf. Series **1040** (2018) 012008 doi:10.1088/1742-6596/1040/1/012008

The formation of mini illustrated dictionary of rare plants in the village forest of Penglipuran, Bangli, Bali

N Wijana and I G A N Setiawan

Universitas Pendidikan Ganesha.

Abstract. The objective of this study is to formulate a mini illustrated dictionary of rare plants in the village forest of Penglipuran, Bangli, Bali. The population of this research was the entire plant species in the village forest of Penglipuran, Bangli, Bali. The samples in this study were the entire plants covered by 100 quadratics of size 20 x 20 m. From those plant species covered by the quadratics, then conducted document studies, interview and comparative observations to determine the category of rare plant species. The sampling technique used was systematic sampling technique. ADDIE design was used in this dictionary formation. The research results showed that (1) There were 17 rare plant species out of 34 species found in the village forest of Penglipuran, Bangli, Bali; (2) Those 17 rare plant species were furthermore used as the content of the Mini Illustrated Dictionary of Rare Plants; (3) The Mini Illustrated Dictionary of Rare Plants which was developed based on ADDIE design is a dictionary with plants local name, scientific name, picture and the organ, classification, and the description.

1. Introduction

There have been many studies about plant ecology done by Wijana and the Biology FMIPA Undiksha student and other studies done outside Bali by the other researchers. From those studies result, in general they presented the data of plant species composition which constructed the ecosystem in which the study has been conducted. Those studies were conducted on land or terrestrial or in water ecosystem or aquatic. For examples, the studies done by Wijana [1],[2] in the village forests of Tigawasa, Buleleng. In 2008 and 2009 had been done a research in the Penglipuran, Karangasem tradisional forest Wijana & Sumardika [3] & Wijana [4]. The results of the study which was done in Batur Lake area by Wijana & Sumardika [5]. Another study result of terrestrial vegetation analysis in the village of Tenganan Pegringsingan, Sub-district of Manggis, Regency of Karangasem in 2005 Wijana & Sumardika [6]. As well as the study that had been done by Wijana [7] at Tourism Area of Lovina.

The studies of terrestrial vegetation outside Bali had been done also by Arrijani [8], Irwanto [9], Junaedi [10], Sri Hartini [11], Onrizal, et al. [12], Onrizal [13], Purwaningsih [14], and Purwaningsih & Yusuf [15] that examined the species composition, species diversity, and protected forests and National Parks management. Those studies were done in various regions such as Arrijani did the research in Cianjur, Irwanto in Maluku, Junaedi in West Java, Sri Hartini in East Borneo, Sunarti, et al in North Sulawesi, Onrizal in West Borneo, and Purwaningsih did the research in Southeast Sulawesi. The context of these studies were more oriented to the study of vegetation parameters or vegetation analysis.

From those research results showed that the plant species constructing the field research ecosystem were only listed in a form of floristic list. Floristic list is a form of plant species name list both in their scientific and local names. It is not completed yet with the plant pictures or photos, and the plant species description based on the species list composed. If this continues and the students use this as a learning material, then there will be a proverb saying *weruh ring aran tan weruh ring rupa*, or vice versa *weruh ring rupa tan weruh ring aran*, which means the students will only be familiar with the plant species names without knowing the image of them, and vice versa they will know the object without being familiar with the name. Therefore, biology is considered as a rote science.

In general Biology science, and especially in Plant Ecology, the students will be exposed to the field realities. In Plant Ecology, they will be exposed to various plants in an ecosystem. The students will get confused, overwhelmed, and psychologically depressed when they are facing the situation to know the scientific names of those various plant species. There are a lot of unknown plant species, unknown the classification identity, and unknown taxon hierarch. This is one of the problems faced by the students on their field study, even on their thesis preparation.

In order the research results related to Plant Ecology have practical value to be used by junior and senior high school students, and then one alternative that can be done is to compile a Mini Illustrated Dictionary of Rare Plants in The Village Forest of Penglipuran, Bangli Bali. This mini dictionary contains the floristic lists, images and photos of the plant species, their classifications and descriptions. The species description includes the morphological, ecological, and taxonomic descriptions.

The formation of this illustrated dictionary about tree habits means that the dictionary is limited to plant species of tree habits, while plant species of bush and herbaceous habits are not yet explored. The location of this study is limited to the village forest of Penglipuran only, so that the illustrated dictionary is limited only by that ecosystem study area. Thus, this dictionary will be in mini, limited and local form. Although this dictionary is in mini and local form, but in terms of meaning, it can be used globally.

The establishment of this mini dictionary is not only based on some collection of book references, but it is also based on the field exploration results, the field realities on certain ecosystem. Therefore, it is important to do exploration on an ecosystem to acknowledge the plant species that construct the ecosystem. In this study, the exploration is done to the village ecosystem in one of tourism object of Penglipuran Village. From the explanation above then can be presented the objective of this study is to establish a mini illustrated dictionary of rare plants in the village forest of Penglipuran.

2. Methods

This was an explorative research. The location of this study was in the village forest of Penglipuran, Bangli, Bali. The population of this study was the entire plant species in the village forest of Penglipuran. The samples of this study were the entire species plants covered by 100 quadratics of size 20 x 20 m with interval 10 m (Cox [16], Barbour et al. [17], Mueller-Dombois & Ellenberg [18], Wijana [19]). From those plant species covered by the quadratics, then conducted a document study, interview, and comparative observation to determine the rare plant species category. The sampling technique used in this study was systematic sampling technique (Figure 1).

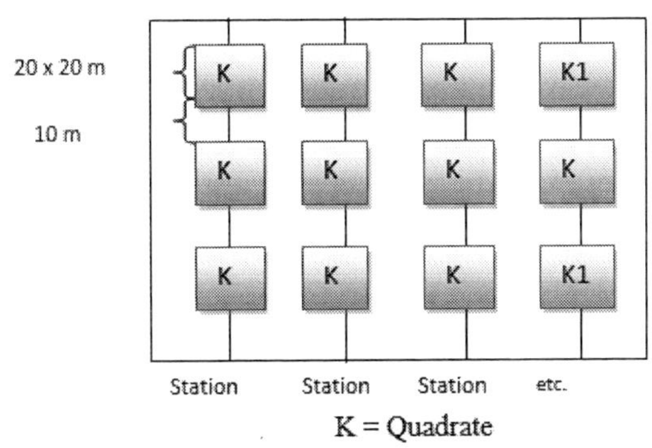

Figure 1. Layout of Quadratic Positioning

ADDIE design was used in the formation of this rare plant mini illustrated dictionary. ADDIE design consists of 5 steps, they are: (1) Analyze, (2) Design, (3) Development, (4) Implementation, (5) Evaluation (Tegeh, et al. [20]) (Figure 2). The data were analyzed descriptively.

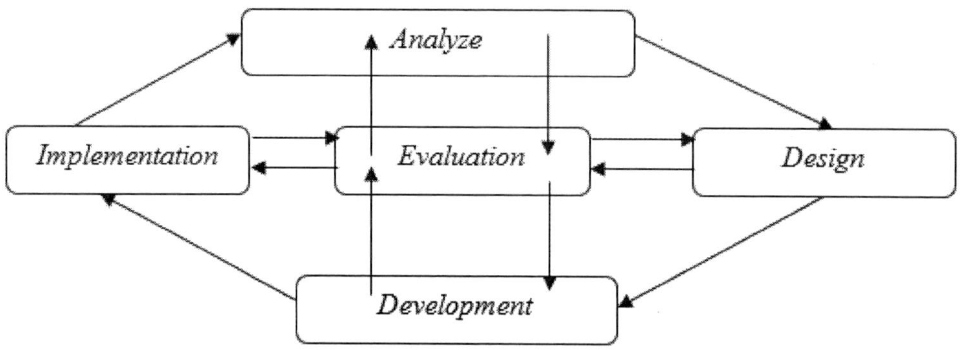

Figure 2. ADDIE Design was used in the Formation of This Rare Plant Mini Illustrated Dictionary

3. Results and Discussion

The research results related to rare plants show that there are 17 rare plant species out of 34 plant species found in the village forest of Penglipuran. The data of rare plants in the village forest of Penglipuran are presented in Table 1.

Table 1. The List of Rare Plant Species in the Village Forest of Penglipuran

No.	Family	Plant Species Name	
		Local/Indonesian Name	Scientific Name
1	Magnoliaceae	Base-base	*Elmerillia ovalis*
		Cempaka	*Michelia alba*
2	Anacardiaceae	Bejaran	*Lannea coromandelica*
3	Rutaceae	Bila	*Aegle marmelos*
4	Euphorbiaceae	Boni	*Antidesma bonius*
5	Elaeocarpaceae	Genitri	*Elaeocarpus ganitrus*
6	Myrtaceae	Juwet batu	*Syzygium cumini*
7	Myristicaceae	Kayu jelema	*Knema laurina*
		Pala jiwa	*Myristica* sp
8	Phyllanthaceae	Kepundung	*Baccaurea racemosa*
9	Moraceae	Mabi	*Ficus rasemosa*
		Nangka-nangka	*Artocarpus* sp
		Beringin	*Ficusbenjamina*
10	Meliaceae	Majegau	*Dysoxylum densiflorum*
11	Clusiaceae	Mundeh	*Garcinia dulcis*
		Peredah	*Garcinia cerlebica*
12	Sapotaceae	Nyaman	*Manilkara* sp

From Table 1 above it is found that there were 12 families with 17 species which were categorized as rare plant in the village forest of Penglipuran. Based on the results of the document study, the interview and the comparative observation, it was stated that the above plant rarity level could be explained that the plants with national rarity level were the family of Euphorbiaceae (*Antidesma bonius*) and the family of Meliaceae (*Dysoxylum densiflorum*). The plants with province rarity level were the family of Magnoliaceae (*Michelia alba*), the family of Elaeocarpaceae (*Elaeocarpus ganitrus*), the family of Myrtaceae (*Syzygium cumini*), the family of Myristicaceae (*Knema laurina*), the family of Phyllanthaceae (*Baccaurea racemosa*), the family of Moraceae (*Ficus benjamina*), and the family of Clusiaceae (*Garcinia dulcis* and *Garcinia cerlebica*). The plants with regency rarity level were the family of Magnoliaceae (*Elmerillia ovalis*), the family of Anacardiaceae (*Lannea coromandelica*), the family of Rutaceae (*Aegle marmelos*), the family of Myristicaceae (*Myristica sp*), Moraceae (*Ficus rasemosa* and *Artocarpus sp*), and Sapotaceae (*Manilkara sp*). So, there were 2 (11.76%) plant species categorized as rare plants at national level, 8 (47.05%) plant species as rare plants at provincial level, and 7 (41.17%) plant species as rare plants at regency level.

From the rare plant exploration in the village forest of Penglipuran, it could be developed into a dictionary. This dictionary had been formatted in 3 types of dictionary, they were type A dictionary, which contained scientific and local names only; type B dictionary which contained scientific name, local name and description; and Type C dictionary which contained images/photos of the plants completed with the organs, scientific name, local name, classification and description. The dictionary content displays of each type are presented in Figure 3, Figure 4 and Figure 5.

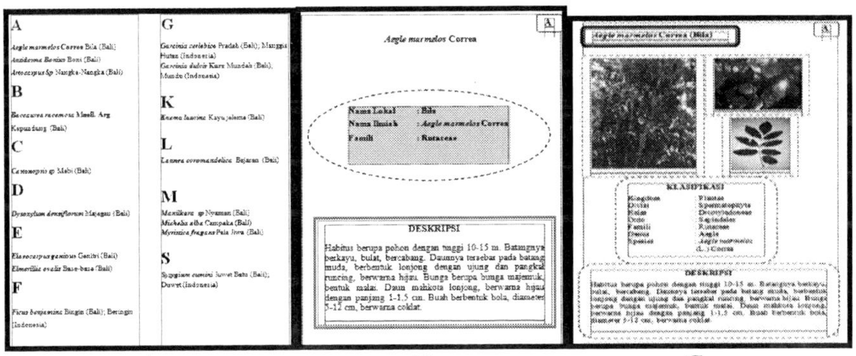

A B C

Figure 3. The Content Display of Dictionary Type A, B dan C

Type A, type B, type C dictionaries were created to analyze a better dictionary type to be developed. Therefore, a survey was conducted to 5 students of 8th semester majoring at Biology Education, Faculty of Mathematics and Natural Sciences, Universitas Pendidikan Ganesha as the samples. The results of the survey showed that the students chose the dictionary type C because the dictionary type C was improved with the plant species images of trunk, leaf, flower and fruit that are the essential content of this dictionary. Therefore this mini illustrated dictionary was developed. The dictionary development result can be seen in the figure below (Figure 4, 5, 6, 7).

Figure 4. The Front and Back Covers of Rare Plants Mini Illustrated Dictionary

The development of this Rare Plants Mini Illustrated Dictionary in this study used ADDI development design. ADDIE design consists of five steps which are Analysis, Design, Devolepment & Production, Implementation, Evaluation (Tegeh, et al. [20]). On the early stage, there are 3 dictionary types composed, they are dictionary type A, type B and type C. The survey results explained the weakness and the strength of each dictionary type. Type A dictionary is good enough and easy to understand, but its weaknesses are that type A dictionary is only suitable to be developed if the species number used is huge, alphabetically completed, and it only has the species scientific and local names. Thus, it is difficult to identify the plants in the field. Type B dictionary is more interesting and more complete than type A one because it contains the species description and easier to use as a learning material. While type C dictionary has the advantages with the images of the species included the images of stems, leaves, flowers and fruits. Therefore this type C dictionary is developed.

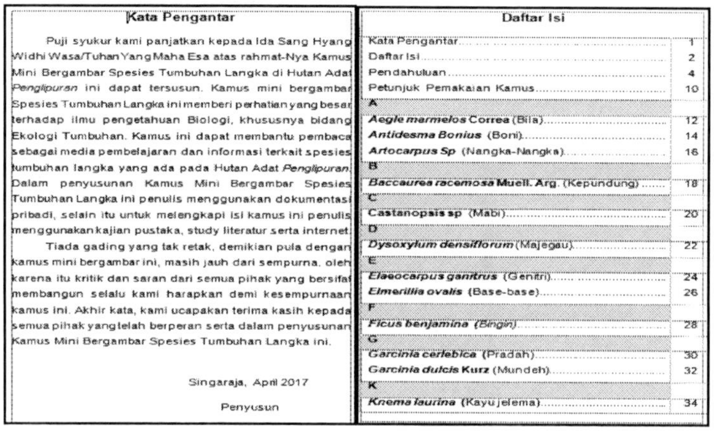

Kata Pengantar

Puji syukur kami panjatkan kepada Ida Sang Hyang Widhi Wasa/Tuhan Yang Maha Esa atas rahmat-Nya Kamus Mini Bergambar Spesies Tumbuhan Langka di Hutan Adat *Penglipuran* ini dapat tersusun. Kamus mini bergambar Spesies Tumbuhan Langka ini memberi perhatian yang besar terhadap ilmu pengetahuan Biologi, khususnya bidang Ekologi Tumbuhan. Kamus ini dapat membantu pembaca sebagai media pembelajaran dan informasi terkait spesies tumbuhan langka yang ada pada Hutan Adat *Penglipuran*. Dalam penyusunan Kamus Mini Bergambar Spesies Tumbuhan Langka ini penulis menggunakan dokumentasi pribadi, selain itu untuk melengkapi isi kamus ini penulis menggunakan kajian pustaka, study literatur serta internet.

Tiada gading yang tak retak, demikian pula dengan kamus mini bergambar ini, masih jauh dari sempurna. oleh karena itu kritik dan saran dari semua pihak yang bersifat membangun selalu kami harapkan demi kesempurnaan kamus ini. Akhir kata, kami ucapkan terima kasih kepada semua pihak yang telah berperan serta dalam penyusunan Kamus Mini Bergambar Spesies Tumbuhan Langka ini.

Singaraja, April 2017

Penyusun

Daftar Isi

Figure 5. The Preface and Table of Content Designs

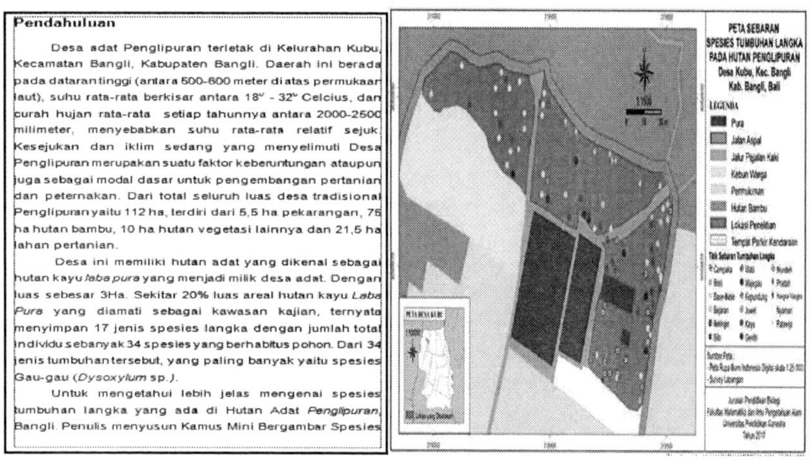

Pendahuluan

Desa adat Penglipuran terletak di Kelurahan Kubu, Kecamatan Bangli, Kabupaten Bangli. Daerah ini berada pada dataran tinggi (antara 500-600 meter di atas permukaan laut), suhu rata-rata berkisar antara 18° - 32° Celcius, dan curah hujan rata-rata setiap tahunnya antara 2000-2500 milimeter, menyebabkan suhu rata-rata relatif sejuk. Kesejukan dan iklim sedang yang menyelimuti Desa Penglipuran merupakan suatu faktor keberuntungan ataupun juga sebagai modal dasar untuk pengembangan pertanian dan peternakan. Dari total seluruh luas desa tradisional Penglipuran yaitu 112 ha, terdiri dari 5,5 ha pekarangan, 75 ha hutan bambu, 10 ha hutan vegetasi lainnya dan 21,5 ha lahan pertanian.

Desa ini memiliki hutan adat yang dikenal sebagai hutan kayu *laba pura* yang menjadi milik desa adat. Dengan luas sebesar 3Ha. Sekitar 20% luas areal hutan kayu *Laba Pura* yang diamati sebagai kawasan kajian, ternyata menyimpan 17 jenis spesies langka dengan jumlah total individu sebanyak 34 spesies yang berhabitus pohon. Dari 34 jenis tumbuhan tersebut, yang paling banyak yaitu spesies Gau-gau (*Dysoxylum* sp.).

Untuk mengetahui lebih jelas mengenai spesies tumbuhan langka yang ada di Hutan Adat *Penglipuran*, Bangli. Penulis menyusun Kamus Mini Bergambar Spesies

Figure 6. Introduction Design and The Rare Plant Species Distribution Map

On the field implementation, these species images/photos and description available in the Mini Illustrated Dictionary of Rare Plants can be used to identify and to look for the rare plants. The description on the Mini Illustrated Dictionary of Rare Plants clearly shows the exact species location, so by using a GPS tool the rare plants location will be able to find. This mini illustrated dictionary is already worth to be used as guidance in a field practice, as a learning material for the students, and as an additional information for the next researchers and the community who visit the village forest of

Penglipuran, Bangli. This can be seen from this dictionary implementation in the field on the evaluation step to see the plant identification results by the students with this Mini Illustrated Dictionary of Rare Plants. The results of evaluation done in the field showed that from 17 rare plant species, there was 1 plant species could not be clearly identified, that was Nyaman plant species (Manilkara sp). It could be concluded that 99% this Mini Illustrated Dictionary of Rare Plants can be utilized to identify the rare plants in the village forest of Penglipuran.

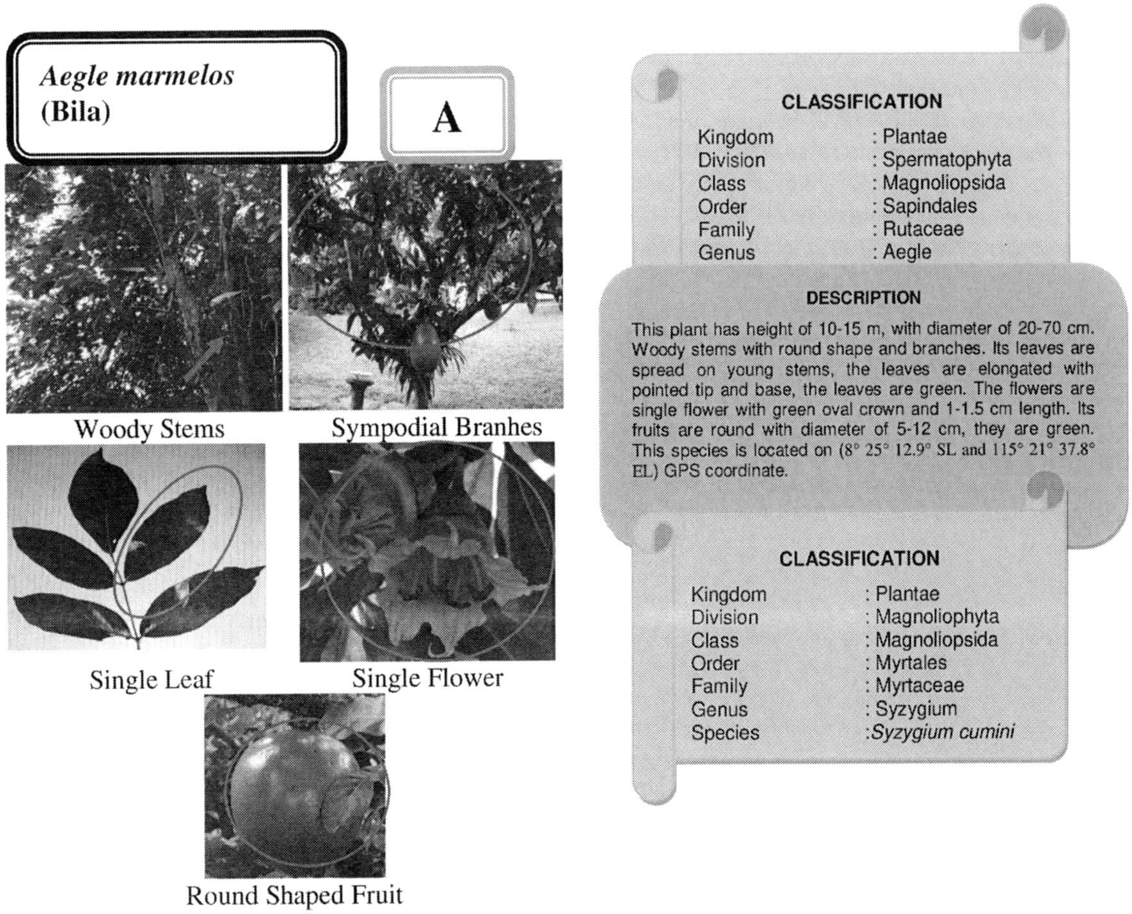

Figure 7. The examples of the dictionary content from alphabetical A and S

4. Conclusions

From the results of this study can be concluded that (1) There were 17 rare plant species out of 34 plant species found in the village forest of Penglipuran; (2) Those 17 rare plant species were furthermore used as the content of the Mini Illustrated Dictionary of Rare Plants (3) The Mini Illustrated Dictionary of Rare Plants was developed by using ADDIE design, contains the local name, scientific name, images/photos and organs, classification and description.From the conclusions, here are some recommendations (1) The educator may use this dictionary as a learning material, both in the classroom or outside the classroom, especially when inviting the students to the field (Village Forest of Penglipuran, Bangli, Bali); and (2) The management of Penglipuran village forest may use this dictionary as an additional information for the visitors.

IOP Conf. Series: Journal of Physics: Conf. Series **1040** (2018) 012008 doi:10.1088/1742-6596/1040/1/012008

References

[1] Wijana N 2013 *Analisis Vegetasi Hutan Adat, Upaya Pengelolaan Berbasis Kearifan Lokal dan Pemberdayaan Masyarakat Melalui Pendekatan Ergologi di Desa Bali Aga Buleleng-Bali* Research Report. Unpublished.

[2] Wijana N 2014 *Jurnal Sains dan Humaniora Lemlit Undiksha* **1** (1) 55-65

[3] Wijana N and Sumardika I N 2009 *Pelestarian Jenis-Jenis Tumbuhan Berguna Melalui Kearifan Lokal di Desa Adat Tenganan Pegringsingan, Kabupaten Karangasem, Bali.*Dimuat dalam prosiding *Konservasi Flora Indonesia dalam Mengatasi Dampak Pemanasan Global.Kebun Raya "Eka Karya – LIPI.Hal. 724 - 731. ISBN 978-979-799-447-1.*

[4] Wijana N 2008 *Jurnal Wahana Matematika dan Sains* **5** (10) 17-34

[5] Wijana N and Sumardika I N 2004 *Penentuan Kualitas Air Danau Batur (Kajian dari Sisi ABC Envirotment)*. Research Report. Unpublished.

[6] Wijana N and Sumardika I N 2005 *Analsis Vegetasi Hutan Bukit Kangin Desa Adat Tenganan Pengringsingan, Kabupaten Karangasem. Research Report*. Unpublished.

[7] Wijana N 2012 Analisis Dampak Lingkungan Dan Upaya Pengelolaan Berbasis Ergologi Kawasan Wisata Lovina, Buleleng – Bali. Research Report. Unpublished.

[8] Arrijani D S, Edi Guhardja,Ibnul Qayim 2006 *Jurnal Biodiversitas* **7** (2) 147-153

[9] Irwanto 2007 Analisis Vegetasi untuk Pengelolaan Kawasan Hutan Lindung Pulau Marsegu, Kabupaten Seram Barat, Provinsi Maluku.*Thesis*. Fakultas Kehutanan Universitas Gajah Mada. Retrieved from http://www.irwantoshut.com (on 6 Maret 2011).

[10] Junaedi D I and Mutaqien Z 2010 *Biodiversitas* **11** (2) 75-81

[11] Hartini S 2007 *Jurnal Biodiversitas* **8** (1) 67-72

[12] Onrizal, Cecep Kusmono, Bambang Hero Saharjo, Iin P. Handayani, Tsuyoshi Koto 2006. Analisis Vegetasi Hutan Hujan Tropika Dataran Rendah Sekunder di Taman Nasional Danau Sentarum, Kalimantan Barat *Jurnal Biologi* **4** (6) 359-371

[13] Onrizal 2010 *Jurnal Biologi Indonesia* **6** (2) 163 – 172

[14] Purwaningsih 2006 *Jurnal Biodiversitas* **7** (1) 49-53

[15] Purwaningsih dan Yusuf R 2008 *Jurnal Biologi Indonesia* **4** (5) 385-399

[16] Cox G W1976 *Laboratory Manual of General Ecology* (USA: WM. C. Brown Company Publisher)

[17] Barbour M G, J. H. Burk and W. D. Pitts 1987 *Terrestrial Plant Ecology* (California : The Benjamin/ cummings Publishing Company)

[18] Mueller-Dombois D, H Ellenberg 1974 *Aims and Methods of Vegetation Ecology*. (Sanfransisco : W. H. Freeman and Company)

[19] Wijana N 2014 *Metode Analisis Vegetasi* (Yogjakarta: Graha Ilmu)

[20] Tegeh M, Jampel N, Pujawan K 2014 *Model Penelitian Pengembangan* (Yogyakarta: Graha Ilmu)

Cytotoxic activity of andrographolide in colon cancer through inhibition cox-2 by in silico study

N P L Laksmiani, K R Reynaldi, M I Widiastari, I P W Nugraha, I M K Suyadnya and R A I K Maharani

Departement of Pharmacy, Faculty of Mathematics and Science, Udayana University

E-mail: rio.reynaldi28@gmail.com

Abstract. Colon cancer is a cancer with characteristics of cyclooxygenase (COX-2) protein overexpression. Increased COX-2 expression may inhibit regulation of apoptotic mechanism. Andrographolide was known has activity as an anticancer. The aimed of this study was determined the mechanism and affinity of andrographolide with COX-2 protein as anti-colon cancer using molecular docking. In silico assay used molecular docking method with Chimera 1.10.1 (use for protein preparation COX-2), Hyperchem 8 (use for optimization 3D structure of andrographolide), and Autodock 4.2 programs (use for molecular docking).The results in molecular docking was a binding energy value of and hydrogen bonds. Data were compared with native ligand, celecoxib, and meloxicam. The binding energy values between COX-2 protein with andrographolide, native ligand, celecoxib, and meloxicam were respectively -8.66; -9.94; -10.12; and -7.68 kcal/mol. This indicate that andrographolide had a stronger affinity and more stable than meloxicam for COX-2 protein with hydrogen bonds on Arg120 amino acid. Based on the results obtained, andrographolide had an anti-colon cancer activity caused it can bind to the protein COX-2. The mechanism of andrographolide as anti-colon cancer by inhibited the overexpression of COX-2 in order to increase apoptotic regulation.

1. Introduction

Cancer is a disease characterized by the unchecked division and survival of abnormal cells. One of the other cancer's types that often cause of death was colon cancer. Colon cancer is a disease that includes the deadly disease cause not known until a more severe level [4]. Colon cancer is the third most common cancer in men (746,000 cases, 10.0% of the total) and the second in women (614,000 cases, 9.2% of the total) worldwide [3]. Cancer Country Profiles (2014) reported colon cancer was the third cancer in men (10,516 cases, 10.2% of the total) and the second in women (7,837 cases, 8.5% of the total) in Indonesia.

Cyclooxygenase (COX)-2 is an 80 % overexpression in colon adenocarcinomas. COX-2 is normally absent in most cells, this immediate-early response gene is rapidly induced by a variety of pro-inflammatory and growth associated stimuli, resulting in increased PGE_2 synthesis. The overexpression of COX-2 can inhibit apoptotic, stimulate adhesion, invasion, and induce angiogenesis [9]. Based on this, it is necessary to develop a new anti-colon cancer's drug from natural product.

Andrographolide has several pharmacological activities, for instance anticancer [8]. Rajagopal *et al.* [6] studied the cellular processes and targets modulated by andrographolide treatment in human cancer and immune cells. Andrographolide treatment inhibited the in vitro proliferation of different tumor cell lines, which representing the various types of cancers. The compound exerts directly in anticancer activity in cancer cells by arresting cell-cycle in G0/G1 phase. Furthermore, through the induction of cell-cycle inhibitory protein p27 and the inhibition expression of Cyclin-dependent Kinase 4 (CDK4).

Generally, by discovering and developing the anti-colon cancer's drug, it needs very long times progress, costs, and the possibility of the results are not appropriate yet. In silico molecular docking have benefit, for example safe, free from chemical waste, cost-effective and the time progress is short [5]. The aim the molecular docking is to achieve the optimized conformation and relative orientation for both the protein and ligand. Therefore, we want to know about the affinity and mechanism between andrographolide with COX-2 protein through in silico as anti-colon cancer.

2. Method

The three dimensional of COX-2 (6COX) were obtained from Protein Data Bank (http://www.rcsb.org). The 3D structure of andrographolide (5318517) downloaded from Pubchem (https://pubchem.ncbi.nlm.nih.gov) which provides information on the biological activities of small molecules. The initial preparation of the pdb files to select the need chains, remove water, add polar hydrogen, delete multiple ligands and non-protein parts in using Chimera 1.10.1 program. Open Babel GUI version 2.3.2 program was used to convert the pdb file format. The three-dimensional structure of andrographolide was optimized by using HyperChem 8 semi-empirical AM1 (Austin Model 1) method and calculating with a single point and geometry optimization. Autodock 4.2 program used for validation of molecular docking method with Root Mean Square Deviation (RMSD) parameter value $\leq$ 3 Å. Next running docking protein target and ligand in file Autodock 4.2 configuration then the ligands binding energy to target protein score was compared to native ligand.

3. Result and Discussion

COX-2 database (6COX) downloaded from http://www.rcsb.org/pdb/home/home.do. Preparation of COX-2 protein aimed to separate protein with native ligand. Therefore, it is available pocket/cavity during docking progress with andrographolide.

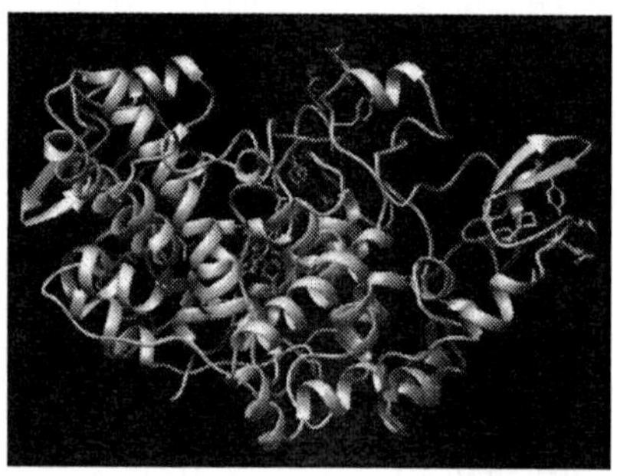

Figure 1. COX-2 Protein Structure without Native Ligand

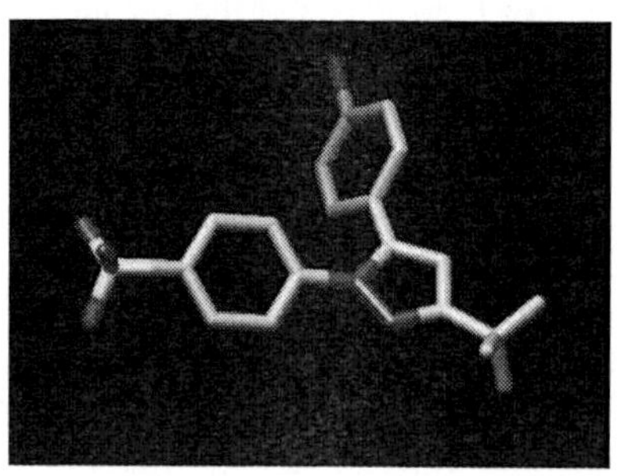

Figure 2. S58 Native Ligand

The protein preparation is performed by removing water molecules (H_2O), so the only one that interact is compound test with amino acids. Those preparations were done by using Chimera 1.10.1 program that obtained protein (without native ligand) and S58 native ligand.

The optimization of 3D structure andrographolide was done by using semi-empirical AM1 on Hyperchem 8. The first step was by calculating single point that used to determine the total molecular energy of the structure. The energy obtained at this single point calculation is -5471.48 kcal/mol. Moreover, to employ the energy minimization algorithms that locate the stable structure was using geometric optimization calculation. The energy obtained in here is -5509.50 kcal/mol.

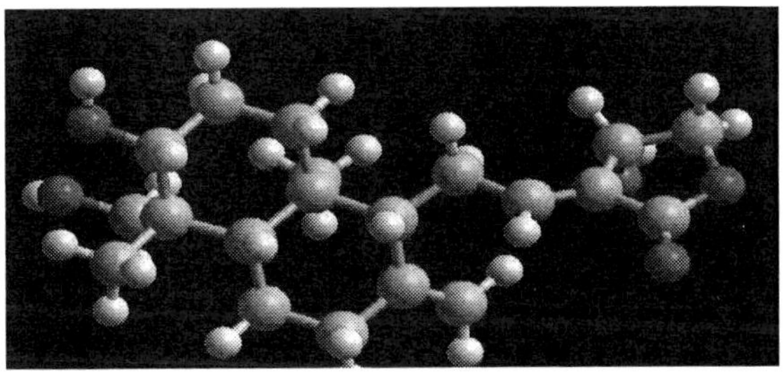

Figure 3. The 3D Structure Andrographolide after Single Point Calculation

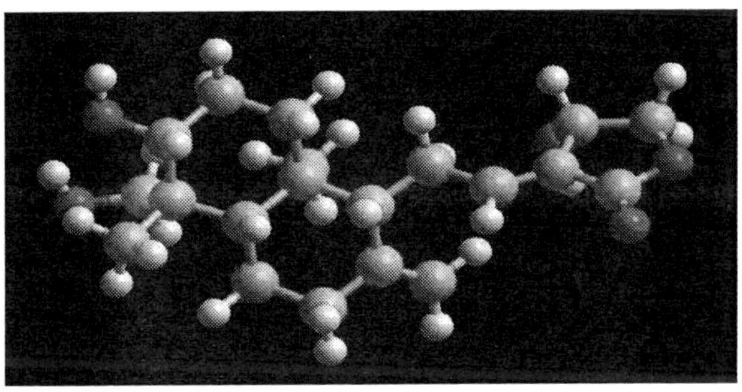

Figure 4. The 3D Structure Andrographolide after Geometry Optimization Calculation

Molecular docking was performed to elucidate the molecular mechanism underlying andrographolide inhibits COX-2 receptor by seeing the drug-receptor binding energy. Score docking shows the bond strength between the active compound and the target protein. Lower score docking shows stronger and more stable bond. When the bond grows stronger, the affinity between the compounds and the target protein increases. Higher its affinity will be caused higher its biological activity [5].

Validation method aims to prove the parameters on the method are appropriate to use for the requirement where it's used. The RMSD value is a parameter used to indicate that the molecular docking method is valid. RMSD value is a measurement media to compare between two poses atomic position, those are the experimental and the predicted structure. Smaller the RMSD value that indicates the predicted ligand poses better as it approaches the native conformation [1]. Validation result from COX-2 protein showed molecular docking protocol for that protein could be accepted with RMSD value < 3 Å (table 1) and molecular docking andrographolide to COX-2 could be continued.

Table 1. Docking results between COX-2 Protein and S58 Native Ligand

Protein	Conformation	RSMD (Å)	Energy Binding (Kcal/mol)	Hydrogen Bond
COX-2	1	0.83	-10.55	Phe518
	2	0.88	-10.39	Arg120
	3	0.83	-10.23	His90
	4	0.85	-10.09	-
	5	1.01	-9.69	-
	6	0.64	-9.94	Arg120
	7	0.96	-9.44	His90
	8	0.94	-8.92	-
	9	1.42	-9.29	Phe518
	10	2.91	-9.74	-

Based on the results of re-docking in table 1 show, the value of RMSD is below 3 Å. Those results that indicate the method are valid. Binding energy values obtained from the docking of COX-2 protein with S58 native ligand, were compared to determine the andrographolide potential of anti-colon cancer by inhibiting mechanism of COX-2 protein.

In silico molecular docking between andrographolide and COX-2 protein are performed by using the Autodock 4.2 program. The docking result showed in the table 2. Conformation with the lowest binding energy value was selected from the conformation which has the most stable conformation. The potential of andrographolide as an anti-colon cancer which has an inhibition mechanism COX-2 can be determined by comparing the value of binding energy for drugs that have the same mechanism, for example celecoxib and meloxicam. Based on the comparison binding energy value between native ligand and andrographolide with COX-2 protein in table 3, it shows that andrographolide has greater binding energy value than S58 native ligand. However, in the binding energy value obtained in here, it got negative value that indicates the formation of bond between COX-2 and andrographolide.

Table 2. Docking results between COX-2 Protein and Andrographolide

Protein	Conformation	Energy Binding (Kcal/mol)	Hydrogen Bond
COX-2	1	-8.66	Arg120
	2	-8.27	Ser530 Unl1
	3	-7.92	-
	4	-8.09	Arg120
	5	-7.8	Arg120
	6	-7.31	Arg120
	7	-6.48	Arg120
	8	-5.64	Arg120
	9	-4.54	-
	10	-4.26	Arg120

Table 3. Docking results between COX-2 protein with Native Ligand, Andrographolide, Celecoxib, and Meloxicam

Ligand	COX-2 Protein	
	Energy Binding (Kcal/mol)	Hydrogen Bond
Native Ligand (S58)	-9.94	Phe518
Andrographolide	-8.66	Arg120
Celecoxib	-10.12	His90 Unl1
Meloxicam	-7.68	Arg120 Ser530

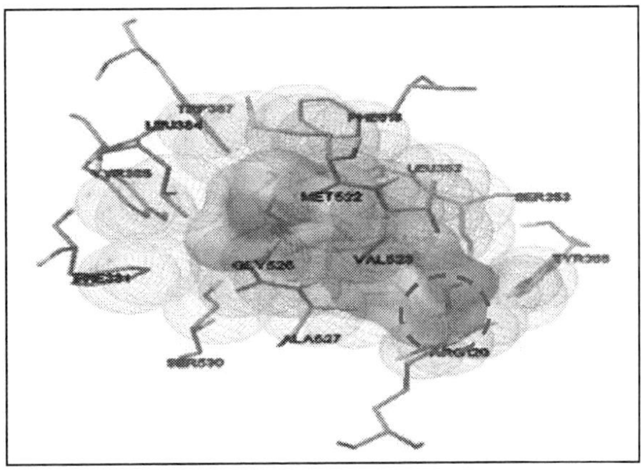

Figure 5. Interaction between Andrographolide and COX-2 Protein

The result that occur in docking of andrographolide compound against COX-2 receptor are energy binding and affinity. Based on the table 3, andrographolide has greater binding energy value than celecoxib, but it is lower than meloxicam. This outcome indicates that andrographolide has stronger affinity and more stable bond than meloxicam. The interaction between proteins and ligands are indicated the form of hydrogen bonds. The hydrogen bonds showed figure 5. Hydrogen bonds are electronegativity that can affect the bond strength of a compound. The hydrogen bond is formed in the same active side between the andrographolide and meloxicam in the COX-2 protein, which it is in the Arg120 amino acid. Those indicates that andrographolide has potential as an anti-colon cancer by inhibiting the expression of COX-2 with a stronger affinity and a more stable bond than meloxicam. Higher its affinity will be caused higher its biological activity.

4. Conclusion

The *in silico* test with molecular docking showed that andrographolide is potentially as anti-colon cancer through the increasing apoptotic mechanism by inhibiting the activity of the COX-2 protein. The molecular docking demonstrate high affinity interaction between andrographolide and COX-2

Acknowledgment
This research was supported by DITJEN DIKTI and Departement of Pharmacy, Mathematics and Science Faculty, Udayana University.

References

[1] Agistia, D.D., H. Purnomo, M. Tegar, and A.E. Nugroho. 2013. Interaction between Active Compounds from Aegle Marmelos Correa. As Anti Inflammation Agent with Cox-1 and Cox-2 Receptor. *Trad. Med. J.* Vol. 18 (2), p 80-87.

[2] Cancer Country Profile. 2014. An Essential Tool for Informing NCD and Cancer Control Planning. Access 17/8/2017. Available: http://www.who.int/cancer/country-profiles/en/Indonesia.

[3] GLOBOCAN. 2012. *Estimated Cancer Incidence, Mortality and Prevalence Worldwide in 2012.* Access 15/04/2017. Available:http://globocan.iarc.fr/Pages/factsheetscancer.aspx

[4] Labianca, R. B., G. D. Nordlinger, A.Beretta, A. Brouquet, and Cervantes. 2010. Primary colon cancer: ESMO Clinical Practice Guidelines for diagnosis, adjuvant treatment and follow-up. *Annals of Oncology*. Vol. 21 (5).

[5] Mukesh, B. And K. Rakesh. 2011. Molecular Doking: A Review. *International Journal of Research in Ayurveda & Pharmacy*, Vol. 2 (6).

[6] Rajagopal, S., R. A. Kumar, D. S. Deevi, C. Satyanarayana, and R. Rajagopalan. 2003. Andrographolide, a potential cancer therapeutic agent isolated from Andrographis paniculata. *Journal of Experimental Therapeutics and Oncology*. Vol. 3: 147–158

[7] Terzic, J., S. Grivennikov, E. Karin, M. Karin. 2010. Inflammation and Colon Cancer. *Gastroenterology*. Vol. 138 pp. 2101-2114.

[8] Tung, Y.T., H. L. Chen, H. C. Tsai, S. H. Yang,Y. C. Chang, and C. M. Chen. 2013. Therapeutic Potential of Andrographolide Isolated from the Leaves of Andrographis paniculata Nees for Treating Lung Adenocarcinomas. *Evidence-Based Complementary and Alternative Medicine*. ID 305898.

[9] Wang, R.N. Dubois. 2010. The role of COX-2 in intestinal inflammation and colorectal cancer. *Oncogene*. Vol. 29:781.

[10] Yang, S.Y., K. M. Sales, B. Fuller, A. M. Seifalian dan M. C. Winslet. 2009. Apoptosis and colorectal cancer: implications for therapy. *Trends in Molecular Medicine*. Vol.15 (5).

The inhibitory activity of peonidin purple sweet potato in human epidermal receptor-2 receptor (her-2) expression by in silico study

N P L Laksmiani, M I Widiastari and K R Reynaldi

Departement of Pharmacy, Faculty of Mathematics and Science, Udayana University

Email: irmawidiastari2@gmail.com

Abstract. An acetylated anthocyanin (peonidin) is one of the flavonoid group compounds contained in purple sweet potato (Ipomoea batatas) which has anticancer activity. Overexpression of Human Epidermal Receptor-2 receptor (HER-2) could induce spontaneous dimerization and autophosphorylation, and trigger the occurrence of focal adhesion kinase (FAK) activation in order to induce migratory process and breast cancer cell metastasis. Lapatinib is one the drug which used to inhibit over-expression of HER-2. The purpose of this study was to determined the inhibition mechanism of over-expression HER-2 protein of peonidin by using in silico molecular docking method. In silico method such as molecular docking is done through stages such as validation of molecular docking method, optimization of peonidin 3D structure, docking of peonidin with HER-2 protein which refers to the binding energy parameter where the lower binding energy value indicates the stronger and stable bond between peonidin with HER-2 protein. The binding energy of peonidin was compared with native ligand's binding energy and lapatinib's binding energy. The result of this study was peonidin's binding energy with HER-2 protein was -7,54 kcal/mol, while the native ligand's binding energy and lapatinib's binding energy with HER-2 protein were -5,77 kcal/mol and -0,56 kcal/mol. The binding energy indicates peonidin as a purified anthocyanin of purple sweet potato had potential activity as anti-breast cancer because it suppressed the excessive expression of the HER-2 protein.

1. Introduction

Breast cancer has relatively high incidence in the worldwide, approximately 20% of all malignancies [1]. Breast cancer in Indonesia got the second rank after cervical cancer. It is estimated that 10 out of 100,000 populations are affected by breast cancer and 70% of patients check themselves at an advanced stage [2].

In 30% of cases of breast cancer is caused by Human Epidermal Receptor-2 (HER-2) overexpression [3]. Invasion of cancer cells can occur due to the overexpression of HER-2 protein. The overexpression of HER-2 protein could induce migratory process and metastasis process of cancer cell through spontaneous dimerization induction, autophosphorylation, and the occurrence of focal adhesion kinase (FAK) activation [4]. Several treatments have been done for breast cancer, but the results have not been satisfactory, even the effects of surgical failure can cause cancer to spread to the other part of body with severe conditions [5]. Because of that, we need to developing a new drug which has a good therapeutic effect by excavating natural compounds derived from plants.

Purple sweet potato (*Ipomoea batatas* L.) is one of the most abundant commodities in Indonesia [6]. One of the phytochemical properties in *Ipomoea batatas* is anthocyanin [7]. Purple swet potato contains 10 types of anthocyanins with differences in R1, R2, and R3 are: sianidin-3-sophorosid-5-glucosid, peonidin-3-sophorosid-5-glucosid, sianidin-3- (6 "-6 " - dicoffeoysophoroside) -5-glucosid, sianidin3- (6 "-caffeoyl-6 " - p-hidroxybenzoylsophorosid) -5-glucosid, sianidin 3- (6" - caffeoylsophorosid) -5-glucosid, sianidin-3- (6 '- caffeoyl-6 " - feruloylso-phorosid) -5-glucosid, peonidin-3- (6 ', 6' '- dicoffeoylsophoroside) -5-glucosid, pionidin 3- (6 "-caffeoylsophorosid) -5-glucosid, pionidin 3 - (6 "-caffeoyl-6 " - p-hidroxyibenzoylsophorosid) -5-glucosid and peonidin-3- (6" -caffeoyl-6 " 'feruloylsophorosid) -5-glucosid [8]. Anthocyanin has been widely used as a dye, especially beverages, as many synthetic dyes are known to be toxic and carcinogenic. According to JEFCA (Joint FAO / WHO Expert Committee on Food Additives) has stated that extracts containing anthocyanin have low toxicity effects. Attention to intensive anthocyanin pigment in recent years due to its health benefits, including reduce the risk of coronary heart disease, stroke risk, anticancer activity, anti-inflammatory effects, improving eye acuity, and improving cognitive behavior. Attention to intensive anthocyanin pigment in recent years due to its health benefits as anticancer [9]. One type of anthocyanin which can be developed as a new drug is an acetylated anthocyanin (peonidin) which has better stability during storage than the other type of anthocyanin.

In silico molecular docking techniques can be used to predict how a protein interacts with a ligand, so the activity of the bioactive compound and its synergistic action with other drugs can be known. This method will improve the effectiveness and efficiency of research on new drug development. Therefore, we want to know about the inhibition mechanism of overexpression of HER-2 protein by an acetylated anthocyanin from purple sweet potato through in silico simulation.

2. Method

Molecular docking was performed to get the drug-receptor binding energy. The PDB files obtained from Protein Data Bank (PDB) (www. rcsb. org) is a worldwide repository for processing and distribution of 3D biological macromolecular structure data. The structure of HER-2 (3PP0) proteins were downloaded from PDB and prepared for docking by removing the heteroatoms and water molecules in them. The software used were as follows: Autodock 4.2. for molecular docking, Chimera 1.10.1 for protein and ligand preparation. The three-dimensional structure optimization by using HyperChem 8 by semi-emphiric AM1 computation and calculate with a single point and geometry optimization. The downloaded PDB file of HER-2 were first read in Chimera 1.10.1, added waters removed, and polar hydrogens were added. The structure of peonidin was drawn by using Hyper Chem 8. Next running docking protein target and ligand in file Autodock 4.2. The configuration then the ligand's binding energy to target protein score was compared to native ligands.

3. Result And Discussion

The result of protein preparation can be seen in Figure 1. The aim of protein preparation is obtaining protein structure without ligand and its native ligand (available space / pocket / cavity) in which chain is selected chain A. At the time of preparation of H_2O molecule on chain A protein without native ligand is removed so that at the time of docking peonidin can interact perfectly with HER-2 proteins [10].

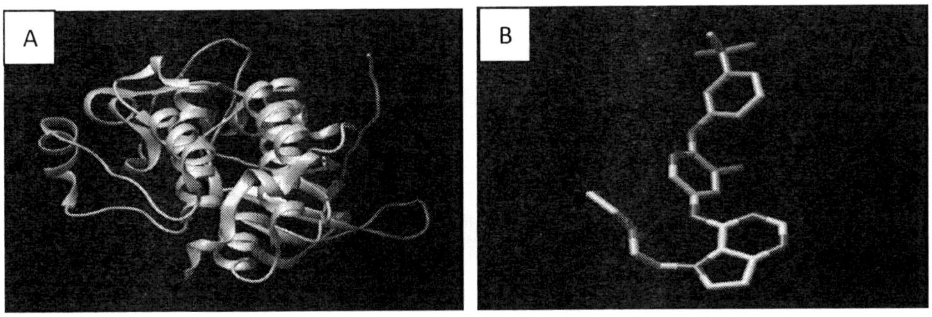

Figure 1. Structure of HER-2 protein without Native Ligand (A) and Native Ligand (B)

Table 1: Molecular Docking Method Validation of HER-2 Protein

Conformation	RMSD (Å)	Energy Binding (kkal/mol)	Hydrogen Bond
1	1.42	-5.77	-
2	1.88	-4.97	-
3	5.41	-4.53	ASP863 THR862
4	5.12	-4.14	-
5	2.68	-3.03	-
6	2.90	-2.33	MET801
7	2.48	-1.57	ARG849 SER728
8	2.58	-1.09	-
9	3.23	-0.90	MET801
10	4.62	-0.25	-

The results and interactions that occur in the validation of the molecular docking method can be seen in table 2. The aim of molecular docking validation is to determined the conformational similarities between native ligand-protein complex in crystallography compared with experimental results [11]. Validation parameter for docking is Root Mean Square Deviation (RMSD) value which shows the ligand irregularity rate experimentally against the crystallographic ligand on the same binding site. The RMSD value ≤ 3Å indicates that the method used is valid.

Preparation is done by first drawing the structure of peonidin 3D and then done optimization on HyperChem 8 program to get a stable structure. The energy produced by peonidin compounds after calculating the single point of energy is 3529,799 kcal / mol and decreased to -3634.471 kcal / mol after geometry optimization aimed to minimize the energy to obtain a more stable compound [12].

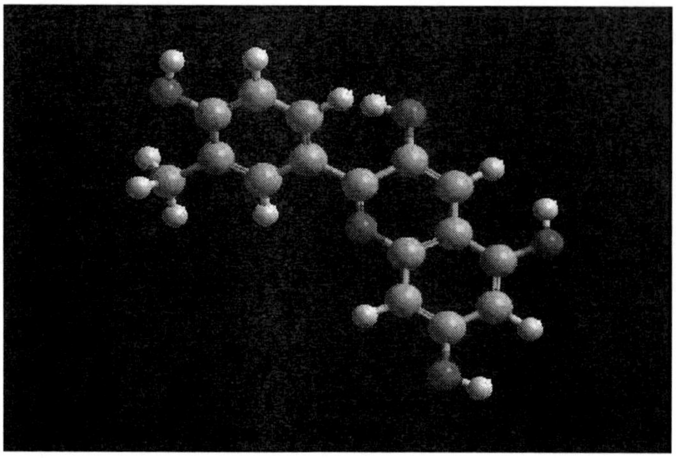

Figure 2: The Structure of Peonidin after Geometry Optimization

Table 2: Docking of Peonidin and HER-2 Protein

Conformation	Energy Binding (kcal/mol)	Hydrogen Bond
1	-7,54	-
2	-7,51	-
3	-7,49	-
4	-7,42	MET801
5	-7,31	ASP863
6	-7,29	ASP863
7	-7,29	ASP863
8	-7,23	-
9	-7,20	MET801
10	-7,10	ASP863

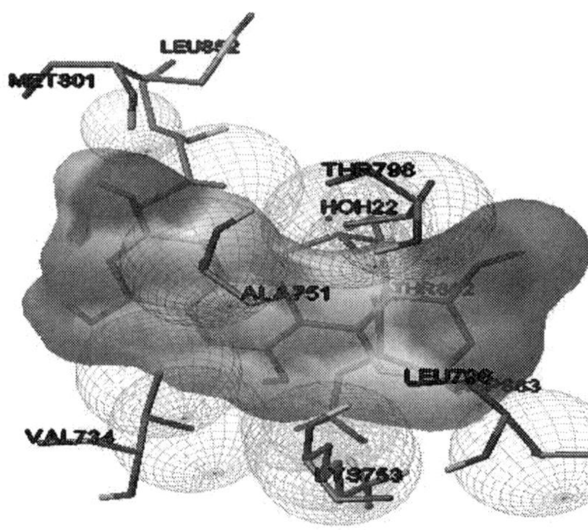

Figure 3: Interaction Between Peonidin and HER-2 Protein

The results and interactions that occur in docking of peonidin compounds against HER-2 protein can be seen in table 2 and figure 3. Docking of peonidin and HER-2 protein results is energy binding of -7.54 kcal/mol which can be seen in Table 2. When compared with the native ligand and lapatinib as a control to determined how strong and stable the bond, the binding energy of peonidin is lower (more negative) than native ligand's binding energy of -5.77 kcal/mol and lapatinib's energy binding of -0.56 kcal/mol which indicating that peonidin has a stronger and more stable bond compared to the native ligand. In the peonidin docking results no hydrogen bonds are found and it was same with native ligand. The amount of number of hydrogen bonds can determine the strength of the interactions, as can be seen from the physical and chemical properties between H_2O and HF which are much different even though they have hydrogen bonds. H_2O has a higher boiling point than HF. This indicates that it takes a higher energy to break the H2O molecule bond, the explanation is because the H_2O molecule has two hydrogen bonds while the HF (Jefrey, 1997). Based on the comparison of binding energy result between peonidin, native ligand, and lapatinib, it can be seen that peonidin as an acetylated anthocyanin in purple sweet potato has anticancer activity by inhibiting the overexpression of the HER-2 protein. Overexpression of HER-2 protein can lead to the process of migration and metastasis of breast cancer cells.

4. Conclusion

Peonidin as an acetylated anthocyanin of purple sweet potato has anti-breast cancer activity through inhibition of overexpression of HER-2 protein with a binding energy of -7.54 kcal / mol.

Acknowledgment

The authors would like to thank DITJEN DIKTI and all those who have helped this research.

References

[1] Tjahjadi G 1995 *Patologi Tumor Ganas Payudara, Kursus Singkat Deteksi Dini dan Pencegahan Kanker* (Jakarta: FKUI-POI)
[2] Ana K 2007 *Panduan Lengkap Kesehatan Wanita* (Yogyakarta: Gala Ilmu Semesta)
[3] Gibbs J B 2000 *Science* 287 1969-1973
[4] Johnson S C, Dweck C S, Chen F S, Stern H L, Ok S J, and Barth M 2010 *Cognitive Science A Multidisciplinary Journal* Vol 34 pp 807-825
[5] Nafrialdi dan Gunawan S G 2007 *Antikanker, Farmakologi dan Terapi* Edisi ke-5 (Jakarta: Departemen Farmakologi Dan Terapeutik Fakultas Kedokteran Universitas Indonesia)
[6] Koeswara S 2008 *Teknologi Pengolahan Umbi-umbian. Bagian 5: Pengolahan Ubi. Research and Community Service Institution* (Bogor: Agricultural Technology)
[7] Pochapski M T et al 2011 *Pharmacogn Mag* 7 pp 165-170.
[8] Montilla E C, Hillebrand S, and Winterhalter P 2011 *Fruit, Vegetable and Cereal Science and Biotechnology* 5 19-24.
[9] Wrolstad R E 2004 *Journal of Food Science* 69 C419 – C442.
[10] Kitchen D B, Decornez H, Furr J R, and Bajorath J 2004 *Nat. Rev. Drug Discov* 3 935-949.
[11] Luthfiyah Y dan Santoso B 2014 *Pemodelan Tiga Dimensi (3D) Ikatan Hasil Docking Molekular Turunan Diketopiperazin (DKP) dengan Bcl-2 pada Sel MCF-7* (Surakarta: Fakultas Farmasi Universitas Muhammadiyah Surakarta)
[12] Fitriasari A, Wijayanti N K, Ismiyati N, Dewi D, Kundarto W, Sudarmanto B S A and Meiyanto E 2008 *Pharmacon* 9.
[13] Jefrey G A 1997 *An Introduction to Hydrogen Bonding* (New York: Oxford University Press)

IOP Conf. Series: Journal of Physics: Conf. Series **1040** (2018) 012011 doi:10.1088/1742-6596/1040/1/012011

Identification of coral recruitment in Lembongan Island, Nusa Penida, Bali

I N D Prasetia[1], Supriharyono[2], S Anggoro[2] and L Sya'rani[2]

[1]Fisheries and Marine Departement, Ganesha University of Education, Bali
[2]Program Doctor of Coastal Resource Management, Faculty of Fisheries and Marine Sciences, Diponegoro University, Semarang

E-mail: indprasetia@gmail.com

Abstract. Coral recruitment is the increasing number of individual coral in reef ecosystems. Coral recruitment is generated from generative reproduction that produces coral planula and migration.The success of coral recruitmen is determined by environmental factors. Coral recruitment beneficial are survived from environmental impact, maintaining the potential for recovery and reproduction. Lembongan Island coral reefs are categorized as fringing reef as coral reefs live along the coast and surround the island. Coral reef Lembongan Island is the main environmental carrying capacity for the life of this island. The coral reef of Lembongan Island bioeconomically serves as a tourism object and fishery area. The aim of this research is to know the species, abundance, and survival of coral planula. The research was conducted in Lembongan Island, Bali in 2015 until 2016. The research method is manta tow survey, line intercept transect and coral recruitment. Research shows the coral reefs condition of Lembongan Island in medium to excellent category. Recruitment of coral attached to the substrate consists of Acropora millepora, A. palifera, A. tenuis, Fungia fungites, Montipora digitata, Pocillipora damicormis, Porites sp, Seriatophora hysterix, and Stylophora pistillata. The numbers of individual coral in 2015 are 63 individuals and in 2016 are 50 individuals.

1. Introduction

Lembongan Island coral reefs has a high value for people in this area. Coral reefs also provide millions of people with ecosystem services such as shoreline protection from wave action and erosion, generation of coralline sands, fisheries, tourism revenue, raw material for construction, raw natural products for pharmaceuticals, recreational and cultural values [1] [2]. Their total economic value has been estimated at $352,249 per hectare per year; higher than any other biome on earth [3].

Coral reef ecosystems comprise a complex array of habitats and communities that differ in biodiversity, ecological function, and associated organisms [4]. Most reefs are separated into distinct regions or zones that differ in topography, bottom type, and wave energy [5] [6].

Corals have sexual and asexual reproductive methods. Asexual reproduction is done by forming new shoots that will become a new individual on its parent. The formation of these shoots is done continuously which is a mechanism of increasing the size of the colony, but not forming new colonies.

Sexual reproduction is done by producing one of the sperm or egg to be released into the free waters. The egg will be equipped with a coelenterons that will be fertilized by sperm cells. This prose produces free-swimming planetary larvae, and will settle on the bottom or on hard subsets to form new colonies [7] [8] [9].

Corals can be hermaphrodite or dioeciously with an average of 7-10 years of sexual maturity. Conception will take place inside "the mother's" gastro vascular chamber and the sperm released into the waters will enter this room. The eggs will be fertilized and held until they reach the stage of the larvae of the planula. The planula will be released and swimming in open water for an indefinite period, generally lasting several days or weeks to settle and start a new colony [8] [9].

Recruitment is the process by which young individuals (coral larvae) undergo larval settlement and become part of the adult population. Some species tend to settle close to their "mother colony", others drift for great distances before settling down. Theoretical and empirical work has taught us that recruitment is influenced by a wide range of ecological processes such as competition, succession, facilitation, predation, and disturbance [10] [11].

Recruitment is an operational term and has been defined in several ways, but is commonly defined as when individuals become established in the local population [12] [13] [14]. Thus, recruitment combines both settlement and post-settlement Mortality prior to census. Recruitment is affected by different processes throughout the early life-history stages of an organism, and there are frequent trade-offs that influence success [15].

The environmentalconditions that allow new individuals to colonize an area successfully and become established in the local population has been defined as an individual's recruitment niche [10], similar to Grubb's regeneration niche [16]. If the recruitment niche of an organism is parameterized according to its multiple environmental influences, trade-offs can be quantified and recruitment success predicted under multiple scenarios of biophysical forcing [17].

Theoretical and empirical work has taught us that recruitment is influenced by a wide range of ecological processes such as competition, succession, facilitation, predation,and disturbance [10, 11]. Demographic bottlenecks during recruitment can impact adult populations and occur in a variety of benthic organisms at different stages throughout early ontogeny [15] [18] [19] [20].

The rate, scale, and spatial structure of larval dispersal drive population replenishment, and therefore have significant implications for population dynamics, marine reserve orientation, and resilience of a system. The purpose of study was to know the species, abundance, and coral survival. The present study documented patterns of coral recruitment in Lembongan Island.

2. Theoretical Framework

Coral recruitment is a critical process that helps maintain coral populations and facilitates recovery after a disturbance [21] [22] [23]. Hard corals have a bipartite life cycle consisting of a pelagic dispersive larva and a sessile benthic adult stage. During this life cycle they pass through three demographic bottlenecks that influence coral recruitment: (1) larval supply, (2) settlement, and (3) post-settlement survival [24] [25].

Like most sessile marine invertebrates, scleractinian (reef-building) corals have a bipartite life history that involves a dispersive larval phase followed by settlement (i.e., attachment and metamorphosis) of minute individuals ($\leq$1 mm) onto the reef substratum [13].

Coral larvae generally settle within several hours to weeks [26]. Certain larvae are competent to settle minutes after being released [27] [28], whereas others survive for over 244 days before settlement. While some larvae will settle close to their parent colony [29], others will be dispersed by currents and waves before they are ready to settle [30].

A consistently result in higher recruitment rates in this microhabitat nor does it reflect juvenile community structure *in situ*. When the distribution of coral recruits has been compared among microhabitats on the reef, they are usually found in higher abundances on exposed surfaces [31] [32] [33], rather than crevices [34] [35], creating an important and unresolved disparity between patterns of recruit distributions among microhabitats on settlement tiles and the reef [36].

3. Research methods

The research was conducted with the general condition began observations of coral reef ecosystems in the region by using the method of Manta Tow Survey. Manta Tow survey was conducted along the coast at intervals of 2 minutes were recorded on the percentage of coral cover from closing scoring system between live coral, dead and soft [37]

Manta Tow Survey data are analyzed and used a basis general condition of coral reefs mapping. This map will be used as the basis for determining the observation station of coral reefs with corals recruitment methods and Line Intercept Transect. Determination of research stations with randomly conducted with safety and represents the East, North, and West of Lembongan Island.

Line Intercept Transect method carried out to assess the benthic community based on the characteristics of life forms, especially the morphology of the coral reef community, so it can be known diversity of coral species in the area. Observations with Line Intercept Transect done by SCUBA diving at a depth of 3 m and 10 m [37]. Measurements were taken at two depths is assuming the two depths is considered to represent the condition of coral reefs because it usually grows well and a high diversity of coral species were also obtained at these depths.

Observation and identification of coral recruitment in accordance with English methods [37]. Observations were made at a depth of 5 meters, with 3 pieces of research racks on each research station. One unit that has a shelf research 24 pieces tile substrate placed at 4 different positions. Substrate has the size 10x10 cm with the manufacture of following the natural substrate reefs. Observations made with coral recruitment methods such as Figure 1.

4. Result and discussion

4.1. Coral reef ecosystem

Coral reef formation in Lembongan Island can be categorized as fringing reefs constituting the ones living along the beach and surrounding an island. The formation of coral reef surrounds Lembongan Island along the beach and surrounding an island.

Manta tow survey method result in North island of Lembongan Island had coral covering category between 4 - 5, it mean the life coral covering between 51 - 100 % (Figure 1).

Area residing in Tanjung Ental, Pura Sakenan until Tanjung Pemaroan represent area which high exploited as tourism area because their beauty and natural underwater view. Strong current in this area result coral covering species of branching, massive, and coral soft ware dominant. Sea weed farming activity conducted in this area from the beach until 3 meter deepness.

Coral reef condition in South Island of Lembongan Island had category between 1 - 3 with 0- 50 % for life coral covered. Sea weeds farming activity in Ceningan Strait to be depressing factor existence of coral reef in this area.

West part of Lembongan Island had life coral covering between 31% until 100% with category 3 until 5. This area represented as benefit tourism area, such as tourism activity and tourism accommodation. Besides exploiting for tourism this area also conducted by sea weed farming activity.

IOP Conf. Series: Journal of Physics: Conf. Series **1040** (2018) 012011 doi:10.1088/1742-6596/1040/1/012011

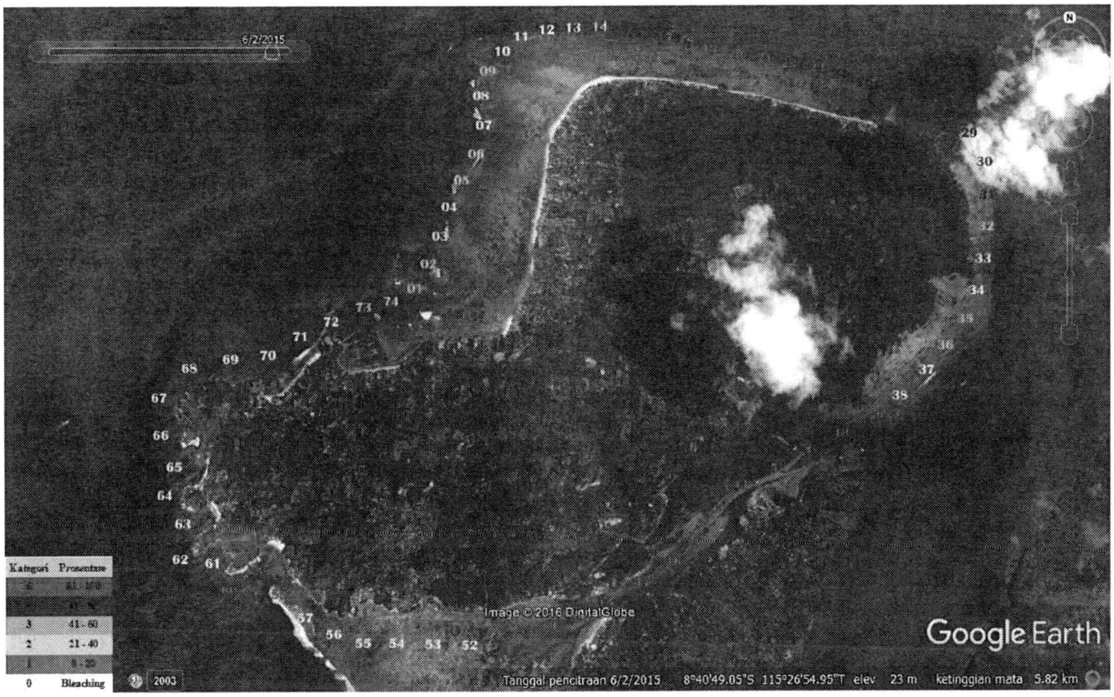

Figure 1. Manta Tow Survey

East part of Lembongan Island had life coral covering with category range from 3 - 5, with percentage between 31 - 100%. These areas represent Lembongan Island mangrove forest. Existence of mangrove in this area given double meaning to coral reef, one side mangrove protect coral reef from continent input and on the other when low tide substrate from the mangrove distributed to coral reef area. Sea weeds farming activity conducted by people in front of mangrove forest.

After observation with Manta Tow Survey, observation was done by using Line Intercept transect method.

Observations of November and 2015 and 2016 show the percentage of live coral cover at a depth of 3 meters ranging from 43.61 - 78.8% and depth of 10 meters between 43.61 - 78.8%. Research shows the coral reefs condition of Lembongan Island in medium to excellent category.

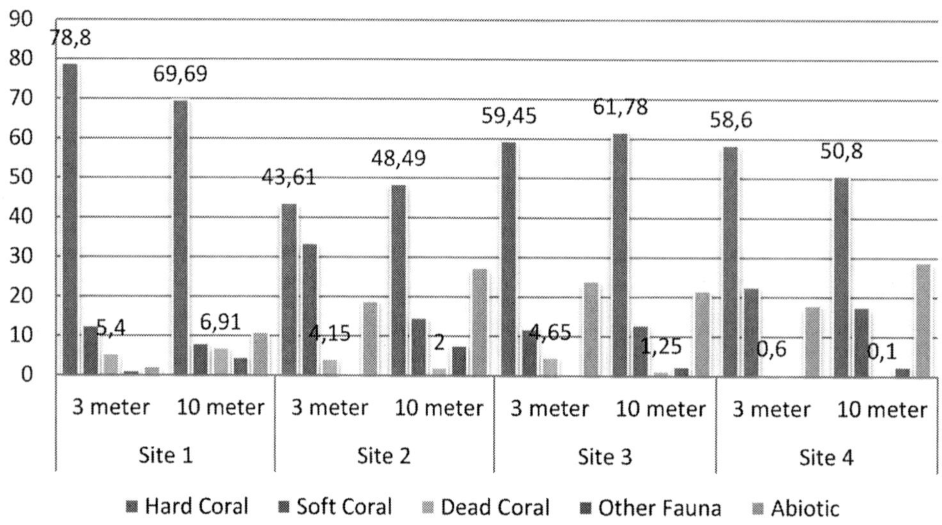

Figure 2. Line Intercept Transect November 2016

Height of live coral covering can be indication condition of environment supporting continuity coral reef in this area. Strong current bringing substrate as coral food also can be human being activity minimization function in coral reef Lembongan Island.

4.2. Coral recruitment

This studies use settlement tiles to quantify recruitment and the early dynamics of coral recruits. Settlement tiles have consistently been attributed to higher recruitment rates [38] [39] [40] [41]. It is also well established that coral larvae generally settle in much higher abundances on the cryptic undersides of flat settlement tiles [30] [42] [43].

Recruitment of coral attached to the substrate consists of *Acropora millepora, A. palifera, A. tenuis, Fungia fungites, Montipora digitata, Pocillipora damicormis, Porites sp, Seriatophora hysterix,* and *Stylophora pistillata.*

Table 1. Coral Recruitment in Lembongan Island

Species	Coral recruitment in November 2015 and 2016							
	Site 1		Site 2		Site 3		Site 4	
	2015	2016	2015	2016	2015	2016	2015	2016
Acropora millepora	2	1	4	2	1	2	3	2
Acropora palifera	2	2	1	2	2			2
Acropora tenuis	2	1	1	1	2	1	2	2
Fungia fungites	1	1	3					
Montipora digitata	3	3	4	3	1	2	2	2
Pocillopora damicormis	3	2	1	1	2	2	3	1
Porites sp	3	2		1				
Seriatophora hysterix	3	2	1	1	2	1	3	2
Stylophora pistillata	1	2	2	2	1	1	2	1
TOTAL	**20**	**16**	**17**	**13**	**11**	**9**	**15**	**12**

The number of available larvae for settlement larval supply therefore depends both on local larval availability and the number of larvae that arrive from more distant sources and is a combined result of reproduction, fecundity and larval dispersal [44].

Settlement refers to the metamorphosis process during which the coral larvae undergo morphological changes and attach to the substratum. Different cues can induce coral larvae to descend in the water column and start probing the benthos for a suitable nursery habitat [30].

The recruitment success throughout the earliest life-history stages of corals and uncover some intriguing trade-offs between growth, competition and predation, highlighting how these change and even reverse during ontogeny and under alternate disturbance regimes [45].

Recruits were categorised to each of the three major families Acroporidae, Pocilloporidae and Poritidae. The number of fecund adult corals and their reproductive output define larval production. Coral larvae travel on the water surface and have to survive a wall of mouths consisting of planktivorous fishes during the day and hard corals, zoanthids and anemones at night [46] [47].

The age of pocilloporid recruits with one polyp was determined by comparing their skeleton to skeletogenesis patterns [48]. Pocilloporid recruits of the age of 12 hours, 24 hours, 48 hours and 3 days or older could be identified. The *Pocillopora damicornis* recruits surveyed in each site, developed a second polyp after 11 to 15 days. Pocilloporid recruit size is constant during the first week after settlement and increases thereafter, supporting my result that a second polyp is formed after 11-15 days.

Porites sp constituted a majority of the small corals sampled within the plots, but only approximately 6 individual of the overall coral composition across the site. While *P. porites* has reportedly been a dominant recruiter in certain reef habitats [49], its congener *P. astreoides*, in addition to *Agaricia* spp. and *Siderastrea siderea*, more commonly dominate coral recruit assemblages in nearby Florida reefs [50] [51], as well as adult coral assemblages [52].

For a given supply of coral larvae, a suite of ecological interactions influence the success of coral recruitment and include: (1) settlement behavior and interactions with the substratum [30] [53]; (2) post-settlement Competition with algae and sessile invertebrates [54] [55]; and, (3) incidental and targeted predation from a variety of fish and mobile invertebrates [55] [56].

Reef formation in Lembongan Island can be categorized as fringing reefs constituting the ones living along the beach and surrounding an island. Conclusion research shows the coral reefs condition of Lembongan Island in medium to excellent category. Recruitment of coral attached to the substrate consists of *Acropora millepora, A. palifera, A. tenuis, Fungia fungites, Montipora digitata, Pocillipora damicormis, Porites sp, Seriatophora hysterix,* and *Stylophora pistillata*. The numbers of individual coral in 2015 are 63 individuals and in 2016 are 50 individuals.

References

[1] Conservation International 2008 *Economic Values of Coral Reefs, Mangroves, and Seagrasses: A global compilation*

[2] Wilkinson C 2008 *Status of coral reefs of the world: 2008*. Global Coral Reef Monitoring Network and Reef and Rainforest Research Center, Townsville, Australia. 296pp.

[3] de Groot R, et al 2012 *Ecosyst Serv* **1** pp 50–61

[4] Cameron C M, Rachel E P and Margaret W M 2015 *Bull Mar Sci.* **92** (1)

[5] Lalli C and Parsons T 1997 *Biological oceanography: an introduction*. Elsevier Butterworth-Heinemann. 2nd ed.

[6] Castro P and Huber M 2000 *Mar Biol*. 3rd ed. (New York: McGraw Hill)

[7] Bengen D 2001 *Sinopsis Ekosistem dan Sumberdaya Alam Pesisir dan Laut*. Pusat Kajian Sumberdaya Pesisir dan Lautan Institut Pertanian Bogor. Bogor

[8] Morton J 1990 *The Shore Ecology of The Tropical Pasific* (Jakarta: Unesco Regional Office for Science and Technology for South East Asia)

[9] Nybakken J W 1988 *Biologi Laut, Suatu Pendekatan Ekologis*. Alih Bahasa : H. M. Eidman, Koesoebiono, D. G. Bengen, M. Hutomo dan S. Sukardjo (Jakarta: PT. Gramedia)

[10] Young T P, Petersen D A and Clary J J 2005 *Ecology Letters* **8** 662–673.
[11] Pineda J, Reyns N B and Starczak V R 2009 *Population Ecology* **51**17–32.
[12] Keough M J and Downes B J 1982 *Oecologia* **54** 348–352.
[13] Harrison P L and Wallace C C 1990 *Reproduction, dispersal and recruitment of scleractinian corals* pp 133–207 *in* Z. Dubinsky, editor. Coral reefs. Elsevier Science, Amsterdam,the Netherlands.
[14] Caley M J, Carr M H, Hixon M A, Hughes T P, Jones G P and Menge B A 1996 *Annual Review of Ecology and Systematics* **27** 477–500
[15] Werner E E and Gilliam J F 1984 *Annual Review of Ecology and Systematics* **15** 393–425.
[16] Grubb P J 1977 *Biological Reviews* **52** 107–145
[17] Doropoulus C, Roff G, Bozec Y M, Zupan M, Werminghausen J and Mumby P J 2016 *Ecological Monographs* **86** (1) 20 – 44
[18] Wahle R and Steneck R 1991 *Marine Ecology Progress Series* **69** 231
[19] Arnold S N, Steneck R S, Mumby P J 2010 *Mar Ecol Prog* Ser **414** 91–105
[20] Doropoulos C, Ward S, Roff G, Gonzalez-Rivero M, and Mumby P J 2015 *PLoS ONE* **10**:e0128535
[21] Vermeij M J A and Sandin S A 2008 *Ecology* **89** 1994–2004
[22] Graham N a. J, Nash K L, Kool J T 2011 *Coral Reefs* **30** 283–294
[23] Gilmour J P, Smith L D, Heyward A J, Baird A H and Pratchett M S 2013 *Science* **340** 69–71
[24] Arnold S N and Steneck R S 2011 *PLoS ONE* **6**:e28681.
[25] Chong-Seng KM, Graham N a. J and Pratchett M S 2014 *Coral Reefs* **33** 449–461
[26] Ritson-Williams R, Arnold SN, Fogarty ND, Steneck RS, Vermeij MJ, Paul VJ 2009 *New perspectives on ecological mechanisms affecting coral recruitment on reefs.* Smithson Contrib Mar Sci
[27] Best B A and Resing J M 1987 *Active settlement of coral planulae-the effect of flow and swimming behavior on local recruitment* **27** A103–A103
[28] Carlon D B and Olson R R 1993 *J Exp Mar Biol Ecol* **173** 247–263
[29] Swearer S E, Shima J S, Hellberg M E, Thorrold S R, Jones G P, Robertson D R, Morgan S G, Selkoe K A, Ruiz G M, Warner R R 2002 *Bull Mar Sci* **70** 251–271
[30] Raimondi P T and Morse A N C 2000 *Ecology* **81** 3193–3211
[31] Edmunds P J, Bruno J F, and Carlon D B 2004 *Marine Ecology Progress Series* **278** 115–124.
[32] Roth M S and Knowlton N 2009 *Marine Ecology Progress Series* **376** 133–142.
[33] Trapon M L, Pratchett M S and Hoey A S 2013 *PLoS ONE* **8**:e57788.
[34] Gleason, M. G 1996 *Journal of Experimental Marine Biology and Ecology* **207**:79–101.
[35] Brandl S J, Hoey A S and Bellwood D R 2014 *Coral Reefs* **33** 421–430.
[36] Mizrahi D, Navarrete S A and Flores A A 2014 *Journal of Experimental Marine Biology and Ecology* **452** 22–30.
[37] English S C, Wilkinson and Baker V 1994 *Survey Manual for Tropical Marine Resources.* (Townsville: Australia Institute of Marine Science)
[38] Carleton J H and Sammarco P W 1987 *Bulletin of Marine Science* **40** 85–98.
[39] Harriott V J and Fisk D A 1987 *Marine Ecology Progress Series* **37** 201–208.
[40] Nozawa Y 2008 *Journal of Experimental Marine Biology and Ecology* **367**127–130.
[41] Edmunds P J, Nozawa Y and Villanueva R D 2014 *Journal of Experimental Marine Biology and Ecology* **454** 78–84.
[42] Babcock R C and Mundy C 1996 *Journal of Experimental Marine Biology and Ecology* **206** 179–201.
[43] Mizrahi D, Navarrete S A and Flores A A 2014 *Journal of Experimental Marine Biology and Ecology* **452** 22–30
[44] Harrison P L and Wallace C C 1990 *Ecosyst World* **25** 133–207
[45] Doropoulus C, Roff G, Bozec Y M, Zupan M, Werminghausen J and Mumby P J 2016 *Ecological Monographs* **86** (1) 20 – 44

[46] Hamner W M, Jones M S, Carleton J H, Hauri I R and Williams D M 1988 *Bull Mar Sci* **42** 459–479
[47] Fabricius K E and Metzner J 2004 *Coral Reefs* **23** 245–248
[48] Baird A H and Hughes T P 2000 *Journal of Experimental Marine Biology and Ecology* **251** 117–132
[49] Chiappone M and Sullivan K M 1996 *Bull Mar Sci.* **58**(2) 555–569
[50] Miller M W and Barimo J 2001 *Bull Mar Sci.* **69**(2) 395–405
[51] Miller M W, Weil E and Szmant A M 2000 *Coral Reefs.* **19**(2) 115–123. http://dx.doi.org/10.1007/s003380000079
[52] Wagner D E, Kramer P and Van Woesik R 2010 *Mar Ecol Prog Ser.* **408** 65–78. http://dx.doi.org/10.3354/meps08584
[53] Harrington L, Fabricius K, De'Ath G and Negri A 2004 *Ecology* **85** 3428–3437
[54] Day R W 1983 *Bulletin of Marine Science* **33** 597–605
[55] Box S J and P J Mumby 2007 *Marine Ecology Progress* Series **342** 139–149
[56] Wolf A T and Nugues M M 2013 Coral Reefs **32** 227–231

The structure of the *Meiozoobenthos* community and its contribution to demersal fishery in mangrove forest ecosystem in Banyuwedang Bay, Buleleng, Bali

I B J Swasta

Biology Department, Faculty of Matematic and Natural Science,
Ganesha University of Education, Bali - Indonesia

E-mail : swastajelantik@yahoo.com

Abstract. Meiozoobenthos are small animals of less than 1.0 millimeter in length that live in mud substrate. Meanwhile, demersal fish that live in the seabed is one of the predatory groups that allegedly exploit the existence of meiozoobenthos by preying on them. Some interesting issues to be studied are the ecological aspects of meiozoobenthos, especially meiozoobenthos community structures and their contribution to the existence of demersal fish. To answer this problem, field research has been conducted by taking a location in the mangrove ecosystem in Banyuwedang Bay. In this study, the researcher collected biotic and abiotic data, taken on 20 sample points spread over 4 (four) sub locations. The collected data is then analyzed by quantitative and qualitative descriptive approach. The results of this research are: (a) the diversity of meiozoobenhos communities in Banyuwedang Bay is high with the number of species of 53 species and diversity index of 3.78, (b) the diversity of demersal fish in Banyuwedang Bay is moderate with the number of species as many as 13 species with diversity index of 2.50, (c) meiozoobenthos community contributes greatly to the presence of demersal fish in Banyuwedang Bay through meiozoobenthos's role as the food for demersal fish, and (d) ecologically, the condition of mangrove ecosystem in Banyuwedang Bay is quite feasible as a place to develop a demersal fishery.

1. Introduction

As an aquatic ecosystem, the sea has a complex biological component that according to its form (life form) can be divided into 6 (six) kinds which are neuston, plankton, nekton, periphyton, benthos, and microbes Each of these six forms of aquatic life occupies its own place according to its role and position within the marine biological community. Phytoplankton and phytobenthos take on the role of producers producing organic materials that can be used by other components. Meanwhile, zooplankton, zoobenthos, neuston, and nekton take the role of the consumer at various levels depending on the food source. Microbes as microorganisms are components that take on the role of decomposers that decompose the bodies of other components and its group into more simple organic particles either stored in the substrate (deposit) or in the water column (suspension) [1].

About benthos, one of the most important benthos groups in marine ecosystems is the meiozoobenthos community, a small zoobenthos group (<1.0 mm) inhabiting a muddy and sandy seabed [1] [2]. Ecologically, meiozoobenthos are an important link in marine ecosystems that take on the role of consumers that utilize organic particles trapped in sediment (sand and mud) as food [3]. This organic material is then continued by meiozoobenthos to the higher trophic level through predation events when the meiozoobenthos itself is preyed on by other consumers [3] One group of consumers that used to prey on meiozoobenthos is a group of fish commonly associated with basic habitats commonly called demersal fish [4]. Therefore the existence of meiozoobenthos for the demersal fish community is very important especially in the waters that the decomposition and sedimentation process is quite large, as in the mangrove forest ecosystem. So it is very rational if the trophic relationship between meiozoobenthos and demersal fish becomes an inspiration for the development of demersal fisheries with mangrove-based silvo fisheries system

As an important resource for the development of demersal fisheries, meiozoobenthos is very important to learn, especially from the ecological aspect so that the understanding of the ecological aspects of this unique meiozoobenthos community becomes clearer. Talking about the ecological aspects of the meiozoobenthos community, the important aspects to be studied are aspects of community structure and habitat requirements and a good environment for meiozoobenthos's life. The study of meiozoobenthos community structure can provide an overview of diversity, dominance, relative abundance, trophic structures and the life forms of meiozoobenthos.. This study of meiozoobenthos community structure can be used as a material to assess the potential of meiozoobenthos resources in an area. Meanwhile, the study of habitat and environmental requirements for pelos can provide an overview of the physical and chemical variables appropriate for meiozoobenthos' life. Then, it can be used as a material to assess whether an area has a condition suitable for pelos' life. Thus, these findings can be used as a consideration for making decisions about whether an area with certain conditions and meiozoobenhos resources is suitable to be used as a site for developing demersal fisheries.

In relation to study the ecological aspects of meiozoobenthos and demersal fish communities, the mangrove forest ecosystem is a very good form of ecosystem used as the media to study it. It is because the ecosystem of mangrove forest is very rich with meiozoobenthos community, which in the end can invite the presence of various demersal fish in search of food in mangrove forest [5] [6] [7].. The existence of mangrove forests does not only support the existence of meiozoobenthos through the sedimentation and decomposition process, it also sustains the life of demersal fish through the provision of food and shade of canopies. Leaf detritus, fruit, stems, twigs, and mangrove flowers are food for meiozoobenthos and some demersal fish species, while the shade of mangrove canopy can cover the forest floor including meiozoobenthos in it, and also shelter fish communities in the mangrove forest. Thus, meiozoobenthos and fish communities are safe from the hot sun [8].

In determining the mangrove forest as the media to study the ecological aspects of meiozoobenthos community in relation to the demersal fish community, the mangrove ecosystem in Banyuwedang Bay is a representative place to be used as the media. It is because the mangrove forest in this region is quite dense with a high diversity of mangroves (6 species), and with a thick sediment of forest (about 30 cm). In addition, the condition of mangrove ecosystems in this region is relatively good and natural. Thus, mangrove forests in this region are estimated to be rich of meiozoobenhos and demersal fish, so it is appropriate to be used as the media to examine the ecological aspects of meiozoobenhos and demersal fish communities. Based on the above description, there are a number of interesting issues to be studied, namely ; (a) how is the structure of meiozoobenthos and demersal communities that inhabit mangrove ecosystem in Banyuwedang Bay ; (b) whether pelos community has contributed to the existence of demersal fish community in mangrove ecosystem in Banyuwedang Bay, and in what form is the contribution given by pelos and; (c) based on existing conditions, is the mangrove ecosystem in Banyuwedang Bay suitable for developing demersal fisheries?

2. Research Method

This study is categorized as a descriptive exploratory study that takes place in Banyuwedang Bay, Gerokgak Sub-district, Buleleng Regency, Bali. The subjects in this study are meiozoobenthos and demersal fish communities that inhabit the base substrate of mangrove forest in Banyuwedang Bay, Buleleng, Bali. While the object of this research is the structure of meiozoobenthos and demersal fish community, and the contribution of meiozoobenthos community to the existence of demersal fish community in the mangrove forest area of Banyuwedang Bay, Buleleng, Bali. The population in this study is the whole meiozoobenthos and demersal fish that inhabit the base substrate of mangrove forest in Banyuwedang Bay, Buleleng, Bali, while the sample of this research is a number of pelos and demersal fish that have been caught in the sampling done in accordance with sampling rules that have been set.

The study used a number of tools such as ; (a) a set of mud sampling tools consisting of Petersen Grab and Mud Core, (b) a set of meiozoobenthos extraction tools consisting of UME (Uhlig Mud Extractor), MEA (Mud Elutriation Apparatus) [9] ; c) a set of fishing gear consisting of spreading nets, fishing rods, and traps ; (d) a set of sea water filters comprising plankton net, bucket and Beker glasses ; (e) a set of seawater quality checker called Water Quality Checker and Hand Refractometer ; (f) a set of Rh Meter ; (g) binocular stereoscopic microscope and ; (h) a number of flakon bottles and plastic bags as the place to put the samples of meiozoobenthos, mud and seawater. Materials needed in this study were ; (a) Rose Bengal coloring ; (b) preservatives of 5% formaldehyde, and (c) seawater ice. All of these materials are in the laboratory (Biology and Aquaculture).

The data collection process consists of several steps, namely ; (1) division of research sites into four sub locations covering sub location around Avicennia mangrove, sub location around Sonneratia mangrove, sub location around Rhizophora mangrove, and sub location around hot springs ; (2) determination of sampling points in which there are five sampling points in each sub location, making 20 sampling points in total ; (3) sampling of meiozoobenthos, demersal fish, substrate and seawater ; (4) identification of meiozoobenthos samples and demersal fish as well as measurement of abiotic variables of seawater and substrate, and ; (5) analysis of demersal fish stomach contents.

For the data analysis, this research uses 2 (two) approach that are quantitative approach (statistic, descriptive) and qualitative approach (descriptive, inductive and deductive). The issues that require quantitative analysis (statistics) are ; (a) determination of diversity index of meiozoobenthos and demersal fish ; (b) determination of abundance value of meiozoobenthos and demersal fish, and ; (c) determination of meiozoobenthos percentage in the whole contents of demersal fish stomach. Meanwhile, the problems that require qualitative analysis are ; (a) the composition of pelos and demersal fish species ; (b) meiozoobenthos species consumed by demersal fish, and ; (c) the feasibility level of the mangrove ecosystem in Banyuwedang Bay for the development of demersal fisheries in terms of physical and chemical variables of the substrate and its waters. In this qualitative analysis, inductive and deductive approaches are used where the results are described in narrative with the help of graphics and illustrations.

3. Results and Discussions

3.1 Results

From the research conducted and from the results of data analysis, the following results have been obtained.

a. Diversity of *Meiozoobenthos* community in Banyuwedang Bay is very high with the number of species as many as 53 species with diversity index of 3.78. The 53 meiozoobenthos species mentioned are shown in the following table (Table 1).

Table 1. The composition of *Meiozoobenthos* species that live throughout the Bay of Banyuwedang

No	Species Names	Groups	Abundance
1	*Paramonohystera sp*	Nematoda	65
2	*Microlaimus sp*	Nematoda	50
3	*Leptolaimus sp*	Nematoda	50
4	*Comacolaimus sp*	Nematoda	45
5	*Metalinhumous sp*	Nematoda	30
6	*Enoplus sp*	Nematoda	30
7	*Enoploides sp*	Nematoda	30
8	*Bathylaimus sp*	Nematoda	25
9	*Halalaimus sp*	Nematoda	25
10	*Anticoma sp*	Nematoda	20
11	*Draconema sp*	Nematoda	20
12	*Axonolaimus sp*	Nematoda	15

13	*Pionoxyllis sp*	Polichaeta	40
14	*Protodrilus sp*	Polichaeta	35
15	*Phryatrocha sp*	Polichaeta	35
16	*Saccocirrus sp*	Polichaeta	30
17	*Manayunkia sp*	Polichaeta	20
18	*Meganerilla sp*	Polichaeta	20
19	*Nerillidium sp*	Polichaeta	15
20	*Mesonerilla sp*	Polichaeta	10
21	*Nematoderma sp*	Turbelaria	25
22	*Nematoplana sp*	Turbelaria	25
23	*Anoplodiera sp*	Turbelaria	20
24	*Polichoirus sp*	Turbelaria	10
25	*Mesostoma sp*	Turbelaria	10
26	*Urotricha sp*	Ciliophora	15
27	*Halophrya sp*	Ciliophora	15
28	*Tracheloraphis sp*	Ciliophora	10
29	*Aktedrillus sp*	Oligochaeta	20
30	*Grania sp*	Oligochaeta	15
31	*Heterodrillus sp*	Oligochaeta	15
32	*Enchytraeus sp*	Oligochaeta	10
33	*Batillipes sp*	Tardigrada	10
34	*Stygarctus*	Tardigrada	10
35	*Hypsibius sp*	Tardigrada	5
36	*Turbanella sp*	Gastrotricha	15
37	*Dalichodasys sp*	Gastrotricha	10
38	*Meioherpia sp*	Molusca	20
39	*Prochaetoderma sp*	Molusca	15
40	*Falsiden sp*	Molusca	15
41	*Gnathostomula sp*	Gnathostomulida	15
42	*Haplognatha sp*	Gnathostomulida	10
43	*Aspidosiphon sp*	Sipuncula	5
44	*Pascolion sp*	Sipuncula	5
45	*Archinemertinea sp*	Nemertinea	20
46	*Tubiluchus sp*	Priapulida	10
47	*Cypridina sp*	Crustacea	15
48	*Macrocypris sp*	Crustacea	10
49	*Herpectocypris sp*	Crustacea	10
50	*Harpacticus sp*	Crustacea	10
51	*Laophonte sp*	Crustacea	10
52	*Muna sp*	Crustacea	5
53	*Colomastix sp*	Crsutacea	5
∑	53 Species	13 Groups	1030

These 53 meiozoobenthos species were extracted from four sub locations with the following details

Table 2. Number of Species, Abundance and Diversity of Living Meiozoobenthos in Each Sub Location in Teluk Banyuwedang

No	Sub location	∑ Species	Abundance	Diversity
1	Avicennia	39	325	3.55
2	Sonneratia	36	285	3.47
3	Rhizophora	32	260	3.34
4	Hot Spring	27	160	3.25

Graphically, species richness, individual abundance and diversity index of *Meiozoobenthos* in each sub location can be described as follows.

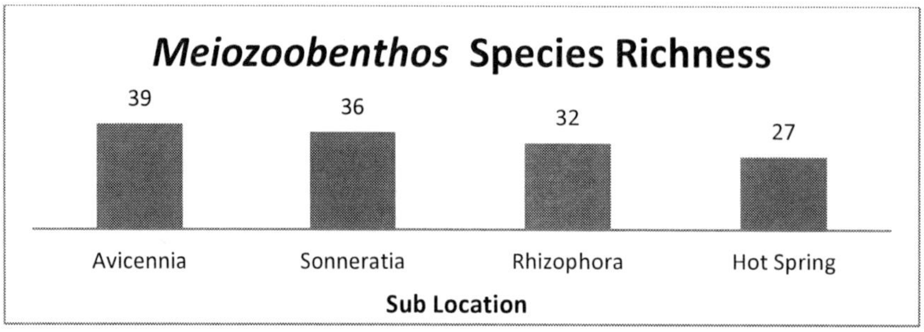

Figure 1. Meiozoobenthos species richness in each sub location in Banyuwedang Bay.

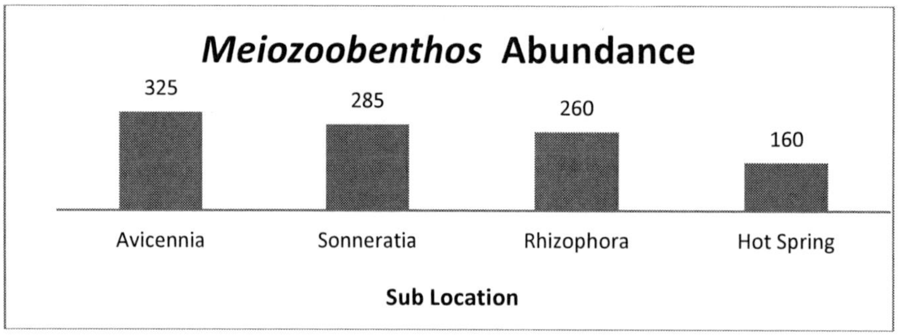

Figure 2. Meiozoobenthos abundance in each sub location in Banyuwedang Bay.

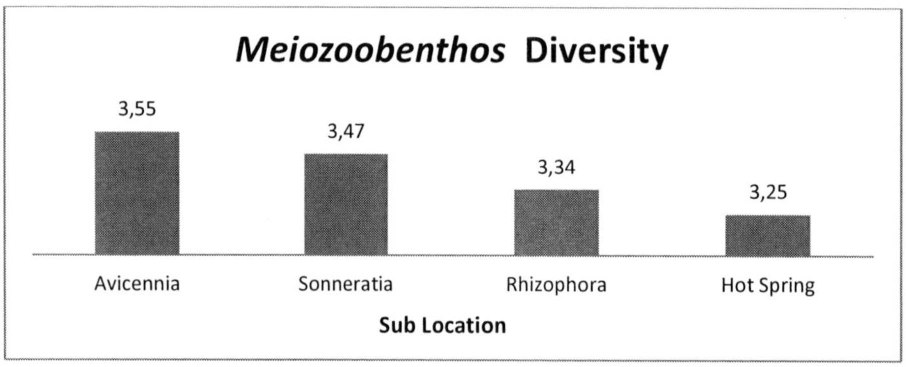

Figure 3. Meiozoobenthos diversity index in each sub-location in Banyuwedang Bay.

b. The demersal fish community in Banyuwedang Bay is quite diverse with the number of species as many as 13 species with a diversity index of 2.50. The 13 species of demersal fish mentioned are shown in the following table (Table 3).

Table 3. The Spesies Composition of Demersal Fishes in Banyuwedang Bay

No	Species	Native name	Abundance
1	*Mugil labiosus*	Ikan Belanak	45
2	*Terapon jarbua*	Ikan Kerung	35

3	*Epinephelus lanceolatus*	Kerapu Lumpur	35
4	*Lutjanus argentimaculatus*	Ikan Mengarat	30
5	*Lutjanus johnii*	Ikan Badur	30
6	*Chanos chanos*	Ikan Bandeng	35
7	*Scorpaenopsis sp*	Ikan Lepuh	15
8	*Tetraodon hispidus*	Ikan Buntel duri	25
9	*Tetraodon reticularis*	Ikan buntel pasir	15
10	*Pseudorhombus arsius*	Ikan Lidah	20
11	*Leiognathus splendens*	Ikan Cotek	40
12	*Periopthalmus sp*	Ikan Gelodok	55
13	*Plotusus canius*	Ikan Sembilang	25
∑	13 Species		405

These 13 demershal fish species live in four sub-district as shown in the table bellow.

Table 4. Species richness, Abundance and the diversity of demersal fish that live in each sublocation in Banyuwedang Bay.

NO	SUB LOCATION	∑ SPECIES	ABUNDANCE	DIVERSITY INDEX
1	Avicennia	12	140	2.43
2	Sonneratia	13	115	2.47
3	Rhizophora	13	115	2.43
4	Hot Spring	5	35	1.48

Graphically, the abundance, species richness, and diversity of the demersal fish that live in those four sub-locations in Banyuwedang Bay are shown bellow

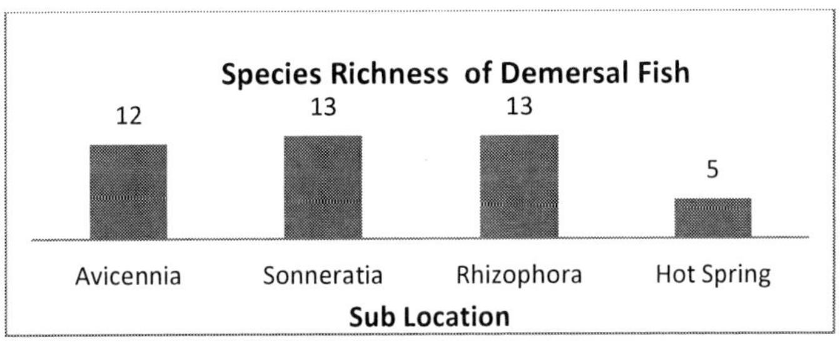

Figure 4. Species richness of demersal fish in each sub location in Banyuwedang Bay

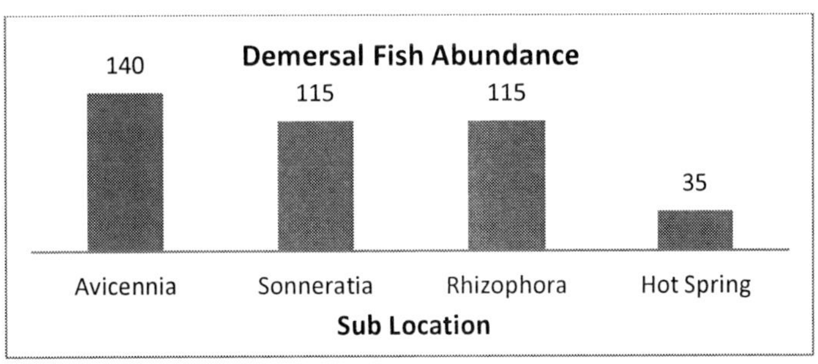

Figure 5. The abundance of demersal fish in each sub-location in Banyuwedang Bay.

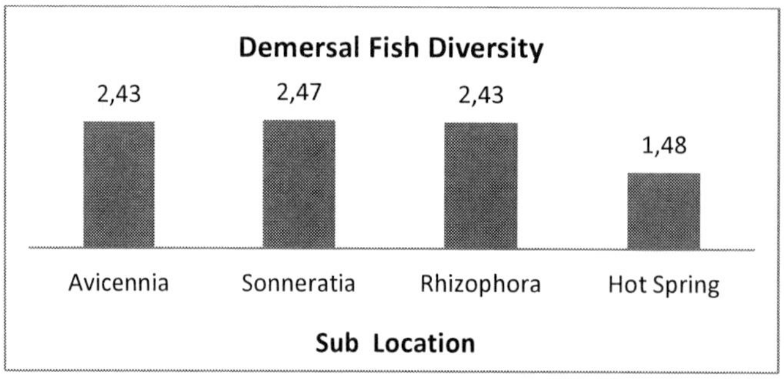

Figure 6. Demersal fish diversity index in each sub-location in Banyuwedang Bay.

c. Meiozoobenthos has a major contribution to the life of demersal fish. The contribution is given by acting as a food for demersal fish. This is proved by the results of the analysis of stomach contents of some demersal fish caught like Mud Grouper (*Epinepelus lanceolatus*), Kerung fish (*Terapon jarbua*), Mengarat fish (*Lutjanus argentimaculatus*), Sembilang fish (*Plotusus canius*), and blisters (*Scorpaenopsis sp*). From those five fishes stomach, it was obtained a percentage of average pelos as follows: (a) Crustaceans as much as 24%, (b) Polichaeta as much as 26%, (c) Tardigrada as much as 12%, (d) Nematodes as much as 14%, and (e) others as much as 24%. Identifying the other contents in the demersal fish stomach (probably meiozoobenthos) is difficult because it was completely destroyed.

d. Judging from the ecological conditions, the mangrove forest ecosystems in the Bay of Banyuwedang is pretty decent to be used as a place for demersal fisheries development. It is based on the results of monitoring and measurement of the ecological condition of the mangrove ecosystem, where all physical and chemical variables showed pretty good condition, viewed from the requirements of life for demersal fish community and the meiozoobenthos community. In addition, the water is not polluted and guarantees the survival of demersal fish and meiozoobenthos in the region. More than that, mangrove vegetation condition are still relatively intact and natural, and has enough diverse types so that it can guarantee the life and development of demersal fish and meiozoobenthos that live in the region. For sustainability, the ordinance of West Bali National Park is quite a guarantee that there will be no destruction of mangrove forest ecosystems in this region, considering this area is still in the area of West Bali National Park.

3.2 Discussion

By seeing all the data above, we can conclude that the diversity of meiozoobenthos in Banyuwedang Bay is very high. This can be seen from the number of species reaching 53 species, and the value of the Diversity Index reaching 3.78. The high diversity of meiozoobenthos in this region is estimated to be caused by 3 (three) important factors namely ; (a) the variety of meiozoobenthos habitats in this region ; (b) the good condition of meiozoobenthos habitats in the region, and ; (c) the stability of the mangrove ecosystem in the region.

Seen partially for each sub-location, it appears that the diversity of meiozoobenthos in each sub-location is slightly varied. The results showed that the highest meiozoobenthos diversity was found around the mangrove of the Avicennia family (3.55), followed by the meiozoobenthos diversity in the vicinity of the Sonneratia cluster mangrove (3.47), the diversity of meiozoobenthos around the mangroves of the Rhizophora clan (3.34) , and the variety of meiozoobenthos around the hot springs (3.25). This variation is suspected to be closely related to the tannin content of each mangrove species. According to Lemmen and Wulijarni (1992) the content of tannins in Sonneratia ranged from 9 to 15%, whereas in Rhizophora ranged between 8 - 40% [10]. It is known that tannin has properties that suppress meiozoobenthos population, so the higher the tannin in mangrove, the greater the pressure on the meiozoobenthos population [5]. The diversity of meiozoobenthos in sub-hot springs that show the smallest number compared to the other sub-sites is probably caused by the relatively warm water temperature (31° C) and high sulfur content. The water temperature is relatively warmer than the temperature of the surrounding sea water. it tends to be suppressing meiozoobenthos population, except the population of certain meiozoobenthos types that have adapted well in hot water. In relation to sulfur substances in the form of H2S, these sulfuric substances can be toxic to most meiozoobenthos types, even in low concentrations [2]. It is said to be toxic, because H2S compounds can block the enzyme cytochrome oxidase acting on the cellular respiration chain [2]

In relation to the diversity of demersal fish which is not too high (H '= 2.50) in Banyuwedang Bay area, this is rationally acceptable because Banyuwedang Bay waters are almost isolated from seawater around it. Therefore the demersal fish in this bay is almost entirely a permanent species settled in this bay. In other words, it is very difficult to find marine migrant species that come to Banyuwedang Bay. Thus it can be understandable that the number of species and their diversity are in the moderate category.

Seen partially for each sub location, it appears that demersal fish diversity in each sub location is almost the same that is classified as moderate, except for sub hot springs location where the diversity is low. The factors that may cause the high diversity of demersal fish in the other three sub-sites (Avicennia, Sonneratia, and Rhizophora) are because there are no dividing boundaries between the sub-sites so that demersal fish can still move from one sub-location to another sub-location. Thus the diversity of demersal fish in the three sub locations is almost the same. The low diversity of demersal fish in hot springs sub-locations is probably due to a relatively higher water temperature factor and the presence of sulfuric substances in the form of H2S. The relatively warm water temperature is generally avoided by the fish because it is less convenient, also because the oxygen solubility is lower than the water with the lower temperature. Low oxygen content in the water will make it difficult for the fish to breathe.

The fact that meiozoobenthos is a food source for demersal fish, especially the carnivores, is very much in line with the previous researches conducted around the world that conclude meiozoobenthos as an important source of energy and material for the demersal fish community in addition to other sources such as detritus, carcasses, phytobenthos and zoobenthos from the epifauna group. These are the previous researches that are very much in line ; (1) The 1976 Ankar and Elmgren studies in the Asko estuary around the Baltic Sea, and; (2) Warwick and Radford's 1979 study in the Lynher estuary, southern England [11].

In terms of ecological conditions, the mangrove ecosystem in Banyuwedang Bay is suitable to be used as a place to develop mangrove-based demersal fisheries. It is because the measurement results of all physical and chemical variables of the waters as well as the observation of the condition of the mangrove vegetation generally indicate the condition is good. The salinity, temperature, pH, DO, and turbidity show a range within feasible limits for aquatic biota, except for physical and chemical variables in hot springs sub-location that appear less friendly. Then, the vegetation of the mangrove is quite diverse and still in relatively good condition and natural. Therefore, everything can certainly support the existence of demersal fisheries in this bay, except in sub hot water spots that appear to be less suitable for demersal fisheries development. This is because the water temperature is warm and the content of sulfur is less favored by most species of fish, including demersal fish. The existence of mangrove vegetation for demersal fish is as a place to find food, a place to reproduce, a place to live as well as to take a shelter [12]

4. Conclusions and Suggestions

4.1 Conclusion

There are some findings after the research was conducted: (a) the diversity of meiozoobenthos communities living within the mangrove ecosystem in Banyuwedang Bay is very high where 53 species were identified with the Diversity Index of 3.78, (b) the diversity of demersal fish communities living within the mangrove ecosystem of Banyuwedang Bay is moderate with 13 species were identified, with a Diversity Index of 2.50, (c) a meiozoobenthos community in Banyuwedang Bay has a major contribution to the presence of demersal fish in this bay, because meiozoobenthos can be a food source for demersal fish, especially carnivorous and omnivorous, and (d) based on its good ecological condition, the mangrove ecosystem in Banyuwedang Bay is feasible as a place to develop demersal fisheries.

4.2 Suggestions

Based on the research results, there are some suggestions: (1) to ensure the survival of demersal fishery in Banyuwedang Bay, it is suggested not to damage or convert the existing mangrove forest area, and (2) to ensure the survival of demersal fishery in Banyuwedang Bay, it is suggested to use an environmentally friendly way to exploit the yield and pay attention on its sustainability.

References

[1] Jenkin M 1983 *Seashore Studies* (London: George Allen and Unwin Publisher)
[2] Giere O 1993 *Meiobenthology, the Microscopic Fauna in Aquatic Sedimen* (Berlin: Springer-Verlag)
[3] Mann K H 1982 *Ecology of Coastal Waters* (London: Blackwell Scientific Publications)
[4] Burhanuddin M, Djamali and Muljanto 1984 *Perikanan Demersal di Indonesia* (Jakarta: Lembaga Oseanologi Nasional, LIPI)
[5] Gwyther J 2000 Meiofauna in Phytal Based and Sedimentary Habitat of Temperate Mangorve Ecosystem *Proceeding of the Royal Society of Victoria* **112** (3) 137 – 151
[6] Gwyther J and Fairweather P G 2002 Colonization by Epibiont and Meiofauna of Real and Mimic Pneumatophores in a Cool Temperate Mangrove Habitat *Marine Ecology Progress Series* **229** 137 – 149
[7] Gwyther J 2002 Nematode Assemblages from Avicennia marina Leaf Litter in a Temperate Mangrove Forest in Southeastern Australia *Marine Biology* **142** 289 – 297.
[8] Jelantik S 2010 *Bioekologi Endopsammon dan Metode Penelitiannya* (Yogyakarta: Yogya Media Utama)
[9] Holme N A and McIntyre A D 1984 *Methods for the Study of Marine Benthos* (London: Blackwell Scientific Publications)
[10] Lemmens R H M J and Soetjipto N W (Ed) 1992 *Dye and Tannin Producing Plants* (Bogor: PROSEA)
[11] McLusky D S 1989 *The Estuarine Ecosystem* (London: Blackie Academic and Professional)
[12] Hutchings P and Saenger P 1987 *Ecology of Mangroves* (London: University of Queenslands Press)

Bioprospecting of the Balinese marine sponges and nudibranchs

I W Mudianta

Universitas Pendidikan Ganesha, Bali – Indonesia, 81116

E-mail: mudianta@undiksha.ac.id

Abstract. Secondary metabolites play an invaluable role as a starting point in drug discovery process. Among marine organisms, sponges and nudibranch mollusc (Mollusca: Gastropoda: Nudibranchia) have been the prolific sources of bioactive secondary metabolites with diverse chemical structures and promising bioactivities. This report presents the isolation and structure elucidation of bioactive natural products from the Balinese marine sponge Aplysinella strongylata and the nudibranchs Hypselodoris infucata and Glossodoris hikuerensis. The samples were collected from Tulamben beach Bali, an unexplored site but rich in marine biodiversity. The separation and purification of the metabolites were achieved by mean of NP-flash column chromatography and HPLC. Structure elucidations of the isolated compounds were completed by analysis of spectroscopic dataset including ^{1}H and 2D NMR. A series of bromotyrosine-derived alkaloid possessing a spirooxepinisoxazoline moiety were identified from the extract of A. strongylata. One of the metabolites, 19-hydroxypsammaplysin E (**1**), showed the best antimalarial activity, with an IC_{50} value of 6.4 μM. The absolute configuration of psammaplysin series compounds has been assigned as (6R, 7R) using experimental and calculated electronic circular dichroism (ECD) data and NMR analysis of MPA esters prepared from the acetamide derivative of psammaplysin A. A furanosesquiterpen, (–)-furodysinin (**2**), was identified for the first time from the nudibranch H. infucata. The compound inhibited the growth of HeLa cell line at IC_{50} 102.7 μg/mL. Two new scalarane sesterterpenes (**3-4**) were characterised from an organic extract of a single specimen of the nudibranch G. hikuerensis.

1. Introduction

Marine animals have been known to produce myriad of secondary metabolites (or natural products) with diverse chemical structures and an array biological properties [1,2]. Studies of the chemical properties of these secondary metabolites have led to discovery of lead compounds some of which have entered clinical trials, particularly as anticancer and antimicrobial agents [3]. Currently, the number of bioactive natural products isolated from marine organisms now exceeds 28,000 with hundreds of new compounds being discovered every year [4].

Among marine organisms, sponges and nudibranchs (Mollusca: nudibranchia) have been prolific sources of bioactive secondary metabolites, several of which have inspired the development of new classes of therapeutic agents. Recently, the active compounds of four out of seven new FDA-approved drugs were originally collected from sponges and nudibranchs. For example, the drug marketed as Cytarabine (anticancer), Vidarabine (antiviral), and Eribulin Mesylate (anticancer) were originally isolated from Caribbean sponge *Cryptotethia crypta* and *Halichondria okadai* respectively [5].

Here we presented chemical studies of marine natural products from sponges and nudibranchs collected from Bali, an under-exploited marine hot spot but rich in species diversity. Although there have been several reports on chemical studies of marine sponges and sponge-associated fungi collected from West Bali [6-7], this is an early study carried out of specimens from East Bali including Tulamben. A single specimen of sponge *Aplysinella strongylata* and two nudibranch extracts including *Hypselodoris infucata* and *Glossodoris hikuerensis* were investigated.

The isolation and purification of secondary metabolites from the extracts of these specimens were achieved by means chromatography including flash column chromatography and high performance liquid chromatography (HPLC). Accordingly, the structure elucidations of the isolated metabolites were completed by analysis of 1D and 2D nuclear magnetic resonance (NMR) data. Furthermore, biological activities of the metabolites were tested in vitro against chloroquine-sensitive (3D7) *P. falciparum* malaria parasites and HeLa cell line.

Herein reported the isolation and structure elucidation of 19-hydroxypsammaplysin E (**1**) from the extract of Balinese marine sponge *Aplysinella strongylata*. It showed the best antimalarial activity, with an IC_{50} value of 6.4 μM. The absolute configuration of psammaplysin series compounds were assigned as (6*R*, 7*R*) using experimental and calculated electronic circular dichroism (ECD) data and NMR analysis of MPA esters prepared from the acetamide derivative of psammaplysin A. A furanosesquiterpen, (–)-furodysinin (**2**), was identified for the first time from the nudibranch *H. infucata*. Two new scalarane sesterterpenes (**3-4**) were characterised from an organic extract of a single specimen of the nudibranch *G. hikuerensis*.

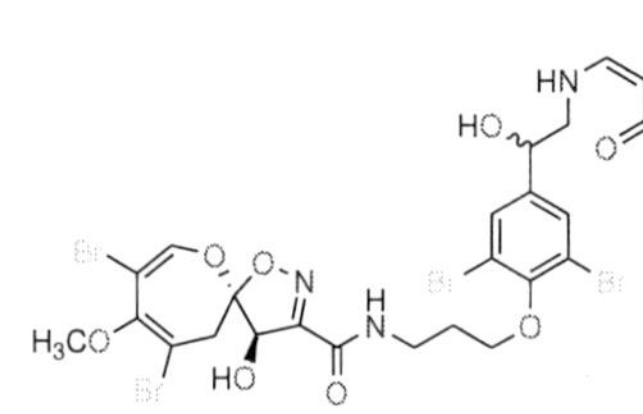

1

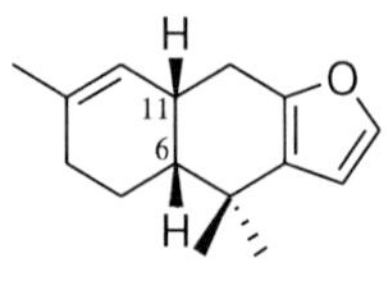

2

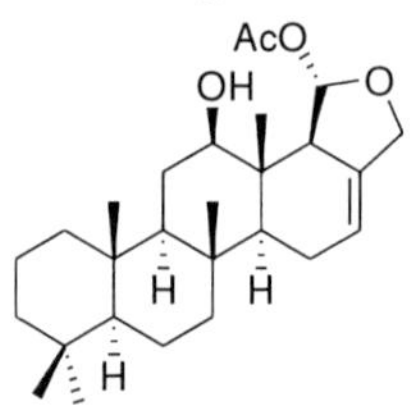

3

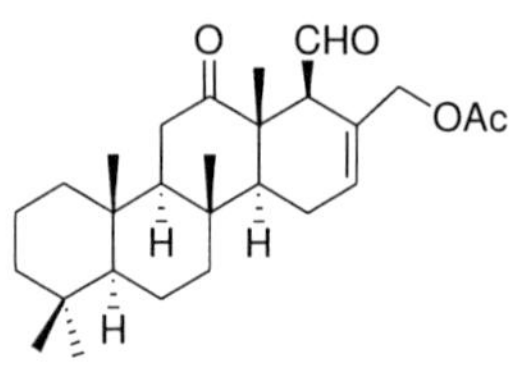

4

2. Methods

General Experimental Procedures. All NMR spectra were referenced to solvent signals as follows: δ 7.26 and 77.16 ppm for $CDCl_3$; δ 2.05 and 29.84 for acetone-d_6; and δ 3.31 and 49.00 ppm for methanol-d_4. 1D and 2D NMR spectra were acquired using a Bruker Avance 400 or a Bruker Avance 500 spectrometer at 298 K. Optical rotations were obtained using a Jasco P2000 polarimeter. Positive ion electrospray mass spectra (LRESIMS) were determined using a Bruker Esquire HCT or (HRESIMS) using a MicroTof Q instruments each with a standard ESI source. Reversed-phase HPLC was carried out on an Agilent 1100 Series instrument fitted with a variable-wavelength UV and refractive index detector, an Agilent D1311A quaternary pump, and a semipreparative Phenomenex C18 Gemini 5 μm 110 Å column (10 mm × 250 mm). Gas chromatography was carried out on a Shimadzu GCMS-QP2010 Plus. Initial temperature was 100 °C, isothermal for 3 min, then ramped 16 °C/min for 10 min. The final temperature was 270 °C, injection temperature 250 °C, and flow rate 1.5 mL/min.

Biological Material. The sponge *Aplysinella strongylata* was harvested in Tulamben Bay, Bali, at a depth of approximately 20 m by hand using scuba in November 2010. The sponge was spherical in shape, fleshy to the touch, and compressible. The color in life was milky-colored inside and gray outside, compressible with mucus secretion, turning deep brown in 70% EtOH preservative. The sponge identification was undertaken at the Research Center of Oceanography, Indonesian Institute of Science, where a voucher specimen (TL-20) is deposited. One specimen of *Glossodoris hikuerensis* (crawling length 2 cm) was collected during fieldwork at Tulamben Bay, Bali, in July 2012. Two specimens of *Hypselodoris infucata* (crawling length 2cm) were collected during fieldwork at Tulamben Bay, Bali in November 2014. The specimens were identified by comparing the surface pictures with that in the encyclopaedia of nudibranch as well as with the online database such as nudipixel (www.nudipixel.net) and WoRM (www.marine-species.org).

Extraction and Purification. The freshly collected sponge (147 g wet wt) was frozen before being extracted with 1:1 DCM-MeOH (3×200 mL) at room temperature. The combined extracts were dried under vacuum to produce 64.5 g (wet wt) crude extract. The crude extract was sequentially partitioned between hexanes (3×200 mL) and H_2O, followed by EtOAc and H_2O (3×200 mL), and finally BuOH and H_2O (3×200 mL). This gave hexanes (1.2 g), EtOAc (3.1 g), and BuOH (0.84 g) fractions, respectively. A portion of the EtOAc fraction (3.1 g) was resolved by normal-phase VLC (3×10 cm in diameter) with step gradient elution from 100% hexanes to DCM, EtOAc, and 100% MeOH and gave 11 fractions. Subsequently, each of the fractions was purified by HPLC to give the known and new metabolites.
A single specimen of *Hypselodoris infucata* was diced, extracted in acetone (3×10 mL), and sonicated for 2 min. The combined extracts were partitioned between water and dichloromethane (3×5 mL) and subsequently the organic layer was dried with Na_2SO_4. Dried organic layer was evaporated to dryness to give a crude extract (5 mg). The extract was passed through a normal-phase Sep-Pak cartridge eluting with 100% hexane (10 mL) to provide compound **2** (1.3 mg).

A single specimen of *Glossodoris hikuerensis* was diced and extracted in acetone (3×10 mL). The combined extracts were dried under reduced pressure to give a crude extract (50 mg). The extracts were passed through a NP flash column eluted with hexanes (100%), hexanes→chloroform (3:1 to 1:3), chloroform→EtOAc (3:1 to 1:3), and EtOAc (100%) to give two terpene-rich fractions coded as F3-4 (17.6 mg) and F5-6 (14.4 mg). The less polar fraction (F3-4) was subjected to NPHPLC with hexanes: EtOAc (80:20) to provide the two new sesterterpenes **3** (3.8 mg) and **4** (~0.1 mg) together with the known compounds.

3. Result and Discussion
Twenty-one new psammaplysin derivatives exhibiting a variety of side chains, as well as six previously known psammaplysins including psammaplysins A, B, D, and E and ceratinamides A and B were identified from the extract of the Balinese sponge *Aplysinella strongylata* (order *Verongida*, family *Aplysinellidae*). Detail spectroscopic data analysis of each of the new metabolites was discussed in the previous publication [8]. Here we only presented an example of structure elucidation of a new compound that possesses the highest bioactivity named as 19-hydroxypsammaplysin D (**1**).

The presence of the psammaplysin carbon framework in **1** could be deduced from ^{1}H and ^{13}C NMR chemical shifts (MeOH-d_4) compared to those of psammaplysin E. A signal at δ_H 7.15 (1H, s) assigned to H-1 had HMBC correlations to signals at δ_C 103.5 (C-2), 148.9 (C-3), and 122.1 (C-6). Two distinctive geminal protons at δ_H 3.06 and 3.38 could be attributed to H-5a and H-5b, respectively. This AB system correlated to signals for C-4 (δ_C 105.7) and C-6 as well as to C-7 at δ_C 79.3, diagnostic for a hydroxymethine carbon. The H-7 signal appeared at δ_H 4.99 as an isolated singlet and correlated to an amide carbon atom at δ_C 159.1 (C-9). These assignments secured the presence of the spirooxepinisoxazoline system in **1** [9]. Furthermore, three mutually coupled methylenes at δ_H 3.62, 2.13, and 4.07 suggested the presence of a -$CH_2CH_2CH_2O$- moiety attached to the amide bond of the spiro ring system. A singlet at δ_H 7.60 was designated to the symmetric aromatic ring protons H-15 and H-17. HMBC correlation between H-12 and C-13 (δ_C 153.2) secured the connection between the propyloxy side chain and the substituted aromatic ring.

The ^{1}H NMR spectra of **1** was consistent except for the signals corresponding to H-19 and H-20. In psammaplysin E, H-19 appeared at δ_H 2.83 (2H) and H-20 at δ_H 3.55 (2H); in contrast for **4**, H-19 was shifted to δ_H 4.77 (1H, dd, J = 3.7, 7.2 Hz) while H-20 appeared as an AB system [δ_H 3.42 (1H, dd, J = 7.2, 13.4 Hz) and δ_H 3.59 (1H, dd, J = 3.7, 13.4 Hz)]; since the H-20 signals were partially obscured by those of H-5b and H-10, they were further resolved by a 1D TOCSY experiment involving irradiation of H-19. The chemical shift of C-19 was also shifted downfield from δ_C 36.0 in psammaplysin E to δ_C 71.8 in **1**, consistent with a C-19 hydroxy substituent as found previously in both psammaplysins B and D [9-10].

The absolute configuration of psammaplysin A has been assigned as (6*R*,7*R*) using experimental and calculated electronic circular dichroism (ECD) data and NMR analysis of MPA esters prepared from the acetamide derivative of psammaplysin A. Detailed conformational analyses of a truncated model compound of psammaplysin A with an in vacuo method and with the PCM solvent model for MeOH have identified the major conformers and factors governing the ECD spectrum of psammaplysin A. The correlation of the ECD data with the stereochemistry of psammaplysin A allows configurational assignment of related psammaplysin analogues on the basis of their ECD spectra (Figure 1)[11].

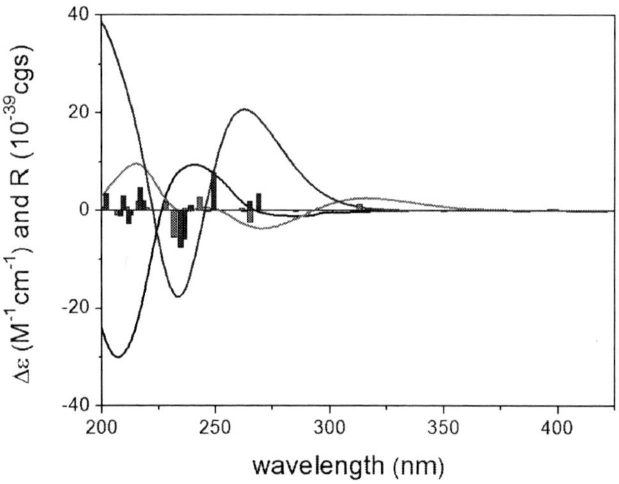

Figure 1. Experimental ECD spectrum of psammaplysin A recorded in MeOH (black curve) and PBE0/TZVP PCM/MeOH ECD spectra of (6*S*,7*S*)-truncated model for the two lowest-energy conformers obtained by B97D/TZVP (PCM/MeOH) reoptimization (blue and red curves). Bars represent rotational strength values of conformers A and B.

Compound **2** was isolated as a colorless oil ([α]$_D$ −51) and showed a single GCMS peak at *m/z* 216 (M^+) with a strong diagnostic base peak at *m/z* 122 (C_8H_8O) arising from a retro Diels-Alder reaction of the carbocyclic ring adjacent to the furan ring [12]. The ^{1}H NMR ($CDCl_3$, 500 MHz) spectrum of **2** revealed the presence of diagnostic furan protons at δ_H 7.21 (1H, d, J = 1.7 Hz, H-1) and δ_H 6.23 (1H, d, J = 1.7 Hz, H-2) together with an olefinic proton at δ_H 5.61 (1H, brs, H-10). There were also two geminal methyl groups resonated at δ_H 1.19 (3H, s, H-15) and δ_H 1.18 (3H, s, H-14) along with a methyl attached to an on olefinic carbon at δ_H 1.66 (3H, s, H-13). These ^{1}H NMR data of **2** were in close agreement with those of the synthetic sample of furodysinin reported by Vaillancourt *et al.* in 1991 [13].

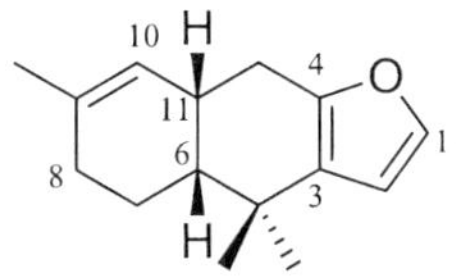

2

The tricyclic furanosesquiterpene (+)-furodysinin, ([α]$_D$ +64) was first reported by Wells *et al.* in 1978 from an Australian *Dysidea herbacea*[14] without relative stereochemical assignment detail. Guella *et al.* subsequently characterised its enantiomer (–)-furodysinin ([α]$_D$ –47) [15] from a specimen of *D. tupha* collected from the Mediterranean. The absolute configuration of (+)-furodysinin has been established as as 6*R*,11*R* when optically pure (–)-furodysinin ([α]$_D$ –54) was synthesised by Vaillancourt *et al.* from (+)-9-bromocamphor in four steps [13]. Recently, (–)-(6*R*,11*R*)-furodysinin (**2**) has been isolated from *D. herbacea* (Fiji) [16], from *H. bayeri* collected from Cuba [17], and from *H. jacksoni* from South-East Queeensland, Australia. (–)-Furodysinin isolated from *H. infucata* in the current study has consistently been associated with a 6*R*,11*R* configuration, whereas the positive counterpart shows a 6*S*,11*S* configuration [18].

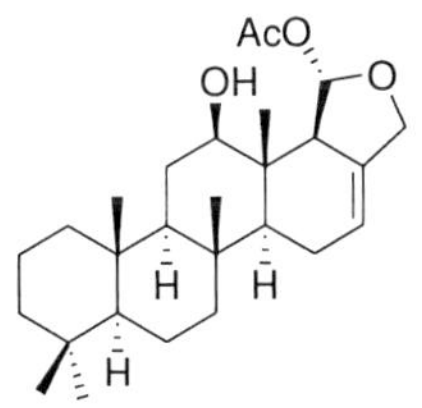

3

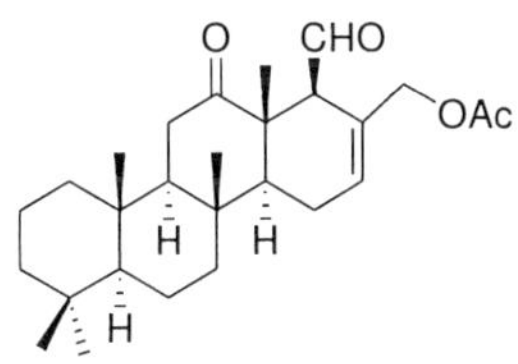

27

Compound **3** was isolated as a colorless oil, the molecular formula of which was established to be $C_{27}H_{42}O_4$ from the sodiated (+)-HRESIMS ion at *m/z* 453.2965. The presence of a 12-hydroxy scalarane framework in **3** was inferred from proton NMR signals (Table 1) for an acetal (δ_H 6.49, d, J = 2.7 Hz, H-19), an olefinic signal (δ_H 5.60, brs, H-16), and a carbinolic proton (δ_H 3.44, dt, J = 4.6, 11.3 Hz, H-12) as well as an AB system [δ_H 4.38, d, J =11 Hz (H-20a); 4.24, d, J = 11 Hz (H-20b)]. There were also five methyl singlets at δ_H 0.89, 0.85, 0.84, 0.80, and 0.76, each linked to their respective carbons at δ_C 16.3, 16.5, 33.4, 21.4, and 9.1 by HSQC data. The ^{1}H and ^{13}C NMR data of **3** were identical to those of 12-deacetyl-12-*epi*-deoxoscalarin isolated in the current study as well as to those reported by Fontana *et al.* , with the exception of the addition of an acetate methyl signal at δ_H 2.09. The acetate moiety was placed at C-19 based on a key HMBC correlation observed from the signal for H-19 (δ_H 6.49, d, J = 2.7 Hz) to the carbonyl at δ_C 171.3. The proton chemical shifts along with the coupling constants of the signals for H-18 (δ_H 2.43, s) and H-19 (δ_H 6.49, d, J = 2.7 Hz) in **3** were consistent with those for scalarane isolated in the current study. Further COSY and HMBC correlations secured the gross structure of **3** as 12-deacetyl-12-*epi*-deoxoscalarin-19-acetate [21].

Compound **4** was isolated in trace amount and gave a molecular formula of $C_{27}H_{40}O_4$ by HRESIMS. The ^{1}H NMR spectrum displayed an aldehyde signal (δ_H 10.02, d, J = 2.5 Hz), an olefinic proton (δ_H 5.89, br s, H-16) as well as a distinctive AB system [δ_H 2.57, t, J = 14.2 Hz (H-11a); 2.39, dd, J = 13.9, 2.1 Hz (H-11b)]. There were five methyl singlets at δ_H 0.85 (CH_3-21), 0.82 (CH_3-22), 1.11 (CH_3-23), 0.87 (CH_3-24), and 1.16 (CH3-25) that were correlated with their respective carbons at δ_C 33.4 (C-21), 21.5 (C-22), 16.4 (C-23), 15.8 (C-24), and 15.1 (C-25) in the HSQC spectrum. These NMR data were consistent with the presence of a tetracyclic 12-keto scalarane skeleton, such as in 12-deacetoxy-12-oxo-scalaradial from *Glossodoris rufomarginata*. The 12-keto moiety in **4** was secured from HMBC correlations from H2-11 and from H3-25 to the carbon at δ_C 215.5 (C-12). However, the signals for the aldehyde proton at C-17 of 12-deacetoxy-12-oxo-scalaradial were absent in **4**, instead replaced by oxymethylene signals [δ_H 4.48 (s); δ_C 66.6 (C-20)] [19].

4. Conclusions

The extract of Balinese sponge *Aplysinella strongylata* contained twenty-one new psammaplysin derivatives as well as six previously known psammaplysins. The new metabolite 19-hydroxypsammaplysin D (**1**) found to be the best antimalarial activity, with an IC_{50} value of 6.4 μM. The absolute configuration of psammaplysin A has been assigned as (6*R*,7*R*) using experimental and calculated electronic circular dichroism (ECD) data. Chemical investigation of the extract of Balinese nudibranch *Hypselodoris infucata* yielded the known (–)-furodysinin (**2**) that exhibited *in vitro* cytotoxic activity of against HeLa cell line with an IC_{50} of 102.7 μg/mL. Two new scalarane sesterterpenes (**3**, **4**) were characterized from an organic extract of a single specimen of the nudibranch *Glossodoris hikuerensis*, and no bioactivity was currently obtained.

ACKNOWLEDGEMENT

We thank to the International Postgraduate Research Scholarship (to I.W.M.). We also appreciated for the financial support provided by Indonesian Directorate General of Higher Education (DIKTI) under *Hibah Bersaing* research grant scheme 2015-2016 (contract number: 51/UN48.15/LT/2016).

REFERENCES

[1] Jaspars M, De Pascale D, Andersen JH, Reyes F, Crawford AD, Ianora A. 2016 *J. Mar. Biol. Assoc. U.K.* **96** 151-8. https://doi.org/10.1017/s0025315415002106
[2] Yong KWL, Mudianta IW, Cheney KL, Mollo E, Blanchfield JT, Garson MJ. 2015 *J Nat Prod.* **78** 421-30. https://doi.org/10.1021/np500797b
[3] Dias DA, Urban S, Roessner U. 2012 *Metabolites* **2** 303-36. https://doi.org/10.3390/metabo2020303
[4] Blunt JW, Copp BR, Keyzers RA, Munro MHG, Prinsep MR. 2016 *Natural Product Reports* **33** 382-431. https://doi.org/10.1039/c5np00156k
[5] Gerwick WH, Moore BS. 2012 *Chem Biol* **19** 85-98. https://doi.org/10.1016/j.chembiol.2011.12.014
[6] Setyowati EP, Jenie UA, Sudarsono, Kardono LBS, Rahmat R. 2009 *J Biol Sci* **9** 29-36. https://doi.org/10.3923/jbs.2009.29.36
[7] Edrada RA, Heubes M, Brauers G, Wray V, Berg A, Gräfe U, et al. 2002 *J Nat Prod* **65** 1598-604. https://doi.org/10.1021/np020085c
[8] Mudianta IW, Skinner-Adams T, Andrews KT, Davis RA, Hadi TA, Hayes PY, et al. 2012 *J Nat Prod* **75** 2132-43. https://doi.org/10.1021/np300560b
[9] Yang X, Davis RA, Buchanan MS, Duffy S, Avery VM, Camp D, et al. 2010 *J.Nat. Prod.* **73** 985-7. https://doi.org/10.1021/np900834g
[10] Roll DM, Chang CWJ, Scheuer PJ, Gray GA, Shoolery JN, Matsumoto GK, et al. 1985 *J. Am. Chem. Soc.* **107** 2916-20. https://doi.org/10.1021/ja00296a014
[11] Mándi A, Mudianta IW, Kurtán T, Garson MJ. 2015 *J. Nat. Prod.* **78** 2051–6. https://doi.org/10.1021/acs.jnatprod.5b00369
[12] Dunlop RW, Kazlauskas R, March G, Murphy PT, Wells RJ. 1982 *Aust. J. Chem.* **35** 95-103. https://doi.org/10.1071/CH9820095
[13] Vaillancourt V, Agharahimi MR, Sundram UN, Richou O, Faulkner DJ, Albizati KF. 1991 *J. Org. Chem.* **56** 378-87. https://doi.org/10.1021/jo00001a069
[14] Kazlauskas R, Murphy PT, Wells RJ. 1978 *Tetrahedron Lett.* **49** 4951-4. https://doi.org/10.1016/s0040-4039(01)85779-8
[15] Guella G, Mancini I, Guerriero A, Pietra F. 1985 *Helv. Chim. Acta.* **68** 1276-82. **https://doi.org/10.1002/hlca.19850680523**
[16] Horton P, Inman WD, Crews P. 1990 *J. Nat. Prod.* **53** 143-51. https://doi.org/10.1021/np50067a019
[17] Fontana A, Gimenez F, Marin A, Mollo E, Cimino G. 1994 *Experientia* **50** 510-6. https://doi.org/10.1021/np50106a011

[18] Mudianta IW, Martiningsih NW, Prasetia IND, Nursid M. 2016 *J. Pharm.* **27** 104-10. https://doi.org/10.14499/indonesianjpharm27iss2pp104
[19] Mudianta IW, White AM, Garson MJ. 2015 *Nat. Prod. Commun.* **10** 865-8.

Synthesis and characterization of hydrogel composite based on bacterial cellulose-gambir leaf extract (*Uncaria gambir* Roxb.)

A Putra, B A Nugraha and A Amran

Department of Chemistry, Faculty of Mathematics and Natural Sciences, Universitas Negeri Padang, Padang, Indonesia 25131

E-mail: anandap@fmipa.unp.ac.id

Abstract. Synthesis and characterization of hydrogel composite based Bacterial Cellulose - Gambier Leaf Extract (BC-GLE) has been done. The purpose of this research was to obtain an elastic hydrogel composite BC-GLE that can be applied as a substitution material for cartilage. Preparation of this hydrogel composites conducted by immersing BC in GLE for 1, 2, 3 and 4 days. The effects of immersion time on the physical and mechanical properties of BC-GLE were studied by measuring its water content, compressive and tensile strength. The structures of BC-GLE were analyzed by using FTIR and XRD. The results showed an improvement of physical and mechanical properties of BC-GLE hydrogel composites in line with the duration of immersion. Absorption of GLE into BC matrix caused decreasing the water content of BC. The best water content percentage result was obtained from the BC-GLE composite which was immersed for 4 days, 98.211%. The best compressive strength result was also obtained from samples on the immersion time of 4 days, 2.633 MPa. The tensile strength of the BC-GLE hydrogel composite increased from day 1 to 3, but decreased in day 4. The best tensile strength result of 0.17 MPa was obtained on the day 3 immersion time. Physical and mechanical properties improvements of BC-GLE hydrogel composite were only caused by absorption of GLE to BC matrix without forming new bond. The FTIR results showed no new functional group was formed, but only showed vibration shift to a lower wave number. BC-GLE hydrogel composite structure was still in a cellulose type 1. Combination of BC and GLE produced BC-GLE hydrogel composite material that having physical and mechanical properties better than pure BC. However these properties values were not reached the standard to be applied as a substitute material for cartilage yet.

1. Introduction

Design and fabrication of a suitable material for soft tissue substitutes is an important aspect for biomedical applications. This encourages the need to develop a material that not only exhibits the same mechanical properties as the soft tissue to be replaced, but also exhibits an improvement of functionality, biocompatibility and durability. A material that is promising and appropriate for this purpose is hydrogel. Hydrogel is a network of three-dimensional hydrophilic polymers formed from the reaction of one or more monomers or by the incorporation of bonds between polymer chains. Hydrogel can absorb water from 20% to thousands of times its dry weight. This unique property may cause hydrogel to be used in various medical, pharmaceutical and prosthetic applications such as contact lens, synthetic wound cover, drug delivery system, organs, and soft tissue [1].

Bacterial cellulose (BC) is one of the hydrogels produced by bacteria from the genus Acetobacter such as *Acetobacter xylinum*. BC has many advantages such as high purity, excellent network structures, unique mechanical strength and high water content (98-99%), and can be sterilized safely without changing its characteristics [2]. BC has high tensile strength along the fiber layer direction, but has low compressive modulus value. Therefore, BC is less elastic [3]. Utilization of BC in the biomedical field as a substitute material for soft tissue is constrained by BC's low elasticity properties. To improve the BC elasticity, it is necessary to combine BC with other materials that produce a composite material with better physical and mechanical characteristics [1]. One alternative way to produce a good composite material is to combines BC with natural materials such as gambir leaf extract.

Gambir plants (*Uncaria gambir* Roxb.) are herbs that are widely found in Indonesia. About 90% of Indonesia's gambir production is produced in West Sumatra. Gambir plants contain compounds that are high in antioxidants that have been exploited in the field of pharmaceuticals as anti-inflammatory drugs, astringent, and even potentially anti-cancer drugs [4] [5]. Gambir leaf is commonly used by people to treat various diseases such as dysentery, diarrhea, and canker sores. The main content of gambir leaf extract (GLE) is Catechin compound (7-33%), katechu tannat acid (20-55%), Pyrocatechol (20-30%) and quersetin (2-4%) [7].

Based on the above description, a research was done by combining BC with GLE which is expected to produce a material of Bacterial Cellulose-Gambir Leaf Extract (BC-GLE) which can be applied in the biomedical field.

2. Method

2.1. Materials

The equipment used were glassware (pyrex), stainless steel container, plastic container, mixer/blender, vapor plate, plastic, iron (Sanyo), analytical balance (KERN ABS 220-4), oven (Memmert Model 300), filter, thermometer, pH meter (Seven Easy S20 METTLER-TOLEDO), FTIR (PerkinElmer Frontier Optica), and Compressive tester (Tony Technik Compressive Test Bauform model 2020). The material used in this research were waste coconut water, $CO(NH_2)_2$ (PT. Pupuk Sriwijaya), $C_{12}H_{22}O_{11}$, CH_3COOH 25% (PT. Brataco Chemica), starter of the *A.xylinum* obtained from Biochemistry Laboratory of Universitas Negeri Padang, aquades, and gambir leaves was obtained from Siguntur Gambir plantation, Pesisir Selatan, West Sumatera, Indonesia

2.2. BC fermentation

Four liters of filtered waste coconut water was poured into a 5 L stainless steel container, then heated and added 400 g of $C_{12}H_{22}O_{11}$, 40 g of $CO(NH_2)_2$, and stirred. Then, the mixture was acidified with 25% of CH_3COOH to reach pH 4-4.5 (± 80 mL), and then heated. In hot conditions, 600 mL of the mixture was poured into a sterilized fermented container and closed using sterile pore paper. The mixture was cooled to a temperature of ± 28°C (room temperature), and then aseptically added 10% v/v (60 mL) *A.xylinum*. This mixture was fermented for several days to form BC with a thickness of ± 1 cm.

2.3. Washing and purifying BC

The obtanined BC was washed with running water for ± 24 hours, and then was immersed in NaOH 2% (w/v) solution for ± 24 hours to remove impurities and bacterial cells. The pure BC was then washed again by using running water for ± 24 hours for further storage, and ready to use.

2.4. Gambir Leaf Extract preparation
The clean old gambir leaves were cut into small pieces and mashed with a blender. Then, the leaves were soaked for ± 15 minutes in hot water (temperature ± 90°C) with a ratio of leaf and water of 1: 4. The mixture was stirred and filtered until a brownish extract was obtained. This brownish extract was used for the preparation of BC-GLE hydrogel composite.

2.5. BC-GLE hydrogel composite preparation
BC was cut in size of 2x2x1 cm and 15x2x1 cm, and inserted into a plastic container. Into this plastic container, 500 mL GLE was added and placed on the shaker. BC in GLE was shaking for 1, 2, 3 and 4 days. Each day, the resulting BC-GLE hydrogel composite was removed from the plastic container and the sides of the surfaces were dried by using tissues. The BC-GLE samples were then used for characterization

2.6. BC-GLE hydrogel composite characterizations
2.6.1. Water Content of BC-GLE hydrogel composite measurement
A 2x2x1 cm BC-GLE hydrogel composite sample was weighed using an analytical balance to obtain wet weight (W_w). Then, it was flattened by using glass, and fed into an oven with a temperature of ± 105°C. The BC-GLE hydrogel composite was weighed every 15 minutes to obtain a constant final weight (W_d). The same method was used to calculate the water content of pure BC. The percentage of water content can be calculated using the formula:

$$W_c(\%) = \frac{W_w - W_d}{W_w} x100$$

where, W_c = water content (%), W_w = wet weight (g), and W_d = dry weight (g).

The water content of GLE was also measured by putting 10 mL of GLE into hot vapor plate and weighed to obtain the initial weight (W_1). Then, the vapor plate containing GLE was fed into the oven and weighed every 1 hour until a constant weight (W_k) was obtained. The percentage of GLE water content can be calculated using the formula:

$$W_c(\%) = \frac{W_1 - W_k}{W_1} x100$$

where, W_c = water content (%), W_1 = initial weight (g) and W_k = dry weight (g).

2.6.2. Compressive strength of BC-GLE hydrogel composite measurement
A 2x2x1 cm BC-GLE hydrogel composite sample was placed on a 4x4x4 cm concrete block that serves as a supporting material. BC-GLE hydrogel composite sample and concrete block was placed inside Tony Technik Compressive Tester's test chamber, and then the equipment was operated. The BC-GLE hydrogel composite sample was observed until the sample was flattened. The compressive force required to flake the sample would be visible when the pressure device began to press the surface of the concrete block resulting in a dramatic increase in compressive forces and in the event of an increase in this process the suppression is stopped. The magnitude of the compression force seen before the increment was calculated as the value of the compressive force of the BC-GLE hydrogel

composite sample. To obtain the value of compressive strength of BC-GLE hydrogel composite sample used formula:

$$P = \frac{F}{A}$$

where, P = Compressive strength (Pa), F = Pressure (N), and A = Area (m^2).

2.6.3. Tensile strength of BC-GLE hydrogel composite measurement
The 15x2x1 BC-GLE hydrogel composite sample was flattened using glass. The sample was then pressed using an iron until it reaches a thickness of ± 0.3 mm. Next, the flattened sample was cut to the size of 15x1.5 cm, and put it on the sample holder of horizontal tensile tester. Then, the sample was pulled up, and the value of tensile and strain was obtained.

2.6.4. Fuctional group of BC-GLE hydrogel composite analysis
The functional group of BC-GLE hydrogel composite was analyzed by using FTIR. BC-GLE hydrogel composite sample was dried prior to analysis.

2.6.5. Crystallinity of BC-GLE hydgrogel composite analysis
Crystallinity of BC-GLE hydrogel composite was analysed by using XRD. A dry sample of BC-GLE hydrogel composite was cut to 0.5x0.5 cm in size to obtain a flat sample portion. Sample was placed in the sample holder and placed in the XRD equipment. The obtained diffractogram was analyzed to obtain the degree of crystalline.

3. Results and Discussions
In the process of BC preparation, *A.xylinum* produced cellulose fibers starting from the surface to the bottom of the media solution which over time will overlap until the bacteria grow out. Illustration of BC formation can be seen in figure 1:

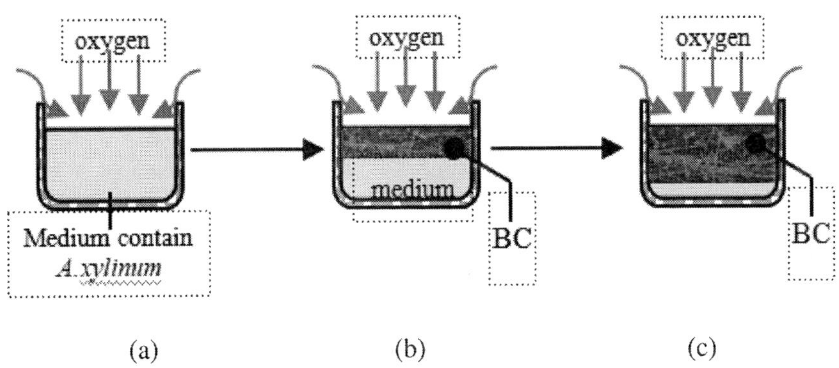

Figure1. Illustration of the formation of BC, a) initial state of the fermentation process, b) BC begins to form from the surface of the medium, c) the formation of BC continues to the bottom until the medium is depleted.

Washing and Purifying of BC was done to remove non-cellulose components and residual bacteria which was still present in BC. The remaining non-cellulose components in BC could block the hydrogen bonds that occur between the cellulose molecular chains and caused decreasing the mechanical strength of the cellulose. The remaining bacteria also decreased the mechanical strength because these bacteria could still move on by consuming the nutrients present in the BC. Therefore, the resulting BC became fragile and could not be used to produce hydgrogel composite materials that have good physical and mechanical properties [8].

The old gambir leaves used were from the Cubadak variety. One of the main compounds found in gambir leaves was Catechin. Catechin which was warmed up long enough will become catechu tannat acid due to its own condensation process. So, it becomes easily soluble in hot water [4]. Catechu tannat acid was a brownish compound commonly used as a natural dye [9]. This was what caused the resulting GLE colour brown.

The result of immersion accompanied by stirring caused colour changed in BC due to the absorption of catechu tannat acid derived from GLE. The changes can be seen in figure 2.

(a) (b) (c)

Figure 2. Figure of pure BC hydrogel and BC-GLE hydrogel composite: a) BC hydrogel before immersion to GLE, b) BC-GLE hydrogel composite formed after immersion to GLE, c) Side view of BC-GLE hydrogel composite

3.1 Water Content of BC-GLE hydrogel composite

The water content is the total amount of water contained in a material compared to the dry weight of the material. The water content was a physical properties test parameter to determine how much water was contained within the BC-GLE hydrogel composite during the immersion time and how much GLE was absorbed to replace water in BC. Water content measurements were also performed on GLE to determine the percentage of extracts obtained during the extraction process. The effect of immersion time on water content of BC-GLE hydrogel composite can be seen in figure 3.

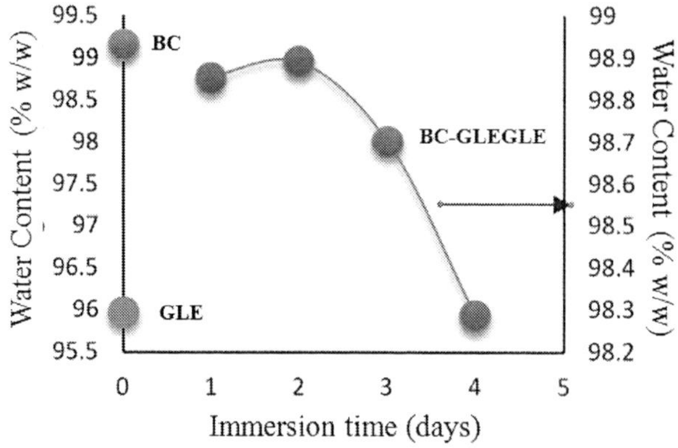

Figure 3. The effect of immersion time on water content of BC-GLE hydrogel composite

BC is a material that has a high water content exceeding 90% of its total weight [10]. From the figure 3, the water content of BC was 99.156%, while the water content in GLE was 95.970%. The water content in the BC-GLE hydrogel composite tends to decrease with the length of the immersion time. Water content reduction was caused by the absorption process of GLE which was acting as absorption filler replaces the water within the BC (matrix).

3.2. Compressive strength of BC-GLE hydrogel composite
Compressive strength is one of the test parameters of mechanical properties performed to determine the durability of a material when given the compressive force. It is necessary to know the compressive strength of the resulting BC-GLE hydrogel composite to be comparable to the compressive strength of the cartilage. The effect of immersion time of BC-GLE hydrogel composite on the compressive strength value can be seen in figure 4.

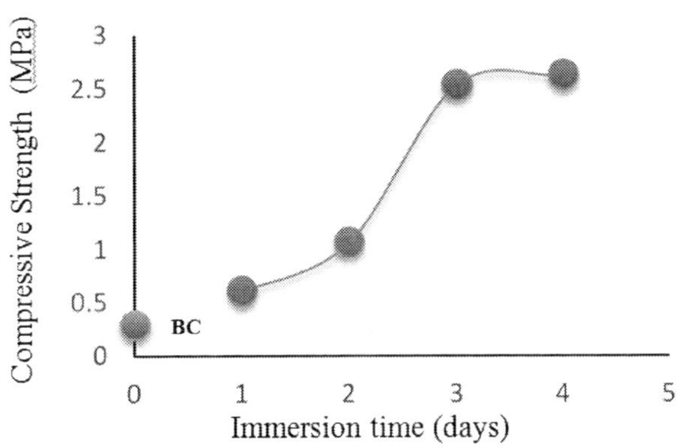

Figure 4. The effect of immersion time on the compressive strength of BC-GLE hydrogel composite

Figure 4 showed an increasing in compressive strength value along with the BC-GLE hydrogel composite immersion time. The best compressive strength value reached 9 times higher than that of pure BC which was obtained on the 4th day of BC-GLE hydrogel composite (2,633 MPa). This is mean that the incorporation of BC with other materials into composite material proved to increase the compressive strength of BC [1].

The increasing value of BC-GLE hydrogel composite compressive strength was caused by the amount of filler absorbed into the BC matrix. Filler replaced the position of water that filled the cavity on the BC network. It could be said that the longer the immersion time, the more filler was absorbed, so that greater compressive force was needed to flake the BC-GLE hydrogel composite. The occurrence of absorption or absorption process was also evidenced by testing the water content. The compressive strength value possessed by the BC-GLE hydrogel composite was still below the cartilage compressive strength value of 14-59 MPa [11].

3.3. Tensile strength of BC-GLE hydrogel composite
Tensile strength testing is the maximum tensile force a material can retain during the test until the material is disconnected. Tensile strength testing is useful to determine the elasticity of BC-GLE hydrogel composite which can be determined through the comparison of the value of tensile strength (stress) with strain value. The elasticity value is directly proportional to the tensile strength value of a

IOP Conf. Series: Journal of Physics: Conf. Series **1040** (2018) 012014 doi:10.1088/1742-6596/1040/1/012014

material and is inversely proportional to the strain value. The graph of time immersion ratio to the value of tensile strength, strain and elasticity can be seen in figure 5.

Figure 5 showed that the value of BC's tensile strength was 0.170 MPa. The value of tensile strength of BC-GLE hydrogel composite increased with increasing immersion time. The highest value of tensile strength was found in the 3rd day of immersion time. However, the value of tensile strength decreased during submersion entering the 4th day. This phenomena happen was due to the amount of GLE as a filler contained in the BC on the 1st and 2nd days was still low, so it did not significantly affect the resistance of the sample when it was withdrawn. Meanwhile, on the 3rd day, the amount of GLE absorbed substantially replaced the water in the BC matrix cavities, causing increased of tensile strength value in the BC-GLE hydrogel composite.

Because the interaction between filler and matrix was not accompanied by a bond formation between the two, but only the process of physical absorption, then on the 4th day, filler contained in the matrix regardless. It was also affected by the shock generated by the shaker that helps the release of the filler from the matrix so that the sample tensile strength drops again. Drop in tensile strength value was evidenced by the significant reduction of water content from the 3rd day BC-GLE hydrogel composite to the 4th day which means that the amount of filler in BC was decreased considerably.

A long with the value of tensile strength, the elasticity of BC-GLE hydrogel composite on the 1st and 2nd day tends to decrease, but increased on the 3rd day (1.3 M.Pa) and again decreased back on the 4th day. This was related to the amount of filler present in the sample matrix.

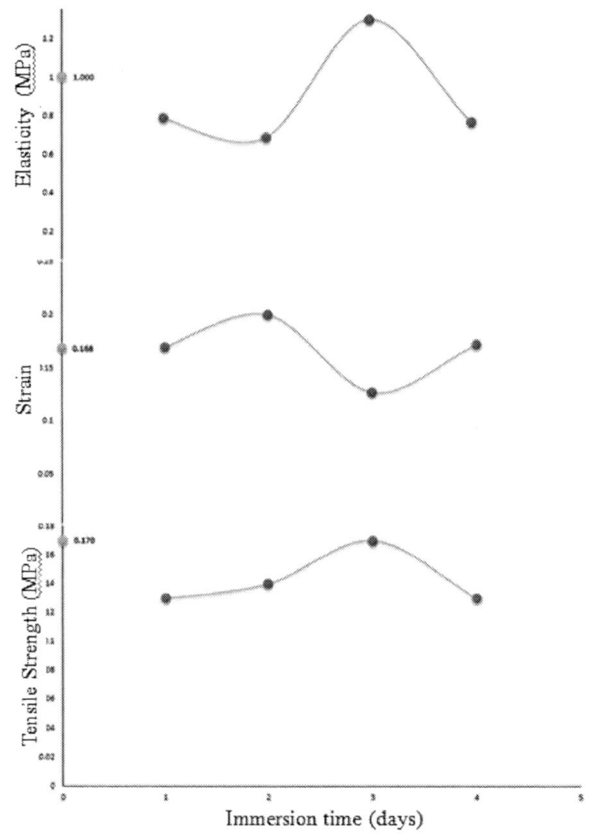

Figure 5. The effect of immersion time on the tensile strength of BC-GLE hydrogel composite

3.4 Fuctional group of BC-GLE

The BC-GLE hydrogel composite fucntional group test was performed using FTIR instrument. The spectra generated from the test were analyzed qualitatively to see if a new functional group was formed after merging BC with GLE. It aimed to know the interaction that occurs between the two during the immersion process that involved changing in physical and mechanical properties characteristics of BC-GLE hydrogel composite. IR spectra of BC, GLE and BC-GLE hydrogel composite can be seen in figure 6.

The peak cellulose characteristics were shown in the wave number range of 4000-600 cm^{-1} as shown in Figure. 6. The characteristic peak of the functional group of BC was in the strain vibration OH (3100-3800 cm^{-1}), CH (2901 cm^{-1}), CO (1370 cm^{-1}) and COC (1163 cm^{-1} and 1068 cm^{-1}) [12] [13].

Figure 6 showed that in BC, there was a wide peak at a wavelength of 3335.85 cm^{-1} indicating a vibration of a group of OH, a vibration at a wavelength of 1635.11 cm^{-1} indicating the existence of a six cycle ring of glucose monomers, 1033.81 cm^{-1} indicating the presence of a β-1,4-glycosidic COC bond. This was the characteristic peaks of cellulose [14].

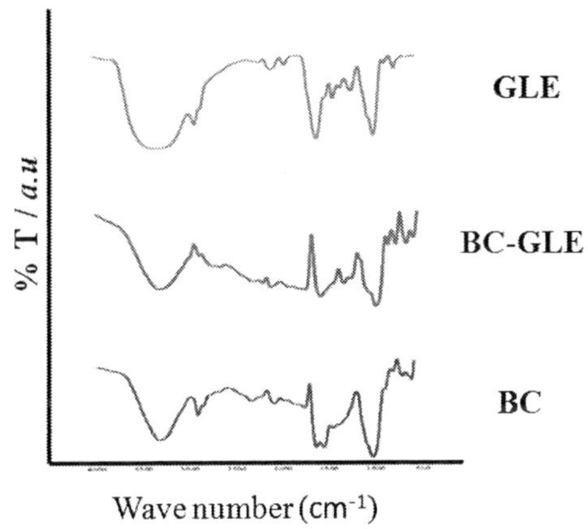

Figure 6. FTIR spectra of BC, GLE and BC-GLE hydrogel composite

Based on the data obtained, the interaction between GLE and BC did not produce a new peak, but only caused shifting in the functional group vibration to a lower wave number. This shift proved that GLE appeared in BC-GLE hydrogel composite. So, it could be said that the interaction that occurs between GLE and BC was only physical interaction that was not accompanied by the formation of bond between the two.

3.5 Degree of crystallinity of BC-GLE hydrogel composite

The degree of crystalline was performed using XRD. The Diffractogram of BC-GLE hydrogel composite showed that the type cellulose produced was cellulose type I which was present at 2Θ 14°, 16°, 22° and 22.48° [15]. Diffractrogram of BC and BC-GLE hydrogel composite can be seen in figure 7.

Based on figure 7, there was no significant difference occurred from the BC and BC-GLE hydrogel composite diffractogram. The absence of new emerging peaks indicated the absence of bonding between BC and GLE, but only the absorption of physics. To determine the degree of crystalline, the difractogram was analyzed manually by transferring the difractogram onto the graph paper and measuring the peak height as well as weighing the crystal and amorphous portions to obtain the weight ratio of crystals and total weight (amorphous + crystals) [16]. The result of calculation of crystalline degree of BC-GLE hydrogel composite can be seen in table 1.

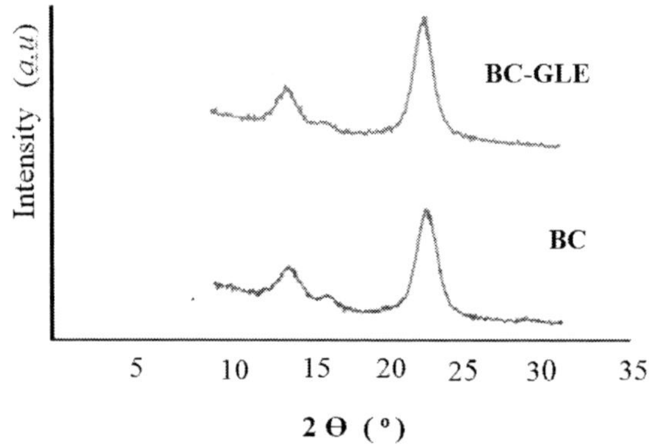

Figure 7. Diffractogram of BC and BC-GLE hydrogel composite

The results obtained that the BC-GLE hydrogel composite had a degree of crystalline greater than BC with an amorphous structure of only 32.414%. The measurement of the degree of crystalline showed that BC had a larger amorphous structure, so that BC was slightly more elastic compared to BC-GLE hydrogel composite.

Table1. Degree of Crystalline of BC-GLE hydrogel composite

Samples	Total weight (g)	Crystalline (g)	Amorphous (g)	Degree of crystalline (%)
BC	0.222	0.149	0.073	67.117
BC-GLE	0.203	0.137	0.066	67.586

4. Conclusion

The composite hydrogel based on BC-GLE can be synthesized. The Water content of BC-GLE hydrogel composite was found lower than pure BC. Immersion time of BC in the GLE to produce BC-GLE hydrogel composite improved the physical, mechanical (compressive and tensile strength) properties of the BC-GLE composite hydrogel compare to pure BC. However, these improvements were not suitable for replacement of cartilage yet.

Acknowledgment

This work was supported by Research Grants provided by Universitas Negeri Padang under Applied Product Research Scheme year 2017 with the contract no.1746/UN.35.2/PG/2017.

References

[1] Nakayama B A, Kakugo A, Gong J P, Osada Y, Takai M and Erata T 2004 *Journal Advanced Functional Materials* **14** (11)

[2] Ciechańska D *Journal Fibres and Textiles in Eastern Europe* **12** (4) 69-72

[3] Putra A, Kakugo A, Furukawa H, Gong J P, and Hagiwara Y 2010 *Journal of Cellulose* **17** (1) 93-101.
[4] Hayani E 2003 *Buletin Teknik Pertanian* **8** (3) 31-33.
[5] Anggraini T, Tai A, Yoshino T and Itani T 2011 *African Journal of Biochemistry Research* **5** (1) 33–38
[6] Redaksi Agromedia 2008 *Buku Pintar Tanaman Obat (I)* (Jakarta: PT. Agromedia Pustaka)
[7] Isnawati A, Raini M, Sampurno O D, Mutiatikum D, Widowati L, and Gitawati R 2012 *Buletin Penelitian Kesehatan* **40** (4) 201-208.
[8] Widyaningsih S and Diastuti H 2008 *Molekul* **3** 28–33.
[9] Green C L 1995 *Non-Wood Forest Products 4: Natural Colourants and Dyestuffs. Food and Agriculture Organization of the United Nations* (Rome: Food And Agriculture Organization Of The United Nations)
[10] Putra A, Kakugo A, Furukawa H, Gong J P, Osada Y, Uemura T and Yamamoto M 2008 *Polymer Journal* **40** (2) 137-142.
[11] Doulabi A H, Mequanint K and Mohammadi H 2014 Materials **7** (7) 5327-5355.
[12] Gayathry G and Gopalaswamy G 2014 *Indian Journal of Fibre and Textile Research* **39** (1) 93-96.
[13] Yue Y, Han G and Wu Q 2013 *BioResources* **8** (4) 6460-6471.
[14] Rahayu T and Rohaeti E 2012 *Prosiding Seminar Nasional Penelitian, Pendidikan Dan Penerapan MIPA* **53** (9) 1689-1699.
[15] Yue Y, Han G and Wu Q 2013 *BioResources* **8** (4) 6460–6471
[16] Terinte N, Ibbett R and Schuster K C 2011 *Lenzinger Berichte* **89** 118–131

Characterization and analysis of the molecular weight of corn corbs microcrystalline cellulose (MCC) fiber using mass-spectrometry methods

W R Kunusa, H Iyabu , M Taufik and D N Botutihe

Universitas Negeri Gorontalo, Indonesia

E-mail: rewinikunusa2014@gmail.com

Abstract. Microcrystalline cellulose(MCC) was isolated from corncobs waste. It was activated by NaOH solution with varied concentration in range from 4%, 6%, 8%, 10%, 12%, 14% and 17%. Characterization of physicochemical properties of microcrystalline cellulose (MCC) molecular weight was conducted by using Mass Spectrofotometer (MS).Preconditioning of MS used 20 μl injection volume, 10 μl concentration, acethonitrile: ultra water (50:50) as eluent. The result showed that NaOHconcentration affect to the type of oligosaccharides fragmentationwith an increased abundance. MS spectra for NaOH 4% showed four major peaks at 834 m/z, 623 m/z, 425 m/z, 383 m/z that indicated 2-5 glucose polymers respectively, with an increased abundance withadditional Na^+ and K^+ ions.NaOH8%, 12%, 14%, 17% indicated the type of oligosaccharide fragmentation with equal three major peaks at 827 m/z, 623 m/z, 425 m /z. The fragmentatition type of NaOH 10% and avicel standard pH 102, indicated 2-5 glucose polymers respectively,with an increased abundance with additional Na^+ and K^+ ions, there are peak at 827 m/z and 425 m/z, and there is no peak at 623 m/z because there are no Na^* and K^+ ions.

1. Introduction

Cellulose is an interesting and sought of material due to current demands for green chemistry, more environmentally friendly resources and renewable raw material. The use of cellulose is therefore highly atractive because of it being renewable, biodegradable and non-toxic. Cellulose is a homopolysaccharide consisting of unbranched units of D-glucose (anhydroglucose) connected to 1 and 4 C atoms with β-glycosidic binding (β-1,4 glycosidic bond) [1]. Microcrystalline Cellulose (MCC) one of the cellulose derivative that can be obstained by enzimatik hydrolysis [2] extracted by acid hydrolysis with H_2SO_4, hydroclorid acid [3] [4]. MCC is a purified, partiallyde-polimeryzed form of cellulose occurring as a fine, freeflowing crystalline powder. MCC has been used for many years in different industries like pharmaceuticals, cosmetics, plastic, food etc [4]. Cellulose has four different crystalline forms named cellulose I-IV.The crystal structure of cellulose I in native cellulose can be converted to cellulose II by dissolution and regeneration or merserization [5].

Polymer characterization is generally performed to evaluate the quality of the synthesis or substitute product one of which is the determination of molecular weight. Molecule weight affects the cellulose dissolution process in the IONCELL-P process using 1-etil-3-methylimidazolium asetat ([emim] OAc) and water solvent. Degree of the crystanillity, accessibility and reactivity of the cellulose chains are important properties of the cellulose [6]. Determination of molecular weight (BM), average molecular weight (Mw) and polydispersity index (D) depending on biomass source, solvent usage pretreatment conditions and isolation methods. Cellulose has a high molecular weight (MW), with polymerization (DP) levels up to 20,000.The molecular mass value describes the polymeric molecular size distribution, an important variable determining the polymer's chemical properties. High molecular weight polymers have stronger properties. In other polymerizations, polymers with very low molecular weight distributions can be obtained [7].

The properties of polymers, such as viscosity depend on molecular weight and molecular weight distribution. Therefore, it is important to determine the average molecular weight associated with the distribution of specific molecular weights. The polymerization reaction produces a weight distribution and a molecular shape. The molecular weight of the polymer can be expressed as the average molecular weight, and polydispersity/Molecular Weight Distribution (MWD). The polydispers index represents the quality of degradation. Molecular weight is highly influential on the physicochemical properties of cellulosic material [8] [9]. Some methods for determining these parameters are Gel Permeation Chromatography (GPC) and Mass Spectrometry (MALDI TOF-MS) and Size-Exclusion Chromatography, SEC is a technique of determining the molecular weight of polymers with a short time. The molecular weight can be expressed by the average molecular weight (Mn), average molecular weight (Mw), average molecular weight (Mz), average molecular weight of viscosity (Mv) [10] [11] [12]. Aplication of size exclusion chromatography (SEC) combinated with electrospray ionization mass spectrometry (ESI-MS) for the determination of accurate moleculer weight distribution corrected for chromatography brand broadening [13].

LC-MS-technieque is used to determine the molecular weight of microcrystalline cellulose (MCC). Liquid chromatography (LC) is eminently suistable for separating soluble polymers. In Mass-Spectrometry (MS) the molecules in a sample are first ionized, than separated according to their mass to charge rasio (m/z) and finally detected. MS provides accurate information on the molecular weight of the separated and detected ions [10].

2. Research Method

2.1. Material, Chemicals and Tools

Corncob from Milango Village, Pohuwato District, Gorontalo Province, Indonesia. Corncob waste was cut into small pieces ± 2cm, washed with aquadest and in oven temperature 60^0C-70^0C for 24 hours. The dried sample is blended and in the shaked to obtain corncob powder particles 180 micron by size [4].The reagents were of analytical grades, NaOH (reagent grade, 98%), HCl (ACS reagent, 37%), H_2O_2 (technical grade) were supplied by Sigma-Aldrich, $K_2Cr_2O_7$, H_2SO_4, Avicel 102 (technical grade) a commercial brand of MCC, Ethanol (95% pa), Tholuena (95% pa) and distilled water used as solvent. Instruments using Shaker Digital ASTM E11 180 micron, pH meter (Thermo SCIENFIC ORION), LC-MS waters UPLC MS G-2 QTQF. System : ESI (Electrospray Ionization). Instrument condition of volume injection 20 μl, concentration 10 μl, acethonitrile:ultra water (50:50) as eluent.

2.2. Method

Cellulose isolation, purification of α-cellulose and the preparation of microcrystalline cellulose (MCC) is a modified procedure [4] that is : dewaxing stage: 50 gr of 180micron corncob powder was soxhleted in ethanol :tholuene (2:1) for 4hours. Sample was filtered and washed until neutral pH, then it was oven-dried at 60^0C for 8 hours. De-hemicellulose stage: 10 gr of 180 micron corncob powder was dissolved into 200 mL of 4%, 6%, 8%,10%, 12%, 14% and 17% NaOH solution. The mixture was heated on hot plate at 100^0C for ± 2 hours. Then it was filtered and washed until neutral pH, oven-dried at 60^0C. De-lignification and bleaching stage using H_2O_2 4% for ± 1 hour at room temperature. Then sample was filtered and washed until neutralpH. It was oven-dried at 60^0C for 8 hours. Further purification of α-cellulose was done by 5 gr of cellulose dissolved in 500 ml of 17.5% NaOH at 80^0C for 30 min. Furthermore the precipitate was filtered and washed until neutral pH, and oven-dried at 60^0C for 8 hours. The preparation of microcrystalline cellulose (MCC) that is: α-cellulose product was hydrolysed with 0.1N HCl ratio 1: 2 and refluxed at 80^{0}C for ± 2 hours. The precipitate was filtered and washed with aquadest until neutral pH and oven-dried at 60^{0}C for 8 hours.

Microcrystalline Cellulose (MCC) product characterized by molecular weight using MS Analysis was performed using waters MS G-2 QTQF. Full scan mode from m/z 100 to 1200 was performed with a source temperature at 140^{0}C. Solvent was Acethonitrile 50%.

3. Result and discusion

3.1. Effect of NaOH concentration on α-cellulose acquisition

Degradation of the cellulose under alkalline condition starts from the reducing and group of the cellulose at moderate temperature 80-100^{0}C [6]. Pretreatment process with NaOH was to lysis and reduce lignin and hemicelullose, to break the crystal structure of celullose and to improve the porosity of sample [17]. The result showed that conversion ratio of cellulose is high than enzymatic hydrolysis [19]. Produce isolation of cellulosewith heating NaOH 4%, 6%, 8%, 10%, 12%, 14%, 17% and preparation of microcrystallinecellulose (MCC) at Figure 1.

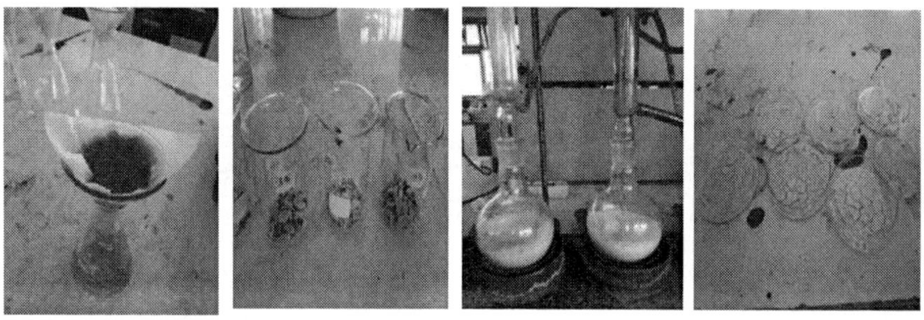

(a) (b) (c) (d)

Figure 1.(a) produce of cellulose, (b) purification of α-cellulose, (c-d) preparation MCC

Here is the curve of purified α-cellulose content in variations of NaOH concentration of 4%, 6%, 8%, 10%, 12%, 14% and 17%.

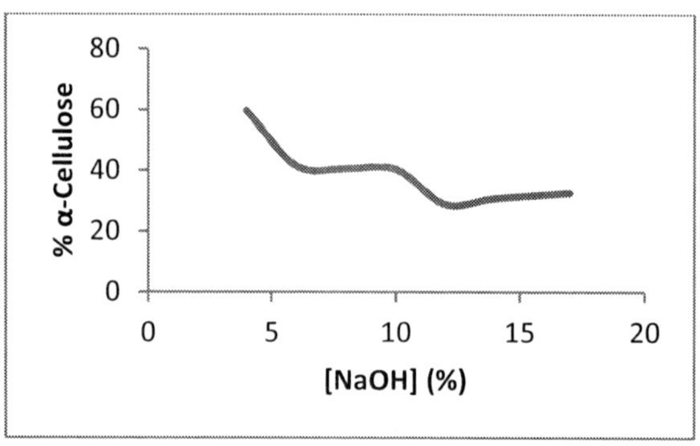

Figure 2. Effect of NaOH concentration on α-cellulose acquisition

NaOH 4% gave higest α-selulosa content i.e 60% (figure 2). When NaOH concentration was higher, the α-selulosa content decreased. It is because the bonds of celullose chain are not tightly , therefore celullose molecules had dispersed freely in NaOH solutions. This allows celullose to pass when filtering and washing.[15]The addition of NaOH 5% (w/w) on *Chlorella vulgaris* at 50^{0}C for 24 hour gave 9,8% of carbohydrate solubility compared with no NaOH added.When NaOH concentration that add to *Scenedesmus sp* is higher, the solubility of carbohydrate increased. In addition, NaOH cause autohydrolysis reaction that make microorganism release hyrdrolyze enzyme. Celullose content of biomass was expected to increase in pretreatment process [18].

3.2. Effect of NaOH Concentration on Fragmentation Type of Oligosaccharida

Polymers are materials with chains of varying length, and each chain consists of monomer residues that affect its properties and thus require the characterization of several parameters [2]. There is influence of NaOH concentration to fragmentation type. If NaOH concentration increased, the spectra showed an increased abundance. The cellulose II crystallites in regenerated celluloses increase in size to the longitudinal direction by the alkaline treatment and acid hydrolysis. There are increase cristallinity size with alkaline treatment in 20% NaOH and acid hydrolysis at 105 °C[12]. Breakdown of celullose structure will make celullose lysis more easier. Simply sugar was the product of lysis, then will be fermented by certain microorganisms [16]. The following is a type of oligosaccharide fragment in several variations of NaOH concentration.The following types of oligosaccharide fragmentation in 4% NaOH;

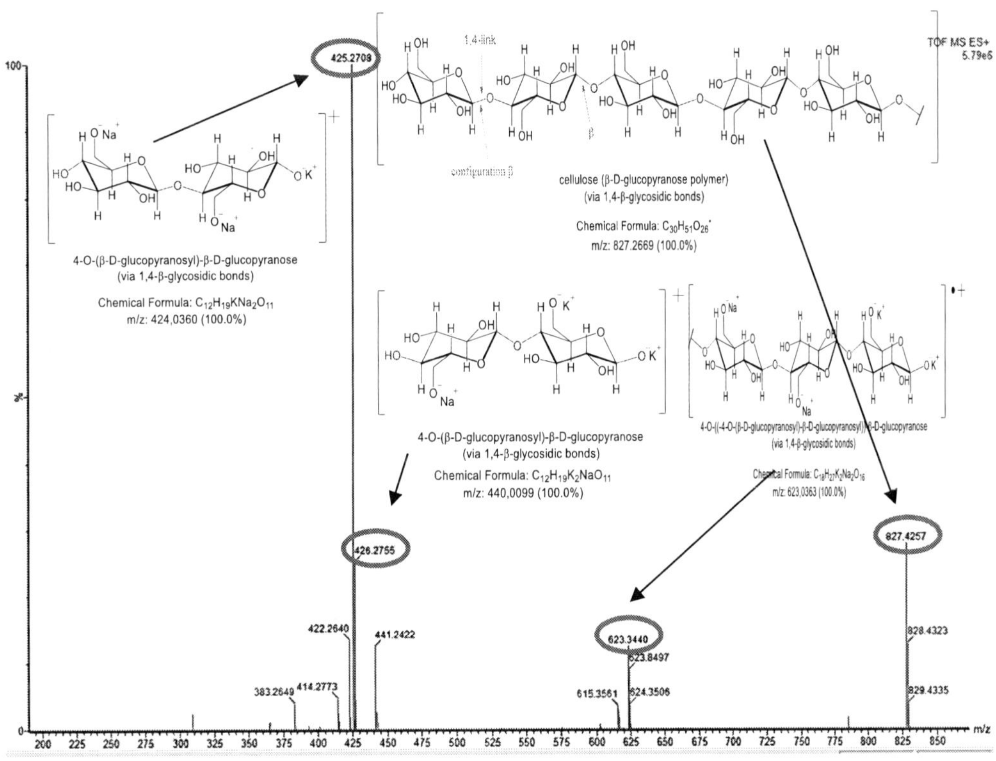

Figure 3. Fragmentation type of oligosaccharida with NaOH 4%

Based on MS spectra of NaOH 4%showed that corncob samples were olighoshacaride. Oligosaccharides consist of short chains of monosaccharide units combined by covalent bonds, including disaccharides having two monosaccharide units, eg sucrose or sugar cane consisting of 6-carbon D-glucose sugar and D-fructose.Most oligosaccharides have three or more non-free units but are combined as polypeptide side chains in glycoproteins and proteoglycans [14]. The ES-MS spectra of 4%NaOHshow peak of ion $[M]^+$ at 827,4257 m/z, andappropriate with molecular mass (Mr) of oligosccharide ($C_{30}H_{51}O_{26}^{\cdot}$) 827,2669 g/mol. NaOH 4% gave higher α-selulosa content ii.e 60% (Fig.2)This indicated the presence 5 polimer of glucose. The determination of the molecular weight of the lignocellulose in the original biomass depends on the source of the bioresources used and the procedure and purification used.The average weight (Mn), average molecular weight (Mw) and polydispersity index (D) all vary greatly depending on biomass sources, pre-treatments and methods [6][7]. Here is (Figure. 4) the fragmentation of oligosaccharide:

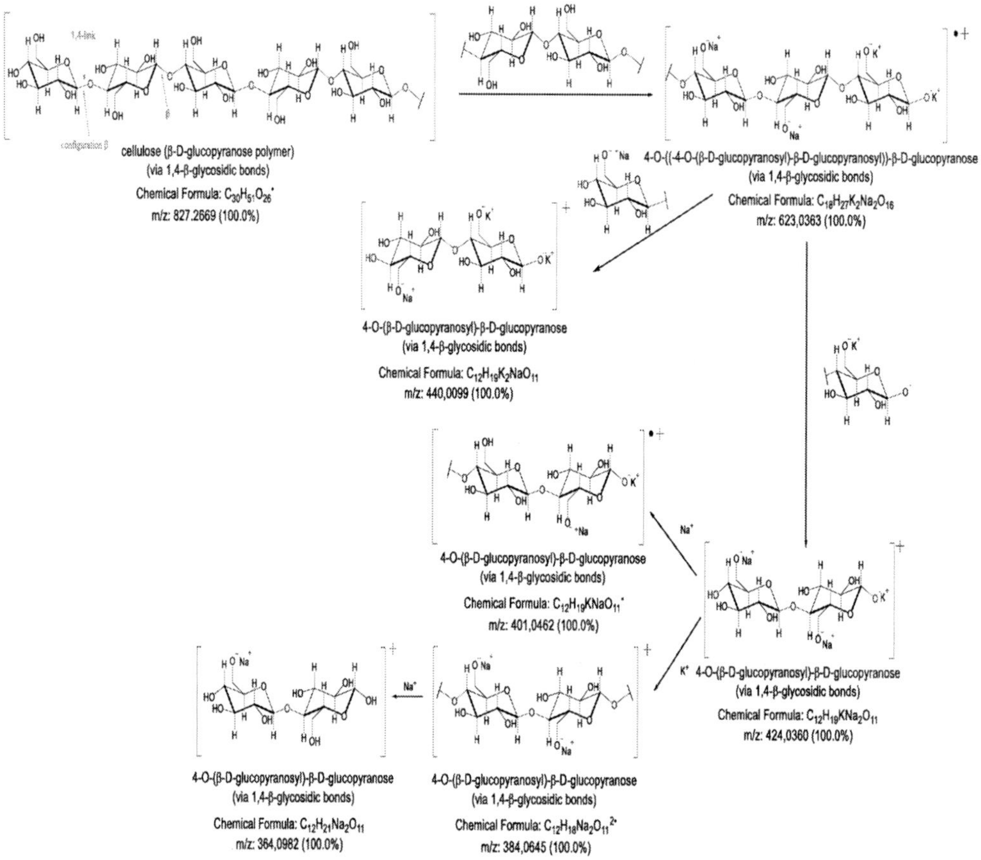

Figure 4. Oligosaccharide fragmentation

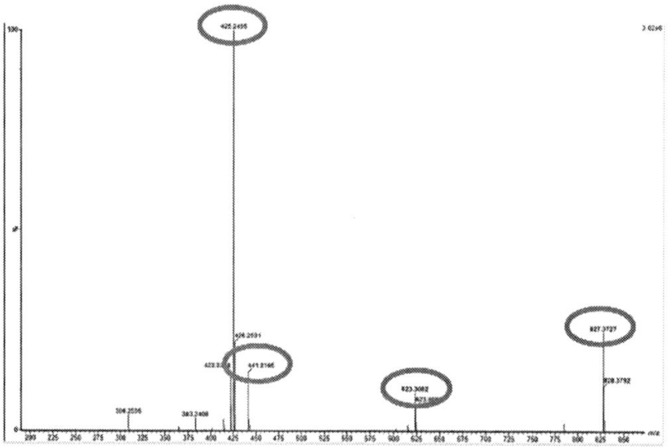

Figure5 . Fragmentation type of oligosaccharida with NaOH 8%

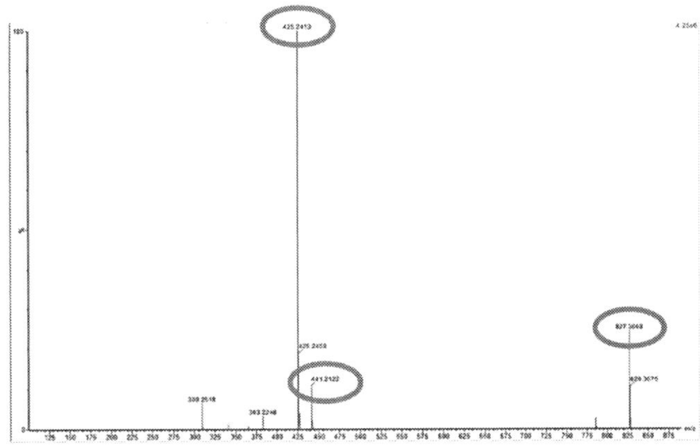

Figure6 . Fragmentation type of oligosakarida with NaOH 10%

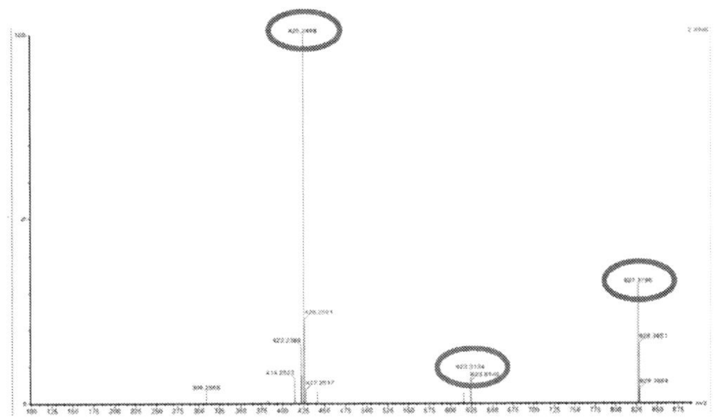

Figure7. Fragmentation type of oligosakarida with NaOH 12%

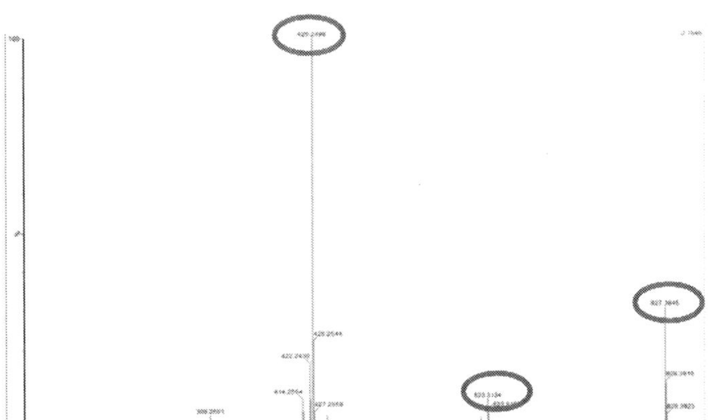

Figure8 . Fragmentation type of oligosakarida with NaOH 14%

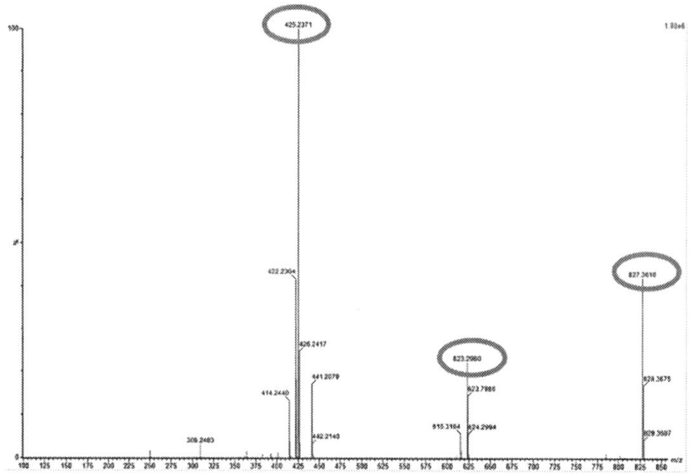

Figure 9 .Fragmentation type of oligosakarida with NaOH 17%

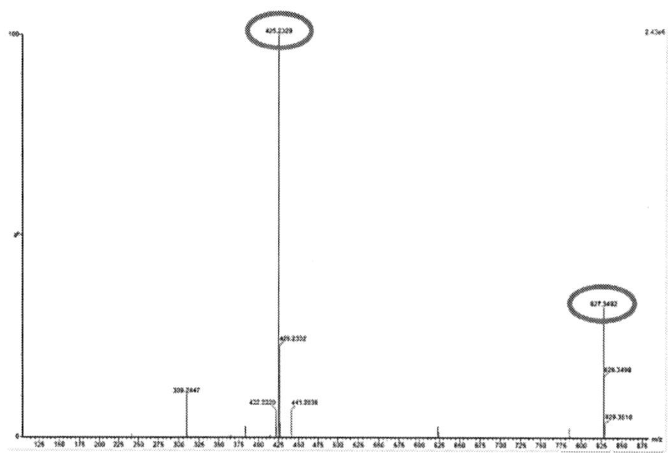

Figure10. Standard Avicel 102

Based on MS spectra of NaOH 8%,10%,12%,14%,17%showed that corncob samples were olighoshacaride. Oligosaccharides consist of short chains of monosaccharide units combined by covalent bonds, including disaccharides having two monosaccharide units, eg sucrose or sugar cane consisting of 6-carbon D-glucose sugar and D-fructose.Most oligosaccharides have three or more non-free units but are combined as polypeptide side chains in glycoproteins and proteoglycans [14]. The ES-MS spectra of 8%,10%,12%,14%,17%NaOHshow peak of ion $[M]^+$ at 827,4257 m/z, andappropriate with molecule mass (Mr) of oligosccharide ($C_{30}H_{51}O_{26}^{\cdot}$) 827,2669 g/mol. This indicated the presence 5 polimer of glucose. High radiation on celluoce biomssa will increase surface area, reduce polymerization degree and celullose crystallinity, hydrolize hemicellulose and cause lignin de polimerizaation. When surface area is high, the glucose from celullose increased [20].

Spectraof ion $[M]^+$at 623,3440 m/z indicated the presence 3 glucose with additionalion–ion $2Na^+$and $2K^+$ion, with molecular mass (Mr) 623,0363 g/mol. There are 3 possibilities i.eNa^+ and $2K^+$ ions from ultra water still have mineral contaminant, Na^+ and K^+ ions as intraseluler ions in plant that carried away during extraction of cellulose with NaOH and $K_2Cr_2O_7$ as activator. Spectra peak $[M+H]^+$ ion at 441,2422 m/z show 2 glucose with additional $2K^+$ and Na^+ ions, that appropriate with molecular formula Mr440,0099 g/mol and molecular formula i.e$C_{12}H_{19}K_2NaO_{11}$. Spectra peak $[M+H]^+$ion at 425,2708 m/z show 2 glucose with additional K^+ dan 2Na ions, that appropriate with molecular formula Mr424,0360 g/mol and molecular formula i.e$C_{12}H_{19}KNa_2O_{11}$.

Based on the result, compound inside concorbi.ecellobiose oligosaccharide. Cellobiose is non reducing glucose that composite matrix of cellulose [14].There is elimination reaction at β-alkoxy group, if NaOH concentration increase8%,10%,12%,14%,17%. This is causing the dissolved monosaccharide units and shortening the polysaccharide chain [1].The MS spectra of Avicel standard pH 102 indicate the type of oligosaccharide fragmentation with same two main peaks i.e at 827 m/z, that indicated the presence of 5 Glucose polymers. The absence of spectra at 623 m/z peak showed that there is no additional Na^+ and K^+ in 3 glucose polymer.

Table 1. Interpretation oligosaccharide fragmentation with NaOH

No Peak	m/z						Structure
	4%	8%	10%	12%	14%	17%	
1	827,4257	827,3727	827,3668	827,3786	827,3845	827,3610	[G5]
2	623,3440	623,3082	-	623.3134	623.3134	623.2980	[G3 + 2K + 2Na]
3	441,2422	441,2165	441,2122	-	-	441.2079	[G2 + 2K + Na]
4	425,2708	425,2955	425,2413	425,2498	425,2498	425,2371	[G2 + K + 2Na]

4. Conclusions

If NaOHconcentration increased, this is affect the type of fragmentation with increasing abundance. NaOH 4% shows the four main peaks of 834 m/z, 623 m/z, 425 m/z, 383 m/z respectively indicating the presence of 2 - 5 glucose polymers with an increased abundance i.e the addition of $2Na^+$ and $2K^+$. The concentration of NaOH 8%, 12%, 17% showed the type of oligosaccharide fragmentation with the same three main peak at 827 m/z, 623 m/z, 425 m/z. For NaOH10%, the oligosaccharide fragmentation type indicated the presence of 2 - 5 Glucose polymers with an increased abundance of the addition of $2Na^+$ and K^+ ions, with absence peak at 623 m/z.

Acknowledgements

Based on the research and the desired results for subsequent researchers to perform the measurement and analysis.

References

[1] Linnea N 2015 *Preparation Methods for Nanocrystalline Cellulose: Acid hidrolysis and various cellulose sources* (Gothenburg-Sweden: Chalmers University of Tecnology,)

[2] Hermin S, Sutriyo, Hasty RS and Dianah R 2017 *Preparation of Microcrystalline Cellulose from Water Hyacinth Powder by Enzymatic Hydrolysis Using Cellulose of Local Isolate .Journal Young Parm* (Jakarta : Universitas Indonesia) 99(1)

[3] Kunal D, Dipa R, N. R. Bandyopadhyay and Suparna S 2010 *Journal of Polymers and the Environment* 18(3) pp 355–363

[4] Lalduhsanga P, David C V, Shri K T, H. Lalhlenmawia 2014 *Journal of App. Pharm. Sci.*4(11) pp. 087-094

[5] Morgado DL, Frollini E, Possidonio S, El Seoud OA Ciacco GT 2010 *Journal Brazil Chem. Soc.* pp. 71-77

[6] Alireza M 2015 *Influence of the molecular weight of cellulose on the solubility in ionic liquid-water mixtures Master's thesis* (Espoo-Finland : Faculty of Chemistry and Materials Sciences Department of Forest Product Technology)

[7] Allison T, Hannah A, Ratayakorn K, Amit K N, Arthur J R 2014 *Characterization and analysis of the molecular weight of lignin for biorefi ning studies Journal Biofuels, Bioprod. Bioref.* (Atlanta,GA, USA : Georgia Institute of Technology) DOI: 10.1002/bbb.150

[8] Kari V, Tom L, Asko K, Marcus L, Olli D 2016 *Journal of Mat. Sci.*51(12) pp 6019–6034

[9] Zoia L, King AW, Argyropoulos DS 2011 *Journal Agric. Food Chem.* 59(3) pp 829-38

[10] Peter J S and Chris G. de K 2002 *LC–MS of Synthetic Polymers: The Parable of the Lame and the Blind* (Amsterdam : Swammerdam Institute for Life Sciences,University of Amsterdam)

[11] Oluwasina O, Lajide L and Owolabi B 2014 *Microcrystalline Cellulose from Plant Wastes*

through Sodium Hydroxide-Antraquinone-Ethanol Pulping Journal Bioresource (Akure,Nigeria : Federal University Of Technology, Department of Chemistry 9(4) pp 6166-6192

[12] Takuya I, Masahiro Y, Akira I 2008 *Degrees of polymerization (DP) and DP distribution of dilute acid-hydrolyzed products of alkali-treated native and regenerated celluloses Journal Cellulose* (Tokyo Japan: The University of Tokyo Agricultural and Life Sciences) Volume 15, Issue 6, pp 815–823

[13] Satish N. Patkar[a] and Prasad D. Panzade[*a] 2016 *Fast and efficient method for molecular weight analysis of cellulose pulp, in-process and finished product Journal Analytical Method* (Mumbay India : Aditya Birla Science and Technology Company) 8(15) Page 2981 to 3272

[14] Isaksen T, Westereng B,J 2014 *Journal of Biological of Chem(JBC)* (Europe PMC) 289(5)pp 2632-2642

[15] Mahdy A, Lara M, Mercedes B, Cristina GF 2014 *Journal Energy* 78 pp 48-52

[16] Mosier NS, Wyman C, Dale B,RichardE, Y Y Leee, MarkH, MichaelL 2005 *Bioresource Tech.*96(6) pp 673-686

[17] Sun Y and Cheng J 2002 *Journal Bioresource Tech* 83(11) pp 1–11

[18] Moh. Risal S, Yusuf H, and Wahyunanto A N 2014 *Jurnal Teknologi Pertanian* (Universitas Brawijaya : Indonesia 15 (2)

[19] Zhao X, Zhang L, and Liu D 2007 *Journal Bioresource Tech.* 99(9) pp 3729-3736

[20] Zheng Y, Pan Z, and Zhang R 2009 *nt. Journal Agric & Biol Eng* 2(3) pp 51 – 68

IOP Conf. Series: Journal of Physics: Conf. Series **1040** (2018) 012016 doi:10.1088/1742-6596/1040/1/012016

Silicon chemistry for sustainable development of rice agriculture

I W Karyasa

Department of Chemistry, Faculty of Maths and Natural Sciences, Universitas Pendidikan Ganesha, Singaraja, Bali, Indonesia

E-mail: karyasa@undiksha.ac.id or karyasa.undiksha@gmail.com

Abstract. The literature reviewed paper was aimed to explain the role of paddy's plantation for reserving silica powder, to explain the current developing of advanced materials from the high siliconeous rice plant biomasses, to propose a scheme of value addition of paddy's biomasses for sustainable rice agriculture development. A comprehensive literature review through meta analysis on articles published in last few decades was conducted. Because of high silicon content of rice biomasses especially rice straws and husks and their abundance in availability, the rich silicon biomasses can be used as raw material for producing a high purity reactive fine powder silica, and unexpesive and environment-friendly production processes. Current development results on advanced materials from the rich siliconeous biomasses were reported, namely silicon carbides, amorphous reactive silica, nanosilica, and silica-carbon nanocomposite. The implementation of silicon chemistry takes important role for enhancing the added value of rice husks and straws as the side products of paddy's agriculture to produce some advanced materials inexpensively and environment-friendly. The added values can be considered to overcome the problems of lack paddy's farmer income and the land conversion tendency of rice field, thus the paddy's agriculture can be developed sustainably.

1. Introduction

Silicon and silica powders, as well as their generative advanced materials todays take more attention industrially. Worldwide market of silica powder, for instance, has increased sharply from 27.8 million metric tonnes at 2004 [1] to ten times about 278 million metric tonnes in year 2016 as predicted by PR Newswire [2]. Mainly, the silica powder was reserved from the earth deposite of high silica content of hardrock such as *taconite, granite,* and *quartzite* as well as from quart sand, where both kinds of raw materials needed to be broken down, milled, and physically as well as chemically purified respectively [3]. Those steps of producing silica powder from the rock and sand were high costs for machine investment and energy. It affected the relative expensive of the silica powder price, for instance, in Europe at 2004 the price silica powder reached USD 927/tonne [1]. The mining of rocky silica deposites or quartz sands was also environmental unfriendly, for example, purification processes from metallic impurities used clean water about 4,500 to 6,000 gallons/minute and the throwing away some impurties solvents in the water body [4]. Thus, the production of silica powder from rocks or quartz sands is not only high cost but also environment-unfriendly.

The finding of high silicon content of the rice plant biomasses and the high silica content of rice husk ash triggered further investion intensively among their silica based advanced materials. It is also as a potential alternative solution for supplying silica powder in enough quantity, high purity, and low cost because the avaiability of the raw rich siliconeous biomasses as reviewed previously [5]. The high potency of rice biomasses can be seen from the annual world production of rice grain about 600 million tonnes, rice husk about 120 million tonnes, and the rice straw about ten times of the rice production [6]. Indonesia it self produces rice husk about 13 million tonnes annually [7]. When the rice husk was burned until it became a white powder called white rice husk ash, the randemen of the ash was 22-29 %[8] and the content of silica about 87-97% [9]. The related finding was also reported that the temperature controlled burning at 800°C along 4 hours of rice husks taken from Tabanan Regency of Bali Province Indonesia produced about 20.6% amorphous silica with purity of 99% [10]. Thus, there is a high potency of high purity amorphous silica powder from the rice biomasses.

The high potency of high purity amorphous silica powder from the high siliconeus tropical plants especially the rice biomasses causes today the development of the biosilica based advanced materials taken into account intensively and a comprehensive literature review on that is needed for future looking of the competitive beneficial direction on researches and developments. This paper described the role of paddy's plantation for reserving silica powder unexpensive and environment-friendly, to explain the current developing of advanced materials from the high siliconeous rice plant biomasses, to propose a scheme of value addition of paddy's biomasses for sustainable rice agriculture development. A comprehensive literature review through meta analysis on articles published in last few decades was conducted.

2. Discussion

2.1. The role of rice agriculture for providing high purity silica powder

Paddy's plants are one of the 15 tropical plants containing rich of silicon [5]. The rice plant biomasses, namely its straw and husk, containing siliconeous cellulose and lignocellulose, when it burned, it produced ashes with high content of silica. The high abundance of paddy's biomasses is supported by the increase of the rice need by Indonesian people that pushed people to open new rice field to produce more rice because almost Indonesian people eat rice. However the rice biomasses, such as straw and husk, have been yet added their value optimally. For instance, the rice husk was only burned by the rice farmers or it used as biofuel for brick burning and as a raw material for organic fertilizer.

The use of rice straw and husk to produce high purity biosilica powder can add the economic value of the biomasses. Every tonne of rice husk can produce high purity silica powder about 200 kgs or every 5 tonnes of rice husk can produce 1 tonne of silica powder, where its price at USA was USD 500/tonne and in Europe at USD 927/tonne [1]. If it is assumed that the cost production is about 40% from the market price, the added value of the silica powder production from the paddy's biomasses is about USD 300/tonne. Thus the paddy's farmers can gain more income and they will retain the rice field from land conversion and the the paddy agriculture can develope sustainably.

It is not only a simpler process in preparing and purifying silica powder from the ashes of rice husks and straws but also it need much lower energy than the preparation and purification processes of silica powder from rocky quartz deposit or quartz sand. The biosilica powder production does not need a lot energy for stone breaking and milling. It is not need also a lot of water and chemicals for purification processes. The biosilica from rice husk ash has very high purity about 87-97% w/w[9] with impurities of oxides K_2O, Na_2O and Fe_2O_3 that those are easily dissolved in dilute mineral acid solutions to become a reactive amorphous silica powder with purity up to 99 % w/w [11]. If the burning at lower temperatures is, for instance at 400°C, a black silica powder consisting mainly black carbon ash and silica is produced, where the black silica powder of rice husk was converted into cubic silicon carbide nanosize powders [10]. Moreover, the burning of those paddy's biomasses produce heat energy about 3,600 kcal/kg comparing to the burning of coal producing 6,000 kcal/kg [12]. For instance, the villagers used the rice husk for brick or clayware burning and todays it used for heating a boiler in a turbine of an electric generator. The excess smoke of the burning rice husks can be collected through a distillation process to get a liquid smoke. The burning of 1 kg of dried rice husk produced about 810 mL liquid smoke containing volatile compounds such as aldehides, acids and phenols that those can be used as preservatives [13]. Figure 1 shows four usefullneses of rice husk and straw producing white purely silica powder, carbonized black silica powder, heat energy and liquid smoke.

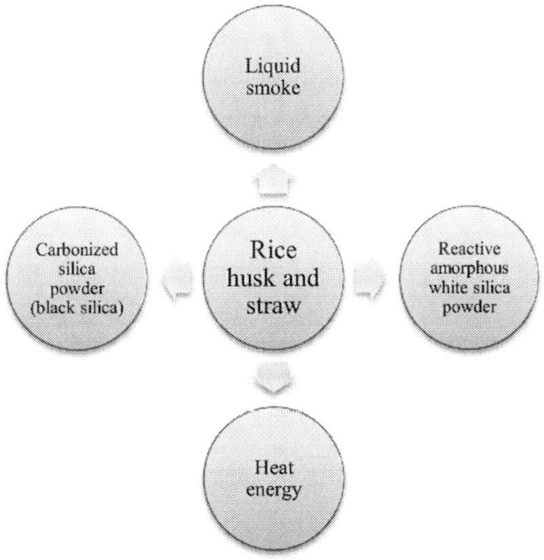

Figure1. The usefullnes of the burning of rice husks and straws

2.2. Development of Advanced Materials from Biosilica

The development of rice biomass ash biosilica based advanced inorganic materials are reviewed in terms of two raw materials namely black biosilica and white biosilica powder as well as their production processes of both kinds of biosilica and their application for producing prospective industrial materials. The black biosilica powder was produced by burning rice biomasses at a controlled temperature 400°C and the white biosilica powder was prepared by burning the biomasses at temperature about 800-1000°C [10]. Both powders were removed their impurities namely metall oxides such as K_2O, Na_2O, MgO, CaO, and Fe_2O_3 by washing the powders using dilute mineral acid solutions [10]. The black biosilica based advanced materials discussed further were focused on silicon carbide, silica-carbon nanocomposite, and their application. Meanwhile, the white biosilica based advanced materials discussed furthermore are ultrafine reactive amorphous silica, nanosilica, and their application.

A research on developing of advanced materials from the carbonized rice husk ash called black biosilica for synthesizing silicon carbide was reported [10]. The research showed that controlled heating under vacuum at temperature of 1500°C along 4 hours, the rice husk ash black silica was completely converted into nanosize cubic silicon carbide (β-SiC) powder[10]. This preparative method is relatively simpler and more unexpensive than the Acheson process that used quarzt and coal under argon at heating a temperature above 2000°C. Furthermore, the β-SiC powder was reacted with aluminium metal powder in form of pelletes under vacuum at a temperature range of 1000-1500°C along 4 hours [14]. The series of reaction produced a superlight material of aluminium silicon carbide Al_4SiC_4 [14]. The compound of Al_4SiC_4 overcame the formation of aluminium carbide Al_4C_3 which its existance on aluminium alloy could promote the formation of the easily burning methane gas. Thus the aluminium silicon carbide is more pavourable than the aluminium-carbon alloy as a superlight material for aircrafts. The finding of the simple preparation process of the nanosize cubic silicon carbide powder from black rice husk ash silica powder contributed to solve the problem of providing unexpensive and environment-friendly silicon carbide powder for industrial purposes. Moreover, the black biosilica was successfully converted into a reactive silica-carbon nanocomposite powder in a simple way [15]. A forming of composite matrix between carbon and silica has a prospective uses as reinforcer because of its ability to enhance bulkily the microstructure of a target materials, for instance, it can hold tighting together the filler and matrix in the cellulose fiber – resin matrix system. The use of silica-carbon nanocomposite as reinforcer material for producing a synthetic wood from non-wood lignocellulose fiber was proofed to improve significantly the mechanical properties of resulted synthetic wood comparing with the synthetic wood without using the silica-carbon nanocomposite [15]. These findings may contribute to develope bio-inorganic material based on non-metallic matrix composites.

White biosilica powder from paddy's agriculture wastes was taken into attention n terms of the development of preparation and characterization methods and its industial application. The pyrolysis methods of rice husks and straws for producing a reactive biosilica powder, its chemical composition analysis, its purification and its characterization were reported by a lot of researchers [16] [17] [18] [19] [20] [21] [22] [23] [24] [25] since the last two decades. Some studies on directlly applying of the reactive amorphous biosilica powder were reported for absorbing gases[26] [27] [28] [29] [30]; for absorbing liquid or solution [27] [31] [32]; as a mixing component in a concrete or mortar [33] [34] [35] [36] [37] [38] [39]; as a precursor [40]; as a ceramic component [41]; and as soil stabilisator [42]. Some researches in improving the rice husk ash biosilica were reported namely, as ultrafine amorphous reactive biosilica powder [43] [44]; as coloidal/gel silica [45]; as meso/nanoporous silica [31] [46] [47]; as nanosilica powder [48] [49]; as glass/membrane silica [50] [51]; for synthesizing zeolites or zeolite-like materials and their application[52-58]; for synthesizing silicon carbide based materials and their applications [10] [59] [60] [61] [62]; for synthesizing silica based composites and their applications [15] [63] [64] [65] [66]. An ultrafine amorphous silica [67] was used as a part composition of cementious binder which reported, was applied for making quake resistant building bricks [68]. The ultrafine amorphous biosilica was also implemented for making calsium silicofosfate having Si-O-P ternary bonding for teeth and bone restorative materials [69].

Structure engineering on the biosilica, namely amorphization and particle size reduction, causes the biosilica powder more porous and reactive. It can take important role in developing inorganic solid state advanced materials. Hence the change of the structure and particle size of the materials can shift their properties. For instance, the biosilica powder with particles in nanosize has distinctive properties comparing to that having particle size in nanometer. Thus, a brief review on the developing of methods and their target of methods for advancing the biosilica based materials is also reported. In term of developing preparation routes, identification and characterization, some articles were reported by some researchers [16] [17] [18] [19] [20] [21] [28] [31] [43] [44] [45] [46] [47] [48] [49] [50] [51]. Some studies were also conducted in order to develop the biosilica based advanced materials and their industrial applications [28] [31] [49] [50] [51] [52] [53] [54] [55] [56] [57] [58] [59] [60] [70] [71].

2.3. A proposed scheme of sustainable development on rice agriculture

Some issues todays have been faced by paddy's farmers, especially in Bali Island namely (1) the high production costs including the treating of rice field, planting, weeding, fertilizing, pest avoiding, and harvesting, (2) the increase of fertilizers and pesticides, (3) the low income of the farmers because of the difference between the harvesting grain rice price and the total production costs is so small, (3) the lack of farmers young generation interesting in working at rice field, and (4) the lack of rice field land because of massively land conversion. Those problems treat the retaining the rice agriculture and its 'Subak" culture, where Subak is a famous rice farmer organization, thus the sustainable development on providing food for our country is also treated.

A key factor is proposed to solve those faced paddy's farmer problems, namely a scheme of adding value of rice husks and straws that those todays as side products or called paddy's agriculture wastes. The proposed scheme focusses on the post harvesting process at a milling instalation and its surrounding area. As seen on Figure 2, the proposed scheme can be described as follows. Firstly, a milling machine will converted the dried milling husked rice grains into unhusked rice grains and rice husks. The rice husks are used for heating resources of the drying maschines by controlled burning them, their heat is used for drying the wet rice grains. The resulted black ashes can be converted into white rice husk ashes by further burning and its excess heating can be also used for drying process. The black rice husk ashes can be used also for producing silicon carbide powders or silica-carbon nanocomposite powders. The white rice husk ashes can be used for generating further advanced materials such as nanosilica, zeolites, silica fumed, and soon. The excess of rice husk as well as rice straws can be used for providing linocellulosic fibers. The fibers can be used futher for synthesing wood-like materials by combining them with silica-carbon nanocomposites. The excess of smokes at the burning processes of rice husk to become black ashes as well as the burning of black rice husk ashes into white ashes can be distillated for producing a liquid smoke containing active compounds for food preservative purposes. The proposed scheme is believed to give impacts in increasing the farmers income and retaining the rice field as well as irrigation system and its Balinese traditional farmer organization "*Subak*". Thus the sultainable development in rice agriculture will come true.

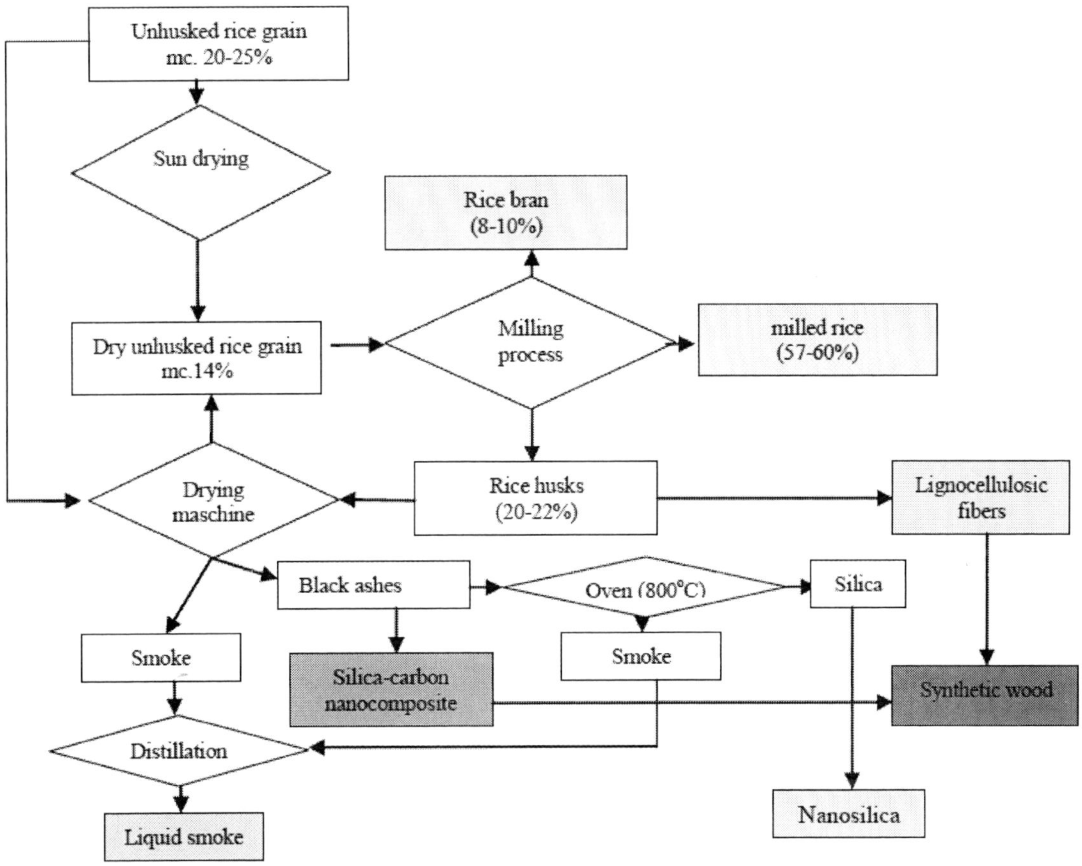

Figure 2. A scheme of value adding for developing paddy's agriculture sustainably

3. Conclusion

Silicon chemistry takes important role for adding values of the rice biomasses namely rice husk and straws by converting them into black as well as white ashes containg rich silica and their generative products such as nanosilica powders, silica-carbon nanocomposite and liquid smoke. The rice straws and husks can be also converted into lignocellulosic fibers which can be combined with the silica-carbon nanocomposite and other components such as resin and catalist to produce a synthetic wood.

The added values can be predicted to enhance the farmers income significantly, thus the paddy's agriculture can be developed sustainably.

Acknowledgement
Direktorat Riset dan Pemberdayaan Masyarakat, Direktorat Jenderal Penguatan Riset dan Pengembangan, Kementerian Riset, Teknologi dan Pendidikan Tinggi is acknowledged for the Competitive National Strategic Research Grant Year 2016 with Grant Number of 67/UN48.16/LT/2016.

References
[1] Dolley T P 2012 *Silica* Online resource: http://www.cdc.gov/niosh/nas/rdrp/appendices/chapter3/a3-53.pdf. Accessed on August 3, 2017
[2] Newswire 2012 P R *World Industrial Silica Sand Market* Online resource: http://www.bizjournals.com/prnewswire/press_releases/2012/11/08/SP08905 Accessed on September 1, 2013
[3] Environmental Quality Board 2013 *Report on Silica Sand* Online resource: http://www.eqb.state.mn.us/documents/23.%20March%20Final%20Silica%20Sand%20repor t. pdf Accessed on August 3, 2017
[4] McCurdy S E 2012 Silica Sand Processing Water Quantity and Quality *Conference on Silica Sand Resources of Minnesota and Wisconsin* Precambrian Research Center, University of Minnesota, Duluth, MN and Society of Mining, Metallurgy, and Exploration. 1-3 Oct 2012.
[5] Karyasa I W 2012 Meta-analisis Terhadap Material Berbasis Silika Terbarukan dari Sekam Padi dan Pemetaan Biomassa Tropis Kaya Silikon *Prosiding Seminar Nasional MIPA II* Undiksha, 30 Nopember 2012
[6] Giddel M R and Jivan A P 2007 Waste to wealth, potential of rice husk in India a literature review. *International Conference on Cleaner Technologies and Environmental Management PEC* pp 586-590
[7] Balipost 2003 *Dulu Bahan Baku Pakan Ternak Kini Sekam Jadi Energi Hijau* 3 September 2003
[8] Muthadhi A, Anitha R and Kothandaraman S 2007 *Journal of the Institution of Engineers* **88(5)** 50
[9] Rozainee M, Ngo S P and Salema A A 2008 *Bioresource Technology* **99** 703
[10] Karyasa I W and Walter D 2004 X-ray Powder Diffraction Study on High Temperature Vacuum Sintered Silicon Carbide (SiC) Synthesized from Carbonized Rice Husk *PubScieAEIF* **9** 1.
[11] Real C, Alcala M D and Criado J M 1996 *J. Am. Ceram. Soc.* **79[8]** 2012
[12] Wada S, Mosungnoen W, Hemachandra K and Jarunworaluck H 2005 Survey of the Research on the Utilization of Rice Husk and Rice Husk Silica *Proceeding of the First Workshop on the Utilization of Rice Husk and Rice Husk Silica* September 19, 2005 Chulalangkorn University Thailand
[13] Karyasa I W, Vivi Oviantari N M and Artawan K 2010 *Jurnal Penelitian dan Pengembangan Sains & Humaniora* **4(3)** 1
[14] Walter D and Karyasa I W 2005 *Journal of Chinese Chemical Society* **52(5)** 873
[15] Karyasa I W, Muderawan I W, Gunamantha I M and Vivi Oviantari N M 2014 *Paten* registered number P00201407241
[16] Chiang W, Fang H, Wu C, Chang C, Chang Y and Shie J 2008 *J. Environ. Eng.* **134(4)** 316
[17] Markovska I G and Lyubchev L A 2007 *Journal of Thermal and Calorimetry* **89(3)** 809
[18] Worasuwannarak N, Sonobe T and Tanthapanichakoon W 2007 *Journal of Analytical and Applied Pyrolysis* **78(2)** 265
[19] Kumar B and Godkhindi M M 1996 *Journal of Materials Science Letters* **15(5)** 403
[20] Genieva S D, Turmanova S C, Dimitrova A S and Vlaev L T 2008 *Journal of Thermal Analysis*

and Calorimetry **93(2)** 387
[21] Azizi S N and Yousefpour M 2008 *Eurasian Journal of Analytical Chemistry* **3(3)** 298
[22] Oidde M R, Dutta J and Jadhav S 2009 Comparative Adsorption Studies on Activated Rice Husk and Rice Husk Ash by using Methylene Blue as Dye *International Congress On Environmental Research at Bits Pilani Goa* 08-09
[23] Rukzon S, Chindaprasirt P and Mahachai R 2009 *International Journal of Minerals, Metallurgy and Materials* **16(2)** 242
[24] Omatola K M and Onojah A D 2009 *International Journal of Physical Sciences* **4(4)** 189
[25] Olawale O and Oyawale F A 2012 *Journal of Science and Technology* **2(4)** 210
[26] Kim K S and Choi H C 1998 *Wat. Sci. Tech.* **38[4-5]** 95
[27] Naeem S, Zafar U and Amann T 2011 *Bangladesh J. Sci. Ind. Res.* **46(1)** 101
[28] Bhagiyalakshmi M, Yun L J., Anuradha R and Jang H T 2010 *Journal of Hazardous Materials* **175 (1-3)** 928
[29] Yalçin N and Sevinç V 2001 *Ceramics International* **27(2)** 219
[30] Dahlan I, Mei G M, Kamaruddin A H, Mohamed A R and Lee K T 2008 *Journal of Engineering Science and Technology* **3(2)**109
[31] Kim II J, So S J and Han C S 2010 *Journal of Nanoscience and Nanotechnology* **10(5)** 3705
[32] Lakshmi U R, Srivastava V C, Mall I D and Lataye D H 2009 *Journal of Environmental Management* **90(2)** 710
[33] Osinubi K J, Edeh J E and Agada J O 2012 *Journal of ASTM International* **9(1)** JAI 103752
[34] Oyetola E B and Abdullahi M 2006 *Leonardo Electronic Journal of Practices and Technologies* **8** 58
[35] Ramezanianpour A A, Mahdi khani M and Ahmadibeni G 2009 *International Journal of Civil Engineerng* **7(2)** 83
[36] Habeeb G A and Mahmud H B 2010 *Materials Research* **13(2)** 185
[37] Kartini K, Mahmud HB and Hamidah M S 2010 *Journal of Engineering Science and Technology*. **5(1)** 1
[38] Kartini K 2011 *International Journal of Applied Science and Technology* **1(6)** 169
[39] Ephraim M E, Akeke G A and Ukpata J O 2012 *Scholarly Journal of Engineering Research* **1(2)** 32
[40] Bondiolia F, Andreola F, Barbieri L, Manfredini T and Ferrari A M 2007 *Journal of the European Ceramic Society* **27(12)** 3483
[41] Haslinawati M M, Matori K A, Wahab Z A, Sidek H A A and Zainal A T 2009 *International Journal of Basic & Applied Sciences* **9(9)** 111
[42] Brooks R M 2009 *International Journal of Research and Reviews in Applied Sciences* **1(3)** 209
[43] Chandrasekhar S, Satyanarayana K G, Pramada P N, Raghavan P and Gupta T N 2003 *Journal of Materials Science* **38(15)** 3159
[44] Javed S H, Naveed S, Feroze N, Zafar M and Shafaq M 2010 *Journal of Quality and Technology Management* **6(1)** 81
[45] Halimaton H 2007 *USPTO Patent* Application 20070276051.
[46] Shelke V R, Bhagade S S and Mandavgane S A 2010 *Bulletin of Chemical Reaction Engineering & Catalysis* **5(2)** 63
[47] Bhagiyalakshmi M, Yun L J, Anuradha R and Jang H T 2010 *Journal of Hazardous Materials* **175(1-3)** 928
[48] Conradt R, Pimkhaokham P and Leela-Adosorn U 1992 *J. Non-Crystalline Solids* **145** 75
[49] Thundaij N and Nuntiya A 2008 *Chiang Mai J. Sci.* **35(1)** 206
[50] Wasanapiarnpong T, Vorajesdarom B, Rujirakamort E, Nilpairach S and Mongkolkachit C 2011 *IOP Conf. Ser.: Mater. Sci. Eng.* **18** 222028
[51] Lee S Y and Han C S 2006 *Journal of Nanoscience and Nanotechnology* **6(11)** 3384
[52] Panpa W, Sujaridworakul P and Jinawath S 2005 Synthesis of ZSM-5 Zeolite from Rice Husk Ash. *Proceeding of the First Workshop on the Utilization of Rice Husk and Rice Husk Silica*

Chulalangkorn University Thailand
[53] Naskar M K, Kundu D and Chatterjee M 2012 *Journal of the American Ceramic Society* **95(3)** 925
[54] Wong J T, Eng-Poh Ng and Adam F 2012 *Journal of the American Ceramic Society* **95(2)** 805
[55] Teabpinyok N, Chareonpanich M, Samingprai S and Limtrakul J 2012 *The Canadian Journal of Chemical Engineering* **90(4)** 881
[56] Saiintawong K, Wadal S, and Jarunworaluck H (2005). Low Cost Process for Mullite Utilizing Industrial Waste as Starting Materials, *Proceeding of the First Workshop on the Utilization of Rice Husk and Rice Husk Silica*, September 19, 2005. Chulalangkorn University Thailand.
[57] Chumee J, Grisdanurak N, Neramittagapong S and Wittayakun J 2009 *Brazilian Journal of Chemical Engineering* **26(2)** 367
[58] Ghasemi Z and Younesi H 2011 *Journal of Nanomaterials* **2011** 1-8
[59] Narciso-Romero F J and Rodríguez-Reinoso F 1996 *Journal of Materials Science* **31(3)** 779
[60] Gorthy P and Pudukotta M 1999 *J. Am. Ceram. Soc.* **82[6]** 1393
[61] Qadri S B, Imam M A, Fliflet A W, Rath B B, Goswami R and Caldwell J D 2012 *Appl. Phys.* **111** 073523
[62] Sujirote K and Leangsuwan P 2003 *Journal of Materials Science* **38(23)** 4739
[63] Wang W, Martin J C, Huang R, Liu A, Han A and Sun L 2012 *RSC Adv.* **2** 9036
[64] Rout A K and Satapathy A 2012 *CompMechComput Appl Int J.* **3(2)** 95
[65] Lee S H, Weinmann M N and Aldinger F 2008 *Acta materilia* **58** 1529
[66] Rahman I A 1994 *Ceramic International* **20(3)** 195
[67] Karyasa I W 2014 *Jurnal Sains & Teknologi* **3(1)** 263
[68] Karyasa I W 2013 Pengembangan Cementious Binder dari Limbah Batu Pipih, Silika Abu Sekam Padi dan batu Gamping serta Pemanfaatannya dalam Pembuatan Batako Interlocking *Prosiding Seminar Nasional MIPA* 30 Nopember 2013, 178
[69] Beni Widana G A and Karyasa I W 2012 Sintesis dan Karakterisasi Biomaterial Restoratif Kalsium Silikofosfat dari Limbah Tulang Sapi dan Silika Amorp Abu Sekam Padi *Prosiding Seminar Nasional Kimia XIII* ITS Surabaya 22 Januari 2013 A-15
[70] Karyasa I W, Muderawan I W and Gunamantha I M 2015 Silica-Carbon Nanocomposite from Bamboo Stems and Its Use for Reinforcing Synthetic Woods Produced From Lignocellulosic Agrifibers *1st-International Conference on Inovative Research Across Descipline (ICIRAD)*, 18 Nopember 2015, Kuta Bali
[71] Karyasa I W, Muderawan I W and Gunamantha I M 2016 *KnE Engineering*

IOP Conf. Series: Journal of Physics: Conf. Series **1040** (2018) 012017 doi:10.1088/1742-6596/1040/1/012017

Classification of breast cancer using Wrapper and Naïve Bayes algorithms

I M D Maysanjaya[1,2], **I M A Pradnyana**[1] **and I M Putrama**[1]

[1] Department of Informatics Education, Faculty of Engineering and Vocational, Universitas Pendidikan Ganesha, Bali, Indonesia – 81116
[2] Department of Electrical Engineering and Information Technology, Faculty of Engineering, Universitas Gadjah Mada, Yogyakarta, Indonesia – 55281

E-mail: dendi.ms@undiksha.ac.id

Abstract. Breast cancer is defined as a malignant neoplasm disease originating from the parenchyma. This disease is most commonly afflicted by women and is the second killer after lung cancer. By the end of 2012, the WHO shows the prevalence of breast cancer worldwide reaching 6.3 million spread across 140 countries. Symptoms of the disease are diagnosed using mammogram. The results will be examined by paramedics, to find out if the cancer is malignant or not. However, the accuracy of the diagnoses obtained is directly proportional to the level of paramedical expertise. The general objective of this research aims to classify breast cancer types based on the extracted breast cancer features, and the specific purpose is to offer a hybrid machine learning method. To help increase the accuracy of the diagnosis, through this study, a CAD-based scheme was developed by applying a combination of Wrapper and Naive Bayes algorithms. The Wrapper algorithm is used at the feature selection stage, while the Naive Bayes algorithm is used at the classification stage. There are 683 data taken from the UCI Knowledge Repository consisting of two class, namely 444 benign cancers and 239 malignant cancers, with 9 types of attributes. The data is divided into two groups, with the various amount for each class in each group. The first group was used as a training group for the feature selection stage, and the second group as the testing group for the classification stage. The classification results show the accuracy up to 99.27%, sensitivity and specificity on each up to 99.30%. Based on these results, the proposed scheme is expected to contribute in the development of CAD for the diagnosis of breast cancer.

1. Introduction

Breast cancer is defined as a malignant neoplasm disease originating from the parenchyma tissue. This disease commonly affects women and becomes the second killer after lung cancer. By the end of 2012, the World Health Organization (WHO) shows that the number of cases of breast cancer in the world reaches 6.3 million scattered in 140 countries. By 2015, the American Cancer Society [1] estimates about 231,840 new cases occur, and from that number is estimated around 40,290 women die. Breast cancer is not only suffered by women, but men too, although the number is not as many as the number cases in women. By 2015, cases of breast cancer that occurred in men is estimated to be 2,350 cases, and 440 of them died [1]. There are two types of breast cancer, i.e. benign and malignant. The illustration can be seen in Figure 1, and the brief explanation of the differences can be seen in Table 1.

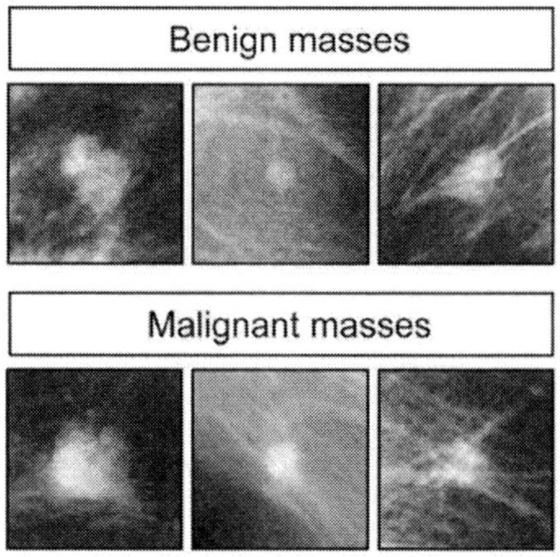

Figure 1. Types of Breast Cancer

Table 1. Differences of Benign and Malignant Cancer

Feature	Benign	Malignant
Size	< 2 cm	> 2 cm
Consistency	Soft	Hard, firm, or rubbery
Mobility	Fixed	Mobile
Rate of Growth	Slow	Rapidly
Boundaries	Circumscribed	Irregular
Metastasis	Absent	Frequently present

In the examination of breast cancer, paramedics commonly use mammography. Mammography is a process of breast examination in humans by using low-dose X-rays. With proper use, mammography can reduce mortality caused by breast cancer. Certainly, the case requires knowledge and paramedical skills in handling it. A false diagnosis on mammography examination can be affected by some factors particularly the expertise level of paramedics, the acquisition method, and the quality of mammography equipment used. Hence, several studies have been conducted to develop computer-aided breast cancer diagnosis based on digital image processing to reduce the error possibility.

Akay [2] applied a combination of Support Vector Machine (SVM) and feature selection to diagnose breast cancer. However, the data used requires further exploration, so the results can be more convincing and interesting. Chen [3] employed a hybrid intelligent model that uses the cluster analysis technique with feature selection for analyzing clinical breast cancer diagnosis. Eltoukhy *et al.* [4] presented a method for breast cancer diagnosis in digital mammogram images using multiresolution representations (wavelet and curvelet), and validated using 5-folds cross-validation. Based on the result, the experiment using wavelet gain the accuracy of 96.56%, and achieved the accuracy of 97.30% by curvelet-based.

Furthermore, Zheng *et al.* [5] introduced combination of K-Means clustering and SVM algorithms. According to the result, the proposed method reduces the computation time significantly without losing diagnosis accuracy. However, by using K-Means algorithm, the training sample size has not been decreased by filtering the similar samples in the data, which could be a potential way of reducing the dimension of training set further.

Based on the previous research, we assess the accuracy of breast cancer classification results can still be improved. Thus, the main contribution of this study aims to develop hybrid machine learning methods to classify the types of breast cancer by using a combination of Wrapper and Naïve Bayes algorithms. In this study, we proposed hybrid method because of its advantages. Wrapper is one type of feature selection method that uses a controlled algorithm as part of the evaluation function, while

Naïve Bayes has several advantages, such as fast training process, unaffected by irrelevant features, and is also capable of handling real and discrete data. By combining their respective advantages, it can improve the accuracy of the classification process.

To complete the identification study of breast cancer, this paper proposes a scheme to classify breast cancer types. The classification is categorized into two types, i.e. benign and malignant. The structure of this paper is organized as follows Section 2 explained the methods and procedures. The results and discussion are presented in Section 3 followed by conclusion and future work in Section 4.

2. Methods and Procedures

2.1. Approach

The methodology consists of four main processes, namely data cleansing, modelling the training-test data partition, feature selection, and classification as depicted in Figure 2.

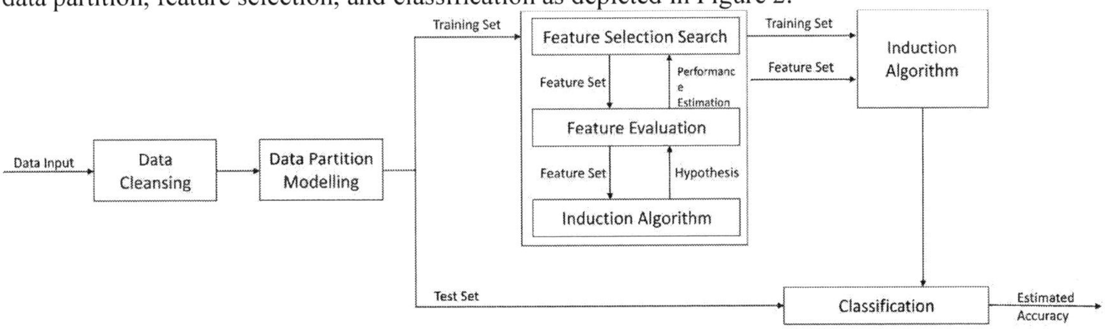

Figure 2. Scheme diagram of the approach

After the dataset taken from the open source repository, firstly, we applied data cleansing process, especially removed the data with missing values. The second step is randomly divide data into two types, training and testing data, based on three types of partition (50%-50%, 70%-30%, 80%-20% of training-testing). The purpose of this partition modelling is to see its effect on the feature selection and classification process, and to compare which partition pattern is capable of producing the highest accuracy. Furthermore, Wrapper algorithm was conducted on feature selection process, which aims to gain the relevant feature for the classification process, and the last step is the classification process which is provided by Naïve Bayes algorithm.

2.2. Dataset

The dataset in this study was taken from the University of California – Irvine (UCI) Machine Learning Repository, created by Dr. Wolberg [6], from University of Wisconsin Hospitals, Madison, Wisconsin, USA. As a brief information, UCI Machine Learning is an open source repository, which like a bank of dataset, database, or domain theory for many case studies that are used by the machine learning community for the empirical analysis of machine learning algorithm. It is commonly used by anyone as a primary source of machine learning dataset [7].

The number of breast cancer dataset is 683 instances, consisting of 9 attributes, i.e. 1) Clump of Thickness, 2) Uniformity of Cell Size, 3) Uniformity of Cell Shape, 4) Marginal Adhesion, 5) Single Epithelial Cell Size, 6) Bare Nuclei, 7) Bland Chromatin, 8) Normal Nucleoli, and 9) Mitoses. This samples arrive periodically as Dr. Wolberg reports his clinical cases. Originally, the number of dataset is 699 instances, but since there are 16 instances have missing values, in this study only 683 instances are used.

The dataset is divided into two classes namely 444 instances of benign class, and 239 instances of malignant class. Furthermore, the dataset is split into two groups namely the training group and the test group, where each group consists of a benign and malignant class. There are three types of training-test partition, i.e. 50-50%, 70-30%, and 80-20%.

Training groups are used to get feature patterns in the feature selection process. The result of this process is used as a reference for determining the number of features in the test group, and would be tested in the classification process. To measure the effectiveness level of the feature selection algorithm, it can be seen from the level of accuracy, sensitivity, and specificity, resulting from the classification process. The higher level of accuracy, sensitivity, and specificity gained, indicates that the feature selection algorithm used is effective, and vice versa.

2.3. Framework

The framework used in this work is the Waikato Environment for Knowledge Analysis (WEKA). This framework was developed at University of Waikato [8]. WEKA is open source (free license or freeware), designed, and developed based on JAVA, and special purpose is used for pattern recognition methods. WEKA supports various data mining functions, i.e. data pre-processing, classification, clustering, association, regression, feature selection, and visualization. Data points are explained by a number of fixed attributes, such as nominal, numeric, normally, and other attribute types.

2.4. Algorithms

2.4.1. Wrapper Subset Evaluation. Wrapper Subset Evaluation, commonly called Wrapper, is one type of feature selection method that uses a supervised learning algorithm as part of its evaluation function. This supervised learning algorithm is used as a "black box" system during the feature selection process [9,10]. The evaluation function of each feature will provide an approximate quality to the model created by the learning algorithm, so this would result in better accuracy forecasts. Besides using learning algorithm, this method also uses *k*-folds cross-validation evaluation. Although the Wrapper method can produce better accuracy, this method has a weakness, which causes the computation process is relatively longer and slower [9,10]. In general, the algorithm of Wrapper described as follows:

```
Wrapper Algorithm
I.S.:   D(F0,F1,...,Fn-1) //a training dataset with N features
        S0                //a subset from which to start the search
        δ                 //a stopping criterion
F.S.:   Sbest             //an optimal subset
begin
    initialize: Sbest = S0;
    γbest = eval(S0,D,A);//evaluate S0 by a mining algorithm A
    do begin
        S = generate(D);//generate a subset for evaluation
        γ = eval(S,D,A);//evaluate the current subset S by A
        if(γ is better than γbest)
            γbest = γ
            Sbest = S;
    end until(δ is reached);
    return Sbest;
end;
```

2.4.2. Naïve Bayes. Naïve Bayes is the algorithm developed from Bayesian theorem, which has a number of advantages, i.e. relatively fast training process, able to handle real and discrete data, uninfluenced by irrelevant features, as a strong assumption towards the independence of each features [11–13]. In simply, the formula for Naïve Bayes can be seen through the Equation (1).

$$Posterior = \frac{Prior \times Likelihood}{Evidence} \tag{1}$$

Posterior is a probability of the presence of the sample with a specific characteristic in a class. It is obtained by multiplying between the probability of the class appearance (*Prior*) and the probability of the emergence of sample characteristic in class (*Likelihood*), and then divided by the probability of the sample characteristic emergence in general (*Evidence*). Application of the Naïve Bayes algorithm was done as follows:

```
Naïve Bayes Algorithm
Learning Phase:
    for each class value C_i
         compute P(C_i)
         for each attribute X_j
              //compute distribution D_ij
              if attribute X_j is discrete
                        D_ij = categorical distribution for C_i and X_j
              else
                        D_ij = normal distribution for C_i and X_j
Testing Phase:
    given unknown X' = [x'_1, x'_2, ..., x'_n]
    estimate class = argmax_{C_i} P(C_i) · ∏_j D_ij
```

2.5. Performance Evaluation Parameters

2.5.1. Accuracy, Sensitivity, and Specificity. To evaluate the test, the statistic parameter such as accuracy, sensitivity, and specificity were used through Equation (2), Equation (3), and Equation (4) on each. Accuracy is a measurement to see the level of achievement of the classification being done. Sensitivity refers to the ability of classifier in predicting the positive class as the True Positive (TP), while specificity refers to the ability classifier in recognizing the negative class as the True Negative (TN).

$$Accuracy = \frac{TP + TN}{TP + TN + FP + FN} \times 100\% \tag{2}$$

$$Sensitivity = \frac{TP}{TP + FN} \times 100\% \tag{3}$$

$$Specificity = \frac{TN}{TN + FP} \times 100\% \tag{4}$$

2.5.2. K-Folds Cross-Validation. To gain better classification result, *k*-folds cross-validation was utilized. This method has several advantages, such as minimizing bias produced during the training process by using random sampling [14]. This method initiated by dividing all data randomly into *k* group, known as fold, with each fold is assumed to have the same number of members. The classification algorithm has executed the training and testing *k* times. In each process, one-fold is entered as data test while the other folds serve as training data. Each process will give different results. These values are then averaged and the average value is used as the accuracy value of the corresponding algorithm.

3. Result and Discussion

In the feature selection process, Multi-layer Perceptron (MLP) is utilized as part of the evaluation function of the Wrapper algorithm. MLP is chosen because it has several advantages, such as the ability to support parallel implementation, generalization capacity, fault tolerance, and able to handle the missing value [15]. Thus, it is suitable to deal with breast cancer dataset characteristics. To evaluate the effectiveness of our proposed method, we conducted several experiments. The feature selection and classification result can be seen in Table 2 and Table 3 respectively, while Table 4 shows the confusion matrix for each classification experiment.

In general, based on the result of Table 2, Table 3, number of accuracy and features increase with the increase of the training set size, while as we can see from Table 4, number of false negative (FN) decrease with the increase of the training set size. For the comparison purposes, Table 5 gives the classification accuracies of our proposed method and previous methods.

Table 2. Feature Selection Result

Features	Training-Test Partitions (%)		
	50-50	70-30	80-20
Types	Clump of Thickness Uniformity of Cell Size Marginal Adhesion Bare Nuclei	Uniformity of Cell Size Uniformity of Cell Shape Single Epithelial Cell Size Bare Nuclei Bland Chromatin	Clump of Thickness Uniformity of Cell Size Uniformity of Cell Shape Single Epithelial Cell Size Bare Nuclei Normal Nucleoli
Numbers	4	5	6

Table 3. Classification Result

Parameter	Training-Test Partition (%)		
	50-50	70-30	80-20
Folds	7	7	7
Benign	215	130	89
Malignant	117	72	47
Correctly Classified	332	202	136
Incorrectly Classified	9	3	1
Accuracy (%)	97.36	98.54	99.27
Sensitivity (%)	97.40	98.50	99.30
Specificity (%)	97.40	98.50	99.30

Table 4. Confusion Matrixes of Classification Result

Actual	Predict		Training-Test Partitions
	Benign	Malignant	
Benign	215	7	50-50%
Malignant	2	117	
Benign	130	3	70-30%
Malignant	0	72	
Benign	89	0	80-20%
Malignant	1	47	

Table 5. Comparison the Proposed Method with Previous Methods

Author (Year)	Method	Accuracy (%)
Quinlan (1996) [16]	C4.5	94.74
Hamilton *et al.* (1996) [17]	RIAC	95.00
Ster and Dobnikar (1996) [18]	LDA	96.80
Pena-Reyes and Sipper (1999) [19]	Fuzzy-GA	97.36
Goodman *et al.* (2002) [20]	AIRS	97.20
Abonyi and Szeifert (2003) [21]	Supervised Fuzzy Clustering	95.57
Akay (2008) [2]	F-score + SVM	97.95
Zheng *et al.* (2014) [5]	K-Means + SVM	97.28
This Study (2017)	**Wrapper + Naïve Bayes**	**99.27**

As in Table 3, we can see that the model using the 80%-20% part of the test partition yields better results than the model using other test partitions because according to the result, this partition model provides more useful features during the classification process (based on the result in Table 2). The number of features generated from the model affects the case of classification error as shown in Table 4, where only one instance is classified incorrectly. Obviously, this is because the data used for the training process more than the other partition model. It can therefore improve the ability of the system to identify the characteristics of breast cancer, as well as implicate the testing process, which gives more ability to distinguish the types of breast cancer.

As we can see from the result especially from Table 5, our proposed method using Wrapper and Naïve Bayes algorithms obtains the highest classification accuracy so far. Therefore, we conclude that Naïve Bayes with feature selection using Wrapper obtains promising results in classifying the potential breast cancer patients.

4. Conclusion and Future Work

This study proposes a scheme to classify breast cancer into two types, namely benign and malignant. A total of 9 features are extracted to facilitate the classification process. Feature selection based on Wrapper method is conducted to gain the relevant features which may contribute to improve the rate of classification result.

Three types of training-test partition are conducted, i.e. 50-50%, 70-30%, and 80-20%, which gain four, five, and six of selected features on each. By using Naïve Bayes for the classification process, obtained that the proposed method affords the highest classification accuracies on 80-20% training-test partition, it's up to 99.27%. Considering that result, we believe that the proposed method can be very helpful to the paramedics for their final decisions on their patients. By using such a tool, they can make very precise decisions.

In the next investigation, it will be better to do comparison with the other algorithms, so it is expected to increase the value of accuracy and can increase the number of datasets used.

References

[1] Society A C 2015 *Breast Cancer Facts & Figures 2015-2016* (Atlanta)
[2] Akay M F 2009 Support vector machines combined with feature selection for breast cancer diagnosis *Expert Syst. Appl.* **36** 3240–7
[3] Chen C-H 2014 A hybrid intelligent model of analyzing clinical breast cancer data using clustering techniques with feature selection *Appl. Soft Comput.* **20** 4–14
[4] Meselhy Eltoukhy M, Faye I and Belhaouari Samir B 2012 A statistical based feature extraction method for breast cancer diagnosis in digital mammogram using multiresolution representation *Comput. Biol. Med.* **42** 123–8
[5] Zheng B, Yoon S W and Lam S S 2014 Breast cancer diagnosis based on feature extraction

using a hybrid of K-means and support vector machine algorithms *Expert Syst. Appl.* **41** 1476–82

[6] Wolberg D W H 1992 Breast Cancer Wisconsin (Original) Data Set *UCI Mach. Learn. Repos.*

[7] Center for Machine Learning and Intelligent System U-I 2015 Machine Learning Data Repository [UCI] *Databib*

[8] Hall M, Frank E, Holmes G, Pfahringer B, Reutemann P and Witten I H 2009 The WEKA Data Mining Software: An Update *SIGKDD Explor.* **11**

[9] Hall M A and Holmes G 2003 Benchmarking attribute selection techniques for discrete class data mining *IEEE Trans. Knowl. Data Eng.* **15** 1437–47

[10] Naqvi G 2012 *A Hybrid Filter-Wrapper Approach for Feature Selection* (Sweden: Örebro University)

[11] Bishop C M 2006 *Pattern Recognition and Machine Learning* (New York: Springer)

[12] Duda R O, Hart P E and Stork D G 2001 *Pattern Classification and Scene Analysis* (New York: John Wiley & Sons, Inc.)

[13] Natalius S 2010 Metoda Naive Bayes Classifier dan Penggunaannya pada Klasifikasi Dokumen

[14] Polat K, Şahan S and Güneş S 2007 A novel hybrid method based on artificial immune recognition system ({AIRS}) with fuzzy weighted pre-processing for thyroid disease diagnosis *Expert Syst. Appl.* **32** 1141–7

[15] Isa I S, Fauzi N A, Sharif J M, Baharudin R and Abbas M H 2012 Comparisons of MLP transfer functions for different classification classes *2012 IEEE International Conference on Control System, Computing and Engineering (ICCSCE)* pp 110–4

[16] Quinlan J R 1996 Improved use of continuous attributes in C4.5 *J. Artif. Intell. Res.* 77–90

[17] Hamilton H J, Shan N and Cercone N 1996 *RIAC: A rule induction algorithm based on approximate classification*

[18] Ster B and Dobnikar A 1996 Neural networks in medical diagnosis: Comparison with other methods *Proceedings of the International Conference on Engineering Applications of Neural Networks* pp 427–30

[19] Pena-Reyes C A and Sipper M 1999 A fuzzy-genetic approach to breast cancer diagnosis *J. Artif. Intell. Med.* 131–55

[20] Goodman D E, Boggess L and Watkins A 2002 Artificial Immune System Classification of Multiple-class Problems *Proceedings of the Artificial Neural Networks in Engineering* pp 179–83

[21] Abonyi J and Szeifert F 2003 Supervised fuzzy clustering for the identification of fuzzy classifiers *J. Pattern Recognit. Lett.* **14** 2195–207

α-Sutte Indicator: a new method for time series forecasting

A S Ahmar[1], A Rahman[2] and U Mulbar[2]

[1]Department of Statistics, Universitas Negeri Makassar, Makassar, Indonesia
[2]Department of Mathematics, Universitas Negeri Makassar, Makassar, Indonesia

E-mail: ansarisaleh@unm.ac.id

Abstract. α-Sutte Indicator (α-Sutte) was originally from the development of Sutte Indicator. Sutte Indicator can be used to predict the movement of stocks. As the development of science, Sutte Indicator is developed to predict not only the movement of stocks but also able to forecast the data on finance, insurance, and time series data. This method is called α-Sutte Indicator (α- Sutte). α-Sutte was developed by using the principle of the forecasting method of using the previous data. α-Sutte choose because it have best accuracy in forecasting. To implementation and check the accuracy of this method, the Consumer Price Index and Indices by Regions (Turkey) data used in this research. Data taken from January 2003 to June 2017. This data used because CPI data unstable and sometimes difficult to predict using other method. The data is divided into two parts, namely data training and testing. Data training starts from January 2003 to October 2016, and data testing starts from November 2016 to June 2017. To see the accuracy of α-Sutte, benchmark will be conducted from the results of forecasting with other forecasting methods e.g. Multiplicative Holt-Winters, Additive Holt-Winters, and Automatic Time Series Forecasting: The forecast Package for R (AutoARIMA) developed by Hyndman- Khandakar (2008) for the ARIMA/SARIMA model. The comparison of this accuracy is to compare the value of MSE, RMSE, and MAPE of forecasting the result on data training by using the data as reference data. The results are obtained that MSE value of Indicator of α-Sutte is smaller than other methods that is Multiplicative Holt-Winters, Holt-Winters Additive, and ARIMA $(1,2,2)(0,0,2)^{12}$. Thereby the case for MAPE accuracy value Indicator of α-Sutte better than others method: Multiplicative Holt-Winters, Holt-Winters Additive, and ARIMA $(1,2,2)(0,0,2)^{12}$.

1. Introduction

Based on data from the World Bank, Indonesia is categorized as a low-middle income country together with other developing countries, such as India and Myanmar. To see the main economic indicator performance of a country, it is important to consider consumer price index (CPI). CPI is an index that is used for measuring the level of price change of any goods and services from time to time [1]. Furthermore, according to International Labor Organization, CPI is the main statistics that are used as Policy takers in social, economic, and have wide and substantial effect for governments, businesses, and household [1]. This CPI is related to the inflation of certain area. The higher inflation rate, the higher CPI rate. The lower of certain CPI, the better for them towards a country because a low CPI indicates that life costs is increasingly cheaper. Therefore, Inflation and CPI are one of the indicators to see the economic condition of a country.

CPI and Inflation experience a change from time to time and a stagnant CPI / Inflation sometimes it suddenly changes. To look the change of this from time to time, forecasting data needs to be done.

Forecasting data about CPI has been done by various experts, i.e. modeling and forecasting consumer price index which is done by Norbert, Wanjoya, &Waititu in 2016, a case study in Rwanda, it results that ARIMA (3.1.4) is a suitable model to be used in forecasting the CPI historical data in Rwanda [2]. The other research is done by Alvares-Diaz and Gupta in 2016, they conducted a research about forecasting CPI in US economy using Random Walk (RW) [3], seasonal integrative moving average (SARIMA))[4], autoregressive (AR), genetic programming (GP) [5], [6] and non linear (artificial neural network(ANN) [7], [8]. The result that is obtained from this research is that SARIMA model is more superior if it is compared with other methods. The purpose of this research is to implement new forecasting method (α- Sutte) to predict CPI Turkey data.

2. Literature

2.1. ARIMA and SARIMA

A process Z_t is stated as following a mixed-autoregressive-moving average or ARMA (p,q) model, if it meets [9], [10]:

$$\phi_p(B)Z_t = \theta_q(B)a_t \tag{1}$$

where $\phi_p(B) = (1-\phi_1 B - \phi_2 B^2 - ... - \phi_p B^p)$ (for AR(p))
and $\theta_q(B) = (1-\theta_1 B - \theta_2 B^2 - ... - \theta_q B^q$ (for MA(q))

If there is a non-seasonal difference, so that it becomes ARIMA(p,d,q) model as follows [4]:

$$\phi_p(B)(1-B)^d Z_t = \theta_q(B)a_t \tag{2}$$

where $\phi_p(B) = (1-\phi_1 B - \phi_2 B^2 - ... - \phi_p B^p)$ (for AR(p)), $(1-B)^d$ (for differencing non seasonal) and $\theta_q(B) = (1-\theta_1 B - \theta_2 B^2 - ... - \theta_q B^q$ (for MA(q)).

Seasonal ARIMA model (SARIMA) is the same as non-seasonal ARIMA model. A process Z_t is stated as following a mixed autoregressive-moving average model that is expanded for handling seasonal aspect or ARMA (p,d)(P,D)^s if it satisfies [4]:

$$\Phi_P\left(B^s\right)\phi_p(B)Z_t = \theta_q(B)\Theta_Q\left(B^s\right)a_t \tag{3}$$

If there is a seasonal difference on Seasonal ARIMA model, so the model becomes as follows:

$$\Phi_P\left(B^s\right)\phi_p(B)(1-B)^d\left(1-B^s\right)^D \dot{Z}_t = \theta_q(B)\Theta_Q\left(B^s\right)a_t$$

where

$$\dot{Z}_t = \begin{cases} Z_t - \mu, \text{if } d = D = 0 \\ Z_t, \text{otherwise.} \end{cases}$$

$$\Phi_P(B^s) = 1 - \Phi_1 B^s - \Phi_2 B^{2s} - ... - \Phi_P B^{Ps}$$

$$\text{and } \Theta_Q(B^s) = 1 - \Theta_1 B^s - \Theta_2 B^{2s} - ... - \Theta_Q B^{Qs}$$

2.2. Holt-Winters

There are 2 types of Holt-Winters methods applied on this research namely Multiplicative Holt-Winters (MHW) and Additive Holt-Winters (AHW). The formula of Multiplicative Holt-Winters (MHW) is elaborated in the following equation [11]:

$$L_t = \alpha\left(\frac{Y_t}{S_{t-s}}\right) + (1-\alpha)(L_{t-1} + b_{t-1}) \tag{4}$$

$$b_t = \beta(L_t - L_{t-1}) + (1-\beta)b_{t-1} \tag{5}$$

$$S_t = \gamma\left(\frac{Y_t}{L_t}\right) + (1-\gamma)S_{t-s} \tag{6}$$

$$F_{t+m} = (L_t + b_t m)S_{t-s+m} \tag{7}$$

and Additive Holt-Winters:

$$L_t = \alpha(Y_t - S_{t-s}) + (1-\alpha)(L_{t-1} + b_{t-1}) \tag{8}$$

$$b_t = \beta(L_t - L_{t-1}) + (1-\beta)b_{t-1} \tag{9}$$

$$S_t = \gamma(Y_t - L_t) + (1-\gamma)S_{t-s} \tag{10}$$

$$F_{t+m} = L_t + b_t m + S_{t-s+m} \tag{11}$$

where:
Y_t = data on t time,
s = the seasonal length in a certain time,
m = the amount of data to be predicted.

2.3. α-Sutte Indicator

α-Sutte Indicator is the development of Sutte Indicator that was developed in 2015 by Ahmar [12]. Sutte Indicator initially developed for predicting the movement of stock [12]. Along with its time development and needs, Sutte Indicator is developed to be α-Sutte Indicator [13]–[17]. The development is expected to provide a better accuracy level and not only limited to the prediction of stock movement but also predict the time series of data. The development of this α-Sutte Indicator considers the trend of certain data. Moreover, the formula of α-Sutte Indicator is as follows.

$$a_t = \frac{\alpha\left(\frac{\Delta x}{\frac{\alpha+\delta}{2}}\right) + \beta\left(\frac{\Delta y}{\frac{\beta+\alpha}{2}}\right) + \gamma\left(\frac{\Delta z}{\frac{\gamma+\beta}{2}}\right)}{3} \tag{12}$$

Where :
$\delta = a_{t-4}$
$\alpha = a_{t-3}$
$\beta = a_{t-2}$
$\gamma = a_{t-1}$

$$\Delta x = \alpha - \delta = a_{t-3} - a_{t-4}$$
$$\Delta y = \beta - \alpha = a_{t-2} - a_{t-3}$$
$$\Delta z = \gamma - \beta = a_{t-1} - a_{t-2}$$

a_t = series observations at $t$time
a_{t-k} = series observations at $(t - k)$ time

3. Method

The data of the research is obtained from TURKSTAT (http://www.turkstat.gov.tr). The data used on the research is divided into two parts namely data training and testing. Data training used 166 data (month) and testing data will be tested by 8 data in the following (month). From this data testing, the results will be obtained from data training. This forecasting result is then compared with the actual data. In the process of forecasting data, this research uses the R package for α-Sutte Indicator namely: sutteForecastR package [17] and RcmdrPlugin.sutteForecastR [17] and R package for automatic time series forecasting [18].

The forecasting method that is used in this research i.e. ARIMA, Holt-Winters, SARIMA, and α-Sutte Indicator. To evaluate the accuracy level of forecasting result measurement from the applied method, so that the accuracy measurement method that is used i.e. MSE (Mean Squared Error), MAPE (Mean Absolute Percentage Error), and RMSE (Root Mean Squared Error).

The accuracy level of forecasting measurement that is often used is MSE [10].

$$\text{MSE} = \frac{1}{N}\sum_{t=1}^{N}(Y_t - F_t)^2,\ t = 1,2,3,..,N \tag{13}$$

Y is the actual value, Ft is the forecasting value, and N is the number of samples.

MAPE is relative error size of a forecasting and it is defined as:

$$\text{MSE} = \frac{1}{N}\sum_{t=1}^{N}\left|100\frac{Y_t - F_t}{Y_t}\right|[\%],\ t = 1,2,3,..,N \tag{14}$$

RMSE is alternative method for evaluating the result of a forecasting by rooting the value of MSE. The smaller MSE, MAPE and RMSE value, the better forecasting result is obtained or in other words, the better the accuracy level.

$$\text{RMSE} = \sqrt{MSE} \tag{15}$$

4. Result and Discussion

In this research, data training starts from January 2003 to October 2016, and data are tested from November 2016 to June 2017. The first stage in forecasting is that perform plotting data to data training. Plotting Data is conducted to see the trend of certain data. Moreover, the plotting of CPI Turkey data is as follows.

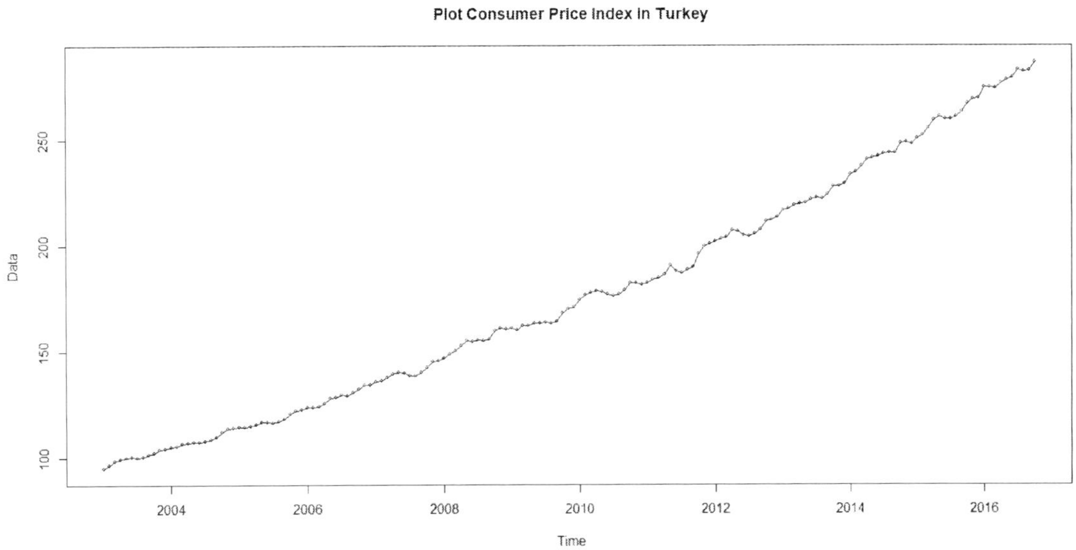

Figure 1. The plot of the consumer price index in Turkey (January 2003 - October 2016)

Based on Figure 1, CPI Turkey experiences an upward trend from year to year so it needs to be considered about the influence of trend in forecasting data. The next stage is forecasting data by using ARIMA/SARIMA method, Holt-Winters and α-Sutte Indicator. This data forecasting uses the R software.

Moreover, the forecasting results of R software from this research by using ARIMA/SARIMA, Holt-Winters and α-Sutte Indicator are as follows.

```
Holt-Winters exponential smoothing with trend and multiplicative seasonal
component.

Call:
HoltWinters(x = data.ts, seasonal = "mult")

Smoothing parameters:
alpha: 0.9044907
beta : 0.03339319
gamma: 1

Holt-Winters exponential smoothing with trend and additive seasonal
component.

Call:
HoltWinters(x = data.ts, seasonal = "additive")

Smoothing parameters:
alpha: 0.8594444
beta : 0.03480424
gamma: 1

Series: data.ts
ARIMA(1,2,2)(0,0,2)[12]

Coefficients:
```

```
          ar1       ma1       ma2     sma1    sma2
      -0.5829   -0.1767   -0.7924   0.2632  0.2616
s.e.   0.1772    0.1329    0.1293   0.0904  0.0733

sigma^2 estimated as 1.496:  log likelihood=-265.62
AIC=543.23   AICc=543.77   BIC=561.83
```

From the output of R Software for Multiplicative Holt-Winters method (MHW or HW-Mul), it is obtained α = 0.9044907, β = 0.03339319, and γ = 1 parameter, while for Additive Holt-Winters method (AHW or HW-Add), it is obtained α = 0.8594444, β = 0.03480424, and γ = 1 parameter. For ARIMA forecasting, it is obtained model namely ARIMA (1,2,2) $(0,0,2)^{12}$. If it is modeled in mathematical form, it will be as follows:
Multiplicative Holt-Winters:

$$L_t = 0.9044907\left(\frac{Y_t}{S_{t-s}}\right) + (0,095509)(L_{t-1} + b_{t-1})$$

$$b_t = 0.03339(L_t - L_{t-1}) + (0.966607)b_{t-1}$$

$$S_t = \frac{Y_t}{L_t}$$

$$F_{t+m} = (L_t + b_t m)S_{t-s+m}$$

Additive Holt-Winters:

$$L_t = 0.85944(Y_t - S_{t-s}) + (0.140556)(L_{t-1} + b_{t-1})$$

$$b_t = 0.0348(L_t - L_{t-1}) + (0.965196)b_{t-1}$$

$$S_t = Y_t - L_t$$

$$F_{t+m} = L_t + b_t m + S_{t-s+m}$$

SARIMA:

$$\phi_1(B)(1-B)^2 \dot{Z}_t = \theta_2(B)\Theta_2(B^{12})a_t$$

$$(1-\phi_1 B)(1-B)^2 \dot{Z}_t = (1-\theta_1 B - \theta_2 B^2)(1-\Theta_1 B^{12} - \Theta_2 B^{24})$$

$$Z_t - (\phi_1 + 2)Z_{t-1} + (1+2\phi_1)Z_{t-2} - \phi_1 Z_{t-3} =$$

$$a_t - \theta_2 a_{t-2} - \Theta_1 a_{t-12} - \theta_1\Theta_1 a_{t-13} + \Theta_1 a_{t-14} - \theta_2\Theta_2 a_{t-24} + \theta_1\Theta_2 a_{t-25} + \theta_2\Theta_2 a_{t-26}$$

$$Z_t = (\phi_1 + 2)Z_{t-1} - (1+2\phi_1)Z_{t-2} + \phi_1 Z_{t-3} + a_t - \theta_2 a_{t-2} - \Theta_1 a_{t-12}$$
$$-\theta_1\Theta_1 a_{t-13} + \Theta_1 a_{t-14} - \theta_2\Theta_2 a_{t-24} + \theta_1\Theta_2 a_{t-25} + \theta_2\Theta_2 a_{t-26}$$

$$Z_t = (-0.5829 + 2)Z_{t-1} - (1 + 2(-0.5829))Z_{t-2} - 0.5829Z_{t-3} + a_t$$
$$-(-0.7924)a_{t-2} - 0.2632a_{t-12} - (-0.1767)(0.2632)a_{t-13} + (0.2632)a_{t-14}$$
$$-(-0.7924)(0.2616)a_{t-24} + (-0.1767)(0.2616)a_{t-25} + (-0.7924)(0.2616)a_{t-26}$$

$$Z_t = 1,4171Z_{t-1} + 0,1658Z_{t-2} - 0.5829Z_{t-3} + a_t$$
$$+0.7924a_{t-2} - 0.2632a_{t-12} + 0,0462a_{t-13} + 0.2632a_{t-14}$$
$$+0,2073a_{t-24} - 0,0462a_{t-25} - 0,2073a_{t-26}$$

Moreover, the forecasting results to 8 months later by using ARIMA/SARIMA, Holt-Winters and α- Sutte Indicator method are displayed in Table 1 and Figure 2.

Table 1. The data of forecasting result for 8 months later

Date	Data	α-Sutte	Auto ARIMA	MHW (HW-Mul)	AHW (HW-Add)
11-2016	286,33	286,33	286,33	286,33	286,33
12-2016	287,81	287,59	280,66	288,04	287,44
01-2017	292,54	289,84	282,16	288,58	287,90
02-2017	299,74	295,99	283,57	291,81	290,84
03-2017	302,17	304,25	285,01	292,34	291,35
04-2017	305,24	307,00	286,44	293,90	292,84
05-2017	309,23	309,51	287,88	296,11	295,04
06-2017	310,61	312,41	289,31	296,77	295,90

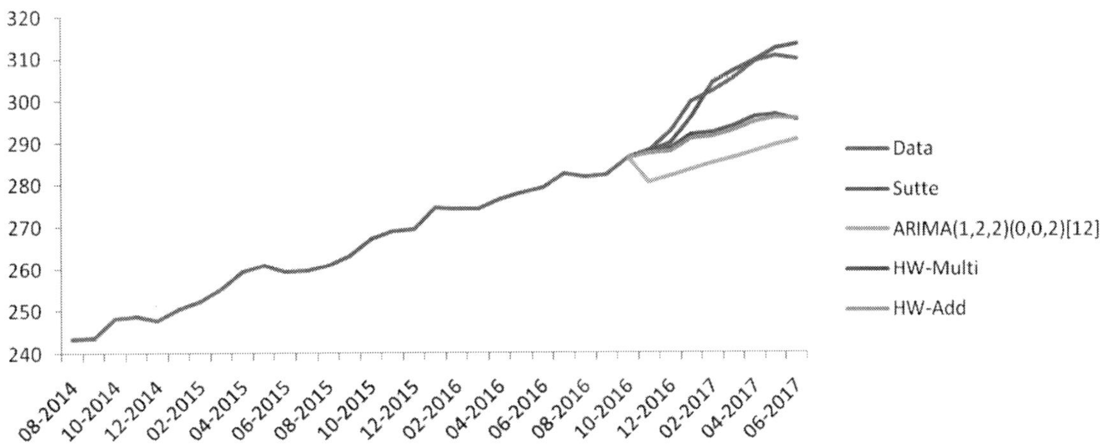

Figure 2. The graph of the comparison from the forecasting result by α-Sutte Indicator and other methods in Consumer Price Index in Turkey (November 2016 - June 2017)

In table 1 and figure 2, it is shown that forecasting result from α-Sutte Indicator is more accurate if it is compared with other forecasting methods (ARIMA/SARIMA and Holt-Winters). The comparison of accuracy level from each forecasting method is displayed in Table 2.

Table 2. The Accuracy Comparison of α - Sutte Forecasting Result with Other Methods

Type	MSE	RMSE	MAPE (%)
MHW (HW-Mul)	109,48	10,46	3,05
AHW (HW-Add)	123,29	11,10	3,27
α-Sutte Indicator	5,70	2,39	0,67
ARIMA(1,2,2)(0,0,2)12	292,51	17,10	5,40

From table 2 and figure 3, it can be seen that the forecasting result of α-Sutte Indicator is far different if it is compared with other methods (ARIMA/SARIMA and Holt-Winters) either from accuracy level of MSE, RMSE, or MAPE.

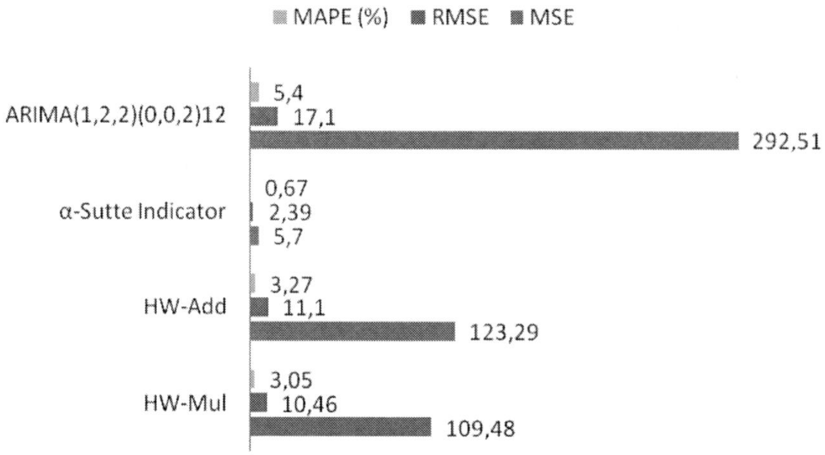

Figure 3. The graph of the accuracy comparison of α-Sutte forecasting result with other methods

5. Conclusion

Based on the analysis about forecasting result to Consumer Price Index data in Turkey, α-Sutte Indicator forecasting method is more suitable if it is compared to other methods i.e. Multiplicative Holt-Winters, Additive Holt-Winters and ARIMA method (1,2,2) $(0,0,2)^{12}$. This can be seen from the data result of forecasting for 8 months later (November 2016 - June 2017) and from this result, it is obtained MSE value for every forecasting method namely: α- Sutte Indicator at 5.70, Multiplicative Holt-Winters at 109.48, Additive Holt-Winters at 123.29, and ARIMA (1,2,2) $(0,0,2)^{12}$ at 292.51. For MAPE value, it is obtained the accuracy as follows: α-Sutte Indicator at 0.67, Multiplicative Holt-Winters at 3.05, Additive Holt-Winters at 3.27, and ARIMA (1,2,2) $(0,0,2)^{12}$ at 5,40.

References

[1] International Labour Organization, 2004 *Consumer price index manual: Theory and practice* International Monetary Fund.

[2] Norbert H Wanjoya A and Waititu A, 2016 Modeling and Forecasting Consumer Price Index (Case of Rwanda) *Am. J. Theor. Appl. Stat.* **5**, 3 p. 101–107.

[3] Álvarez-Díaz M and Gupta R, Oct. 2016 Forecasting US consumer price index: does nonlinearity matter? *Appl. Econ.* **48**, 46 p. 4462–4475.

[4] Ahmar A S *et al.*, 2018 Modeling Data Containing Outliers using ARIMA Additive Outlier (ARIMA-AO) *J. Phys. Conf. Ser.* **954**, 1 p. 01010.

[5] Kaboudan M A, 2003 Genetic Programming Software to Forecast Time Series in *Computing in Economics and Finance*.

[6] Rivero D Rabunal J R Dorado J and Pazos A, 2005 Time Series Forecast with Anticipation Using Genetic Programming in *Computational Intelligence and Bioinspired Systems, 8th International Work-Conference on Artificial Neural Networks, IWANN 2005, Proceedings* **3512** p. 968–975.

[7] Haviluddin Agus F Azhari M and Ahmar A S, 2018 Artificial Neural Network Optimized Approach for Improving Spatial Cluster Quality of Land Value Zone *Int. J. Eng. Technol.* **7**, 2.2 p. 80–83.

[8] Surahman Viddy A Gaffar A F O Haviluddin and Ahmar A S, 2018 Selection of the best supply chain strategy using fuzzy based decision model *Int. J. Eng. Technol.* 7, 2.2.

[9] Ahmar A S *et al.*, 2018 Modeling Data Containing Outliers using ARIMA Additive Outlier (ARIMA-AO) *J. Phys. Conf. Ser.* **954**, 1.

[10] Rahman A and Ahmar A S, 2017 Forecasting of primary energy consumption data in the United States: A comparison between ARIMA and Holter-Winters models in *AIP Conference*

Proceedings **1885**.

[11] Ferbar Tratar L and Strmčnik E, 2016 The comparison of Holt–Winters method and Multiple regression method: A case study *Energy* **109**, Supplement C p. 266–276.

[12] Ahmar A S Rahman A Arifin A N M and Ahmar A A, 2017 Predicting movement of stock of "Y" using sutte indicator *Cogent Econ. Financ.* **5**, 1.

[13] Ahmar A S, 2017 SuIndiWeb: A Web-Based Platform of Sutte Indicator Predicting Movement of Stock in *Proceedings of the 2nd International Conference on Education, Science, and Technology (ICEST 2017)* **149**, Advances in Social Science, Education and Humanities Research p. 237–239.

[14] Sutiksno D U Ahmar A S Kurniasih N Susanto E and Leiwakabessy A, 2018 Forecasting Historical Data of Bitcoin using ARIMA and α-Sutte Indicator *J. Phys. Conf. Ser.* **1028**, 1 p. 012194.

[15] Ahmar A S, 2018 A Comparison of α-Sutte Indicator and ARIMA Methods in Renewable Energy Forecasting in Indonesia *Int. J. Eng. Technol.* **7**, 1.6 p. 20–22.

[16] Kurniasih N Ahmar A S Hidayat D R Agustin H and Rizal E, 2018 Forecasting Historical Data of Bitcoin using ARIMA and α-Sutte Indicator *J. Phys. Conf. Ser.* **1028**, 1 p. 012195.

[17] Ahmar A S, 2018 RcmdrPlugin.sutteForecastR: a plugin in Rcmdr for Forecasting Data *J. Phys. Conf. Ser.* **1028**, 1 p. 012224.

[18] Hyndman R J and Khandakar Y, 2008 Automatic time series forecasting : the forecast package for R *J. Stat. Softw.* **27**, 3 p. 1–22.

IOP Conf. Series: Journal of Physics: Conf. Series **1040** (2018) 012019 doi:10.1088/1742-6596/1040/1/012019

On super (a, d)-edge-antimagic total labeling of Möbius ladder

P K Dewi, I N Suparta and K Y Sumalasia

Mathematics Department, Mathematics and Sciences Faculty, Universitas Pendidikan Ganesha, Bali, Indonesia - 81116

E-mail: kartika.dewi@undiksha.ac.id

Abstract. A Möbius ladder is a simple graph obtained by introducing a twist in a prism graph. A Möbius ladder which has n pairs vertices is denoted by M_n. A graph G is called (a, d)-edge-antimagic total if there exists a bijection f from $V(G) \cup E(G)$ to $\{1, 2, ..., |V(G) + E(G)|\}$ such that the edge-weights, $w_f(uv) = f(u) + f(uv) + f(v)$, $uv \in E(G)$, forms an arithmetic sequence with the first term a and common difference d. Further, it called super if $f(V) = \{1, 2, ..., |V(G)|\}$. In this paper we proved that M_n is super (a, d)-edge-antimagic total with $0 \leq d \leq 2$. Further, we showed that for every even n, M_n is not (a, 0)-edge-antimagic total.

1. Introduction

Let $G(V, E)$ be a graph, where $V(G)$ is the set of vertices and $E(G)$ is the set of edges. A labeling on a graph $G(V, E)$ is an assignment of $V(G)$, $E(G)$, or $V(G) \cup E(G)$ to a set of non negative integers. If the domain is $V(G)$, then it is called the vertex labeling. Respectively, if the domain is $E(G)$, then it is called the edge labeling. While, if the domain is $V(G) \cup E(G)$, then it is called the total labeling.

An example of labeling is antimagic labeling which was introduced by Hartsfield and Ringel [1]. A graph $G(V, E)$ is called antimagic if every $uv \in E(G)$ can be labeled with distinct member of $\{1, 2, ..., |E(G)|\}$ such that the vertex-weights which is the sum of the edge labels' incident to each vertex are distinct. Galian [2] summaries some antimagic graph, such as P_n for $n \geq 3$, C_n, K_n for $n \geq 3$, etc. If the edge-weights form an arithmetic sequence, then the graph $G(V, E)$ is called (a, d)-antimagic. This concept was introduced by Bodendiek and Walther [3]. Bodendiek and Walther [3], [4] and [5] proved that parachutes $P_{g,p}$ is (a, d)-antimagic for some certain classes, while P_{2n}, C_{2n}, and stars are not (a, d)-antimagic. Other concepts of this kind of labeling were proposed in [6], [7] and [8]. There are (a, d)-vertex-antimagic total labeling, super (a, d)-vertex-antimagic total labeling, (a, d)-edge-antimagic total labeling, super (a, d)-edge-antimagic total labeling and (a, d)-edge-antimagic vertex labeling. The Definition of super (a, d)-edge-antimagic total graph was given in [9], [6],[7] and [10].

Definition 1.1. A graph $G(V, E)$ is called (a, d)-edge-antimagic total if there exists a bijection f from $V(G) \cup E(G)$ to $\{1, 2, ..., |V(G) + E(G)|\}$ such that the edge-weight, $w_f(uv) = f(u) + f(uv) + f(v)$, $uv \in E(G)$, forms an arithmetic sequence with the first term a and common difference d. Further, it called super if $f(V) = \{1, 2, ..., |V(G)|\}$.

There were some studies of graphs which are labeled by super (a, d)-edge-antimagic total labeling. Bača, Kovář, Semaničová-Feňovčiková and Shafiq [11] proved that every even regular graph and every odd regular graph with a 1-factor are super $(a, 1)$-edge-antimagic total. While Dafik, Miller, Ryan, and Băca [9] proved that some mC_n are super $(a, 1)$-edge-antimagic total and Băca and Barrientos [12] studied some properties of super (a, d)-edge-antimagic total labeling on mK_n.

Our focus in this paper is proving that Möbius ladder is a super (a, d)-edge-antimagic total and discover its properties. In the next chapter, we introduced the definition and properties of Möbius ladder.

2. Möbius ladder

A Möbius ladder is a simple graph obtained by introducing a twist in a prism graph. A Möbius ladder which has n pairs vertices is denoted by M_n. In the Figure 1 and Figure 2 below, we showed the comparison between a prism graph and a Möbius ladder which have 3 pairs vertices.

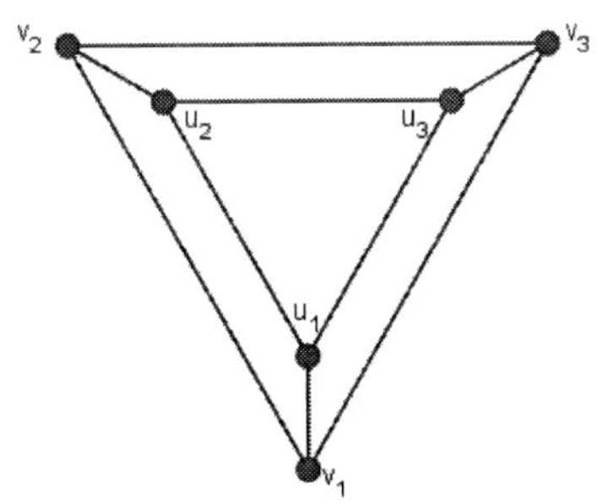

Figure 1. Graph Y_3, a prism graph with 3 pairs vertices.

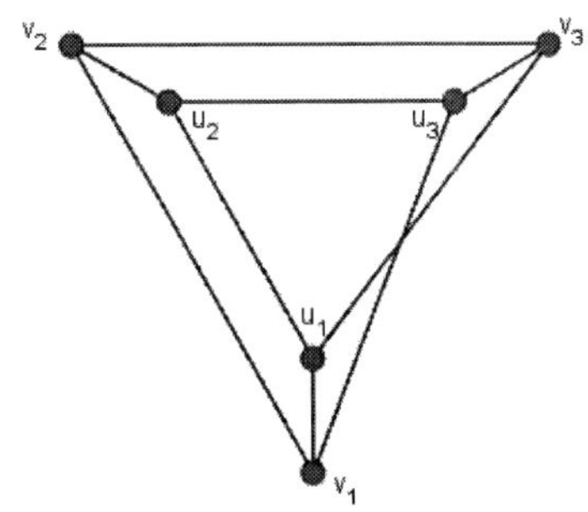

Figure 2. Graph M_3, a Möbius ladder with 3 pairs vertices.

A Möbius ladder which has n pairs of vertices has $3n$ edges and every vertex has degree 3. As we can see in Figure 3, graph C_k is subgraph of M_n for $4 \leq k \leq 2n$. These properties are important to study some properties of the super (a, d)-edge-antimagic total labeling on Möbius ladder.

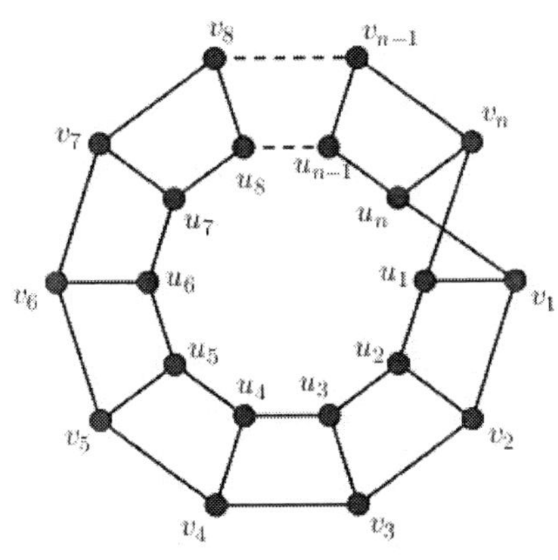

Figure 3. Graph M_n

The application of Möbius ladder can be found in chemistry and chemical stereography. Graph M_3 had been used by Walba to describe synthesis molecular structures [13]. While

IOP Conf. Series: Journal of Physics: Conf. Series **1040** (2018) 012019 doi:10.1088/1742-6596/1040/1/012019

Flapan [14] used Möbius ladder to describe molecular chilarity. Other application also can be found in integer programming [15]. Möbius ladder is used in describing a linear programming relaxation of the problem.

Some labelings on Möbius ladder have been studied, for instance Jayasekaran and Little Flower [16] proved that M_n is edge trimagic total. Pasotti [17] proved that M_n is graceful. While, Chunling, Xiaohui, Yuansheng and Liping [18] showed that when M_n is labeled with irregular labeling, the total edge irregularity strength of M_n is $n + 1$, while the total vertex irregularity strength of M_n is $\lceil \frac{n}{2} \rceil + 1$.

3. On super (a, d)-edge-antimagic total labeling on Möbius ladder

In order to define a labeling on M_n so that it satisfies super (a, d)-edge-antimagic total labeling, we are looking for a pattern by examining some examples.

Example 3.1. The Graph M_3 and M_4 are super $(a, 1)$-edge-antimagic total.

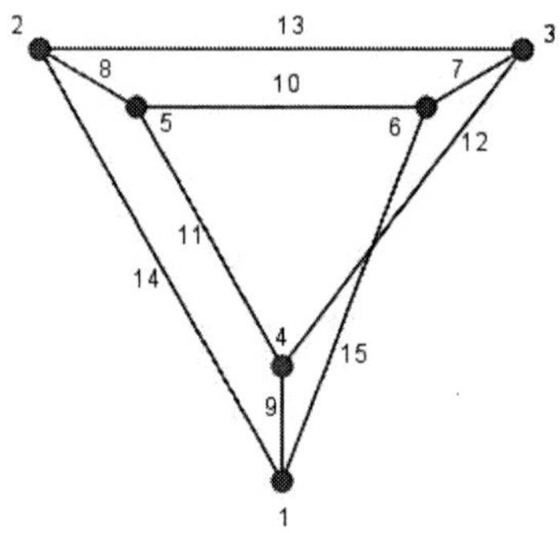

Figure 4. An $(a, 1)$-edge-antimagic total labeling on M_3.

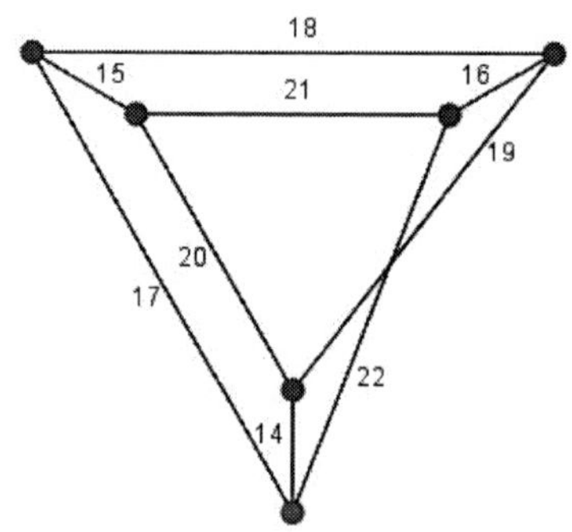

Figure 5. The corresponding edge-weights on M_3.

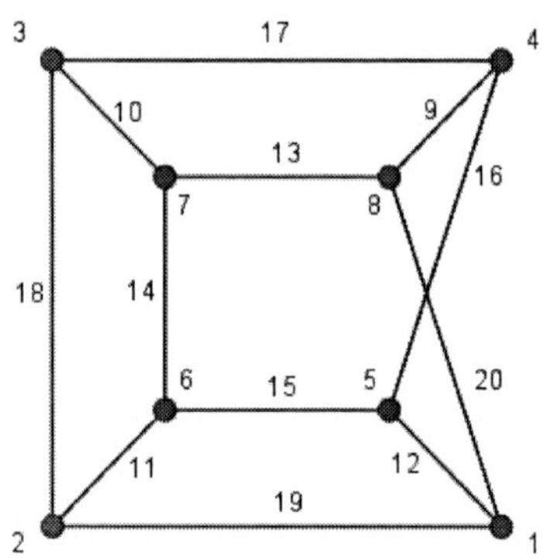

Figure 6. An $(a, 1)$-edge-antimagic total labeling on M_4.

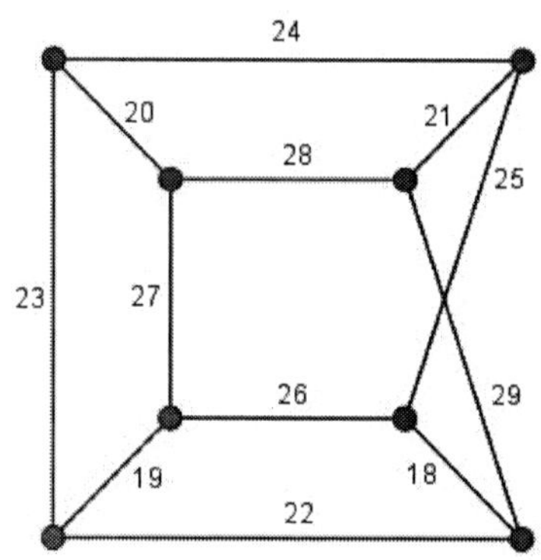

Figure 7. The corresponding edge-weights on M_4.

Let the labeling pattern be as in the Figure 4 and Figure 6. By calculating the edge-weight for each edges, we got arithmetic sequences. In the Figure 5, the edge-weights forms a sequence $\{14, 15, 16, ..., 22\}$, an arithmetic sequence with the first term 14 and common difference 1.

While in the Figure 7, the edge-weights forms a sequence $\{18, 19, 20, ..., 29\}$, an arithmetic sequence with the first term 18 and common difference 1. The set of vertex-labels of M_3 is $f(V(M_3)) = \{1, 2, ..., 6\}$ and the set of vertex-labels of M_4 is $f(V(M_4)) = \{1, 2, ..., 8\}$. So, we can conclude that both M_3 and M_4 are super $(a, 1)$-edge-antimagic total.

Theorem 3.1. *Graph M_n is super (a, d)-edge-antimagic total for every $n \geq 3$.*

Proof. Define a label f on $V(M_n) \cup E(M_n)$.

For $i = 1, 2, ..., n$,

$$\begin{aligned} f(v_i) &= i \\ f(u_i) &= n + i \\ f(u_iv_i) &= 3n - (i - 1) \end{aligned}$$

For $i = 1, 2, ..., n - 1$

$$\begin{aligned} f(v_iv_{i+1}) &= 5n - i \\ f(u_iu_{i+1}) &= 4n - i \end{aligned}$$

and

$$\begin{aligned} f(u_nv_1) &= 5n \\ f(u_1v_n) &= 4n \end{aligned}$$

Considering that on graph M_n, $E(G) = \{u_iu_{i+1}|i = 1, 2, ..., n - 1\} \cup \{v_iv_{i+1}|i = 1, 2, ..., n - 1\}$ $\cup$ $\{u_iv_i|i = 1, 2, ..., n\} \cup \{u_1v_n\} \cup \{u_nv_1\}$, then we calculated the edge-weights for each classes of edges.

For $\{u_iu_{i+1}\}$, $i = 1, 2, ..., n - 1$

$$\begin{aligned} w_f(u_iu_{i+1}) &= f(u_i) + f(u_{i+1}) + f(u_iu_{i+1}) \\ &= (n + 1) + (n + 1 + i) + (4n - i) \\ &= 6n + i + 1 \end{aligned}$$

For $\{v_iv_{i+1}\}$, $i = 1, 2, ..., n - 1$

$$\begin{aligned} w_f(v_iv_{i+1}) &= f(v_i) + f(v_{i+1}) + f(v_iv_{i+1}) \\ &= i + (i + 1) + (5n - i) \\ &= 5n + i + 1 \end{aligned}$$

For $\{u_iv_i\}$, $i = 1, 2, ..., n$

$$\begin{aligned} w_f(u_iv_i) &= f(u_i) + f(v_i) + f(u_iv_i) \\ &= (n + i) + i + (3n - (i - 1)) \\ &= 4n + i + 1 \end{aligned}$$

For $\{u_1v_n\}$,

$$\begin{aligned} w_f(u_1v_n) &= f(u_1) + f(v_n) + f(u_1v_n) \\ &= (n + 1) + n + 4n \\ &= 6n + 1 \end{aligned}$$

For $\{u_n v_1\}$,

$$\begin{aligned} w_f(u_n v_1) &= f(u_n) + f(v_1) + f(u_n v_1) \\ &= 2n + 1 + 5n \\ &= 7n + 1 \end{aligned}$$

Thus, the set of edge-weights $\{w_f\} = \{4n + i + 1 | i = 1, 2, ..., n\} \cup \{5n + i + 1 | i = 1, 2, ..., n - 1\}$ $\cup$ $\{6n + 1\}$ $\cup$ $\{6n + i + 1 | i = 1, 2, ..., n - 1\}$ $\cup$ $\{7n + 1\}$ forms an arithmetic sequence which $a = 4n + 2$, $d = 1$, and $f(V) = \{1, 2, ..., 2n\}$ where $2n$ is the order of M_n. So, we can conclude that M_n is a super $(a,1)$-edge-antimagic total. □

The Theorem 3.1 showed that M_n is a super $(a,\ d)$-edge-antimagic total with d equals 1, then we investigated the possibility of other values of d. We proved the range of d in the next following theorem.

Theorem 3.2. *Graph M_n is super $(a,\ d)$-edge-antimagic total with $0 \leq d \leq 2$.*

Proof. Suppose that M_n is labelled so that M_n is a super $(a,\ d)$-edge-antimagic total. We observed the minimum and the maximum possible edge weights of M_n.

The minimum possible edge weight : $1 + 2 + (2n + 1) = 2n + 4$

The maximum possible edge weight : $(2n - 1) + 2n + 5n = 9n - 1$

Notice that the number of edges is $3n$, thus

$$d \leq \frac{(9n - 1)}{2n + 1} = \frac{7n - 5}{3n - 1}$$

It is clear that $\frac{7(n-1)-5}{3(-1)-1} < \frac{7n-5}{3n-1}$ for all n. Notice that $\lim\limits_{n\to\infty} \frac{7n-5}{3n-1} = \frac{7}{3}$. Thus $\left\{\frac{7n-5}{3n-1}\right\}$ is an increasing sequence which its supremum is $\frac{7}{3}$. So, the maximum d is 2 and the possible interval of d so that M_n is a super $(a,\ d)$-edge-antimagic total is $0 \leq d \leq 2$. □

Regarding to the Theorem 3.1 and Theorem 3.2, if we want to prove the existence of super $(a,\ d)$-edge-antimagic total labeling for the other d, we should prove for $d = 0$ or $d = 2$. Respectively, if we want to prove that there is no super $(a,\ d)$-edge-antimagic total labeling except $d = 1$, we have to prove that it is not possible to have a super $(a,\ d)$-edge-antimagic total labeling for $d = 0$ or $d = 2$. By considering the properties of Möbius ladder, in the next following theorem, we proved that M_3 is a super $(a,\ d)$-edge-antimagic total with a unique $d = 1$.

Theorem 3.3. *Graph M_3 is not a super $(a,\ d)$-edge-antimagic total for $d = 0$ or $d = 2$.*

Proof. Suppose that M_3 is a super $(a,\ d)$-edge-antimagic total, we will proved that $d \neq 0$ and $d \neq 2$. The idea of our proof is checking that there is no possible labeling that satisfies M_3 to be a super $(a,\ d)$-edge-antimagic total with $d = 0$ or $d = 2$. The total sum of the edge-weights of M_3 is 162. Suppose that M_3 is a super $(a,\ 0)$-edge-antimagic total, then the edge-weights of each edge is 18. So, the combinations of edge-labels and vertex-labels are : (7, 5, 6), (8, 4, 6), (9, 3, 6) and these are unique, but we can not find combination for edge-label equals 10, because (10, 4, 4) is not possible since the vertex-labels have to be distinct. Notice that graph C_k is a subgraph of M_n for $4 \leq k \leq 2n$, so triplet (10, 5, 3) is not possible since it will make a C_3 becaused we have fixed the triplet (7, 5, 6) and (9, 3, 6) before. While, (10, 2, 6) is not possible, because the degree of each vertex is 3 and we have fixed triplets (7, 5, 6), (8, 4, 6), (9, 3, 6) before. Thus, M_3 is not a super $(a,\ 0)$-edge-antimagic total.

Suppose that M_3 is a super $(a,\ 2)$-edge-antimagic total. Then the edge-weights are 10, 12, 14, ..., 26. There are unique combinations for edge-weights 10, 12, 14, namely (1, 2, 7), (1, 3, 8), and (1, 4, 9). In the opposite way, there is no suitable combination for edge-weight 16, because triplets (1, 5, 10), (2, 4, 19), (3, 3, 10), (1, 4, 11), (2, 3, 11), (1, 3, 12), (2, 2, 12), (1, 2, 13) are not satisfying some properties of super (a, 2)-edge-antimagic total of Möbius ladder. Thus, M_3 is not a super $(a, 2)$-edge-antimagic total. Therefore we can conclude that there is a unique $d = 1$, so that M_3 is a super (a, d)-edge-antimagic total. □

In the next theorem, we proved that for all even n, graph M_n is not super $(a, 0)$-edge-antimagic total.

Theorem 3.4. *For all even n, graph M_n is not super $(a, 0)$-edge-antimagic total.*

Proof. Suppose that M_n is super $(a, 0)$-edge-antimagic total, thus $f(V(M_n)) = \{1, 2, ..., 2n\}$. Consider that the degree of each vertex is 3, so the total sum of all edge weights is $3(1 + 2 + ... + 2n) + (2n + 1) + (2n + 2) + ... + 5n = \frac{1}{2}3n(11n + 3)$. Since, the number of edges is $3n$ and we assumed $d = 0$, then for each edge, the edge weight is $\frac{1}{2}(11n + 3)$. Notice that n is even, then $(11n + 3)$ is odd, so $\frac{1}{2}(11n + 3)$ is not an integer. Since the edge-weight is always an integer, we can conclude that our supposition is false. □

4. Conclusion and future work

In this paper, we proved that M_n is super (a, d)-edge-antimagic total with $a = 4n + 2$ and $d = 1$. By considering the maximum and minimum edge weights, we showed that $0 \leq d \leq 2$. We proved that M_3 is super (a, d)-edge-antimagic total with a unique $d = 1$, and for every even n, M_n is not super $(a, 0)$-edge-antimagic total. Our future work is investigating whether for every even n, M_n is super $(a, 2)$-edge-antimagic total and the existence of super (a, d)-edge-antimagic total labelings for $d = 0$ or $d = 2$ on M_n, for every odd $n \geq 5$.

References

[1] Hartsfield N and Gerhard R 1990 *Pearls in Graph Theory* (San Diego: Academic Press)
[2] Galian J A 2016 *The Electronic Journal of Combinatorics* **#DS6**
[3] Bodendiek R and Walther G 1993 *antimagische Graphen Graphentheorie* **III**
[4] Bodendiek R and Walther G 1996 *Ars Combin* **42** 129–149
[5] Bodendiek R and Walther G 1998 *Mitt. Math. Ges. Hamburg* **17** 85–99
[6] Bača M and Miller M 2007 *Super Edge-Antimagic Graphs: A Wealth of Problems and Some Solutions* (Boca Raton: BrownWalker Press)
[7] Bača M, Baskoro E T, Miller M, Ryan J, Simanjuntak R and Sugeng K A 2006 *J. Indonesian Math. Soc.(MIHMI)* **12** 113–130
[8] Sugeng K A 2005 *Magic and Antimagic Labeling of Graphs* (Victoria: University of Ballarat)
[9] Dafik, Miller M, Ryan J and Bača M 2009 *Discrete Mathematics* **309** 4909–4915
[10] Bača M, Lin Y, Miller M and Youssef M Z 2007 *Discrete Mathematics* **307** 1232–1244
[11] Bača M, Ková P, Semaničová-Feovčíková A and Shafiq M K 2010 *Discrete Mathematics* **310** 1408–1412
[12] Bača M and Barrientos C 2008 *Discrete Mathematics* **308** 5032–5037
[13] Kawauchi A 1996 *Survey on Knot Theory* (Basel, Boston, Berlin, Birkäuser: Springer Science & Business Media)
[14] Flapan E 2000 *When Topology Meets Chemistry : A Topological Look at Molecular Chirality* (Cambridge, New York, Melbourne, Madrid, Cape Town, Singapore, São Paulo, Delhi: Cambridge University Press)
[15] Newman A and Vempala S 2001 *IPCO Conference Proceeding* pp 333–347
[16] Jayasekaran C and Little Flower J 2017 *Annals of Pure and Applied Mathematicss* **13** 151–163
[17] Pasotti A 2010 *Discrete Mathematics* **310** 3080–3087
[18] Chunling T, Xiaohui L, Yuansheng Y and Liping W 2009 *Indian J. Pure Appl Math* **40** 155–181

Graceful labeling for open superstar of complete bipartite graphs $S^2(t \cdot K_{m,n})$

I N Budayana, I N Suparta and N P H Purnamayanti

Departement of Mathematics, Universitas Pendidikan Ganesha, Indonesia

E-mail: nyoman.budayana@undiksha.ac.id

Abstract. A graceful labeling of graph G with m edges is a one-to-one function f from vertices of G to the set $\{0, 1, 2, \cdots, m\}$ such that the induced function f^* from edges of G to the set $\{1, 2, 3, \cdots, m\}$, which is defined by the difference of its incident vertices labels, is a bijection. An open superstar of complete bipartite graphs is an integration of complete bipartite graphs and superstar graphs. A superstar graph S_t^2 is a graph obtained by replacing each edge of star S_t by a path P_3 of three vertices. An open superstar of complete bipartite graph, $S^2(t \cdot K_{m,n})$, is a graph obtained by identifying each pendant vertex of superstar S_t^2 with exactly one vertex of a complete bipartite graph $K_{m,n}$. In this paper, we have given a graceful labeling for the open superstar of complete bipartite graphs $S^2(t \cdot K_{m,n})$ for all positive integers m, n, and odd positive integers t.

1. Introduction

A graph labeling is a mapping that carries a set of graph elements into a set of integers. There are various types of graph labeling, such as: graceful labeling, harmonious labeling, magic labeling, anti-magic labeling and irregular labeling. More result of graph labeling, can be found in Gallian [1].

The graceful labeling was introduced by Rosa [2] in 1967. A graceful labeling of graph G with m edges is a one-to-one function f from the vertices of G to the set $\{0, 1, 2, \cdots, m\}$ such that the induced function $f^* : E(G) \to \{1, 2, 3, \cdots m\}$, which is defined by $f^*(uv) = |f(u) - f(v)|$, is a bijection. If graph G can be labeled with graceful labeling, we say that G is graceful. In 1979, Bermond [3] conjecture that family of lobsters graph is graceful. In *A Dynamic Survey of Graph Labeling* by Gallian [1], there are a lot of graphs with graceful labeling, such as: trees (if less than 36 vertices), caterpillars, cycles C_n (iff $n \equiv 0, 3 (\mod 4)$), complete graphs K_n (iff $n \leq 4$) and complete bipartite graph $K_{m,n}$. Although numerous families of graceful graphs are known, a general necessary or sufficient condition to classify a graph is graceful has not yet been found.

A star S_t is graceful since it is a special case of caterpillar. Kaneria et. al [4] introduced a open star of graphs and proved that open star of complete bipartite graphs is graceful. Inspired by Kaneria et. al [4], we generalized the star graphs into superstar graph. A superstar S_t^2 is graceful since it is a special case of lobster graph. Then, we proved that open superstar of complete bipartite graphs is graceful, for the odd number of copies of complete bipartite graph, by finding a graceful labeling for this family of graphs.

In this paper, we denote the set of vertices of G by $V(G)$, the set of edges of G by $E(G)$, and uv is the incident edge of vertices u and v. The notation S_t denotes a star graph with t edges and $K_{m,n}$ denotes a complete bipartite graph.

IOP Conf. Series: Journal of Physics: Conf. Series **1040** (2018) 012020 doi:10.1088/1742-6596/1040/1/012020

2. Open superstar of complete bipartite graphs

Definition 2.1. A superstar graph S_t^2 is a graph obtained by replacing each edge of star S_t by a path P_3 of three vertices.

Illustration for Definition 2.1 is given in Figure 1 and Figure 2.

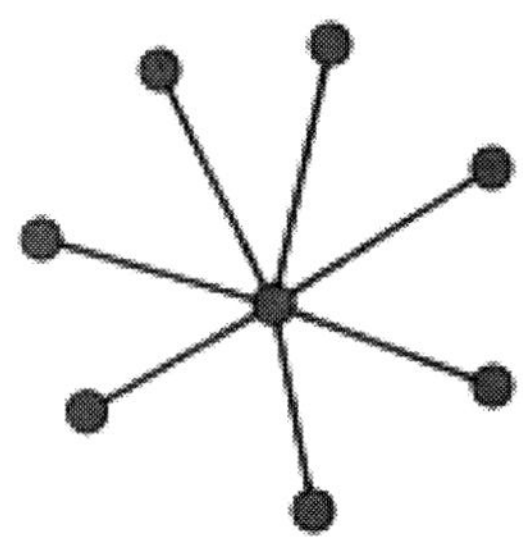

Figure 1. A star S_7.

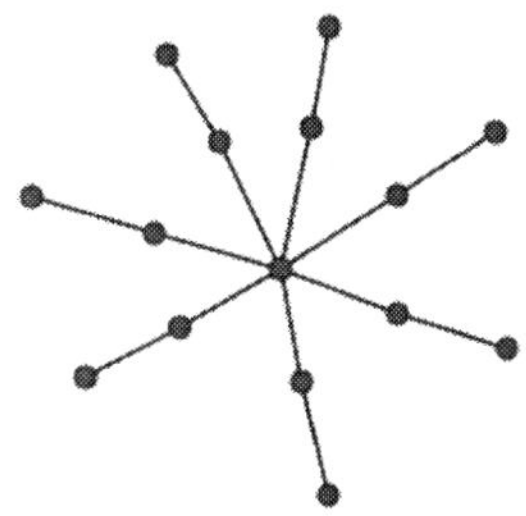

Figure 2. A superstar S_7^2.

It can be seen from the figures that a star S_t (Figure 1) is homeomorphic with a superstar S_t^2 (Figure 2). We can generalize the Definition 2.1 into the definition of superstar S_t^n. Therefore, analogue with definition S_t^2, S_t^n is a graph obtained by replacing each edge of star S_t by a path P_{n+1} of $n+1$ vertices. However in this paper, we just discusses about S_t^2.

Definition 2.2. An open superstar of complete bipartite graph $S^2(t \cdot K_{m,n})$ is a graph obtained by identifying each pendant vertex of superstar S_t^2 with exactly one vertex of a complete bipartite graph $K_{m,n}$.

For more understanding about open superstar of complete bipartite graphs, Figure 3 and Figure 4 show the illustration about how to construct a open superstar $S^2(7 \cdot K_{3,2})$.

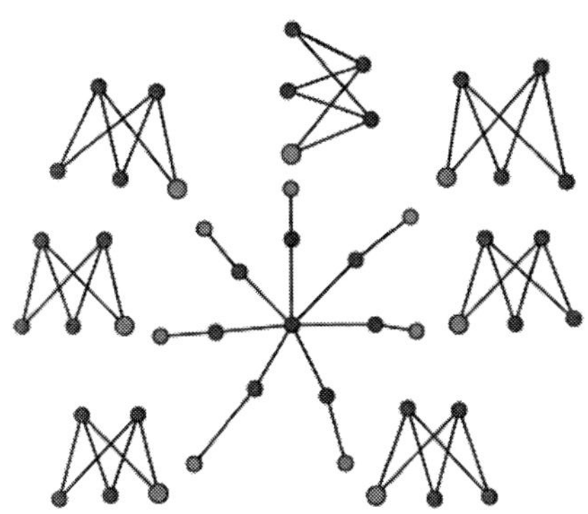

Figure 3. A superstar S_7^2 and seven copies of $K_{3,2}$.

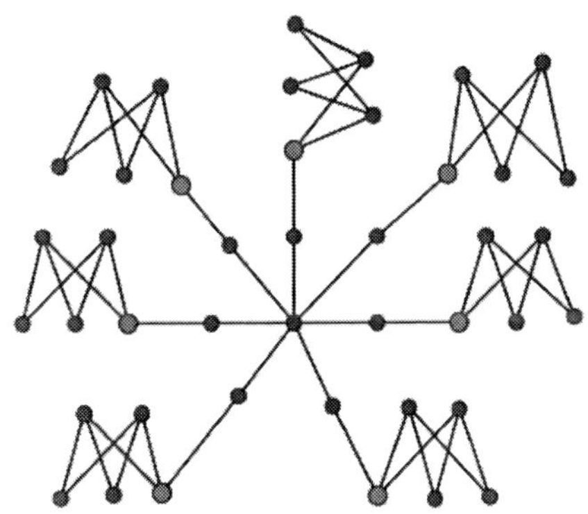

Figure 4. An open superstar $S^2(7 \cdot K_{3,2})$.

Basically, this definition can be generalized into the definition of open superstar of some graphs, by replacing the complete bipartite graph with another graph.

3. Graceful labeling for open superstar of complete bipartite graphs

A graph is graceful if it can be labeled with graceful labeling. To find a graceful labeling for an open superstar of complete bipartite graph $S^2(t \cdot K_{m,n})$, firstly we tried to give a graceful labeling to the easier case, that is for the small positive integers m, n, and t. Then, we looking for a pattern of the labels and generalized it to all positive integers m and n. In this process, we only found a graceful labeling for the odd positive integers t. This result is given in the following Theorem 3.1.

Theorem 3.1. *An open superstar of complete bipartite graphs $S^2(t.K_{m,n})$ is graceful for odd positive integers t and all positive integers m, n.*

Proof. Let G be an open superstar of complete bipartite graph with t copy of complete bipartite graphs $K_{m,n}$. Clearly, G is a graph with $t(m+n+1)+1$ vertices and $t(mn+2)$ edges as shown in the following Figure 5.

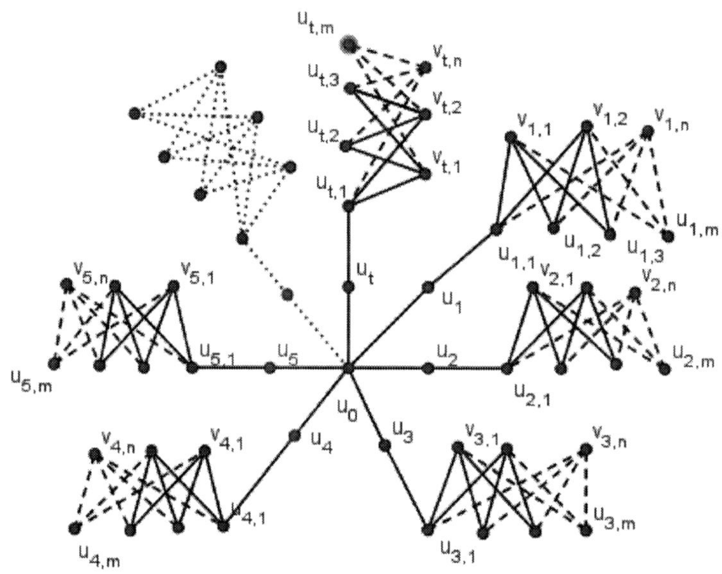

Figure 5. Ordinary labeling of $S^2(t \cdot K_{m,n})$.

Let s be the number of edges of G. So, $s = t(mn+2)$. The labeling function f on the vertices of G is defined as follows:

$$
\begin{aligned}
f(u_0) &= 0 \\
f(u_i) &= \begin{cases} s - \frac{1}{2}(i-1), & i \equiv 1(\mod 2) \\ \frac{1}{2}i, & i \equiv 0(\mod 2) \end{cases} \\
f(u_{i,j}) &= \begin{cases} \frac{1}{2}(t+i) + (j-1)t, & i \equiv 1(\mod 2) \\ s - \frac{1}{2}(t+i-1) - (j-1)t, & i \equiv 0(\mod 2) \end{cases} \quad for\ j = 1, 2, 3, \cdots, m \\
f(v_{i,j}) &= \begin{cases} s - (t + \frac{1}{2}(i-1)) - (j-1)mt, & i \equiv 1(\mod 2) \\ t + \frac{1}{2}i + (j-1)mt, & i \equiv 0(\mod 2) \end{cases} \quad for\ j = 1, 2, 3, \cdots, n
\end{aligned}
$$

To prove that the function f is a graceful labeling, it must be shown that f is one-to-one and the induced function f^* is bijective. First, we will show that f is an injective function from $V(G)$, the set of all vertices of G, to the set $\{0, 1, 2, 3, \cdots, s\}$.

For $i = 1, 2, 3, \cdots, t$, the labels of u_0, u_i, $u_{i,1}$, and $v_{i,1}$ are the consecutive integers $0, 1, 2, \cdots, t - \frac{1}{2}(t-1)$ and $s, s-1, s-2, \cdots, s - (t - \frac{1}{2}(t-1))$. So, there are $s - (t - \frac{1}{2}(t-1)) - t - \frac{1}{2}(t-1) - 1$ integers between $t - \frac{1}{2}(t-1)$ and $s - (t - \frac{1}{2}(t-1))$ which can be used as labels of $t(m+n-2)$ remained vertices. Consider that, $s - (t - \frac{1}{2}(t-1)) - (t - \frac{1}{2}(t-1)) - 1 = t(mn+2) - 3t = t(mn-1) \geq t(m+n-2)$. It indicates that f is may be an injective function.

Now, it must be shown that each label for remained vertices is different. From the function f, we can see that $f(u_{i,j}) \equiv f(u_{i,1}) (\mod t)$ and $f(v_{i,j}) \equiv f(v_{i,1}) (\mod t)$. Furthermore, for i is odd, $f(u_{i,j}) \equiv \frac{1}{2}(t+1) (\mod t)$ and $f(v_{i,j}) \equiv (t - \frac{1}{2}(i-1)) (\mod t)$. For i is even, $f(u_{i,j}) \equiv \frac{1}{2}(t-i+1) (\mod t)$ and $f(v_{i,j}) \equiv \frac{1}{2}i (\mod t)$. Besides that, function f also has these following properties :

(i) for i is odd, $f(u_{i,j}) = f(u_{i,1}) + (j-1)t$ and $f(v_{i,j}) = f(v_{i,1}) - (j-1)mt$.

(ii) for i is even, $f(u_{i,j}) = f(u_{i,1}) - (j-1)t$ and $f(v_{i,j}) = f(v_{i,1}) + (j-1)mt$.

It is clear that for $i \neq k$, $f(u_{i,j})$ and $f(u_{k,l})$ are in different classes of modulo t and it also stand for $f(v_{i,j})$ and $f(v_{k,l})$, but not for $f(u_{i,j})$ and $f(v_{k,l})$. For example, in case $i = 2$ and $k = t-1$, we have

$$f(u_{2,j}) \equiv \frac{1}{2}(t-1) (\mod t) \equiv f(v_{t-1,l}).$$

Since $f(u_{2,j_1}) > f(u_{2,j_2})$ for $j_1 < j_2$ and $f(v_{t-1,l_1}) < f(v_{t-1,l_2})$ for $l_1 < l_2$, we just need to compare the smallest value of $f(u_{2,j})$ and the biggest value of $f(v_{t-1,l})$. The smallest value of $f(u_{2,j})$ is $f(u_{2,m}) = mnt - mt + 2t + \frac{1}{2}(t-1)$. While the biggest value of $f(v_{t-1,l})$ is $f(v_{t-1,n}) = t + \frac{1}{2}(t-1) + mnt - mt$. Therefore, we get

$$f(v_{t-1,l}) < f(v_{t-1,l}) < \cdots < f(v_{t-1,n}) < f(u_{2,m}) < f(u_{2,m-1}) < \cdots < f(u_{2,1}).$$

By generalized this case, we obtained that although $f(u_{i,j})$ and $f(v_{k,l})$ are in the same class of modulo t, the values of $f(u_{i,j})$ and $f(v_{k,l})$, for all j and l, are distinct.

Second, it must be proved that the induced function f^* is bijective. We know that for $uv \in E(G)$, $f^*(uv) = |f(u) - f(v)|$. It is easy to show that for the superstar part of G, we have

$$\begin{aligned} f^*(u_0 u_i) &= \begin{cases} s - \frac{1}{2}(i-1), & i \equiv 1 (\mod 2) \\ \frac{1}{2}i, & i \equiv 0 (\mod 2) \end{cases} \\ f^*(u_i u_{i,1}) &= s - i - \frac{1}{2}(t-1), \qquad i = 1, 2, \cdots, t. \end{aligned}$$

So, the edges labels of superstar part of G are the consecutive positive integers $1, 2, 3, \cdots, \frac{1}{2}(t-1)$ and $s - t - \frac{1}{2}(t-1), s - (t-1) - \frac{1}{2}(t-1), \cdots, s$. There are tmn integers between $\frac{1}{2}(t-1)$ and $s - t - \frac{1}{2}(t-1)$ that can be used as the labels of edges in each copy of complete bipartite graph. There are t copies of complete bipartite graphs $K_{m,n}$. Every complete bipartite graphs $K_{m,n}$ has mn edges. So, there are tmn edges and tmn consecutive integers for labeling it. To prove f^* is bijective, it must be proved that for every i, j, and k, the value of $f^*(u_{i,j}v_{i,k})$ is all distinct and $\frac{1}{2}(t-1) < f^*(u_{i,j}v_{i,k}) < s - t - \frac{1}{2}(t-1)$.

Since $f(u_{i,j})$ is in the same class of modulo t for all j and $f(v_{i,k})$ is in the same class of modulo t for all k, $f^*(u_{i,j}v_{i,k})$ is in the same class of modulo t for all j and k. So, if two edges are located in the same copy of complete bipartite graph, its labels are in the same class of modulo t. From the definition and properties of f, it can be shown that $f^*(u_{i,j}v_{i,k}) = (\frac{1}{2}(t+1) - i) (\mod t)$, for all i, j, and k.

Now, we need to prove that in each copy of complete bipartite graph, there is no edges which has the same label. For the odd i, let $f(u_{i,1}) = a$ and $f(v_{i,1}) = b$. Then, we get $f(u_{i,j}) = a + (j-1)t$, for $j = 1, 2, 3, \cdots, m$, and $f(v_{i,k}) = b - (k-1)mt$, for $k = 1, 2, 3, \cdots, n$. It is clear that for all j and k,

$$f^*(u_{i,j}v_{i,k}) = (b-a) - (k-1)mt - (j-1)t.$$

If there is edge $u_{i,p}v_{i,q}$ with the same label, it must be shown that $u_{i,p}v_{i,q} = u_{i,j}v_{i,k}$. Consider that

$$f^*(u_{i,j}v_{i,k}) = f^*(u_{i,p}v_{i,q}) \Leftrightarrow q - k = \frac{j-p}{m}.$$

Since $1 - m \leq j - p \leq m - 1$ and qk is an integer, then qk must be zero. So, we get $q = k$ and $j = p$, i.e. $u_{i,p}v_{i,q} = u_{i,j}v_{i,k}$. For the even i, the process is analogue.

We already proved that for every i, j, and k, the value of $f^*(u_{i,j}v_{i,k})$ are distinct. Then, we continue to prove that $\frac{1}{2}(t-1) < f^*(u_{i,j}v_{i,k}) < s - t - \frac{1}{2}(t-1)$. Consider that the smallest edges label in each copy of $K_{m,n}$ is $f^*(u_{i,m}v_{i,n}) = \frac{1}{2}(3t+1) - i$. Then, for the biggest i, we have the smallest $f^*(u_{i,j}v_{i,k})$, it is $f^*(u_{t,m}v_{t,n}) = \frac{1}{2}(t+1)$. The biggest label in each copy of $K_{m,n}$ is $f^*(u_{i,1}v_{i,1}) = s - t - \frac{1}{2}(t-1) - i$. Then, for the smallest i, we have the biggest $f^*(u_{i,j}v_{i,k})$, it is $f^*(u_{1,1}v_{1,1}) = s - t - \frac{1}{2}(t-1) - 1$.

Since f is an injective from $V(G)$ to the set $\{0, 1, 2, 3, \cdots, s\}$ and the induced function f^* is a bijective from $E(G)$ to the set $\{1, 2, 3, \cdots, s\}$, we can say that f is a graceful labeling for G and graph G is graceful. So, for positive integers m, n and odd positive integers t, graph $S^2(t \cdot K_{m,n})$ is graceful. □

Illustration 3.1. The open superstar of 5 copy of $K_{2,3}$ and its graceful labeling are shown in Figure 6.

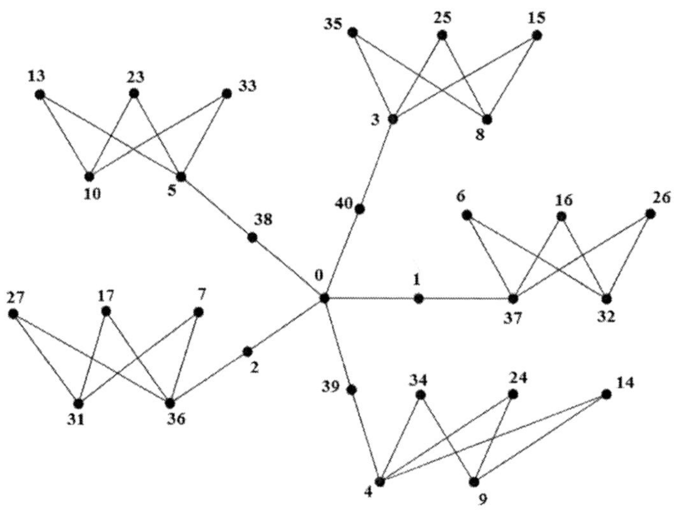

Figure 6. A graph $S^2(5 \cdot K_{2,3})$ and its graceful labeling.

Illustration 3.2. We know that the complete bipartite graphs $K_{m,n}$ and $K_{n,m}$ are isomorphic. However, in the open superstar case, $S^2(t \cdot K_{m,n})$ and $S^2(t \cdot K_{n,m})$ are not isomorphic graphs for $m \neq n$. For example, we can compare $S^2(5 \cdot K_{2,3})$ with $S^2(5 \cdot K_{3,2})$ as can be seen in Figure 7.

4. Conclusion and future work

In this paper, we have introduced a definition of open superstar of complete bipartite graphs. We also have given a graceful labeling for $S^2(t \cdot K_{m,n})$ with odd positive integers t. Therefore, for positive integers m, n and odd positive integers t, open superstar of complete bipartite graphs $S^2(t \cdot K_{m,n})$ is graceful. Our future research is construct a graceful labeling for open superstar of complete bipartite graphs $S^2(t \cdot K_{m,n})$ with even positive integers t and another families of open superstar graphs.

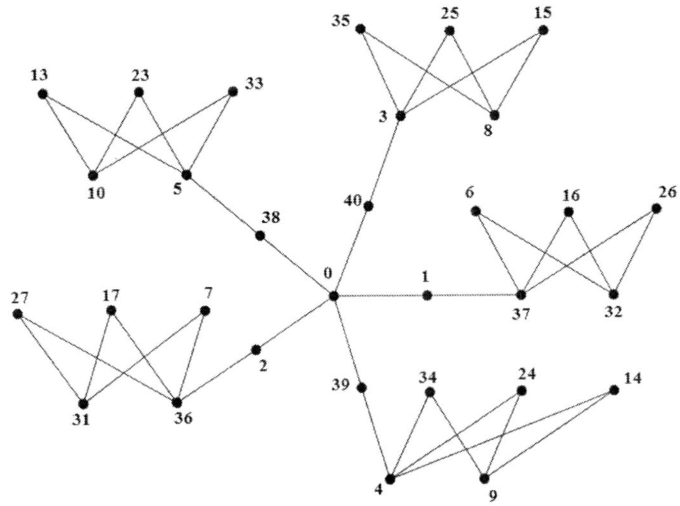

Figure 7. A graph $S^2(5 \cdot K_{3,2})$ and its graceful labeling.

References

[1] Gallian J A 2016 *The Electronic Journal of Combinatoric* **#DS6**
[2] Rosa A 1967 On certain valuation of the vertices of a graph *Theory of Graphs(Int. Symp., Rome, July 1966)* pp 329–335
[3] Bermond J C 1979 Graceful graphs, radio antennae and French windmills *Graph Theory and Combinatorics* (London: Pitman) pp 18–37
[4] Kaneria V J, Meghpara M and Makadia H M 2014 *International Journal of Mathematics and Statistics Invention (IJMSI)* **2** 19 – 23

IOP Conf. Series: Journal of Physics: Conf. Series **1040** (2018) 012021 doi:10.1088/1742-6596/1040/1/012021

Numerical solution of SIR model for transmission of tuberculosis by Runge-Kutta method

S Side, A M Utami, Sukarna and M I Pratama

Department of Mathematics, Faculty of Mathematics and Natural Science, Universitas Negeri Makasar, South Sulawesi, Indonesia

E-mail: syafruddin@unm.ac.id

Abstract. This paper discusses the numerical solution of the SIR model on the spread of Tuberculosis by the Runge-Kutta method. The data used is secondary data of Tuberculosis cases in Makassar city from Health Department of South Sulawesi Province in 2015. The result obtained is a general solution of SIR model of Tuberculosis transmission by fourth-order Runge-Kutta method. The results of model analysis and simulation using data on the number of tuberculosis cases in Makassar showed that tuberculosis cases increased for some time, then decreased. The results have also shown that the fourth-order Runge-Kutta method can be used to predict the transmission and to be considered for the prevention of Tuberculosis in Makassar.

1. Introduction

The development and progress of the modern world today can't be separated from mathematics. Almost all human activity is related to mathematics [1]. Mathematical model is one tool that can help facilitate solving problems in real life. The example is the application to determine the spread of infectious disease model in a particular region or region. To know the spread of infectious diseases, there are known models of disease spread, both models are deterministic, as well as models that are stochastic. The models have their own characteristics, based on the type and shape of the spread of infectious diseases observed, for example the SIR Model of Tuberculosis [2,3].

The World Health Organization states that one-third of the world's population has been infected by M. tuberculosis, 9 million new Tuberculosis patients and 3 millions tuberculosis deaths worldwide, 95% of Tuberculosis cases and 98% death of Tuberculosis [4]. Particularly in Kota Makassar, based on data obtained from the Division of Disease Prevention and Environmental Health of Makassar City Health Office, the findings of new TB, TB (+) patients in 2013 were 72.44% (1,811 people were found to be 2,500), and 2014, increased by 78.12%, so the city of Makassar including the city of concern for tuberculosis [2].

The numerical method found only solutions approaching or nearing true solutions so that numerical solutions are also called approximation [5]. In solving numerical differential equations we can use the one-step method, one of which is the Runge-Kutta method [6].

One of the infectious diseases whose models can be resolved numerically is Tuberculosis. Some researchers have examined these issues as follows: [7] discusses the use of Runge-Kutta method in finding solutions of initial value problems; [8] discusses the numerical solution of the Predator-Prey model by comparing the fourth-order Runge-Kutta method and the Runge-Kutta Gill method, the result of which is the fourth-order Runge-Kutta method will yield more precise accuracy in computation and rounding than Runge-Kutta another order; [9,10] discusses the mathematical modeling of SIR and SEIR on the transmission of dengue fever and tuberculosis by using Lyapunov function method to analyze the models. [2] discusses the process of SIR modeling on the spread of Tuberculosis disease in Makassar starting from determining the point of disease-free equilibrium, the point of equilibrium endemic and its numerical simulation but not examining further the numerical solution of the model already obtained. This paper discusses the numerical solution of the SIR model on the spread of tuberculosis from [2] with the fourth-order Runge-Kutta.

2. SIR Model for Transmission of Tuberculosis

The SIR model for tuberculosis spread [2] is presented in the form of a transfer chart as Figure 1.

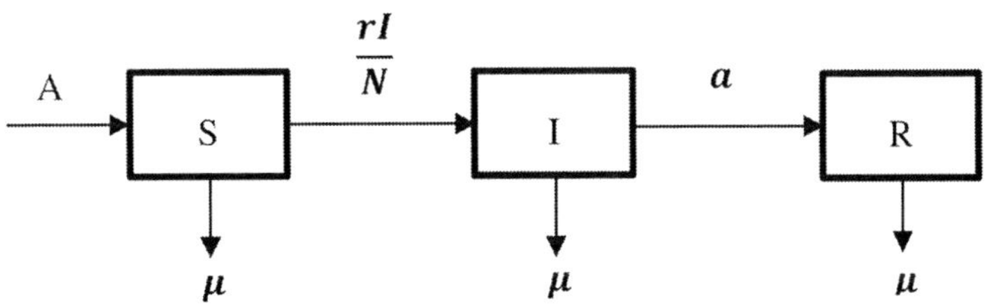

Figure 1. Diagram of SIR Model for Tuberculosis

where : S = Number of vulnerable individuals in the population at time t
I = Number of infected individuals in the population at time t
R = Number of individuals who recover in the population at time t
r = The rate of transmission of the disease from susceptible becomes infectious $(0 \le r \le 1)$
a = The rate of recovery from infectious to recovered
A = Initial value
μ = Death rate

Based on the assumptions and the relationship between the variables and the parameters in Figure 1, we get the SIR model of tuberculosis dispersion as in (1) - (3) follows:

$$\frac{dS}{dt} = A - \mu S - \frac{rI}{N}S \qquad (1)$$

$$\frac{dI}{dt} = \frac{rI}{N}S - (\mu + a)I \qquad (2)$$

$$\frac{dR}{dt} = aI - \mu R \qquad (3)$$

Where: N = S + I + R

3. Fourth Order Runge-Kutta Method

This method was discovered by German mathematician Carl Runge (1856-1927) and Wilhelm Kutta (1867-1944). The rationale behind this method is to defend almost Taylor, but in the settlement of GDP by Taylor method is not practical because the method requires derivative calculations [7]. The fourth-order Runge-Kutta method is the most rigorous method compared to the previous order Runge-Kutta method. Therefore, the fourth-order Runge-Kutta method is often used to solve a differential equation.

The fourth-order Runge-Kutta method has the form as follows equation (4):

$$y_{r+1} = y_r + \frac{1}{6}(k_1 + 2k_2 + 2k_3 + k_4) \qquad (4)$$

with

$$k_1 = hf(x_r, y_r)$$

$$k_2 = hf(x_r + \frac{1}{2}h, y_r + \frac{1}{2}k_1)$$

$$k_3 = hf(x_r + \frac{1}{2}h, y_r + \frac{1}{2}k_2)$$

$$k_4 = hf(x_r + h, y_r + k_3)$$

Source [8].

In addition to having a better accuracy of the solution than the previous order Runge-Kutta method, this method is also easy to program, stable, small cut errors and also small rounding errors.

4. Method

This research is an experimental research by conducting a study of theories about tuberculosis transmission model. The numerical solution of the SIR model is obtained using the fourth-order Runge-Kutta method. The model was then analyzed and simulated using secondary data on the number of tuberculosis cases from the Ministry of Health of South Sulawesi Province in 2015 [2]. The simulation was done using MATLAB software.

5. Results and Discussion

5.1. Numerical Solution of SIR Model forTuberculosis by fourth order Runge-Kutta Method

The SIR model on the spread of tuberculosis diseases present in equations (1) - (3) will be further identified into the form of ordinary differential equations as in equations (5) - (7) as follows:

$$\frac{dS}{dt} = f(t, S, I, R) = A - \mu S - \frac{rI}{N}S \tag{5}$$

$$\frac{dI}{dt} = g(t, S, I, R) = \frac{rI}{N}S - (\mu + a)I \tag{6}$$

$$\frac{dR}{dt} = i(t, S, I, R) = aI - \mu R \tag{7}$$

The equation (5) - (7) above is solved using the fourth-order Runge-Kutta method based on equation (4) by substitution, so that the following equation (8) - (10) is obtained:

$$S_{r+1} = S_r + \frac{1}{6}(k_1 + 2k_2 + 2k_3 + k_4)h \tag{8}$$

$$I_{r+1} = I_r + \frac{1}{6}(l_1 + 2l_2 + 2l_3 + l_4)h \tag{9}$$

$$R_{r+1} = R_r + \frac{1}{6}(m_1 + 2m_2 + 2m_3 + m_4)h \tag{10}$$

with is the step of time

$$k_1 = A - \mu S_r - \frac{r}{N} I_r S_r$$

$$l_1 = \frac{rI_r}{N} S_r - (\mu + a) I_r$$

$$m_1 = aI_r - \mu R_r$$

$$k_2 = A - \mu\left(S_r + k_1\frac{h}{2}\right) - \frac{r}{N}\left(I_r + l_1\frac{h}{2}\right)\left(S_r + k_1\frac{h}{2}\right)$$

$$l_2 = \frac{r}{N}\left(I_r + l_1\frac{h}{2}\right)\left(S_r + k_1\frac{h}{2}\right) - (\mu + a)\left(I_r + l_1\frac{h}{2}\right)$$

$$m_2 = a\left(I_r + l_1\frac{h}{2}\right) - \mu\left(R_r + m_1\frac{h}{2}\right)$$

$$k_3 = A - \mu\left(S_r + k_2\frac{h}{2}\right) - \frac{r}{N}\left(I_r + l_2\frac{h}{2}\right)\left(S_r + k_2\frac{h}{2}\right)$$

$$l_3 = \frac{r}{N}\left(I_r + l_2\frac{h}{2}\right)\left(S_r + k_2\frac{h}{2}\right) - (\mu + a)\left(I_r + l_2\frac{h}{2}\right)$$

$$m_3 = a\left(I_r + l_2\frac{h}{2}\right) - \mu\left(R_r + m_2\frac{h}{2}\right)$$

$$k_4 = A - \mu(S_r + k_3 h) - \frac{r}{N}(I_r + l_3 h)(S_r + k_3 h)$$

$$l_4 = \frac{r}{N}(I_r + l_3 h)(S_r + k_3 h) - (\mu + a)(I_r + l_3 h)$$

$$m_4 = a(I_r + l_3 h) - \mu(R_r + m_3 h)$$

Equation (8) - (10) is a numerical solution of the SIR model on the spread of tuberculosis.

5.2. SIR Model Simulation Using MATLAB Software

Initial values and parameters used in the SIR model on the spread of Tuberculosis disease are presented in Table 1 below:

Table 1. Initial condition of SIR model for Tuberculosis

Initial Condition of SIR Model		
Variable/Parameter	**Value**	**Source**
$S(0) = S(2015)$	$\frac{1446093}{1449401}$	KKRI, 2015
$I(0) = I(2015)$	$\frac{1885}{1449401}$	KKRI, 2015
$R(0) = R(2015)$	$\frac{1423}{1449401}$	KKRI, 2015
r	$\frac{1}{2} = 0.5/month$	KKRI, 2015

Initial Condition of SIR Model

Variable/Parameter	Value	Source
a	$\frac{1}{9} = 0.111111\ /\ month$	
μ	$\frac{1}{71.41 \times 12} = 0.001167\ /\ bulan$	BPS, South Sulawesi, 2015 [11]
$N = A$	1449401	

The SIR model formulation on the spread of Tuberculosis in Makassar is as in equations (11) - (13).

$$\frac{dS}{dt} = 1449401 - 0.001167\,S - \frac{(0.5)}{1449401} IS \tag{11}$$

$$\frac{dI}{dt} = \frac{(0.5)}{1449401} IS - (0.001167 + 0.111111)I \tag{12}$$

$$\frac{dR}{dt} = 0.111111 I - 0.001167\,R \tag{13}$$

The result of simulation equation (8) - (10) by using time interval per month, time step, and initial condition in the form of parameter value and initial value as in Table 1 are shown in Figure 2, Figure 3, Figure 4 and Figure 5 below:

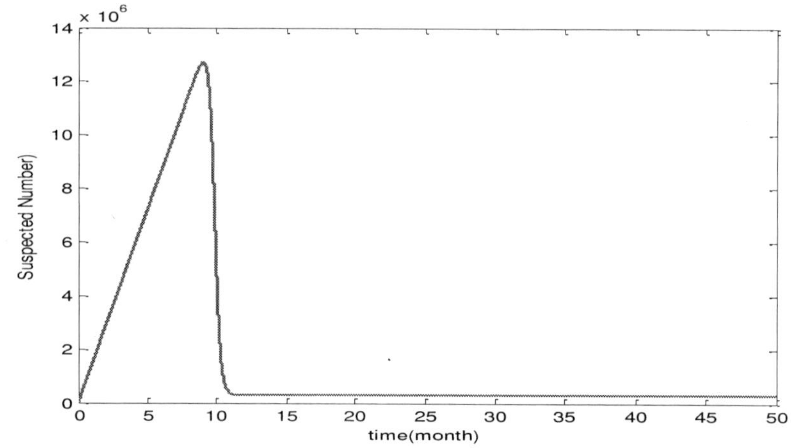

Figure 2. Suspected Population Model SIR for Tuberculosis

According figure 2, the number of susceptible population of 1446094 from 1449401 people, then increased until the ninth month reached the peak of the suspected population about 1,272 x 10^7 population. The increase is due to the amount of population or the initial value used is greater than the amount of population decline in the suspected class in the form of disease transmission rate from suspected to infectious and the rate of death of suspected class. After passing the peak point, there is a continuous decline and the line will converge to a point, indicating that in some time later, the suspected population will decrease.

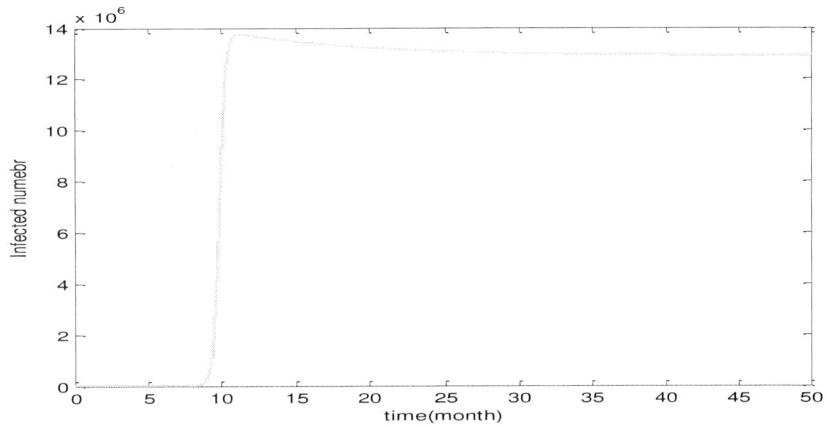

Figure 3. Total Population Infected Model SIR spread of Tuberculosis

According to the Figure 3, the infected population amounted to 1885 from 1449401 inhabitants, which then increased steadily until the eleventh month peaked around 1,375 x 107 populations. The increase is due to the rate of transmission of the disease from suspected to infectious greater than the rate of recovery from infected to recover, and infected class mortality rate. After the hemisphere, the population in the infected class begins to decline and then the line converges to a certain point until sometime later.

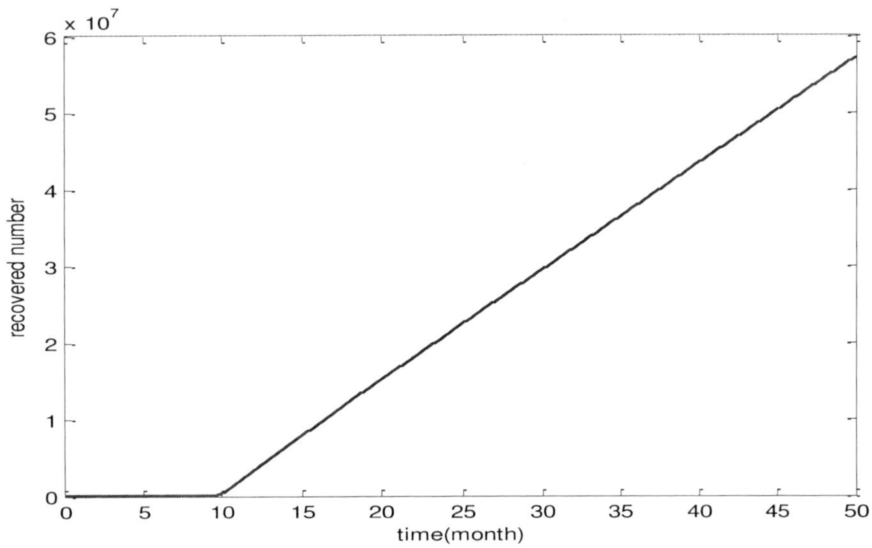

Figure 4. Total Population Recovered SIR Model of Tuberculosis Deployment

Based on Figure 4, Recovered population amounted to 1423 from 1449401 inhabitants, it is seen that the number of healthy population has increased over time. The increase is caused by the rate of recovery rate from infected to recover greater than the rate of death of the recovered class and the inability of the disease transmission rate from suspected to infected.

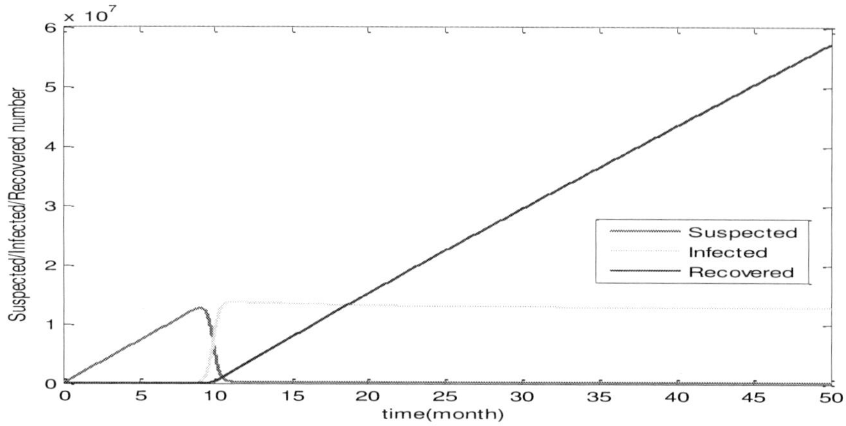

Figure 5. Total Population Suspected, Infected dan Recovered (SIR) Model of Tuberculosis

Based on Figure 5, The number of suspected increased to 9,022 <t <9.069 months later after reaching the peak then the population will decrease and close to zero, this is because some populations move from the suspected class to the infected class, the infected class increases until finally convergent the next time, this is because some populations move to recovered classes through several factors, one of which is healing, and in the recovered class, from the first month to the fiftieth month continues to increase, this explains that the rate of recovery over time has increased

5.3. Discussion

In [7] study only deals with the use of the Runge-Kutta method in finding solutions of initial value problems in ordinary differential equations. [8] discusses the comparative analysis of numerical solutions of the Predator-Prey model by comparing the fourth-order Runge-Kutta method and the Runge-Kutta Gill method. [9,10] discusses the mathematical modeling of SIR and SEIR on the transmission of Tuberculosis and dengue fever disease using Lyapunov function method. The method analyze both models can be used to determine Tuberculosis disease status in a region by using data simulation from certain region.

In [2] describes the SIR model on the spread of Tuberculosis with its assumptions, determining the equilibrium point of the model so as to obtain two equilibrium points, stability and basic reproduction number, and simulation using secondary data of Tuberculosis (+) disease case in Makassar. The simulation results explain that the transmission of Tuberculosis disease has a great influence on the model and explains that Tuberculosis disease in Makassar City decreases or even disappears if its basic reproduction number is less than one.

This paper describes numerical solutions and SIR model analysis using the fourth-order Runge-Kutta method. Then, establish the parameters and initial values used by using real data from the South Sulawesi Provincial Health Office. The model is then simulated to predict and see the trend of the number of tuberculosis cases in Makassar.

6. Conclusion

The numerical simulation results of the SIR model on the dispersion of Tuberculosis by the fourth-order Runge-Kutta method indicates that: a) the larger the time interval used, the clearer the movement of each class. There is an increase in susceptible, infected, and very long increase classes in the recovered class, but for the susceptible class after reaching its peak it will move down while the infected class and the recovered class will move steadily at a later time; b) Infected classes will experience an increase influenced by the rate of transmission of the disease from susceptible to infected. Infected classes will decrease influenced by the rate of recovery of the disease from infected to recovered; c) Tuberculosis cases in Makassar have an incubation period and a longer recovery process than the average data obtained; and d) The fourth-order Runge-Kutta method can be used to find numerical solutions of the SIR model and predict the number of Tuberculosis cases in Makassar.

Acknowledgments
We would like thank to PNBP-UNM for financial support.

References

[1] Belo Y J T 2016 *SIRv Modeling of Swine Flu (H1N1) with Vaccination in Constant Populations* (Thesis) Universitas Negeri Makassar
[2] Hasrina 2015 *SIR (Susceptible, Infectious And Recovered) Model for Spreading Tuberculosis Disease in Makassar* (Thesis) Universitas Negeri Makassar
[3] Tracy A 2008 *Modeling transmission dynamics of tuberculosis inclunding various latent periods* Spring Term
[4] Fardinah 2009 *Solution of Ordinary Differential Equations using Runge-Kutta Method of Old Order and Euler Method* (Thesis) Universitas Negeri Makassar
[5] Side S 2015 A Susceptible-Infected_Recovered Model and Simulation for Transmission of Tuberculosis *Advance Science Letters* **21**(2) 137-139.
[6] Mursalin 2006 *Numerical Solutions for Differential Equations using Ordinary Multi-Step Methods* (Thesis) Universitas Negeri Makassar
[7] Asmah S 2004 *Numerical Solution of Initial Value Problems by Runge-Kutta* (Thesis) Universitas Negeri Makassar
[8] Putri P P 2013 *Analysis of Numerical Solution of Model Predator-Prey with Runge-Kutta Method of Order Four and Gill* (Thesis) Universitas Jember.
[9] Side S, Sanusi W, Aidid M K and Sidjara S 2016 Global Stability of SIR and SEIR Model for Tuberculosis Disease Transmission with Lyapunov Function Method *Asian Journal of Applied Sciences* **9**(3) 87-96.
[10] Syafruddin S and Noorani M S M 2013 Lyapunov function of SIR and SEIR model for transmission of dengue fever disease *International Journal of Simulation and Process Modelling* **8** 177-184.
[11] BPS South Sulawesi 2016 *South Sulawesi Province in Number* (Makassar: BPS)

Identification of chemistry teaching problems of a prospective teacher: A case study on chemistry teaching

I W Redhana, I B N Sudria, I N Suardana, I W Suja and N K N Handayani

Universitas Pendidikan Ganesha, Bali, Indonesia

Email: redhana.undiksha@gmail.com

Abstract. This study was aimed at identifying chemistry teaching problems of a prospective teacher viewed from the conceptual change model on colloid topic. The study was conducted in one class of one senior high school in Buleleng regency, Bali. The study was a qualitative research of case study type. A number of students in the class were 27 people. The study was done in an academic year of 2013/2014 on the even semester. Methods of the study were observation and interview. The results of the study showed that new conceptions related to the necessity aspects were that the prospective teacher did not discuss the colloid concepts comprehensively and profoundly. On the other hand, the chemistry teaching problems related to the intelligibility aspects were that the prospective teacher (1) asked successive questions to students, (2) asked self-answered questions, (3) gave incomplete information, (4) gave unclear information, and (5) did not ask for students' reasons. The chemistry teaching problems related to the plausibility aspects were that the prospective teacher emphasized the less importance of the context. Finally, the teaching problems related to the fruitfulness aspects were that the prospective teacher did not provide the new conceptions that guided the inquiry on the wider fields.

1. Introduction

Students are learners who have the instinct to understand natural phenomena through observation. In an attempt to understand the phenomena they try to construct a conception [1]. This conception will be used to solve problems or explain the new phenomena encountered [2].

Educational research indicates that students' conceptions are referred to as alternative conceptions that are strongly influenced by age and culture [3]. The alternative conceptions of students have been widely reported by some researchers [4]. In general, the alternative conceptions are misconceptions [5] and these misconceptions are resistant or difficult to change using the traditional learning strategies.

In the process of learning teachers should be able to change the alternative conceptions into scientific conceptions. Therefore, they should challenge students' cognitive so that their understanding of the natural phenomena is disturbed. With these disturbances, students can change their conceptions. These changes are called as a conceptual change model [6]. In the conceptual change model, students are not satisfied with their conceptions because the conceptions cannot be used to explain the natural phenomena and/or to solve the problems encountered well. When this happens, students will experience dissatisfaction with their own conceptions. Therefore, they need new conceptions that are able to explain the natural phenomena or solve the problems encountered. Furthermore, the new conceptions must be intelligibly understood (intelligibility aspect). Likewise, the new conceptions must be able to solve problems (plausibility aspect) and the new conceptions must be extended to other events and open up new areas of inquiry (fruitfulness aspect). A number of studies have been conducted regarding the conceptual change model [7] [8]. All science learning is supposed to be able to make conceptual changes when the students' conceptions are unable to cope the challenges.

A colloid is one of the important topic studied in chemistry. Application of the topic in everyday life is very much, for example in food, medicine, cosmetics, and water treatment. With regard to the colloid application, students should be able to understand the concepts of colloids and their application well. By so doing, they will be able to utilize the concepts learned to help themselves and others.

Because of its complexity, chemistry must be taught comprehensively and coherently. Researchers have observed various problems related to the mastery of the chemistry concepts. Some of these problems are students' misconceptions. Some of these misconceptions are found on the topics of chemical equilibrium, acids and bases, electrochemistry, matter and its change, chemical bonds, physical and chemical changes, and solutions [9], hydrocarbons [10], and buffer solutions [11].

SMA Negeri 1 Singaraja is a favorite school in Singaraja city and many of the junior high school graduates want to go to school here. Many achievements have been achieved by SMAN 1 Singaraja, such as Olympic champions either at local, regional, national, and even international level. However, previous research findings suggested that students had misconceptions about basic chemistry concepts. Redhana and Kirna [12] reported that students of SMAN 1 Singaraja had misconceptions about the topics of atomic structures, periodic table, and chemical bonds.

Every year SMAN 1 Singaraja provides teaching practices for prospective teachers, including the prospective teachers from Chemistry Department of Ganesha University of Education. Like teaching in general, the teaching done by the prospective teachers should be able to change the old conceptions of students so that students' understanding of the concepts learned for the better. However, in fact, there are so many problems encountered by prospective teachers when they practice teaching at schools [11]. Some of these problems are as follows. First, they have no mental readiness and feel insecure as they practice teaching in school. Second, they lack chemistry contents well before they practice teaching. Third, they do not consider the prior knowledge of students in making lesson plans so that the plans are not directed to the conceptual change strategy.

In this research, we would like to investigate the chemistry teaching problems faced by a prospective teacher who taught at SMAN 1 Singaraja. These teaching problems were then grouped according to the conceptual change model Posner et al. [6] which included aspects of necessity, intelligibility, plausibility, and fruitfulness.

2. Theoretical framework

The conceptual change model of Posner et al. [6] comes from the history and sociology of sciences of Khun [13] and from the developmental psychology of Piaget [13]. Khun's view of the scientific revolution explaining the scientific discovery will lead to a new scientific methodology [13]. Piaget, on the other hand, describes how students learn through assimilation and accommodation of knowledge [14]. Assimilation processes are an adjustment of knowledge learned by students with existing schemes. In contrast, accommodation processes are the change of the existing schemes by the knowledge learned because the schemes are unable to explain the natural phenomena and/or to solve the problems. Posner et al. [6] explain that the conceptual change model is similar to the scientific revolution of Kuhn and the accommodation of knowledge of Piaget.

In the conceptual change model, Posner et al. [6] propose four aspects of the conceptual change. The four aspects of it include (1) dissatisfaction with the old conceptions, followed by the new conceptions must be (2) intelligible, (3) plausible, and (3) fruitful. In the first aspect, that is, the dissatisfaction with the existing conception, the students must realize that their own conceptions are inconsistent with the existing conditions and these conceptions cannot be used to solve the problems faced. These inconsistencies by Kuhn [13] are called an anomaly. The second aspect is that the new conceptions that are accommodated must be intelligible. It means that they are not only reasonable but also students can explain them well. The third aspect is that the new conceptions are plausible. It means that the new conceptions being learned must be able to be used or applied to solve problems encountered. The fourth aspect is that the new conceptions being accommodated must be fruitful. It means that they must be extended to other events and open up new areas of inquiry.

3. Method

The study was a qualitative research of case study type. The subjects of the study were students in one class from SMAN 1 Singaraja Bali. The selection of the class was done by purposive sampling. The consideration used to select the class was because the class was taught by a chemistry prospective teacher. The study was conducted in the odd semester of the academic year 2013/2014. The number of students involved in the study was 27 people. Identification of teaching problems was done by observing the chemistry teaching processes on colloid topics conducted by the prospective teacher. In addition, the chemistry teaching processes were also recorded with a video recorder. All notes on the board were copied into the field notebook. Video recording played back to make the transcription. The transcription made was also enriched by the notes found on the board made by the chemistry prospective teacher and students. The transcription was read by the researcher and then the sentences were smoothed to obtain a better meaning of the transcription. The results of the transcription were then cross-checked to the prospective teacher to obtain approvals related to the contents of the transcription. If the prospective teacher agreed with the contents of the transcription, he/she was required to give a signature. The transcription results were then analyzed to obtain the chemistry teaching problems. The chemistry teaching problems were further grouped under the aspects of the conceptual change model, including necessity, intelligibility, plausibility, and fruitfulness.

4. Result and discussion

The chemistry teaching problems were collected through observation and recording video. The chemistry teaching problems were further classified into one of four aspects of the conceptual change model of Posner et al. [6], covering necessity, intelligibility, plausibility, and fruitfulness. The results of the study related to the chemistry teaching problems were presented in Table 1.

Table 1. The chemistry teaching problems experienced by the prospective teacher

Characteristics of the new conceptions	Description
Necessity	The colloid concepts discussed by the prospective teacher were not comprehensive and not profound. 1. He/she classified rough suspensions, true solutions, and colloids (Tyndall effect, homogeneous/heterogeneous, and filtering) not based on experiment. The rough suspensions, true solutions, and colloids were not grouped through experiments, but only by providing information in the form of examples. The example of the rough suspensions was a mixture of water and sand, the example of the true solution was a mixture of water and salt, and the example of the colloid was a mixture of water and milk. 2. He/she did not explain the concepts of colloids thoroughly. He/she discussed fewer concepts of colloid. For example, he/she described the colloid type of liquid and solid froth, Tyndall effect, Brownian motion, adsorption, electrophoresis, coagulation and dialysis, protective colloids, colloidal properties, and colloid manufacture. In the colloid manufacture, he/she informed the colloid manufacture through condensation methods such as redox reaction, hydrolysis, double decomposition, and solvent replacement, and the colloid manufacture through dispersion methods such as mechanical method, peptization, and Bredig's arc. He/she did not explain the principles of each process of the methods. He/she briefly explained the role

IOP Conf. Series: Journal of Physics: Conf. Series **1040** (2018) 012022 doi:10.1088/1742-6596/1040/1/012022

	of colloids in everyday life, such as the colloid applications in the fields of cosmetics, foods, and pharmaceuticals.
Intelligibility	The chemistry teaching problems that arose in relation to the aspects of intelligibility were as follows. 1. He/she asked successive questions to students. One question had not been answered by the students, he/she then asked another question. This could be seen in the following examples. • He/she asked: "What are the differences of Tyndall effect on solutions, colloids, and suspensions?" "How does Tyndall effect work on solutions?" "How does Tyndall effect work on colloids?" "How does Tyndall effect work on suspensions?" • He/she asked: "What are the differences between the colloid manufacture through condensation and through dispersion? What are the differences? What is condensation? What is dispersion?" 2. He/she asked a question that is answered by himself/herself. • He/she asked: "What are the differences among solutions, colloids, and suspensions?" He/asked: "Size of the particles." 3. He/she provided incomplete information. • He/she stated: "The solutions are much more stable than colloids and suspensions." He/she does not explain why the solution is much more stable than colloids and suspensions. • He/she asked: "What are the differences between adsorption and coagulation?" "He/she then states that they are very different." • He/she stated: "Brownian motion is irregular motion or zigzag motion of the colloid particles. "This causes the colloids to remain stable, whereas the suspensions are unstable and more easily precipitated than colloids." He/she did not explain further why Brownian motion in the colloid system can cause the colloids to become stable. 4. He/she provided unclear information. • He/she provided unclear information regarding the phase differences among solutions, colloids, and suspensions. He/she explained: "The solution has one phase, whereas colloids and suspensions have two phases." He/she did not explain why these happen. 5. He/she did not ask the students' reasons. • He/she asked: "What are the differences between adsorption and absorption?" The students replied: "The adsorption is on the surface, and the absorption gets inside." He/she did not ask the student's reasons. • He/she asked: "Manufactures of colloids can be divided into two, what are they?" Students answered: "Condensation method and dispersion method." He/she replied: "Yes, correct." He/she did not ask students' opinion about how to make colloids with both methods.
Plausibility	He/she emphasized the less importance of the context. • He/she explained the little usefulness of the colloid system in everyday life, such as the cosmetics, foods, and pharmaceuticals.
Fruitfulness	He/she did not provide a new conception that guides the inquiry on the wider fields through solving complicated problems.

The results of the study showed that there were the many chemistry teaching problems experienced by the prospective teacher in terms of conceptual change models. In the aspect of necessity, he/she classified rough suspensions, true solutions, and colloids not based on experiments, but only explain the classification without the empirical facts. As a result, the colloid concepts learned by students were less meaningful. This will lead to less understanding of students to the colloid concepts. If this happens they were very easy to forget the colloid concepts learnt.

The second problem related to the aspect of necessity was that the prospective teacher did not explain the colloid concepts thoroughly. In other words, the students' understanding was partial. What happens with this problem was that students only know these concepts on "the surface." As a result, when they applied the concepts, they experienced difficulties. The results of the study also indicated that students were not able to solve simple or complex problems of the aspects of plausibility and fruitfulness, respectively.

The chemistry teaching problems related to the intelligibility aspect were the prospective teacher asked successive questions. The point was that he/she did not give the students the opportunity to think the answer to the first question and he/she then asked the next question. In terms of theory of teaching this condition was not good because The students were not given the opportunity to develop their thinking skills or reasoning. Consequently, they will be indifferent and lazy to think. Finally, they would wait for the answers of the questions from the teacher. This fact was seen in the next teaching problems that the prospective teacher answered their own questions. This conditions tended to encourage the teacher-centered teaching, that was, teachers ask, answer, and think actively, while students as passive listeners. In other words, students were positioned as the objects of teaching, not as the subjects of learning. This teaching contradicts the paradigm of learning embraced in the curriculum that was applied recently, Curriculum 2006 and Curriculum 2013. In both curriculum, students are required to learn actively, both mind on and hand on. In Curriculum 2006 the learning that should be done by the teachers/the prospective teachers is the learning cycles, including exploration, elaboration, and confirmation [15]. The learning cycle forms can be problem-based learning, project-based learning, discovery learning, inquiry learning, and etc. Meanwhile, in the Curriculum 2013, the teachers or the prospective teachers should apply a scientific approach, including observing, questioning, collecting data, processing data, and communicating [16]. The models of learning with the scientific approach were the discovery learning, the inquiry learning, the problem-based learning, the project-based learning, and etc. Indeed, both the Curriculum 2006 and the Curriculum 2013 expect the same learning approach, that is the scientific approach.

The chemistry teaching problems related to the next aspect of intelligibility were the prospective teacher provided an unclear information. The lack of clarity of information provided by the prospective teacher might confuse the students with regard to the correct concepts. The students become unable to distinguish the correct concepts and the wrong concepts. These conditions could lead to misconceptions of students.

The chemistry teaching problems related to the other intelligibility are the prospective teacher did not ask students' reasons. By not asking these there are several possibilities that could happen. Firstly, the students might guess the answer. Secondly, the students might know the answer to the questions. In the former possibility happened, they did not know the true reasoning of the given answers. On the other hand, in the latter possibility happened, the students also did not know whether their reasoning was reasonable or not. Both of these did not establish the scientific concepts learned by the students. In other words, students did not understand the concepts deeply.

The chemistry teaching problems related to the aspect of plausibility were that the prospective teacher emphasizes the less importance of the context. This can be seen from at least the discussion about the application of colloids in everyday life. Indeed, colloids have very high applications in everyday life, such in the food industries (e. g. ice cream, jelly, and jellyfish), cosmetic industries (e. g. powders, lipstick, and deodorant), and in pharmaceutical industries (e. g. ulcer drugs, cough and syrup), as well as in the natural phenomena, such as fog, smoke, froth, and foam. The aims to learn chemistry are to become a someone who literates on chemistry, that are they can explain the phenomena that occur in everyday life and solve the problems encountered with regards to the chemistry concepts. These problems are closely related to the chemistry teaching problems related to the fruitfulness aspect, namely the prospective teacher did not confront the students with the problems that guide them to do an inquiry in the wider fields. As a result, they only memorize colloid concepts. This will cause students to easily forget the concepts learned. The further consequences are that they cannot help themselves and even others when they face problems related to colloid concepts.

The importance of identifying the chemistry teaching problems in terms of conceptual change models has been reported by de Jong, Acampo, and Verdonk [17]. By knowing the problems of teaching, it will be formulated appropriate learning strategies to overcome these chemistry teaching problems encountered. Several learning strategies have been developed to address the teaching problems related to the conceptual change model, including (a) the conceptual change texts [5] [18] [19] [20], (b) analogies [21], (c) the holistic mental model confrontation [22], (d) the web-based conceptual change texts [23], (e) the computer simulations [24], (f) the computer-based learning [25], and local culture-based 7E learning cycle model [26]. In preparing this conceptual change learning strategy, the teachers should also pay attention to the students' learning style [27].

5. Conclusion

From the results of the study could be concluded as follows. The conditions of conceptual change model related to the aspect of necessity were as follows. The colloid concepts taught by the prospective teacher were not comprehensive and not profound. The conditions of conceptual change model for the intelligibility aspects included the prospective teacher asked successive questions to students, asked self-answered questions, gave incomplete information, gave unclear information, and did not ask for students' reasons. The condition of the conceptual change model related to the aspect of plausibility was that the prospective teacher emphasized the less importance of the context. Finally, the condition of the conceptual change model related to the fruitfulness aspect was that the prospective teacher did not provide a new conception that guides the inquiry on the wider fields.

References

[1] Driver B 1989 *Int. J. Sci. Educ.* **11** 481-89
[2] Peterson G 2002 Description of cognitive development from a constructivist perspective. *Paper presented in the third European Symposium on Conceptual Change* (Finland: Turku)
[3] Yeo S and Zadnik M 2001 *Phys. Teach.* **39** 495-504
[4] Feher E and Meyer R K 1992 *J. Res. Sci. Teach.* **29** 5 505-20
[5] Cetengul I P and Geban O 2005 *Hacet. U. Egitim. Fak.* **29** 69–74
[6] Posner G J, Strike K A, Hewson P W, and Gertzog W A 1982 *Sci. Educ.* **88** 2 211–27
[7] Liu X 2004 *Sci. Educ.* **88** 373-96
[8] Yang E, Greenhowe J, and Amdre T 2004 *J. Chem. Educ.* 81 4 587-97
[9] Griffiths A 1994 A critical analysis and synthesis of research on students' chemistry misconceptions *Int. Counc. Assoc. Sci. Educ. Pub.* ed H. J. Schmidt 70-99
[10] Redhana I W 2001 Prosiding Seminar Nasional MIPA I (Singaraja, Indonesia: FMIPA Universitas Pendidikan Ganesha)
[11] Redhana I W, Sudria I B N, Hidayat I, and Merta L M 2017 *Jurnal Pendidikan IPA Indonesia* **6** 2 356-64
[12] Redhana I W and Kirna I M 2004 *Identifikasi miskonsepsi siswa SMA Negeri di kota Singaraja terhadap konsep-konsep kimia Skripsi* (IKIP Negeri Singaraja)
[13] Khun T S 1970 *The structure of scientific revolution (2nd ed.)* (Chicago: The University of Chicago)
[14] Santrock J W 2011*Educational psychology (5th ed.)* (New York: The McGraw Hill Company, Inc.)
[15] Permendiknas Republik Indonesia Nomor 41 Tahun 2007 (Jakarta: Depdiknas)
[16] Permendikbud Republik Indonesia Nomor 103 Tahun 2014 (Jakarta: Depdikbud)

[17] de Jong O, Acampo J, and Verdonk A 1995 *J. Res. Sci. Teach.* **32** 1 1097–110
[18] Onder I and Geban O 2006 *Hacet. U. Egitim. Fak.* **30** 166–73
[19] Balci S, Cakiroglu J and Tekkaya C 2006 *Biochem. Mol. Biol. Educ.* **34** 3 199–203
[20] Sinatra G and Broughton S 2011 *Read. Res. Quart.* **46** 4 374–93
[21] Dilber R and Duzgun B 2008 *Am. J. Phys. Educ.* **2** 3 174–83
[22] Calik M, Okur M, and Taylor N 2011 *Sci. Educ. Technol.* **20** 729–42
[23] Tas E, Gulen S, Oner Z, and Ozyurek C 2015 *Inter. Elect. J. Element. Educ.* **7** 2 263–80
[24] Trundle K C and Bell R L 2009 *Comp. Educ.* **54** 1078–88
[25] Yumusak A, Maras I, and Sahin M 2015 *J. Educ. Gifted Young Sci.* **3** 2 23–50
[26] Suardana I N, Redhana I W, Sudiatmika A A I A R, and Selamat I N 2018 *Int. J. Instruct.* **11** 2 399-412
[27] Sudria I B N, Redhana I W, Kirna I M, and Aini D 2018 *Int. J. Instruct.* **11** 1 89-102

Effectiveness of problem-based learning to optimize student learning outcomes in regression analysis course

M Susilawati and D P E Nilakusmawati

Mathematics Department, Faculty of Mathematics and Natural Sciences, Udayana University, Indonesia

E-mail: mdsusilawati@unud.ac.id

Abstract. This study aims to determine the effectiveness of the implementation of Problem-Based Learning (PBL) model in optimizing student learning outcomes in the regression analysis course at Mathematics Department, Udayana University. The type of research used is action research, with a classroom action research design model. Sources of data in this study are students of the fourth semester who took a course of Regression Analysis. Data collection techniques are tests, interviews, and observations. Statistical analysis technique with analysis of variance is used to test student scores on the three cycles performed. The action procedure is designed in three cycles, with stages of planning, implementation, observation and reflection. The results showed that descriptively the average of student learning outcomes in cycle I = 54.15; Cycle II = 75.25; and cycles III = 90.15 which means an increase in students' understanding and reduced misconceptions that occur in using regression analysis. Further analysis using analysis of variance obtained F_{count} = 34,76 with Pvalue = 0,000 indicates the null hypothesis is rejected, it's means that the three cycles have different effects on student learning outcomes, with a tendency to increase students' ability at every stage of the cycle.

1. Introduction

The learning process is the process of interaction of learners with educators and learning resources in a learning environment. In order for the learning process to make changes to the students in the four domains, namely the cognitive, affective, psychomotor, and cooperative, the quality standards of the learning process in each program are required. In addition, to meet the competence of graduates, the standard of assessment is required at every stage of the learning process to achieve the targeted competencies.

This Regression Analysis course is a compulsory subject for mathematics students who take an interest in statistics. Regression Analysis course discusses the relationship between response variables with predictor variables or explanatory variables. In linear regression analysis, the response variable is required to have a minimum interval measurement scale, while the explanatory variables can be either quantitative or qualitative data. Regression analysis is widely used in both exact and non-exact research. Research in pure sciences such as biology, chemistry, physics, as well as applied sciences such as research in agriculture, animal husbandry and other fields. In non-exact research, regression analysis is widely used in economic and social studies.

Regression analysis in its application, there are still many inappproprite/improper application in using regression analysis by researchers. One of the misuse of linear regression analysis is often done on a scale of measurement of response variables that are not scaled intervals or ratios. As explained by Indriantoro [1], gives an example to the accounting student's thesis that examines the factors that affect customer satisfaction. Customer satisfaction is a response variable measured by Likert scale. Likert scale is the process of scoring is done by making the classification and categories of answers to questions on the questionnaire according to the responses of respondents. The respondent answered the question on the questionnaire by cross-marking (x) on the answer provided with the five available possibilities. Scores given on each question are: 1 = Strongly Disagree; 2 = Disagree; 3 = Neutral; 4 = Agree; 5 = Strongly Agree. It is clear that the scale of measurement of response variables here does not meet the proper requirements if linear regression analysis is used.

Other mistakes that are also often performed in research using regression analysis are: (1) Significant on testing hypotheses simultaneously does not indicate that all explanatory variables are significant; (2) The value of coefficient of determination (R^2) is not a measure that the regression model has been obtained is good, because the addition of more explanatory variables will surely increase the value of the coefficient of determination. So a better measure is the corrected coefficient of determination that is the coefficient of determination that has been corrected by the degrees of freedom of explanatory variables; (3) Corrected coefficient of determination (R^2) obtained less than 50% means that the model obtained must be improved, one of them by adding non linear factor; and (4) The need to ensure that the data has met the assumptions required in the linear regression analysis namely: linear relationship, multivariate normality, no or little multicollinearity, no auto-correlation, and homoscedasticity

The real condition in the Mathematics Department, Udayana University shows that the achievement of student learning outcomes for regression analysis course in the last three years shows the students' learning achievement which is not improved. Students' learning outcomes in this course range from 65.00 to 75.00. Based on the evaluation of the learning process conducted by the lecturer of this subject course and the Quality Assurance Team of the department, it is found that the low of student learning result is caused by the misconception experienced by the student during the intrepretation of data analysis and misconception on diagnosis assumptions required in linear regression analysis.

Starting from the above fact and referring from the research cases which is misuse of linear regression analysis, as well as the importance of regression analysis understood by all fields of science, the determination of the appropriate method of learning process is very important to do. This is an effort to optimize the learning process and improve student learning outcomes in this course in particular, and improve the quality and GPA of graduates in general from quality learning process.

One of the methods that can be applied to optimize the learning process that leads to improving student learning outcomes is to implement Problem-based Learning (PBL) combined with the support of e-learning media. Utilization of e-learning media in this research is intended to provide easy access of teaching materials to students of Mathematics Department who take courses regression analysis as well as for general users outside Mathematics Department students who need regression analysis for analysis of research data in the framework of completion of final project, as well as the completion of other structured tasks within Udayana University and the wider environment.

Problem-based learning is not just a teaching method but also a method of thinking, because in problem-based learning can use other methods starting with finding data to draw conclusions[2]. The problem-based learning method is obviously trying to guide students to be able to think logically, to find cause and effect, and find the key to solving the problem and summarize it as an expected answer. Educating students to want to use their own minds and not just to accept something from others is a way that should be developed in the updating and refinement of education in general and teaching methods in particular in college.

Based on some of the above learning problems, as well as to prove scientifically supported by empirical data about implementation of problems-based learning, especially related to the optimization of learning processes that can maximize students learning outcomes, then need to be studied and tested empirically whether the PBL model with the support of e-learning media that will be applied effectively used to optimize the learning process and student learning outcomes in the course of Regression Analysis?

Based on the orientation about the effectiveness of the use of PBLmodel in an effort to optimize the learning process and student learning outcomes in the course of Regression Analysis, can be identified various research questions: what are the factors affecting the mastery of the concepts of regression analysis in students?; how misconceptions experienced by students in regression analysis courses?; what are the factors that influence the use of PBLmodel in the learning process ?, how effective the application of thisPBL in minimizing student misconception?; how is the influence of using PBLmodel on the mastery of the concepts of regression analysis ?. The problem of this research is focused on optimizing the learning process to improve student learning outcomes, with the aim of this research is to know the effectiveness of the implementation of Problem-Based Learning (PBL) model in optimizing the students' learning outcomes in the regression analysis course at Mathematics Department, Udayana University.

Problem-based learning model not just a teaching method but also a method of thinking, because in problem-based learning can use other model starting with finding data to draw conclusions [2]. Nasution [3] states that problem solving is a learning method that requires students to find the answer (discovery) without special help. By solving the problem students find new rules that are higher in degree even though he may not be able to formulate them verbally. Based on some opinions above can be drawn a sense that the problem-based learning model model is a way of teaching that provides opportunities for students to on their own initiative is able to do analysis and synthesis of the problems faced so obtained the solution.

2. Research methods

The research design used is Classroom Action Research. Classroom action research is a research conducted by teachers in their own classes through self-reflection, with the aim of improving teacher performance so that student learning outcomes are further improved[4]. According Arikunto[5], classroom action research is a reflection of learning activities in the form of an action, which deliberately appear and occur in a class together. Classroom Action Research by Kemmis & McTaggrat (Arends [6]), provides limits on implementation in the stages of planning, action, observation/evaluation, and reflection stage. Based on the aforementioned opinions it can be concluded that the meaning of classroom action research is a corrective action by practitioners to improve the tasks and conditions of learning.

This research is conducted in Mathematics Department, Udayana University, with consideration of Regression Analysis subject is compulsory subject of statistics interest field. Therefore, learning in Regression Analysis courses needs to be made more interactive, innovative, and IT-based, so that the misconceptions experienced by students to be reduced. The whole research was conducted for six months, while the implementation of giving treatment in the form of problem-based learning model with e-learning assistant, was done for four months.

Data collected to evaluate the implementation of the learning process are: (1) The result of the student's work in doing the test questions given by the lecturer of the subjects; (2) The result of the student's performance in doing regression analysis, especially the existing cases of misconceptions; (3) The results of interviews with students in accordance with the development of conceptual understanding; (4) Observations during the learning process; and (5) Evaluation results during the learning process. The subject of this research is the fourth semester students who take the regression analysis course at Mathematics Department, Udayana University, in the even semester of 2016/2017.

Data collection techniques in this study were conducted through tests, interviews, and observations. The test is done to see the conception of students about the concept of simple linear regression analysis, which is done before given the learning and to know the change of student conception after following the learning.Tests on student performance results in completing tasks on research cases using regression analysis. This test aims to determine the ability of students to analyse the data and interpret the results of analysis appropriately and correctly.

The results of the students' work on each test or task given will be explored through interviews, to gain a deeper picture of the progress of understanding or the difficulties students experience in learning, expressing their ideas, and feelings.Observation is used to document/record events during the learning activities. The observations contain descriptions relating to teacher activity, student activities, and situations and conditions during the lesson.

Data obtained from test results on learning with Problem-based Learning (PBL) model was analysed by statistical technique with variance analysis. Statistical analysis technique with variance analysis is used to test student scores on the three cycles performed. The hypotheses tested were:

H_0 : $\mu_1=\mu_2=\mu_3 = 0$ (the implementation of PBL model is not effective in improving student learning outcomes in the regression analysis course at Mathematics Department, Udayana University)

H_1: At least one μ is not equal to zero (the implementation of PBL model is effective in improving student learning outcomes in Regression Analysis course at Mathematics Department, Udayana University)

The calculated statistic used to test the above hypothesis is the F test statistic. Ho will be accepted if the value of F_{count} is greater than the value of F table with $\alpha = 0.05$ and Ho is rejected otherwise [7].

The procedure of action in this study begins with an initial reflection. At this stage, the test is given and followed by interviews to the students, with the aim of obtaining a clear picture of student conception in analyzing data using regression analysis. Then summarized the data obtained as a basis for establishing and formulating the design of learning. The next step is the implementation of classroom action research designed in three cycles. Each cycle includes stages: planning, implementation of the action, observation, and reflection.

3. Results and discussion

3.1. Implementation of action reseach in cycles i, ii, and iii

The result of the research shows that atthe beginning of the action, most of the students had problems with the lack of mastery of the application concept of linear regression analysis that was below the value of 65 of the Minimum Criterion Score. After the action is given through the three cycles of action research, the results obtained descriptively the average of student learning outcomes in cycle I = 54.15; Cycle II = 75.25; and cycles III = 90.15, which means an increase in students' understanding, indirectly also activities on the third cycle in addition to applying regression analysis, also improve misconceptions that occur in using regression analysis. The average increase from cycle I to cycle III shows that there has been an increase in the ability of students in solving cases that use regression analysis.

The increase of conceptual mastery between cycle I and cycle II by average 65.74%, between cycle II and cycle III the average increase in concept mastery percentage is 22.42% and after given action in cycle III there is an increase of conceptual mastery of analysis linear regression application by average 93.95% when compared with cycle I. Thus the average percentage increase in student mastery of the material provided between the action cycle I and after the action cycle III has increased significantly by average 93.95%.Descriptively student learning outcomes on the material given with PBL showed results that increased at the end of cycle III compared with the previous cycle.

Implementation of PBL model that is done well, earnestly, and professionally will be able to increase the student's mastery of the material concept given in accordance with expectations. Only with the condition of students who have been motivated to engage in group discussion activities in sol ving problems given by lecturers will get good results as well as will impact on the results achieved. In addition to good preparation by lecturers in facilitating learning, as well as support from e-learning media also contribute to the learning outcomes achieved by students. Provision of on-line learning objects in the form of text, images/photos, audio, video/animation, simulation, assessment (quiz/task/self-test /formative test/summative-test) provided through e-learning media strongly support students in accessing learning resources .

In this research, there is no further study about how this e-learning media contributes to the improvement of student learning outcomes. The use of e-learning facilities in this study solely serves to provide easy access to learning resources to students. Further studies on the impact and effectiveness of using e-learning in assisting learning need to be further investigated by other researchers.

Observations during the learning process can be noted that during the problem-solving activities in cycle I, it appears that some students seem to be good enough in cooperation, participation, initiative, thoroughness and courage to express their opinions, are in sufficient category compared to before cycle I. These aspects still need to be improved in the next cycle. In this first cycle students are given research cases using the correct regression analysis on terms and assumptions, with the results of the analysis is satisfactory. In this case the students have not found any problems that may be contained in the regression analysis.

The result of observation after the learning process of cycle II, student (group) has been improved from enough category to good category, that is cooperation aspect, participation in group, and courage to express opinion, while the aspect of accuracy and initiative remain in enough category. Based on observations of results in this cycle II, the emphasis needs to be given more on both aspects that have not increased. The initiative aspect is very important, besides the aspect of accuracy is needed in solving linear regression analysis problem. Troubleshooting is not done carefully will result in problem-solving that is not in line with expectations. This needs to be emphasized, because in this second cycle students begin to discuss the case of research which is wrong in the application of regression analysis and also misconception in analyzing it. The mistakes made by researchers, which many found such as the measurement scale of research variables are not appropriate.

The result of observation after cycle III is given, the whole aspect reaches the good category. Students have done their job well, cooperate in groups, participate actively in group, be honest, dare to express their opinion, and have increased their accuracy in solving problems and more initiative compared with learning in previous cycle. In the final stages of this cycle, the research cases discussed are more complex, because in addition to the students' mistakes and misconceptions in applying regression analysis, students are also emphasized on the interpretation of regression analysis results.

The results of observation during the learning process with PBL indicate that the PBL applied in this course can improve aspects of doing the task well, cooperate in groups, participate actively in the group, honest, courage to express opinions, improve student's accuracy in solving problems, and increase the initiative students in solving problems.

The findings of this study are in line with Graaff & Kolmos [8] explained that PBL education builds on the students' background, expectations, and interests. Further explained that students to be motivated to work much harder with the PBL model than with traditional teaching methods. With PBL students spend more time on their studies and more participating compared with traditional models.
The authors argues that the PBL models appear to inspire a higher degree of involvement in study activities and, consequently, a higher level of complex comprehension. A possible drawback is the risk of gaps in specific knowledge areas. Therefore, it is crucial that the students in a PBL curriculum become lifelong learners who have learned to take responsibility for their own learning process.

The results of Inel & Balim [9] explain the same thing, that the PBL method is results in positive development in students'academic achievement. PBL helps students learn by discussing their existing knowledge and skills. Further explained that PBL positively affects concept learning. This positive change is arguably caused by the fact that they are found in group environments to construct their knowledge.

3.2. Effectiveness of implementation of Problem-Based Learning Model

Description of student learning outcomes with the application of problem-based learning (PBL)model on the mastery of the application concept of linear regression analysis, on three cycles of action research conducted obtained results as in Table 1.

Table 1. Description of Student Learning Outcomes in Cycle I, II, and III

Cycle	N	Mean	Std. Deviation	Minimum	Maximum
1	20	52.55	22.01070	27.00	97.00
2	20	73.05	11.50229	50.00	98.00
3	20	77.85	1.46089	87.00	92.00

Sources: Primary data (2017)

The mean value of the three cycles is obtained that the mean of cycle III is the highest, and the mean of cycle I is the smallest, with the least diversity level in the third cycle (standard deviation = 1.46089). The standard deviation value shows how varied the student evaluation results are.The greater the standard deviation value, the more varied the evaluation value of the students. In this study it is expected that the evaluation value of students has a small variance, because it implies the ability of students who are almost the same. While the increasing mean indices from cycle I to cycle III indicate that the average student ability increases in each cycle. The minimum value from cycle I to cycle II is 27, 50, 87, based on this value is seen that the student's ability is increasing. However, based on the maximum value of 97, 98, 92, the tendency is not visible, this may be because there are no students who dominate to get the highest score.

The next analysis is to study the PBL model that is applied effectively to improve student learning outcomes on subject of linear regression analysis, the hypothesis tested using analysis of variance.

Table 2. Results of analysis of variance of student learning outcomes

	Sum of Squares	Df	Mean Square	F	Sig.
Between Groups	14340.400	2	7170.200	34.756	.000
Within Groups	11759.250	57	206.303		
Total	26099.650	59			

Sources: Primary data (2017)

The result of variance analysis in Table 2 obtained the value of $F = 34.756$ with the significance value of 0.000. The value of this significance compared with the significance value of $\alpha = 0.05$ is smaller, it indicates that Ho is rejected or H1 accepted, meaning that there are differences in learning outcomes in the application of PBL model assisted by e-learning applied into three cycles.

Table 3. Tukey Multiple Comparison Test Results

(I) cycle	(J) cycle	Mean Difference (I-J)	Std. Error	Sig.	95% Confidence Interval Lower Bound	95% Confidence Interval Upper Bound
1	2	-22.70000*	4.54205	.000	-33.6301	-11.7699
	3	-37.60000*	4.54205	.000	-48.5301	-26.6699
2	1	22.70000*	4.54205	.000	11.7699	33.6301

*. The mean difference is significant at the 0.05 level.

Which cycle is different? then continued with the Tukey multiple comparison test, the results are presented in Table 3. Tukey test results show that in cycle I compared with cycle II and cycle III obtained significant value 0.000, this significance value is smaller than the significancy level that has been determined that is 0,05, this means there are differences in student learning outcomes in cycle I with cycle II and cycle III. Tukey test results between cycle II with cycle III also resulted in a decision that states there are differences in student learning outcomes. Based on Tukey test results can generally be stated that with the implementation of PBL model applied in three cycles of classroom action research effectively can improve student learning outcomes in the course of Regression Analysis.

The findings of this study show that results the application of PBL is effective in improving student learning outcomes in the course of Regression Analysis optimally. Inel & Balim [9] explains that during the process of problem-based learning, students identify the information they lack, make up for these shortcomings in their knowledge about a subject, share information in group environment, and have the opportunity to construct their knowledge in social and cognitive terms. Authors argued that the problem-based learning method can be used to ensure better construction in students' minds for the concepts in science and technology course.

Research on the effectiveness of the application of this PBL has been done by many previous researchers. A number of studies show the pros and cons of the effectiveness of PBL.Strobel & van Barneveld [10] indicated that PBL is superior when it comes to long-term retention, skill development and satisfaction of students and teachers, while traditional approaches are more effective for short-term retention as measured by standardized board exams. Furthermore Kyeong [11] explains that although the background of learners might affect the effectiveness of PBL, the roles of the teacher in designing the problem-based scenarios remain the key to the successfulness of PBL. Author explained that in mathematical study, teacher's instructional ability is critical to engage students in gathering information and apply their knowledge in their respective fields. Simons, Klein, & Brush [12] explains that in general, the PBL process is anchored by a complex, ill-structured problem, the students proceed through a variety of activities to frame their understanding of the problem, access resources, increase understanding, and recommend solutions.

From another perspective, Ribeiro [13] in his article on the pros and cons of problem-based learning from the teacher's standpoint explains that PBL also seems to distribute teaching workload more evenly throughout the semester than traditional methods do. This hinders routinisation and constrains teachers' autonomy. On the other hand PBL appears to foster the teacher 's development of his teaching knowledge base, especially regarding the knowledge of the students, their reasoning modes and interests.

4. Conclusions and recommendations

4.1. Conclusions

After given the action in the form of Problem-based Learning (PBL) model with assisted E-learning, an increase in the average percentage of mastery of the concept between cycles I and II of 24.92%, in the second and third cycle of 16.63% and after the action is given in cycle III there is an increase in the average percentage of mastery of the concept of linear regression analysis of 39.91% compared to cycle I. So the average percentage increase in student mastery of the material given between the action cycle I and after the action cycle III average of 39.91%.

The result of the analysis of variance, shows that there is a difference of value on the implementation of PBL model that is applied into three cycles. Further test with Tukey multiple comparison test, the result of the three cycles gives significant difference of results, so it can be stated that with the implementation of PBL model applied in three cycles of classroom action research, effectively can improve student learning outcomes in Regression Analysis course.

Implementation of PBL model that is done well, earnestly, and professionally will be able to increase student's mastery of the concept of the given material. Only with the condition of students who have been motivated to engage in group discussion activities in solving problems given by lecturers will get good results as well, will also impact on the results achieved. In addition to good preparation by lecturers in facilitating learning, as well as support from E-learning media also contribute to the learning outcomes achieved by students. Provision of on-line learning objects in the form of text, images / photos, audio, video / animation, simulation, assessment (quiz/task/self-test /formative test/sumative test) provided through e-learning media strongly supports students in accessing learning resources .

4.2. Recommendations
In this research, there is no assessment of the extent to which this e-learning media contributes to the improvement of student learning outcomes. The use of e-learning facilities in this study solely serves to provide easy access to learning resources to students. Further studies on the impact and effectiveness of using e-learning in assisting learning need to be further investigated by other researchers.

Acknowledgments
The authors would like to thank theUdayana University, which has funded this Hibah Unggulan Program Studi (HUPS), through funding DIPA BLU Udayana University academic year of 2017

References
[1] Indriantoro S 2002 *Metodologi Penelitian Bisnis untuk Akuntansi dan Manajemen* (*Edisi Pertama)* (Yogyakarta: BPFE)
[2] Sudjana N 1996 *Cara Belajar Siswa Aktif dalam Proses Belajar Mengajar* (Bandung: Sinar Baru Algensindo)
[3] Nasution S 1992 *Berbagai Pendekatan dalam Proses Belajar Menga*jar (Jakarta: Bumi Aksara)
[4] Wardhani I GAK, Wihardit K dan Nasoetion N 2000*Penelitian Tindakan Kelas* (Jakarta: Universitas Terbuka)
[5] Arikunto S 2002 *Prosedur Penelitian Suatu Pendekatan Praktek* (Jakarta: Rineka Cipta)
[6] Arends R 2009 *Learning to Teach*. Sevent edition (New York: McGraw Hill companies,Inc.)
[7] Walpole R E 1995*Introduction to Statistics*. Terjemahan Bambang Sumantri (Jakarta: PT. Gramedia Pustaka Utama)
[8] Graaff E D and Kolmos A 2003 Characteristics of problem-based learning *Int. J. Engng*. **19** 57-62
[9] Inel D and Balim A G 2010 The effects of using problem-based learning in science and technology teaching upon students' academic achievement and levels of structuring concepts *Asia-Pacific Forum on Science Learning and Teaching* **11** 1-23
[10] Strobel J and van Barneveld A 2009 When is PBL more effective? A Meta-synthesis of meta-analyses comparing PBL to conventional classrooms. *Interdisciplinary Journal of Problem-Based Learning* **3** 44-58
[11] Kyeong H R 2003 *Problem-Based Learning in Mathematics* Digest (EDO-SE-03-07) (USA: Educational Resources Information Center)
[12] Simons K D, Klein J D and Brush T R 2004 Instructional strategies utilized during the implementation of a hypermedia, problem-based learning environment: a casestudy *Journal of Interactive Learning Research* **15** 23-33
[13] Ribeiro L R C 2011 The pros and cons of problem-based learning from the teacher's Standpoint *Journal of University Teaching & Learning Practice* **84**

IOP Conf. Series: Journal of Physics: Conf. Series **1040** (2018) 012024 doi:10.1088/1742-6596/1040/1/012024

Scaffolding in problem based learning to increase students' achievements in linear algebra

I N Gita and R A Apsari
Mathematics Department, Mathematics and Science Faculty
Universitas Pendidikan Ganesha, Bali – Indonesia, 81116

email: ra.apsari@gmail.com

Abstract. The aim of this research is to improve the undergraduate students' learning achievements in Linear Algebra course. This research was a classroom action research which has two cycles, each consist of four stages, i.e.: planning, implementation, observation/evaluation, and reflection. The subject of the present study was 23 second semester students of Mathematics Education Department, Universitas Pendidikan Ganesha, in academic year 2015/2016. The data about students' learning achievement was collected from students' written work in solving Linear Algebra test. The gathered data was analyzed descriptively. The results of this research showed that the implementation of scaffolding in problem based learning are able to improve students' learning achievement.

1. Introduction

Linear Algebra is one of the basic course in mathematics education department, which is appear in the second semester for the first year university's students. The course is generally about generalization of a formal theory. Most of the materials of the course are actually discussed in the high school. Therefore the students will be able to understand the content of the course if they mastered the mathematical concepts in senior high school. Unfortunately, the students were struggling in this pre-requisite knowledge, for instance in the use of symbols and in the mathematical problems which asking for proving something.

Stewart & Thomas [1] said that the linear algebra is one of demanding course because it is based on a mathematical theory and included many definitions which is served in a very frequent. Hence, the students hardly see the connection between their prior knowledge about mathematical objects in high school with the materials of linear algebra.

To deal with the condition of the students, the lecturer frequently ask the students to learn before they enter the class. Also, to enhance students' participation, the lecturer give score rewards for the active students. By doing those strategies, it is hope that the students' understanding in the material will be increase.

Another strategy needed to increase students' understanding in linear algebra is by implementing teaching and learning which stimulate creativity and reasoning ability. Therefore in this study, we implemented Problem Based Learning (PBL) with scaffolding. PBL had been chosen to solve the problem, because it is predicted that this model can be very helpful to guide the students from basic to abstract concept. It helps the students to be familiar with the mathematical problems and be confidence to solve it. As is stated in Dolmans, De Grave, Wolfhagen & van der Vleuten [2], PBL is a prospective learning model for the next generation as it is useful to develop the students' concept construction abilities, self-directed learning, cooperative works and contextualize mathematics based on the real problem encountered in society.

The scaffolding also become a support to enhance students' successful in learning without making them stop thinking. In other words, we can say that the combination of PBL and scaffolding can motivate the students to achieve their best.

PBL is a constructivism based learning model. It means, this model encourage the learner to construct their own concept by their own exploration through the given situational and authentic problem [3]. The major characteristic of this model is the use of problems and questions related to the upcoming discussed materials to start the lesson. When the students cannot understand the problem, lecturer will indirectly explain it by using clues and cross questions. There are five steps described in PBL [4], including: (1) posing problems, (2) clarifying and investigating problems, (3) solving problems, (4) organizing report, and (5) presenting the work. From those steps, it can be seen that PBL is a student centered learning which give chance to construct their knowledge.

H.S Barrows (in [5]) stated that Problem Based Learning is a learning model which based on the principal that problem is the start to integrate or to find a new knowledge. Nowadays, PBL has been used in various field of education. In university level, PBL is a learning model in which the students can solve an authentic problem to construct their own knowledge, develop their inquiry skill, train their higher order thinking, and to empowering their self-confidence and self-regulated learning. In this case, the lecturer should provide non-routine problems with various solving methods and approaches for the students. It helps them to think critically and creatively to solve the mathematical problems. Hence, it can be seen that PBL is a student-centered learning model which is in line with the constructivism point of view about learning.

As in constructivism, a new knowledge constructed based on the students' previous experiences [6]. In line with that it is reasonable to acknowledge the students' prior knowledge will affect their learning [7]. Therefore, the lecturer should connect the students' prior concept with the current discussed topic. By that, we can argue that mathematics learning is about building connection and understanding about mathematical knowledge.

However in many cases, PBL alone will be difficult to stimulate the students to learn, especially when the students were not used to think creatively in seeing a problem. Hence, the educator needs to provide a clue or limited guidance that can be used by the students to drive their elaboration. In education, this kind of help is known by scaffolding.

Scaffolding plays an important role for the success of the implementation of PBL. By definition, it can be acknowledged as a support for someone, given by classmate or adult who is more competent or capable in certain materials. Scaffolding means provide certain amount of support for children in the early steps and after that reduce it step by step to give them more opportunities to take over more responsibilities to finish their assignments [8]. In its interaction with PBL, we can explain that the students who get difficulties in construct certain concept will be given certain support until they able to master it. In this study, the support can be given by lecturer or from student to student.

Scaffolding is important especially in the beginning year of higher education, because the students are having an adjustment or transition from how they used to learn in school and how they expect to learn in university. The transition in the earlier stage of students' life in university plays important role that will determine the success of students' academic for the rest of study [9]. Therefore, they suggest to take care the idea of giving enough support for the students to have an optimal transition.

The scaffolding also become a support to enhance students' successful in learning without making them stop thinking. In other words, we can say that the combination of PBL and scaffolding can motivate the students to achieve their best.

The practical values of this research can be described as follows: (a) the participant of the current study will get learning experience by using PBL with scaffolding which hopefully able to increase their learning achievement, and (b) the result of the study can be used as a consideration to improve the quality of linear algebra teaching and learning and also in different courses.

Reflecting to the aforementioned reason, the question of the research is *how can the use of scaffolding in Problem Based Learning help the students to increase their achievement in linear algebra?*

2. Methods

This study was an action research study, which aimed to improve the quality of learning process in linear algebra for the mathematics students of Universitas Pendidikan Ganesha. Kemmis & Tanggart [10] stated that "*action research is a form of collective self-reflective enquiry undertaken by participants in social situations in order to improve the rationality and justice of their own social or educational practices, as well as their understanding of these practices and the situations in which these practices are carried out*" (p.5).

Hine [11] argued that action research can be benefit for the improvement of the teaching and learning in the classroom because it has a genuine setting as the students are keep in their nature and the teacher also serve them as a teacher who also become a researcher in the same time. Hence, it is clear that the teacher can understand her/his class better and can design many improvements based on the findings during the research.

There were 2 cycles employed in this study, each with four steps, including: planning, implementing, observing/evaluating, and reflecting. The design can be seen as in the following Figure 1.

Figure 1. The steps of action research

The subject of this study was 23 students of B-class mathematics education department. From those, only four students who actively engaged in discussion. The rest of the students were passive during the learning activities. In general, the students were not preparing themselves before they enter the class, for instance, by reading the upcoming discussed topic. They also afraid in the type of problems which asking about mathematical proof.

The result of the preliminary reflection before we conducted the study, found that the mean of the students' score in achievement test was 49.09 (from 100). One of the difficult problems according to the students' answer sheets was to determine the determinant of matrix with size 4×4. It was known that there only 6 students whose answer this problem with correctness value 50%, while 6 others answered with correctness value less than 50%, 5 students gained score 0 and the rest incorrectly chose a strategy to solve the problem. Specifically, the students who used incorrect strategy was using strategy in finding determinant for 3×3 matrix to solve the determinant for 4 × 4 matrix. Hence, it can be stated that they had a misconception in this topic.

To solve the aforementioned problem, this study employed the PBL model with the support of scaffolding to conduct a lesson in linear algebra. The PBL was employed by started the class by giving an authentic problem and asked the students to use their prior knowledge to elaborate it. For instance, when deal with the topic of linear equation system, a problem of price of some foods were given where the students need to create the mathematical model of it afterwards. From their experience in the high school, they were familiar with the linear equation with two and three variables, and were able to solve it by using elimination, substitution, or graphical methods. Started with that knowledge, the designed lesson challenged the students to think the general step of it. When the students lost, the scaffolding helped by asking students several questions, such as: "*what is actually happened when the same variable are eliminated? Why can they be substituted? What is the fact showed by the graph?*"

As mentioned before, we employed four steps in this classroom action study. The first step is *Planning*. There are three activities in this step, including: (a) fixing the schedule for face-to-face meeting, (b) preparing the students' worksheet, (c) preparing research instrument which is achievement test in linear algebra in form of essay test.

The second step is *Implementation*. There are four activities in this step, including: (a) distributing the students' worksheet, (b) group discussion, (c) presenting the result of the discussion in front of the class, and (d) making conclusion. During the *Implementation* stage, the researcher observe implementation of PBL with scaffolding and how the participation and activity of the students during the teaching and learning process.

The third step is *Observation/Evaluation* which will be done by checking the students' responses in answering the prepared achievement. The fourth step, *Reflection*, will be conducted in the end of the First Cycle by evaluating the students' difficulties in following linear algebra lesson. The result of Reflection of the First Cycle will be used to improve the lesson plan for the Second Cycle. The Second Cycle has the same steps as the first one, with improvement in the activity during the lesson. The gathered data will be analyzed by calculating the mean of the achievement test' score and comparing with the previous mean of the achievement test' score.

3. Results and Discussion

Before conducted the research, a preliminary reflection was done by giving the students a short explanation, continued by problem example and then an exercise task. In the end of the preliminary reflection, the students were given an achievement test using essay form. Based on the result of the test, it is found that the mean of the students' score was 49.09 (from 100). Specifically, there are 13 students (56.52%) whose score below 55 which means they failed in the test.

During the implementation data were gathered by using the research instrument and in the end of the cycle, a test about students' achievement in linear algebra was given. After the implementation of PBL with scaffolding, it is found that the mean of the students' score in achievement test in the first cycle was 65.17. From those, 8 students was gained A (34.78%), 3 students gained B (13.04%), 4 students gained C (17.39), 3 students was gained D (13.04%) and 5 students was gained E (21.74%).

In the second cycle, the mean score of linear algebra achievement test of the students was 74.74. The details can be described as follows: 12 students achieved A (52.17%), 2 students achieved B (8.70%), 4 students achieved C (17.39%), 2 students achieved D (8.70%) and 2 students achieved E (8.70%). The comparison chart of the students' range of scores during the first and second cycles can be seen in the following Figure 2.

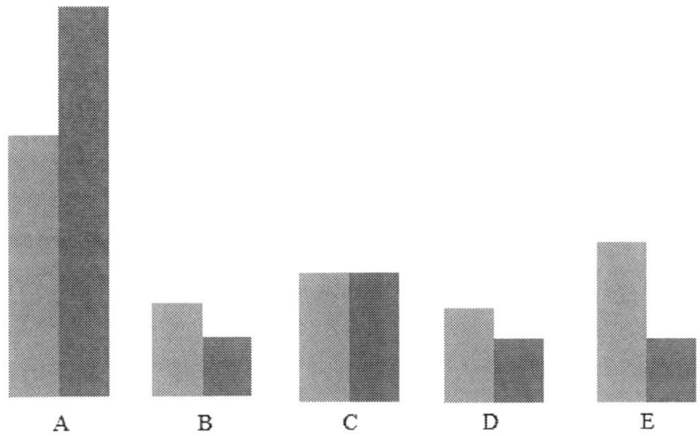

Figure 2. Comparison of students' score in the first and second cycle

Based on the evaluation of the achievement test conducted in preliminary reflection, cycle I and cycle II, it can be seen that there is an increase of the students' achievement in learning linear algebra, even though not really high. The improvement of students' achievement is caused by the implementation of PBL with scaffolding.

In the first cycle, even though the students' score increased, the minimum criteria of students' achievement was not fulfilled. It happened due to the not optimal implementation of PBL with scaffolding. Also, the lesson in the first cycle was dominated by certain group, especially the best one. In the cycle II, this factor was improved to support all students equally. This result confirm another finding which stated that in the learning process, cognitive aspect is not enough, PBL should be implemented because the students need to develop their metacognitive abilities and also their motivation in learning [12].

The analysis of the cycle II indicated that the students' achievement in learning linear algebra improved and the minimum score targeted was fulfilled. This is happened because the optimal implementation of PBL and effective scaffolding which give the students more opportunities to think and work cooperatively with their classmate in solving the given problems.

In general, the scaffolding provides support for the students in term of posing a hint question that can drive them to new connection in their knowledge construction. For instance, in learning Elementary Row Operation, the students got difficulty to understanding the process. The common strategy they used were remembering the steps as it given in the textbook. Without enough understanding, the written steps become inefficient. It is happened because the students ignored the actual goal of elementary row operation is to find the value of each variable and only focused on multiplying numbers. When this happened, the lecturer helped by reminding the students their process of finding the solution of linear equation in two or three variables. For instance, "*what was the aim of eliminating these variable?*" (pointing at the same variable) and after the students gave the correct responses, the question was continued by "*so, what should you do to have the similar answer as you do elimination? It just more variable, but the key idea has no different*".

Another important scaffolding was in helping students to differentiate and in the same time, seeing the similarities, between Gauss Elimination and Gauss-Jordan Elimination. Both were started in the same method and aim to find the solution of linear equation systems as well. Therefore, the students should not be confused why it looks similar. The different only that the Gauss-Jordan have the last step which is finishing the matrix in the reduced-row echelon form. Meanwhile it seems simple case, but from the observation, many students found it unclear, because they unsure whether the steps were same or not. To help the students grasping the method, the lecture ask them to create table of comparison between the steps of Gauss Elimination and Gauss-Jordan Elimination. By trying it themselves and not merely relying by the given procedures on the book, they were able to see that until certain steps, the process is same. They also concluded that the Gauss-Jordan Elimination was developed from Gauss Elimination itself and both are lead to the same answer.

Based on the improvement of students' score in linear algebra achievement test during the research, it can be seen that the implementation of PBL with scaffolding is able to support the students in learning. Besides that, it is powerful to manage more dynamic classroom, where the students actively discuss and construct their meaningful understanding, and the lecturer no longer as the center of the study. The previous study was comparing the effect of PBL and traditional method to the students' achievement in mathematics. It was found that the students who learn with PBL were performing better because in that class the students feel challenged to study and willing to discuss with their group when some obstacles emerged [13].

Moreover, the students who learn mathematics through PBL model also known to be more confidence, because they can making sense the problem and the topic they learned with the reality they have or know previously [14]. It becomes a powerful start for the students to be motivated in learning and solving mathematical problems and later improve the students' mathematical achievement.

The result of this study found that when the students have more chance to manage their own learning construction, they performed better (see the improvement of students in cycle 2 from cycle 1). This improvement can be preserved in future time [15]. The author argue that PBL is able to help students' to achieve their best by enabling the students to develop the ability of solving problem efficiently and develop the intrinsic motivation of the students. Therefore, the students will have a skill to life-long learning which will be meaningful for them.

This is in line with the argument saying knowledge is built in students' knowledge based on their prior knowledge [6]. Besides that, the students can construct their own knowledge through solving a problem which is started from what they already know into what they are going to learn. Therefore, it is necessary for the educator to connect the students' preconception towards a learning material before move to the current material.

4. Conclusion

Based on the results and discussion, it can be concluded that the implementation of PBL with scaffolding can improve the students' achievement in linear algebra. It can be seen from the increase of the mean of students' test score from 49.09 in the preliminary reflection to 65.17 in the first cycle and 74.74 in second cycle. Therefore, it is recommended to the lecturer to use PBL model with scaffolding in teaching linear algebra. To enhance the students' application ability, it also recommend to use authentic problem or at least a related to the-real-world-mathematical problems.

Acknowledgement

The authors would like to express gratitude to the Undergraduate Mathematics Education Study Program of Universitas Pendidikan Ganesha for the permission given for the research and for the students who become the important part of this study.

Refference

[1] Stewart S and Thomas M O J 2003 Difficulties and acquisition of linear algebra concepts *New Zealand Journal of Mathematics* **32** 207-215

[2] Dolmans D H, De Grave W, Wolfhagen I H and van der Vleuten C P M (2005). Problem-based learning: future challenges for educational practice and research *Medical Education* **39** 732-741

[3] Gijselaer W H 1996 Connecting problem based practices with educational theory *New Direction for Teaching and Learning* **68** 12-21

[4] Arends R I, Wenitzky N E and Tannenboum M D 2001 *Exploring teaching: An Introduction to Education* (New York: McGraw-Hill Companies)
[5] Suweken G, Gita I N and Hartawan GNY 2015 *Pelatihan pembelajaran Problem Based Learning (PBL) untuk mengimplementasikan 5M bagi guru matematika SMP Kabupaten Buleleng* Laporan P2M (Unpublished Paper) Mathematics Education Department Universitas Pendidikan Ganesha.
[6] Bodner G M 1986 Constructivism: A theory of knowledge *Journal of Chemical Education* **63**, 873 - 878
[7] Novak J D and Gown D B 1985 *Learning How to Learn* (New York: Cambridge University Press)
[8] Slavin R E 1997 *Educational Psychology: Theory and Practice* Fourth Edition (Needham Heights: Allyn and Bacon Publisher)
[9] Hultberg J, Plos K, Hendry G D and Kjellgren K I 2008 Scaffolding students' transition to higher education: parallel introductory courses for students and teachers Journal of Further and Higher Education **32** 47 – 57
[10] Kemmis W C and Taggart R M 1988 *The Action Research Planner* (Geelong Victoria: Deakin University Press)
[11] Hine G S C 2013 The importance of action research in teacher education programs *Issues in Educational Research* **23** 151-163
[12] Mayer R E 1998 Cognitive, metacognitive, and motivational aspects of problem *Instructional Science* **26** 49–63
[13] Fatade A O, Mogari D and Arigbabu A A 2013 Effect of problem-based learning on senior secondary school students' achievement in further mathematics *Acta Didactica Napocensia* **6** 27-44
[14] Kartikasari A and Widjajanti D 2017 The Effectiveness of problem-based learning approach based on multiple intelligences in terms of student's achievement, mathematical connection ability and self-esteem *IOP Conf. Series: Journal of Physics: Conf. Series* 1-7
[15] Hmelo-Silver C E 2004 Problem-based learning: What and how do students learn? *Educational Psychology Review* **16** 235 - 266

Using puzzle to encourage students to do problem posing

I M Candiasa[1], N Santiyadnya[2] and G K A Sunu[3]

[1]Educational Science Study Program, Post Graduate Faculty, Universitas Pendidikan Ganesha, Bali – Indonesia, 81116
[2]Electronica Technology Education Study Program, Faculty of Technology and Vocational School, Universitas Pendidikan Ganesha, Bali – Indonesia, 81116
[3]Educational Administration, Post Graduate Program Faculty, Universitas Pendidikan Ganesha, Bali – Indonesia, 81116

Abstract. Problem posing is very important in learning mathematics. Many studies have indicated that problem posing can improve the ability to solve mathematics problems. Therefore it is necessary to encourage students systematically to do problem posing. This research tries to pack problem posing in puzzle frame in order to increase student's interest and ability to do problem posing. The puzzle game which is then directed to problem posing is assured can arouse students' enthusiasm, thus encouraging their involvement in problem posing activities. Two groups of junior high school students each consisted of 31 and 32 students selected as samples. The first group follow mathematics class with problem posing learning model that preceded by playing a puzzle but still in a mathematical environment. The second group follow the mathematics class with problem posing learning model only. At the end of the research student's interest and ability to do problem posing are compared by using MANOVA test. Wilks' Lambda test results F = 10.8 on significance (p) = 0.002. Based on these findings, it is concluded that the problem posing learning model in mathematics that is preceded by playing puzzles can increase students' interest and ability to do problem posing.

1. Introduction

Naturally, in the daily interaction people tend to ask a question if they want to know something. Children tend to ask everything they seen, heard or felt. Different things happen in the classroom. Children tend to be reluctant or shy to ask questions. Despite being encouraged to ask questions, the child is still reluctant to ask questions. There are several reasons that allegedly motivate students to be reluctance to ask questions. First of all, the students feel embarrassed to be seen as too stupid by his friend for always asking. For the second one, students feel less confident with the question asked, whether the question is useful. For the third, students are difficult to ask questions because they are not aware of the problems faced.

Conditions above occur also in learning mathematics. Mathematics teachers often complain, why students rarely ask during learning. In addition to the embarrassment and lack of confidence, the inability of students to realize the problems faced is the dominant reason. The low ability of students to be aware of problems in learning mathematics can occur because during the learning students only trained for problem solving. In fact, in learning mathematics, not only problem solving skills that must be pursued but creativity of mathematics must also be obtained, therefore, learning mathematics must consist of two aspects, namely problem posing and problem solving [1].

The problem posing theory was applied by Freire to awaken the oppressed at once against the bank concept of education that became the beloved of the oppressors [2]. Freire's attention focused on the awakening of the oppressed. The oppressed are encouraged to fight for their humanitarian rights that the oppressors oppose. Bank concept of education considers education such as depositing money in the bank. Students are considered as empty boxes to be filled with knowledge. There is no opportunity for students to ask or even to argue. They are only welcome to receive what is given by the teacher. Problem-posing education is an approach to education where teacher and student approach a problem together [3].

Problem posing and problem solving has been identified to be central theme in mathematics education [4]. Problem posing has been used as both a metric of conceptual understanding and also as a pedagogical tool [5]. Activities in mathematics learning are dominated by problem posing and problem solving. The problem solving approach in learning mathematics is used to practice conceptual understanding and application of concepts. Problem posing is mostly used to extract information from students regarding concepts that have not been understood. The information is then used by teachers to provide remedial feedback. Submission of problems from students may also be related to advanced materials. When this happens, the teacher's feedback will become enrichment for the students.

Problem posing is a crucial component for mathematics discipline [6]. The proposed problem is formulated from a concrete situation [7]. Students can pose problems in learning mathematics if they already realize they have not understood the problem. If the proposed problem gets a solution from a friend, then their knowledge will increase and trigger a new problem to be raised. This process will be continued, where the process of problem posing and problem solving will like a cycle that occurs repeatedly. When the process occurs in a sustainable manner, the student's ability in mathematics will increase continuously.

Problem posing refers to creating a completely new problem or revising the problem to be resolved [8]. Problem posing may occur before, during, or after the settlement of a problem. In an effort to solve a problem, students may ask questions before starting to find the solution of the problem. Similarly during problem solving, if deemed necessary, then the student may ask a related question or problem. Even after solution is obtained, students may pose another identical problem. That process occur continuously, so each task can cause one or more other tasks. When students formulate new problems, it can foster the sense of ownership that students need to take for constructing their own knowledge [9].

Problem posing has defined as a means of instruction where the students construct questions in response to different circumstances, namely real life situations, another mathematical problem, or the teacher [10]. The quality of the problem posed by the students in learning mathematics can be used as an instrument to measure the student achievement in mathematics. On the other hand, the process problem posing followed by problem solving that occurs iteratively over and over can be a medium of learning mathematics for students. By utilizing the existing thinking skills, students can take advantage of the existing situation to redefine or revise the given problem. Problem posing skills can be developed by giving students an ill-formulated or partially formulated problem and asking students to redefine it [8]. The main activity of problem posing is to compose new questions based on given mathematical tasks [11].

Problem posing education aims to strengthen the student's skills of thinking critically or contemplating on the object of knowledge and reasons for his/her existence [2]. Students can compose a new problem from existing problem only if they understand the existing problem and can solve it. They will then think creatively and reasonably to formulate the next problem. If this condition occur continuously, the student's critical thinking skill will arise significantly. When this condition is achieved, the student will start acquiring knowledge through a feeling of epistemological curiosity [2]. At a time even they can formulate really a new problem, without analogize from the other problem.

Problem posing has long been viewed as a characteristic of creative activity or exceptional talent [8]. According to their ability students must be create a new problem. On the other hand they may be re-compose an existing problem. It means, students compose the existing problem in their own version. The do it in order to make the problem easier to solve. The students can be use heuristics approach, such as analogy, working back, or may be divide the problem into several parts. This process need not only critical thinking but also creative thinking. In addition, this process refers that the students not to be easy to give up but they must try again and again. They must maintain their tenacity and their adversity to learn.

Empirical experience shows that the problem posing learning model that have been applied in learning mathematics so far have not been able to encourage students to be active and creative to pose a meaningful problem. Students tend to be passive, just wait the problems from their teacher. Posing the problem is often done in a forced way, just to meet the requirements. This is seen from the frequency of posing problem and the quality of the problem posed. It appears that shyness and lack of self-confidence are still the dominant factors that influence the problem posing. In other words, overcoming shyness and a lack of confidence to pose problems should be treated first. A commitment to creating an environment in which problem posing is a natural process of mathematics learning is required [12].

There are several activities that students can do in learning mathematics with problem solving model. As a first step, students can ask questions to complete the previous problem solving. Furthermore, students can raise the problem by analogy of the problem already exists and has been resolved. With slight modifications to the denomination or variables of an existing problem, students can pose new problems. Analogy can also be done by changing the context of the problem. Mathematical problems for application in one field can be modified for application in other fields. At the end, it is expected that students are able to pose complex problems from simpler previous problems.

Well-designed problem-posing tasks for students also require skilled teachers who can manage the complexities of such contexts [13].The role of teachers that is very important in learning with the problem posing is to create a natural atmosphere, so that student interest can be raised. Experienced and skilled teachers begin problem posing by not strictly limiting problems students can ask. What the teacher wants is a student willing to pose a problem. If the student is willing to pose a problem, then the problem limits and the quality of the proposed problem can be corrected while proceeding. Teachers' skills in this situation include the ability of self-management in order to serve students optimally.

This research tries to train students to be active and creative to pose problems in learning mathematics. The tried technique is to package the process of the problem posing into similar activities that are often done in everyday life, such as puzzles. The main target is to increase student interest and student ability to pose a problem. When their interest has grown, then the next goal is to improve the ability to create and pose a problem. There is a common characteristic between puzzles and problem posing, such as one person or may be group raises a problem to another and the other tries to find a solution or answer.

Puzzle means to offer or represent to a person a problem difficult to solve [14]. Puzzle also means about facts or events that need to be searched logically according to chronology or analogy. Puzzles can be made in various fields, such as mathematics, natural science, history, geography, economics, or general things in everyday life. Puzzles can pose a concept, principle, or procedure. At the initial level, the puzzle material is limited to concepts, then gradually increasing to more complex things, such as principles or procedures.

Puzzle can exercise (over oneself, one's brain, etc.) over some problem or matter [16]. It's a question or problem that requires thought, skill, or cleverness to be answered or solved [17]. Puzzles can add knowledge and insight. It needs sufficient reference so someone can ask a question or puzzle to his friend. Conversely, a friend who must answer also needs references and reasoning enough to be able to answer a question or statement of a received puzzle. When students are already engaged in a puzzle game, then they will do their best to answer the puzzle posed by a friend. Furthermore, they will do their best to also find new problems that will be given to their friends. If the puzzle is done in a mathematical frame, then the understanding of the mathematical concepts of students will be improved.

Puzzles are often done by children at home, in the community, or at school during breaks. The puzzle has the same characteristics as the problem posing. During a puzzle game, one party asks a question to the other party and the other party tries to answer. Questions are submitted in accordance with the agreement. Puzzles can be done individually or in groups. The atmosphere of play with the cheerfulness that characterizes children is evident in puzzle activities. Informal impression allows students to engage in puzzle activities freely as without pressure.

The problem posing is considered the student as a formal activity, resulting in frequent shame or lack of confidence to do. The puzzle on the other hand is considered the student as an informal activity, so they are willing to perform casually without any burden. Relaxing conditions and no burden can spur the open mind to bring up new ideas. If the problem posing is preceded by a puzzle, then the student's interest to make a problem can be grown, so that they can then process the problem solving effectively. The change from puzzle activity to problem solving activity is done slickly, so students do not feel it. Conversely, if the filing of the problem is done immediately without the preceding puzzle exercises, then the shyness and lack of self-confidence of students will remain.

This study tried to increase the student's interest and ability to do problem posing in learning mathematics by using puzzle. For this objective, the following problems were researched: 1) how are the student's interest and ability to do problem posing in learning mathematics preceded with puzzle compared with those without preceded without puzzle? 2) what are the obstacles faced by students in learning mathematics by problem posing preceded by puzzle? 3) If there are any obstacles, what are the sugestions given to the students?

2. Methods

The research use quasi experiment approach. Two groups of junior high school students in grade of eight at Singaraja (Bali) were selected as sample by using cluster sampling technique. The first group consisted of 31 students was experiment group that follow mathematics class with problem posing learning model that preceded by playing a puzzle but still in a mathematical environment. The second group consisted of 32 students was control group that follow the mathematics class with problem posing learning model directly.

At the end of experiment, student's interest and ability to do problem posing were measured. Student's interest was measured by using questionnaire while student's ability to do problem posing was measured by evaluating posed problems using rubric that was specially formulated for that purpose. Finally, student's interest and ability to do problem posing from experiment group and control group were compared by using MANOVA preceded by data normality test, data homogeneity test, equality of covariance matrices test, and multicollinearity test.

3. Results and Discussion

Saphiro-Wilks test for data normality test found that all of data were normally distributed and Pearson correlation found that no co-linearity between student's interest and ability to do problem posing. Levene test found that variance of data were homogeneous and Box's M test found that matrices of covariance were homogeneous also.

From MANOVA, Wilks' Lambda test results $F = 10.8$ on significance rate$(p) = 0.002$. It means students with problem posing preceded by puzzle and students with problem posing directly have different interest and ability to do problem posing. Furthermore, tests of between-subjects effects results $F = 28.35$ on significance $(p) = 0.002$ for student's interest and $F = 18.28$ on significance $(p) = 0.005$ for student's ability to do problem posing. The results show that student's interest from group with problem posing preceded by puzzle was better than that from group with problem posing directly. On the other side, ability to do problem posing from group with problem posing preceded by puzzle was also better than that from group with problem posing directly.

Puzzle exercises was able to enhance students' interest in doing problem posing. The relaxed environment that characterizes the puzzle brings them to a relaxed condition when it comes to problem posing. This condition make students can communicate openly without any burden, either with colleagues or with teachers. Such a free condition makes them able to think deeper, analyze more closely, and reason more quickly. In addition their perception also become more opened in relaxed condition. Those situations will arise students' interest to do problem posing. On the other side, those situation can enhance students' ability in formulating and solving problems.

Interview with some students found that the puzzle, even though in the mathematical contents were able to bring them into the playing situation. The exchange of ideas happens smoothly without fear of being wrong. Therefore, they feel faster in finding new problems. Initially they tried to imitate the problem posed by his friend with a little modification. Over time the modifications of the problem they make were increase significantly. One day they were able to formulate their own problems that will be posed. The posed problem was really new, it is not from modification of existing problem.

The role of teachers in packing problem posing and puzzle is crucial. At the beginning, the teacher is not too strict limiting the problems posed by students. Teachers still tolerate, even if the problem that students ask in the game was far from the mathematical content to be discussed. The main objective at this early stage was to foster students' interest in posing a problem. Therefore, as long as the problem was true or reasonable, even if the issue was slightly out of context, the teacher was not paying much attention. If there were any shortcomings or errors in the proposed problem, the teacher will make improvements.

The natural atmosphere that was impressed in the puzzle game was carried away as they entered the problem posing. The impression that the problem posing was formal activity can be forgotten. Teacher expertise in transforming puzzle games into problem posing activities strongly supports the successful increase of student interest in doing problem posing in learning mathematics. Disposition of problem complexity and the adjustment of problems with the context by experienced teachers can make the process of problem posing better and more systematic [13]. At the end, it will lead to the improvement of student interest and ability to do problem posing.

There was no significant obstacle take place during the experiment. The obstacles that were estimated occurred while experiment such as student difficulty to find mathematics puzzle and student difficulty to move from puzzle to problem posing. But until the end of the experiment students were able to find the appropriate mathematics puzzle, despite they were vary in consuming time. In addition, students also moved smoothly from puzzle to problem posing. But once again, they were vary not only in consuming time but also in the quality of the posed problem.

4. Conclusion

Puzzle exercises bring a positive influence on the increased interest and ability of students to do problems posing in learning mathematics. Students who follow mathematics learning with problem posing model preceded by puzzle exercises have better interest and problem posing abilities than students who follow mathematics learning with problem posing directly. Mathematics that is still considered difficult and scary for some students can be arranged by creating a relaxed and informal atmosphere. If students still think that math is difficult and scary, then their learning achievement is difficult to improve.

Puzzle games can create a relaxed and informal atmosphere, so it is suitable to be applied as a first step in mathematics learning, such as in mathematics learning with problem posing. Active student involvement in learning should still be pursued by creating an atmosphere that can attract students' interest. The expected final goal is better learning outcomes, not just cognitive but also affective and psychomotor. Mathematics is learned to be applied, so the various skills that exist in mathematics should be excavated. All that can be achieved if students are actively involved in learning, such as in problem posing.

Acknowledgement
The authors are thank to the all people who contribute to the finalization of this paper. Especially for the colleagues at Mathematics Department Universitas Pendidikan Ganesha, for never ending support and permission to do the research.

References
[1] Xia X, Chuanhan L and Wang B 2008 Research on Mathematics Instruction Experiment Based Problem Posing *Journal of Mathematics Education* **1** 153-163
[2] Durakoglu A 2013 Paulo Freire's Perception of Dialogue Based Education, International Journal on New Trends in Education and Their Implications **4** 102 - 107
[3] Freire P 2005 *Pedagogy of the Oppressed* New York: Continuum
[4] Pittalis M, Christou C, Mousoulides N and Pantazi D A P 2004 Structural Model for Problem Posing, Proceedings of the 28th *Conference of the International Group for the Psychology of Mathematics Education Paulo Freire's Perception of Dialogue Based Education* 49–56.
[5] Yoonsuck C and Timothy A M 2012 From Problem Solving to Problem Posing, *Brain-Mind Magazine* **1** 7 - 8
[6] Sengul S and Katranci Y 2012 Problem solving and problem posing skills of prospective mathematics *Procedia - Social and Behavioral Sciences 69 International Conference on Education and Educational Psychology (ICEEPSY 2012): Elsevier Ltd* 1650 – 1655
[7] Stoyanova E N 1997 *Extending and exploring students' problem solving via problem posing* Retrieved from http://ro.ecu.edu.au/theses/885
[8] Silver E A 1994 On mathematical problem solving *For the Learning of Mathematics.*
[9] Lavy I and Shriki A 2007 Problem Posing as A Means for Developing Mathematical Knowledge of Prospective Teachers Proceedings of the 31st Conference of the International Group for the Psychology of Mathematics Education **3** 129-136.
[10] Ghasempour Z, Bakar M N and Jahanshahloo G R 2013 *Innovation in Teaching and Learning through Problem Posing Tasks and Metacognitive Strategies* International Journal of Pedagogical Innovations **1** 53-62
[11] English L D and Watters J J 2004 Mathematical Modelling With Young Children *Proceedings of the 28th Conference of the International Group for the Psychology of Mathematics Education*, 2004 **2** 335–342
[12] Lin P 2004 Supporting Teachers On Designing Problem-Posing Task As A Tool Of Assessment To Understand Student's Mathematical Learning, *Proceedings of the 28th Conference of the International Group for the Psychology of Mathematics Education* 257–264
[13] Singer F M, Ellerton N and Cai J, Problem-posing research in mathematics education: new questions and directions *Educational Studies in Mathematics* **83** 1-7
[14] Webster A M 1983 *Webster's Ninth New Collegiate Dictionary* (Springfield Massachusetts: Merriam-Webster Inc)
[15] The American Heritage Dictionary 1985 (Boston Massachusetts: Houhton Mifflin Company)
[16] Dictionary.com http://www.dictionary.com/browse/puzzle
[17] Webster M, Learner's Dictionary http://www.learnersdictionary.com/definition/puzzle.

Assessing by doing: A Balinese culture based assessment

N M S Mertasari[1], I M Yudana[2] and N Gita[1]

[1]Educational Research and Evaluation Post Graduate Faculty, Universitas Pendidikan Ganesha, Bali – Indonesia, 81116
[2]Educational Administration, Post Graduate Program Faculty, Universitas Pendidikan Ganesha, Bali – Indonesia, 81116

Abstract. An integrated assessment model for assesing cognitive, affective, and psychomotor simultaniously has been developed for formative assessment in computer class. The assessment model was adopted from a Balinese culture based assessment. Many of the skills in Bali are learned through a collaborative process between skilled workers and young workers who help them. Through observations and in-depth interviews with some community figures it was found that skilled workers taught skills to young workers while they are working together. If the process runs normally, then in the future, young workers will become skilled workers. During the collaboration process, skilled workers evaluate the learning progress of young workers. Evaluation is done through a process similar to analogy. The skilled worker gives an example of performance, then the young worker is asked to imitate the example. Young workers will attempt to implement such performance examples, while expert workers do assessments while working. Assessment is done in an integrated manner including cognitive, affective, and psychomotor components. The assessment model has been tried to be adopted for formative assessment in computer class and developed in the form of performance assessment. Students are given problems to be resolved and similar problems with it's solutions are attached. The student solve the problem by doing analogy with the problem that has been solved. The assessment rubric includes the cognitive, affective, and psychomotor components. Gradually the similarity of the problem is reduced, so that the students are finally able to solve new problems without any examples.

1. Introduction

The development of academic competence and character building are the two main goals of education in schools. These two goals are inseparable in preparing students for entering the workforce, for further education, for lifelong education, and for being good and responsible citizens. The development of academic competence contributes to the improvement of abilities and skills. Character building helps form attitudes and behaviors, such as honesty, integrity, respect, responsibility, self-discipline, self-management, and self efficacy. Character education creates an environment that is expected to improve the effectiveness of learning activities. The hope is that no student can finish school by mastering academic competence but lacking in character. Schools with good quality character education tend to produce graduates with higher academic ability [1].

Recently the school has been pursuing a good and true moral mission. Moral is an essential component in the effort to nurture and develop the ideas and efforts of educators. The moral mission intended here is not to point to religious beliefs, but morals that all peoples may understand as they have responsibilities for each other. Paradigms that are held on this moral mission are including: 1) education is a moral activity; 2) the studentis a youth who can be forged very short and crucial; 3) what is learned and what is not learned is very important; 4) what is customary and what does not become a habit has consequences on the students; and 5) what one person believes to be good and right is in accordance with the general outlook of life [2].

Nowadays, children learn more about the habits and morals from their groups and mass media such as television, magazines, newspapers, or internet, so the attention for experience gained in schools is not as much as in a past decade. The teacher's role has been reduced only as a technician, using various strategies to help transfer information and skills to students. The meaning of the word teacher as someone who helps the child to shape himself better has been reduced to helping students to improve their competencies, skills, or techniques. On the other hand, students perceive that the services provided by teachers are their right, so do not view education as their responsibility. This attitude is clearly not beneficial for the maintenance of good and right relationships between teachers and students.

People feel that the decline in moral values and character is caused by several factors. These factors are including: 1) mass media such as television, film, magazines, or other mass media that present violence, drug use, sexual deviation, theft, and academic fraud; 2) lack of role models because many artists, politicians, or leaders are promoting lifestyles that conflict with moral and ethical principles; and 3) problems in the household. Many students expressed boredom in school, which may be caused by a medium of entertainment that is all beautiful and presented on a large scale or there is a problem at home.

Moral, ethics, politeness, rules, and rights or obligations must be striven for the school to maintain its function. The key to successful education in schools is the quality of interpersonal relationships involved in school. It is increasingly clear that ethics and morality are central issues in children's education. Communities, educators, and parents strongly insist that children learn well to be productive, kind, and useful to humanity. The child must be taught to think rationally and responsibly. In addition, children should be taught to enjoy learning, as long as they want to live in a democratic world, where everyone has the same rights, duties, freedoms, interests and responsibilities.

Character education is integrated within the school environment, whether in the curriculum, teaching strategies, or co-curricular programs. The difficulties experienced by teachers in implementing character education and field studies in an integrated manner occur in the assessment, especially formative assessment. Formative assessment is a process by which teachers apply various instruments and strategies to determine what competencies the students have achieved, identify gaps between the competencies of the students with the competencies of the standards to be achieved, and plan the next learning to improve student learning outcomes [3], [4] &[5]. Formative assessment is a window for teachers to observe student learning progress.

Formative assessment encompasses all the processes that teachers undertake during learning to get feedback about student learning progress [6]&[7]. Teachers are required to be able to apply formative assessment in an integrated manner between character education and field of study. For teachers, information gained from formative assessments is very useful for improving the learning process. In contrast to students, information from formative assessments is very useful as a self-evaluation tool to make changes in learning steps for learning objectives that have been outlined.

There are several elements that must be met in the implementation of the formative assessment to run more effectively, namely: 1) the teacher adjust the learning process with the response obtained from the assessment, 2) the students receive feedback related to the learning outcomes with suggestions on what should be done to improve the learning process, and 3) students participate in formative assessments through self-assessment [8]. The students' responses to formative assessment will be the teacher's guide to adjust the learning process. In the meantime, teacher feedback on student responses will serve as a guideline for students to adapt their next learning steps.

Formative assessment and feedback can be utilized by teachers to improve student self-regulation [4], [9]. Self-regulation can be manifested in the form of setting goals and learning objectives, organizing learning activities, choosing strategies for achieving learning objectives, monitoring learning activities, managing learning resources, regulating reactions to external stimuli, and some other self-activities. The student responds to the tasks assigned during the lesson. Feedback received from teachers is used to monitor the progress of learning. The results of these monitors are used as guidelines to improve the learning process to achieve predetermined learning objectives.

Formative form of assessment can be anything, which is logically important to be done by students and able to provide feedback to students as soon as possible [5]. Material outreach and time allocation are arranged in such a way that there is conformity between the characteristics of the students and the competency standards that the curriculum demands. The character education learning outcomes are more about affective (attitude) and psychomotor (behavioral) domains, such as honesty, responsibility, readiness to work hard, independence, and so on. Therefore, the selected assessment form should be able to measure the cognitive, affective and psychomotor domains well.

The test is an appropriate form of assessment for measuring cognitive domains. Meanwhile, the appropriate form of assessment to measure affective and psychomotor domains is questionnaire, inventory, portfolio, interview, or direct observation. Its difficult for the teachers to apply some form of assessment when conducting a formative assessment. The difficulties can occur either on the preparation of the instrument, the implementation of the assessment, or on the provision of feedback. Therefore a formative assessment is needed that can measure the academic ability and character education as well.

For the need above, a research was conducted to construct assessment that can simultaneously measure learning outcomes in terms of academic ability and character education adopted from Balinese culture based education, hereinafter referred to as integrated character education assessment.The research addressed to solve several problems, namely what are the indicators of integrated character education assessment, how the implementation of the assessment is, and how the validity of the assessment is.

Balinese culture-based education that is adopted is the process of education on the concept of mutual cooperation called *ngayah. Ngayah* means offer or provide social or religious services [10] or to help out with a religious or community activity [11]. The people involved in the mutual cooperative process are from children, adolescents and adults. Therefore, in the mutual cooerative process occurs the process of learning and assessment. While teaching, adults assess teenagers and children, and teenagers assess children. Assessment occurs integrally in knowledge, skills, attitudes, and behavior.

2. Methods

The research aims to construct an integrated formative assessment encompassing a field of study and character education. The data were collected through observations and completed with data obtained from in-depth interviews.The observations were done on places that were doing social activities or religious activities with mutual cooperative. In-depth interviews were conducted on respondents determined by snowball technique. The interview begins with the first selected respondent, then the next respondent is determined by the direction of the first respondent [13]. Interviews with respondents were discontinued when data were considered saturated. The data were analyzed by Miles & Huberman method, consists of data reduction, data display, and conclusion drawing or verification [14].

3. Results and Discussion

The literature review found that from the very beginning, the Balinese learned various skills informally through culture-based education. Carpentry, agriculture, animal husbandry, arts, including the skills of making religious ceremonies tools studied by Balinese through the learning process in various social organizations and religious organizations. Mutual cooperative in Balinese daily life to share among others also includes the learning process. In various activities of social or religious organizations there is always a process of learning from seniors to juniors. Both seniors and juniors work in various organizational activities. Senior work while teaching junior, while junior work while learning from senior.

The learning process includes not only knowledge and skills, but also includes ethics, manners, appreciations, responsibilities, and other forms of character education. The seniors are teaching skills to the juniors. While communicating, seniors also insert moral messages, ethics, manners, responsibilities and so on. When juniors make mistakes, then the senior provides guidance or direction. Conversely, when juniors do the right thing or a good thing, then seniors will give moral awards. Seniors will uses these achievements as an example for other juniors. It can accumulate a sense of brotherhood and build togetherness through the activities undertaken with mutual cooperation without any expectation of money or other materials in return [15].

The commonly applied learning model is giving examples. Senior gives examples of processes, products, attitudes or behaviors that are then imitated by juniors. Furthermore, seniors provide some sort of training to the juniors with analogy techniques. Senior gives an example and provides problems similar to the given example. With various efforts the juniors will solve the problem. Problems or tasks can be individual or group tasks, depending on the type of problem. In the context of social work in Bali, the majority of work is done voluntarily. The seniors volunteer to educate the juniors, on the contrary the juniors are also volunteering and self-awareness learning from seniors.

While working and educating juniors, seniors also conduct assessments for juniors. Assessment is done through direct observation. Assessment includes skills, attitudes, and behaviors. Assessments use unwritten instruments, but are only remembered by seniors. If a junior shows inappropriate skills, then the senior will provide coaching with examples. When a junior makes a mistake in behaving, the senior will provide direction on the correct attitude or behavior. Such attitudes and behaviors include ethics, manners, appreciation, courtesy, and responsibility.

If the above assessment model is adopted in the school, such integrated assessment can be done not only for classroom activities, but also for out-of-class activities. Thus, other school staff may be included as evaluators, as long as their competence is appropriate to the activity. Thus, specifically for character education assessments, other than school staffs, families, churches, and other communities responsible for character education may be involved in the assessment of character education for an established common goal [16]. In this case, the assessment instrument must be well formulated, so that all evaluators can conduct the assessment appropriately.

The relevant instrument for formative assessment of integrated academic education and character education is inventory. Inventory is a collection of statements that state the characteristics of the student's skills, attitudes or behaviors [17]. The statements include a range of scores that the evaluator can select in terms of the student's skills, attitudes or behaviors. The better the students' performance, attitudes, or behaviors are shown, the higher the score will be. In the developed instrument, the range of scores provided 1 to 5, which 1 is at the ugly pole and 5 is at the well pole. The evaluator gives the scores according to the observations made.

Collection of formative assessment results can be used as portfolio because it contains some collection of works done by students. Portfolio as a personal collection that contains evidence of works (artifacts) as well as student reflections on achievement, development, strength, and the best work as a result of learning [18], [19]&[20]. A directed portfolio can improve students' ability to formulate learning needs, select meaningful learning activities, and complete learning tasks to lead to future learning [21]. Portfolio assessments are viewed as alternative assessments, authentic assessments, or performance assessments [20]&[22].

The developed assessment model has been tested in the formative assessment of computer programming. Students are given programming problems. If the problems presented are new, then the teacher includes an example of a solution that students can use to do the analogy. Students' work are then assessed using the inventory already developed. Inventory contains an assessment of process components, products, and attitudes. Assessment results are then given feedback and returned to the student. Students are asked to collect their work in one folder to be viewed and reviewed again whenever needed. At the end of the semester the collection of the results of the assessment is submitted to the teacher as a portfolio. The activity is very similar to the CORP portfolio assessment model (collecting documents, organizing, reflecting, presenting) [18].

Interviews to students indicate that their learning process is greatly assisted by such formative assessment models. On the one hand they can be creative with their abilities, while elsewhere if they encounter problems they can be analogous to the prepared example. Furthermore, an integrated assessment of process components, products, and attitudes gives them a more complete picture of the learning outcomes. Feedback provided gives students guidance to improve themselves on weak components and improve themselves on the already good components. Collection of formative assessment results into the portfolio gives them the opportunity to improve self-management.

4. Conclusion

An integrated formative assessment instrument for academic ability and character education has been attempted to develop. Instruments contain assessment of cognitive, affective, and psychomotor aspects. Assessment is done by giving the problem to be resolved. Assessment is done from the beginning until the completion, including working process, resulting product, and attitude or behavior during working. Besides components of process and products, rubric of assessment including assessment of character education components, such as ethics, manners, appreciation, courtesy, responsibility, and cooperative attitude.

The assessment can be developed anyway for the other fields, such as engineering, building, or fine art. Assessment rubric can be extended including the selection of materials or tools (if needed), choice of methods (if necessary), efficiency (materials, time, or energy), communication methods, and so on. By using assessment rubric, the teacher or the other evaluator performs an integrated assessmentfrom the beginning until the end of the student's work.

The integrated assessment model developed not only can be used in the classroom, but also outside the classroom. The assessment model can be used to assess various activities outside the classroom, such as work services, prayers, wall magazine making, or art performances. Assessment should not only be done by teachers, but may involve other school staff, such as librarians, administrators, parents, and even the community [12]. Thus, schools can be realized as a place for value validation, a place of cooperation between school staff with student's and with student's parents, a place to find out if character education is successful or not. Attempts to maintain character education exist in schools.

The assessment model adopted from the Balinese culture-based learning model has been tested to inherit skills and character from one generation to the next generation. Now a junior with his own consciousness learns from his senior. Someday the junior will become a senior. At that time, they will teach and assess the new juniors. Thus the learning process takes place naturally on self-awareness. For the next time, skills, attitudes, and behaviors are taught to others voluntarily also. Thus a democratic agreement will be established between all parties to a peaceful life.

Acknowledgement

The authors would like to thank all participants who contribute to the finalization of this paper. Especially for the colleagues at Mathematics Department Universitas Pendidikan Ganesha, for the support and many feedbacks for the research.

Refferences

[1] Benninga J S 2003 The Relationship of Character Education and Academic Achievement in Elementary School *Journal of Research in Character Education* **1** 19–32

[2] DeRoche E F and Williams M M 1999 *Educating Heart and Minds: A Comprehensive Character Education Framework* (London: Kogan Page Limited)

[3] Pinchok N and Brandt W C 2009 *Connecting Formative Assessment RESEARCH to PRACTICE: An Introductory Guide for Educators* Learning Point Associates http://www.learningpt.org/pdfs/FormativeAssessment.pdf (Accessed on March 13, 2013)
[4] Nicol D J and Macfarrlane-Dick D 2006 Formative Assessment and Self-Regulated Learning: A Model and Seven Principles of Good Feedback Practice *Studies in Higher Education* **31** 199-218
[5] Orlich D C, Harder R J, Callahan R C and Trevisan M S 2010 *Teaching Strategies: A Guide to Effective Instruction* (Singapore: Wadsworth Cengage Learning)
[6] Heritage M 2010 Formative Assessment and Next-Generation Assessment Systems: Are We Losing an Opportunity? *Paper prepared for the Council of Chief State School Officers*
[7] Anderson O W and Krathwohl D R 2001 *A Taxonomy for Learning Teaching and Assessing* (New York: Addison Wesley Longman Inc)
[8] Black P and Dylan W 1998 Assessment and Classroom Learning *Assessment in Education: Principles, Policy & Practice* **5** 7 - 74
[9] OECD 2005 Formative Assessment: Improving Learning in Secondary Classrooms *Policy Brief* www.oecd.org/dataoecd/19/31/35661078.pdf (Accessed on April 15, 2015)
[10] Sutjaja G M 2006 *Kamus Bali Indonesia Inggris* (Denpasar: Lotus Widya Suari & Penerbit Universitas Udayana)
[11] Hatch V 2010 *What is the difference between 'ngayah' and 'ngo'opin'?*. http://www.balimusicanddance.com/articles/what-is-the-difference-between-ngayah-and-ngoopin (Accessed on October 20, 2016)
[12] Kaler G K 1994 *Butir-butir Terceer tentang Adat Bali* (Denpasar: CV Kayumas)
[13] Creswell J W 2008 *Educational Research: Planning, Conducting and Evaluating Quantitative and Qualitative Research* (Upper Sadlle River NJ: Pearson Education International)
[14] Miles M B and Huberman A M 1994 *Qualitative data analysis: an expanded sourcebook.* (Thousand Oaks CA: Sage Publications)
[15] Input Bali 2015 *Ngayah, Tradisi Yang Harus Tetap Ada di Bali* http://inputbali.com/ budaya-bali/ngayah-tradisi-yang-harus-tetap-ada-di-bali (Accessed on May 18, 2017).
[16] Lickona T 2001 *The Teacher's Role in Character Education* (Boston: Boston University)
[17] Donald A, Jacobs L C and Sorensen C 2010 *Introduction to Research in Education* (Belmont California: Wadsworth)
[18] Wyaatt III R L and Looper S 1999 *So You Have to Have A Portofolio, a Teacher's Guide to Preparation and Presentation* (California: Corwin Press Inc)
[19] Depdiknas 2009 *Kebijakan Nasional Pembangunan Karakter Bangsa 2010-2025* http://pendikar.go.id.
[20] Salvia J and Ysseldyke J E 1995 *Assessment* (New Jersey: Houghton Mifflin Company)
[21] Kicken W 2009 The Effects of Portofolio-based Advice on the Development of Self Directed Learning Skills in Secondary Vocational Educations *Education Tech Research Dev* **57** 439-460 DOI 10.1007/s11423-009-9111-3 Association for Educational Communications and Technology
[22] O'Malley J M and Valdez P 1996 *Authentic Assessment for English Language Learners* (New York: Addison-Wesley Publishing Company)

IOP Conf. Series: Journal of Physics: Conf. Series **1040** (2018) 012027 doi:10.1088/1742-6596/1040/1/012027

Design research on plane figure learning by using picture story and pairing game to improve mathematical communication skills of second grade of primary school students

Lisnani

Primary School Teacher Education. Universitas Katolik Musi Charitas, Jl. Bangau No.60, Palembang 30113, Indonesia

E-mail: Lisnanipcmtk@yahoo.com

Abstract. This research aims to produce the learning trajectory to improve mathematical communication skills of second grade students on the plane figure by using picture story and pairing game. It uses design research which consists of three stages: preliminary design, design experiment, and retrospective analysis. In the stage of preliminary design, researcher designed the Hypothetical Learning Trajectory (HLT). In the stage of design experiment, researcher then applied the HLT in teaching second grade students to improve mathematical communication skills from informal knowledge to formal knowledge through mathematics activities. The activities are picture story by using tangram context and pairing game. Finally, in the of retrospective analysis, the result of HLT application was analyzed to acquire the final result namely; Learning Trajectory (LT) of picture story and pairing game. Second grade students were involved as samples of this research. The data were collected through interview, observation, test, documentation, and field note. The data were analyzed through t*riangulation of data* and interpretation cross. The result of this research is LT of the plane figure by using tangram context picture story and pairing game to improve mathematical communication skills of second grade students.

1. Introduction

Mathematics is a discipline which has a specificity compared with other disciplines. Therefore, one who learns mathematics must notice the nature of mathematics. In the process of learning mathematics, the principle of learning must be chosen at the beginning, so that the process of learning mathematics can run smoothly [1]. Through this process, students are expected to improve the ability to use mathematics in problem solving and communicate ideas by using symbols, tables, diagrams, and other media. However, most students rarely practice their mathematical communication skills due to lack of understanding of the material. Therefore, the students are unable to re-explain the topic that they have been studied before.

This is in line with the result of the interview about mathematics conducted by the researcher with some elementary school teachers. From the interview, the researcher obtained information that some students have difficulties in understanding some topics in mathematics such as fractions, division, plane figure, and 3D dimensional shape. Dealing with the plane figure, some students find it hard to understand it due to the need for visual ability and high imagination. The plane figure is a part of geometry. The theory underlying the geometry is Van Hiele theory which comprises five stages namely: visualization, analysis, informal deduction, formal deduction, and appropriateness. In response to this view, has highly recommended that teacher incorporates some shape puzzles such as tangram, lego, and Froebel's blocks in the plane figure as part of geometry teaching and learning activities [2]. Meanwhile, the tangram is a Chinese game consisting of seven pieces of plane figure namely square, right triangle, equilateral triangle, and parallelogram that form a big square together. Applies tangram in shaped images of animals, people, and objects to improve students' understanding of plane figure [3].

The researcher designed this tangram into a picture story by combining various pieces of the plane figure into the shape of animals, people, and objects. The use of tangram context picture story enhances students' interest in learning due to the use of colorful and unique picture story. The designing of picture story uses the scientific approach which stated in 2013 curriculum. Learning in scientific approach involves observing, asking, gathering information/trying, processing information/ associating, and communicating. Mathematical communication skill cannot be separated from the students' ability to read and present a topic in front of the class.

In relation to it, this picture story can be classified into integrative thematic learning. Here, students are expected to sense the interrelationship between Indonesian and Mathematics subjects. Furthermore, the researcher also uses a game which matches plane figures called "pairing game". Pairing game is a game that pairs up plane figure with the shapes, properties, perimeters, and areas of the plane figure. The use of tangram context picture story and pairing game were in line with the characteristics of *PMRI* (Indonesian Realistic Mathematics Education) which covers context or exploration phenomenology. The context which corresponds to the concept is indispensable in initiating learning and shifting learning approach from teacher centered to student centered [4].

The designing process of picture story and pairing game can be described as follows; (1) designing and developing complete HLT, (2) revising HLT after conducting the pilot test to produce LIT (Local Instructional Theory). Based on the descriptions above, the formulation of the research problem is: How is the role of LT through picture story and pairing game in improving second grade students' mathematical communication skills on plane figure material?

2. Literature Review

2.1 Plane Figure

Plane figure is bounded by straight lines or curves. It has only two dimensions of length and width. For example: square, rectangle, triangle, parallelogram, rhombus, trapezoid, and circle.

2.2 Van Hiele Theory

This theory explains the development of students' way of thinking in geometry learning which consists of five stages [5]. Table 1 describes five thinking level by Van Hiele.

Table 1. Thinking level by van Hiele

Thinking level by Van Hiele	Object thinking
Visualization	Shape and shape types
Analysis	Shape characteristics
Informal deduction	The relationship between the characteristics
Formal deduction	Deduction systems and the characteristics
Appropriateness	Analysis of deductive systems

(1) Ability to express mathematics ideas orally, in writing, and visually, (2) ability to interpret and evaluate mathematics ideas orally, in writing, and visually, (3) ability to use terms, symbols, and structures to demonstrate situation and mathematic problems.

2.3 Tangram

Tangram is an ancient Chinese geometric puzzle which has with seven pieces: two large triangles, one medium triangle, two small triangles, one square, and one parallelogram (shown in figure 1) [6].

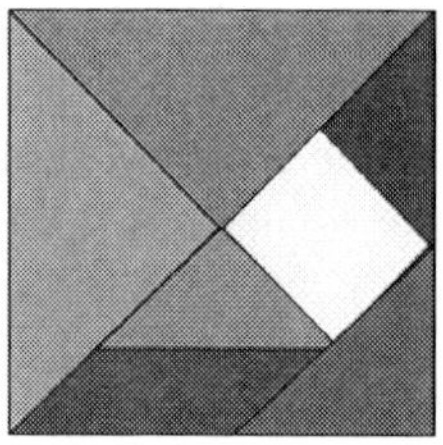

Figure 1. Tangram

2.4 Picture Story

Picture story consists of story and picture. The story is the core and the picture is the complement. Students can recognize the shapes of the plane figure through this picture story. In this research, the researcher used tangram context picture story in the shape of animals, human and objects (shown figure 2a). Figure 2b. describes the pieces of the plane figure in tangram context picture story.

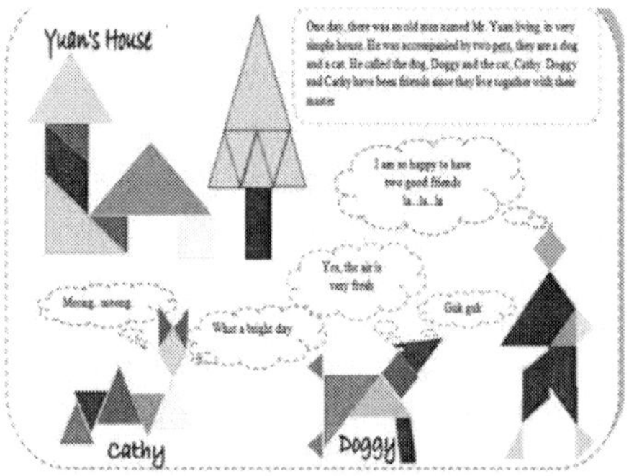

Figure 2a. Picture story using tangram context

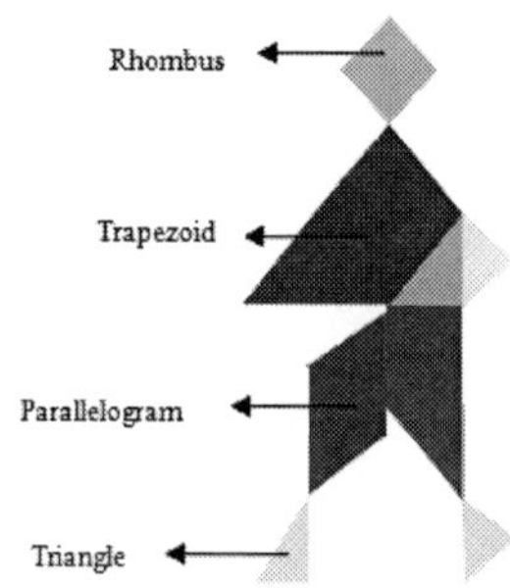

Figure 2b. Shape of plane figure in tangram

2.5 Pairing game

Pairing game is a game which matches plane figure with the shapes, properties, areas, and perimeters of plane figure. This game is played by the students in the group and the purpose is which its purpose to understand and communicate the plane figure within the group. This game uses cards consisting of the plane figure, shapes, properties, areas, and perimeters of plane figure (shown in figure 3).

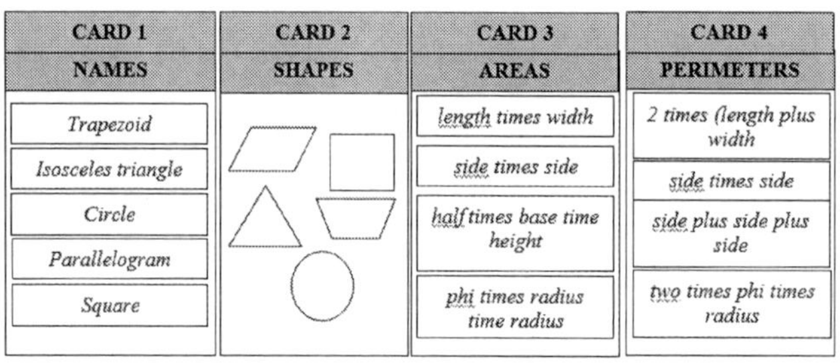

Figure 3. Pairing game cards

2.6 PMRI Approach

PMRI (Indonesian Realistic Mathematics Education) is one of the innovations in the learning of mathematics developed in Indonesia since 2001. *PMRI* is an approach to learn mathematics that starts with the things that are 'real' for students. *PMRI* is implemented by using a real-world context, models, production and construction, interactive, and relatedness (intertwinement) [7]. The use of real-world contexts allows students to utilize their experience of capabilities and browsing new experiences.

The characteristics of PMRI): (1) phenomenological exploration or the use of context; (2) using models and symbols for progressive mathematization; (3) using student's own contribution and production; (4) interactivity; (5) intertwinement [4].

2.7 Mathematical Communication Skills

Communication can be interpreted as an event to convey the message or information to each other within a community [8]. In addition, revealed that the communication of mathematics is an essential skill in mathematic that is the ability to express mathematical ideas coherently to everyone through spoken language [9].

In this research, the indicators of mathematical communication skill which conforms to the needs of the researcher. The indicators can be described as follows; (1) the skills to express mathematical ideas orally and in writing, (2) the skills to interpret and evaluate mathematical ideas orally and in writing, and (3) the skills to use terms, symbols, and structures to model situations or problems of mathematics [10].

2.8 HLT and LIT

HLT is a hypothesis or prediction of how thinking and understanding of learners develop with a context in learning activities (shown in figure 4) [11]. HLT consists of three components: (a) the objectives of mathematics learning for students; (b) learning activities and tools or media used in the learning process; (c) conjecture of learning process, how to know the strategies and understanding of students that arise and develop when learning activities are done in the classroom [12].

LIT is a theory consisting of allegations about the learning process of learners that may be supplemented with assumptions about the aids that support the learning process [13].

Furthermore, HLT is a true relationship between an instruction theory and a true teaching experiment [14]. The conjectures of this relationship can be revised and redeveloped for the subsequent learning activities based on retrospective analysis after teaching experiments performed. Through LIT teachers can design HLT for a mathematics topic by selecting activities that correspond to the allegations that arise in the learning process. LIT is a frame of mind for designing and explaining HLT.

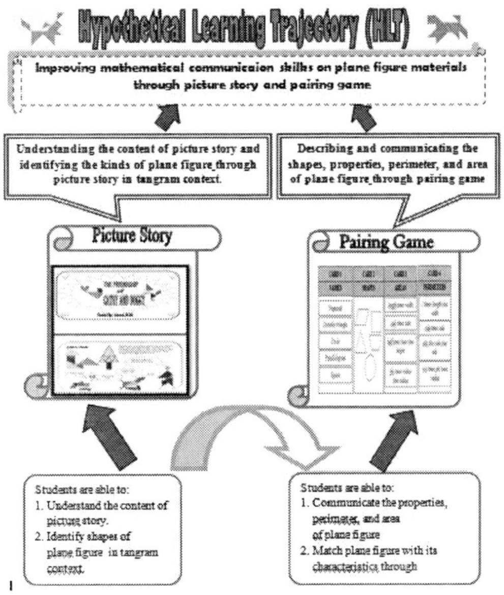

Figure 4. HLT

3. Methodology

3.1 Research Procedure

This research used design research which is appropriate to answer the research question and achieve research objectives. Design research is a research method that aims to develop Local Instructional Theory (LIT) with the cooperation between researcher and educator to improve the quality of learning. A series of student activities consisting of strategy conjectures and student thinking has been developed in this study [15]. There are five characteristics of design research, namely interventionist nature, process oriented, reflective component, cyclic character, and theory oriented [16].

Design research consists of three stages: (1) preliminary design, the main objective of this stage is to formulate local instruction theories that are elaborated and improved during the implementation of the experiment [13]. There is a series of activities that contains the conjectures of student thinking developed by researcher through HLT; (2) the design experiment, the researcher conducts pilot teaching activities that have been designed in the first stage. There are several activities undertaken including implementing the learning design with the aim of exploring, knowing the strategies and thinking of students in the introduction to plane figure material; (3) the retrospective analysis, the data obtained from teaching experiment were analyzed. HLT used in the retrospective analysis is the main reference and hint to answer the research question. In general, the purpose of the retrospective analysis is to develop LIT. The implementation of design research is also a cyclical process of thought experiment and instruction experiment [8]. The research is summarized in the following figure 5.

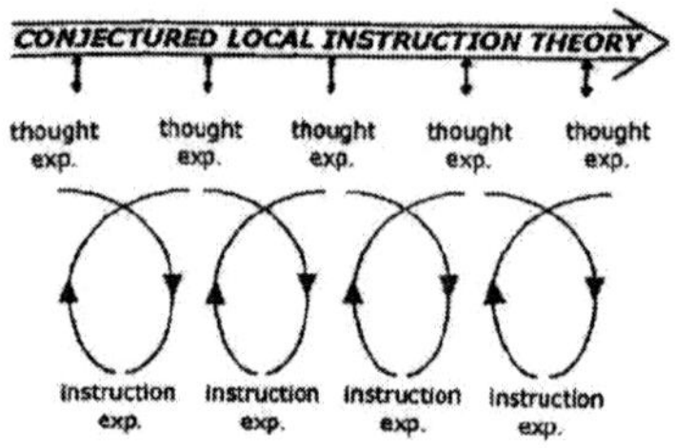

Figure 5. Conjectured LIT

3.2 Sample of Research

The sample of this research was twenty six students of second grade of SD Negeri 42 Palembang.

3.3 Data Collection and Data Analysis

In this research, the data were collected by interview, observation, written test consisting of early test (pre-test) and final test (post-test), documentation, and field note.

Data analysis was carried out with reference to the HLT. The data analysis is done in two ways: (1) triangulation of data is a technique used to see the correlation obtained from the data source in the form of field notes, observation sheets, and video recordings of the HLT that becomes the guide in conducting research; (2) interpretation cross of used to ask the considerations of counselors to advise on data obtained
as observation sheets and video recordings.

4. Result and Discussion

4.1 Pilot Experiment

In the pilot experiment class, the researcher only chose six students in grade IIA for pre-test. The researcher then obtained pretest results from the mathematics teacher in grade IIB that Annisa and Hafidz scored 90 representing high-ability students, Rika and Rafi scored 70 representing moderate level, and Gerry and Reni score 30 representing low level students.

From the pretest, it is obtained hat Gerry and Reni could not answer number 6a (shown in figure 6).

Figure 6. Student's answer for number 6a.

Figure 6 describes Gerry's answers: box and triangle. Based on the answers of the two students, it seems that the students did not understand and imagine the shapes of the plane figure picture on the item test for number 6a.

Then researcher continued first activity about the picture story with the students. Researcher distributed picture stories to the students. One of the students was asked to present an example of the picture in the picture story by communicating the plan figure that is on picture story figure 7.

Figure 7. Hafidz's presentation on the first activity

In figure 7, Hafidz has not been able to distinguish the types of triangles. He only mentioned the triangles.

After having done the pre-test, the students were provided the first evaluation that consists of 2 item tests. It seemed that Reni was less able to answer the questions number 2 related to tangram as shown in figure 8.

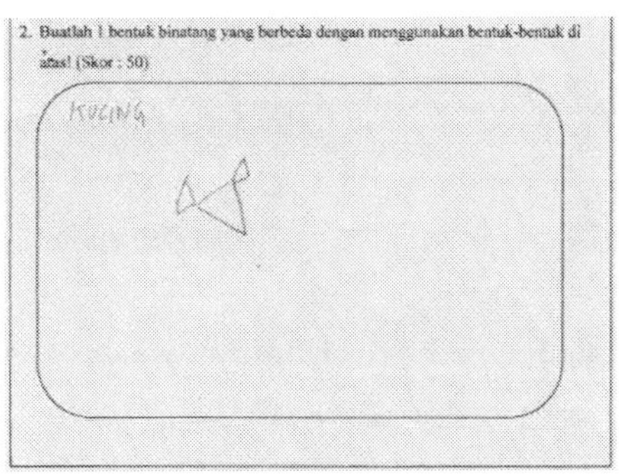

Figure 8. Reni's answer

Figure 8 described that Reni could not make plane figure model to be animal shape, person or object contextual tangram. Her answer was cat because she did not understand about the plane figure in the picture story by using tangram context. She only drew any shape. So, it can be concluded that the indicator of mathematical communication skills is interpreting and evaluating mathematical ideas both orally and in writing have not yet appeared.

In the second activity, that is pairing game, students paired up plane figure with shape, properties, perimeter, and area of the plane figure. Most students have difficulties in pairing plane figure with shapes, properties, perimeter, and area. This is due to the lack of cooperation among groups. Therefore, the researcher did improvement of the second activity through retrospective analysis.

In the second evaluation on question 2, all students could not imagine how to arrange tangram with different shapes. In the pilot experiment class, it was found that some students were very seldom to communicate mathematically.

4.2 Teaching Experiment

In the teaching experiment class, the researcher took all students in grade IIA that consist of twenty six students. After pre-test, *teacher* obtained different results, that is, 30 percent of students obtained the score above 75 while 70 percent obtained the score below 75.

In the first activity students read the picture story. Then, one of the students presented the plane figure that is in the story (shown in figure 9).

Figure 9. Student's plane figure presentation

Figure 9 shows that students are able to mention the shapes of plane figure. It shows the indicator of mathematical communication skills are interpreting and evaluating mathematical ideas both orally and in writing, express mathematical ideas orally, writing, and visual have been able to appear.

Then, the teacher divided the students into five groups consisting of 5-6 students. Each group is given the activity of pairing the plane figure pieces and then presenting it in front of class (shown in figure 10a and figure 10b).

Figure 10a. Result of group 3

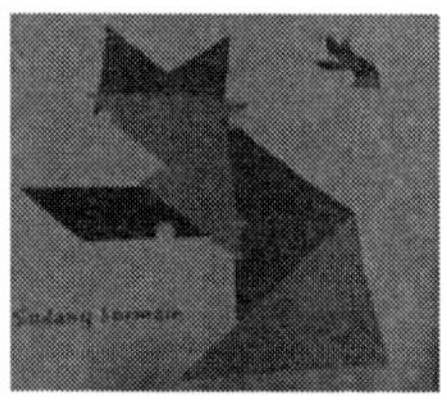

Figure 10b. Result of group 5

Figure 10a shows group 3 understands the shape and color of the plane figure and they can develop the idea. As a result, it shows the indicator of mathematical communication skills is in the form of using terms, symbols, and structures to model situations or problems of mathematics. While in figure 10b, students are less able to integrate the plane figure properly even though the results are almost the same.

In the second activity, that is pairing game, students matched the shapes, properties, perimeter and area of the plane figure. In the second activity, it shows the indicator of mathematical communication skills is interpreting and evaluating mathematical ideas both orally and in writing. Most of the students are able to match properly after the correction in the second activity in the retrospective analysis.

4.3 Retrospective Analysis

At this stage, the researcher corrects the items test on the second activity in the pilot experiment class. The item tests on second activity are less able to improve the skills of mathematical communication in pairing the plan figure with the properties. In addition, to improve students' ability in understanding the interrelationship plane figure with shape, properties, perimeter, and area of the plane figure.

5. Conclusion

On the basis of findings and discussions, it can be concluded that Learning Trajectory of picture story and pairing game generated from this research has an important role in improving students' mathematical communication skills on plane figure material. This role can be identified from a set of learning activities in the classroom.

At the first activity, the picture story aims to guide students in identifying plane figure shapes through a story and present it in front of the class so that the indicators of mathematical abilities arise. The indicators cover ability to express mathematical ideas orally, written, and visual and use terms, symbols, and structures to model situations or problems of mathematics. At the second activity, pairing game trains students' ability to understand the shapes, properties, perimeters, and areas of the plane figure so that the indicators of mathematical abilities arise. The indicators comprise ability to Interpret and evaluate mathematical ideas both orally and in writing. The set of activities above enables the students to understand plane figure both in theory and reality.

Furthermore, it is suggested for the next researcher to design complete HLT in order to generate LT which is not only improving the students' mathematical communication skills but also enhancing other mathematical skills such as mathematical creative thinking, mathematical reasoning, and mathematical concept understanding. In addition, it is better for the next researcher to find an ideal teacher.

Acknowledgement

I would like to thank the headmaster, teachers, and students of SD Negeri 42 Palembang and *Musi Charitas Catholic University Palembang* for the opportunities and supports in my research.

References

[1] Sundayana R 2015 *Media dan Alat Peraga dalam Pembelajaran Matematika* (Bandung: Alfabeta)

[2] Copley J V 2000 *Geometry and Spatial Sense in the Early Childhood Curriculum* (Reprinted from The Young Child and Mathematics) chapter 6 *(CD accompanying Developmentally)*

[3] Gautama P 2010 *Tangram: Melatih Kecerdasan & Kreativitas Anak Edisi: Burung & Kucing* (Jakarta: Elex Media Komputindo)

[4] Zulkardi 2005 *Pembelajaran Realistik Matematics Education (RME) Sebagai Suatu Inovasi Dalam Pendidikan Matematika Di Indonesia* (Palembang: Universitas Sriwijaya).

[5] Van de Walle J A 2008 *Matematika Sekolah Dasar dan Menengah: Pengembangan Pengajaran* (Jakarta: Erlangga)

[6] Bohning G and Althouse J K 1997 *Using Tangrams to Teach Geometry to Young Children (Early Childhood Education Journal vol 24)* pp 239–240

[7] Muttaqin, H Putri R I I and Somakim 2017 *Design Research on Ratio and Proportion Learning by Using Ratio Table and Graph with OKU Timur Context* (Journal on Mathematics Education vol 8) pp. 211–222

[8] Aufa M Saragih S and Minarni A *Development of Learning Devices through Problem Based Learning Model Based on the Context of Aceh Cultural to Improve Mathematical*

Communication Skills and Social Skills of SMPN 1 Muara Batu Students (Journal of Education and Practice vol 7) pp 232–248

[9] Husna Ikhsan M and Fatimah S 2013 *Improving Problem Based Learning and Mathematic Communication of Students Junior High School through Cooperative Learning Model Think-Pair-Share (TPS)* (Jurnal Peluang vol 1)

[10] NCTM 2000 *Curriculum and Evaluation Standard for School Mathematics* (Reston VA: NCTM)

[11] Gravemeijer K 2004 *Local Instructional Theories as Means for Building a Knowledge Base for Teaching in Mathematics Education* (Mathematical Thinking and Learning vol 6) (Lawrence Erlbaum Association, Inc) chapter 6 pp105–128

[12] Simon M A 1995 *Reconstructing Mathematics Pedagogy from a Constructivist Perspective* (Journal for Research in Mathematics Education vol 26) pp 114–145

[13] Gravemeijer K and Cobb P 2006 *Design Research from a Learning Design Perspective* (*Educational Design Research)* eds (New York: Routledge) pp 17–51

[14] Bakker, A 2004 *Design research in statistics education on symbolizing and computer tools.* (Amersfoort: Wilco Press)

[15] Pramana R C I 2017 *Design Research (Teori dan Implementasinya: Suatu Pengantar)* (Jakarta: Rajawali Pers)

[16] Akker J V D, Gravemeijer K, McKenney S and Nieveen 2006 *Educational Design Research.* (London: Routledge Taylor and Francis Group) p 4

IOP Conf. Series: Journal of Physics: Conf. Series **1040** (2018) 012028 doi:10.1088/1742-6596/1040/1/012028

The development of mathematical learning material based on model-eliciting activities (MEAs) approach to improve mathematical problem-solving skill of students of grade X of Senior High School Padang

Irwan, A Asmar and H Syarifuddin

Universitas Negeri Padang, Indonesia

E-mail: irwan.math.165@gmail.com

Abstract. Mathematical problem solving skill is a very important ability in mathematics learning. This skill is used in learning and solving problems. In fact, problem-solving skill of students can be enhanced. Students should get the opportunity to develop their skill in problem solving. One of the efforts to improve problem-solving skill of students is to develop mathematics learning material based on Model-Eliciting Activities (MEAs) approach. The purpose of this study is to produce mathematics learning material based on MEAs approach, which are valid, practical and effective. This developmental research uses the Plomp's model consisting of three phases: preliminary, prototyping, and assessment stage. Subjects of this study were students of grade X of Senior High School Padang state. The results of data analysis from validity of the instrument was showed that mathematics lesson equipment based on MEAs approach has fulfilled valid criteria in terms of content and construct. The lesson equipment has been considered as practical in terms of feasibility, convenience, and time because they can be applied in mathematics learning activities. The lesson equipment has also been effective since they can enhance problem solving skill. Based on these results, it can be concluded that mathematics lesson equipment based on MEAs approach, which were implemented to students of grade X in the first semester was valid, practical, and effective.

1. Introduction

Mathematics is a subject that has significant role in education. By learning mathematics, students can develop logical, critical, analytical, systematic and creative thinking ability as well as cooperative skills [1]. In addition, one of purposes of learning mathematics is to develop students' problem-solving skill both in mathematics context and outside mathematics (real life, science and technology), which consists of some phases, such as: understanding the problem, constructing mathematical models, simplifying the models and interpreting the solution from mathematics into real world [2]. Furthermore, in the process of teaching and learning mathematics in school, teachers should pay attention to five mathematical thinking skills, including problem-solving skill [3]. Therefore, it can be said that problem-solving skill is a skill that has to be possessed and developed by students.

The importance of problem-solving skill was also stated by [4] that problem-solving skill is the heart of mathematics. Problem-solving skill is indispensable not only for someone who will learn mathematics in the future but also for people who will apply it in other fields and daily life [5]. So, problem-solving skill is not only used by students in learning mathematics but also in their daily life to solve any problems or obstacles that they face. By applying some phases of problem-solving skill, such as: understanding the problem, devising a plan, carrying out the plan and looking back the solution [6], students are expected to develop curiosity and motivation and can be a creative, logic and systematic thinking student's in order to solve and face any problems in their daily life.

Problem solving can be seen as a process that integrated in every part of mathematics learning, which also provides skill as a context. Problem-solving skill also can be learned to find out the solution of a problem that cannot be solve immediately [6]. Similarly, [7] and [8] said that problem solving is effort to solve a new problem by using knowledge, skill and understanding that are already possessed.

Old perspective about mathematical problem solving was stated by [9]) that problem solving is a process to develop skills through a learning, which at the beginning consists of concepts and procedures followed by exercises in solving word problems and then transforms to enrich experience by applying competence in solving a new and non-routine problem. Similarly, [10] also stated that problem solving involves high level of mathematical thinking, such as: visualisation, association, reasoning, manipulation, abstraction, analysis, synthesis and generalisation.

There are four indicators of problem-solving skill that are used in this study, namely: (i) students are able to identify unknown aspects that are asked and the sufficiency of aspects that are needed; (ii) students are able to formulate mathematical problems or construct mathematical models; (iii) students are able to apply strategies to solve any problems either in or outside mathematics; and (iv) students are able to explain and interpret the solution of problems [3].

There are many students facing difficulties to solve mathematical problems. These difficulties closely related to mathematical problem-solving skill. The result of interviews with teachers revealed that students always got confused while doing word problems or non-routine problems, which need analysing skill. The students do not understand the problems well so that they cannot explore the steps needed to solve those problems. This leads to the unsatisfactory problem-solving skill. For example, when a teacher presents a mechanistic problem, students can complete it immediately. However, when learners are given questions non routine problem or problems to solve requires analysis then learners look confused in understanding and solving the problem. Only a few people can do the problem. The other students are just waiting for an answer from their friends.

Teaching and learning mathematics in the classroom can be enhanced to develop problem-solving skill, for instance by using Model-Eliciting Activities (MEAs) approach. MEAs is a learning approach to understand, explain and communicate concepts embedded in a problem with the model of [11]. Learning with MEAs approach is started with realistic problems in order to stimulate the interest of the students. This interest will allow them to easily understand the problems because those problems are related with students' daily life. After the students understand problems, it is expected that they can determine and explore the steps to solve problems. There are six principles in learning with the approach of MEAs, namely: 1) The meaningfulness principle, 2) The model construction principle, 3) The self-evaluation principle, 4) The model documentation principle, 5) The simple prototype principle, and 6) The model generalization principle [11].

The learning with MEAs approach needs students to work in a group of 3-4 students [12]. Working in groups allows students to discuss and communicate their ideas as well as to listen to their peers' opinions. The main feature of MEAs is the activity to construct mathematics models. Constructing mathematics models need a strong concept in understanding problems so that students can simplify their thinking [13]. Mathematics models can be defined as representing a situation or things mathematically. By modelling, the students will be used to the modelling steps, such as: simplifying the problem; constructing mathematics models; transforming and finding the results by using the models and interpreting the results. These steps of modelling are in line with Polya's problem-solving steps so that using the phases of MEAs approach is expected to enhance student's problem-solving skill. Previous studies, such as: [14] and [15], both stated that MEAs approach can develop students' problem-solving skill. Similarly, [16] also stated that MEAs approach can develop representing and problem-solving skill. On the other hand, the same opinion is also stated by [17] and [18].

Teaching and learning process by using MEAs approach in the classroom will run well only if the teacher prepares a well-organised lesson plan. Besides that, the handout, such as student's worksheet is needed to support the learning. However, the lesson plan and student's worksheet that are available in the school cannot fully facilitate students to develop their problem-solving skill. Therefore, the development of lesson plan and student's worksheet using MEAs approach is needed.

This study has the purpose to design lesson plan and student's worksheet by using MEAs approach so that lesson plan will be designed by including the steps and principles of MEAs approach. On the other hand, student's worksheet consists of guidances and the collection of problems, which are not only routine problems but also non-routine problems. student's worksheet also consists of problems that relate to students' daily life so that they can easily understand what the problem is asked them to do. Additionally, student's worksheet provides students with instructions that can guide them to solve the problem step by step, such as: understanding the problem and constructing mathematics models of the problem so that students can represent the ideas of the problem into pictures; symbols and mathematical equation, which will lead them to the solution. Using student's worksheet based on MEAs approach is expected to enhance students' mathematical problem-solving skill.

This study will elaborate the characteristics of learning equipment based on MEAs approach to enhance problem-solving skill of X (tenth) grade students, which is valid and practical as well as to investigate the effectiveness of learning equipment based on MEAs approach to enhance students' problem-solving skill.

2. Research Methods

This study is a developmental research. The developmental model used in this study is adapted from Plomp's model, which consists of three phases, such as: preliminary phase; development or prototyping phase; and assessment phase. In the first phase or initial investigation phase, the needs and context analysis are implemented. In the second phase (prototyping phase), a design, development and formative evaluation are implemented. Formative evaluation steps are started by self-evaluation, expert review, one-to-one evaluation, small group evaluation and field test. Next, in the third phase (assessment phase), a semi-summative evaluation in the form of assessment of learning is held [19].

The mathematics learning equipment based on MEAs approach was applied in X (tenth) grade of senior high school in Padang. The subject of this research are students from X grade of SMAN 3 Padang on period of July – December 2017. The characteristic of this research subjects are students that have different learning ability (high, medium, low). There are several instruments used in this study, such as: questions list; check lists; validation sheets; observation sheets; questionnaires and test. Prior to the use, all instruments are validated by experts.

The data as the results of the study is analysed based on its type. For example, qualitative data is analysed based on qualitative paradigm and also quantitative data is analysed based on quantitative paradigm. Finally, the results of the study are categorised and concluded qualitatively.

3. Results and Discussions

3.1. Results of Preliminary Research

In the preliminary phase, the identification and analysis of needs for development of learning equipment based on MEAs approach are held. There are three analysis performed, namely: needs analysis; curriculum analysis and concepts analysis. Needs analysis revealed that students require a new learning approach in addition to conventional learning that is commonly used in the school, which is able to optimise students' mathematical problem-solving skill. Besides, students also need learning equipment for the teaching and learning process runs systematically in order to achieve the learning goals.

The questionnaires revealed that students liked to work in group, had high spirit to study and were more interested to solve problems relating to daily life. Regarding student's worksheet, questionnaires also revealed that students prefered student's worksheet with certain features, such as it had interesting and attractive display, was dominated with blue colour, the paper size used was A4 paper size, and was accompanied with illustrations that was matched with the handouts. Based on the results of the needs analysis, it is required to develop learning equipment based on MEAs approach that will be designed in the development phase.

Next, in the curriculum analysis, data revealed that there was order of the basic competence. There was also an adjustment of teaching materials due to the applicable rules policy. Assessment of teaching materials indicated the needs to do a change of sequence to adjust the interrelationship between each concept and basic competencies.

The last part of the preliminary analysis was conceptual analysis. This analysis revealed information about the content and sequence of teaching materials, total meetings for each basic competency, and indicators of achievement of competence. This information was needed in the designing of student's worksheet.

3.2. Results of Prototyping Phase

There were two prototypes designed in this study, the first was the prototype of lesson plan based on the MEAs approach and the second was the prototype of student's worksheet based on the MEAs approach. The form of lesson plan based on the MEAs approach, was the same as the regular lesson plan, but in the core activities, the learning process refered to the six principles of MEAs approach, such as: reality, model construction, self-assessment, construct documentation, effective prototype, and construct share ability and reusability. Learning activities in the MEAs approach-based lesson plan consisted of preliminary activities, core activities and closing activities. Preliminary activities were initial activities in one meeting, which were aimed at generating motivation and focusing the attention of students to actively participate in the learning process. Learning activities were accompanied by time allocation to facilitate teachers to implement the learning process.

Next, core activities were the learning process to achieve learning objectives. The core activities were started by the teacher while explaining the lesson material briefly, then students were confronted with some contextual problems related to the material being studied. Afterward, teachers asked students to discuss in groups. Students discussed all the information from the problem in student's worksheet and constructed the models based on what was known in the problem. The students must be confident with the models they created. They should be able to measure the feasibility and usefulness of problem solutions without the help of teachers. The students further documented the models and solutions they created and they must be sure that the models should be easily interpreted by others. The last stage of learning was to communicate the models for which they got responses from other groups. Learning ended by concluding activities, which could be done in the form of making a summary or conclusion about the material that has been learned. Lastly, the teacher informed the material to be learned at the next meeting.

At student's worksheet design stage, there were three characteristics that needed to be considered, such as: content, language and display aspect. The content aspect began by providing contextual problems that existed in everyday life. Additionally, student's worksheet provided students with instructions that can help them to solve problems. The first instruction asked students to understand the problem by identifying what was known and asked from the problem. The second instruction asked students to plan the solution by constructing the mathematical models of the problem. The third instruction asked students to solve the problem using the models that have been found. At the end of each student's worksheet, there were exercises to be done by the students.

Regarding the display, student's worksheet consisted of cover, introduction, table of contents, questions and bibliography. Cover contained the identity/title of student's worksheet based on the MEAs approach, subject headings studied, identity columns of students containing names and classes as student's worksheet owners. Type of writing used in the title of student's worksheet was Comic Sans MS with size 20. For the identity of students and class information as well as semester, the type of writing used was Comic Sans MS with size 18.

The third aspect was the language aspect. The language used was adjusted to the rules of writing with enhanced spelling. In addition, the language used was also communicative and in accordance with the level of understanding of junior high school students so that the presentation of the material on student's worksheet can be understood well by students. The questions in student's worksheet were arranged with clear sentences in order to lead students to get the expected answers. The result of this design was called prototype 1. The next activity, after lesson plan and student's worksheet was completed, was to conduct a formative evaluation of the learning equipment. Formative evaluation was preceded by doing self-evaluation to the designed learning equipment. The purpose of this self-evaluation was to see the completeness of the MEAs stages and principles. Another purpose was to check if there were still typing errors and lack of punctuation, capitalisation, spaces between words and so forth.

The results of self-evaluation of lesson plan revealed some mistakes, such as: typing errors, the missing of spaces between each word, and misuse of capital letters. Another mistake was the missing of full stops at the end of the sentence. Furthermore, the size of the letters used was still small that is 11 so that the size was changed to 12. On the other hand, the mistakes found in student's worksheet were typing errors and the missing of spaces between each word. Another mistake was that there was too big space where students writed the solution of the problem. The errors in the lesson plan and student's worksheet were then revised. The revised result of this learning equipment was called prototype 2.

After self-evaluation was done, the next step was to validate prototype 2 to the experts. In this case, the learning equipment was validated by 5 people, consisting of 3 lecturers of mathematics, 1 lecturer of language and 1 lecturer of education technology. There were several suggestions from the validators. Firstly, mathematical validators of lesson plan suggested four points, such as: (1) the learning objective should show the learning process; (2) the assessment has not shown the problem-solving indicators; (3) preferably any problems written in lesson plan are not in bold; (4) it is better to remove "the next" word at the beginning of the activity in lesson plan. Secondly, the suggestion of the language validator was to pay attention to the sentence on the main competences, basic competences and learning goals as there were still the missing of full stop at the end of the sentence. Thirdly, the advice of educational technology experts was to add the source of learning. Meanwhile, the suggestions of a mathematical validator for student's worksheet were to correct the questions on student's worksheet because the questions still did not show the problem-solving indicators, to equate the type of writing used in each student's worksheet, to add attractive illustration relating to the material on the cover of student's worksheet. The suggestion of the language validator was to correct the greetings in the introduction of student's worksheet. The initial greeting should be capitalised and the wording in several sentences also needs to be corrected. Lastly, the advice of educational technology validator was to adjust the type of letters in each student's worksheet.

Learning equipment was then revised in accordance with the suggestions of the validators and re-shown to the validators. Afterward, the validators assigned mark to the existing learning equipment. The validation results of lesson plan revealed that overall average of mark of lesson plan based on the MEAs approach was 3.33, which could be concluded that this lesson plan has met the criteria of the validity. Based on the validation results in terms of content, the average mark obtained was 3.31, which could be interpreted as very valid. In term of the language, the average validity of student's worksheet was 3.25, which also could be interpreted as very valid. Lastly, based on aspect of typography, the average of validity obtained was 3.80 with very valid criteria. So, it can be concluded that student's worksheet based on MEAs approach was valid.

The next stage was to evaluate one-to-one evaluation. One-to-one evaluation was done with 3 students from grade X of Senior High School 3 state of Padang with different learning ability (high, medium, low). The selection of the three students was done by the researchers through the assistance of the grade X mathematics teacher. The purpose of this individual evaluation was to ask students' opinions about possible errors such as poorly understood grammar, misspelling, as well as the use of the term in each problem.

The next stage was to conduct a small group evaluation. This evaluation was conducted on 6 students who had high, moderate, and low ability from grade X-7 Senior High School 3 state of

Padang. The selection of six students was assisted by mathematics teacher at the school. The purpose of a small group evaluation was to identify the lack of learning equipment in terms of timelines and the implementation of learning outcomes. During the learning process in small groups, researchers were assisted by one observer who was tasked to record the implementation of the teaching and learning process by using the MEAs approach and to note the time spent. While doing the trial with small groups, there were some revisions in term of time spent. For example, at the first meeting, time required by students for presentation was only 5 minutes while the time designed on lesson plan was 10 minutes, so that a revision of time was made.

At the second meeting, there was a time revision when the participants worked on the exercises from 10 minutes to 15 minutes, while the time for presentation was changed from 10 minutes to 5 minutes. At the fourth meeting, there was an improvement over time on the lesson plan that was the change of the time to solve the 'problem 1' from 20 minutes to 25 minutes. Learning equipment that has been tested on small groups was revised based on the deficiencies and errors found. In Figure 1. we can see example of part of student's worksheet.

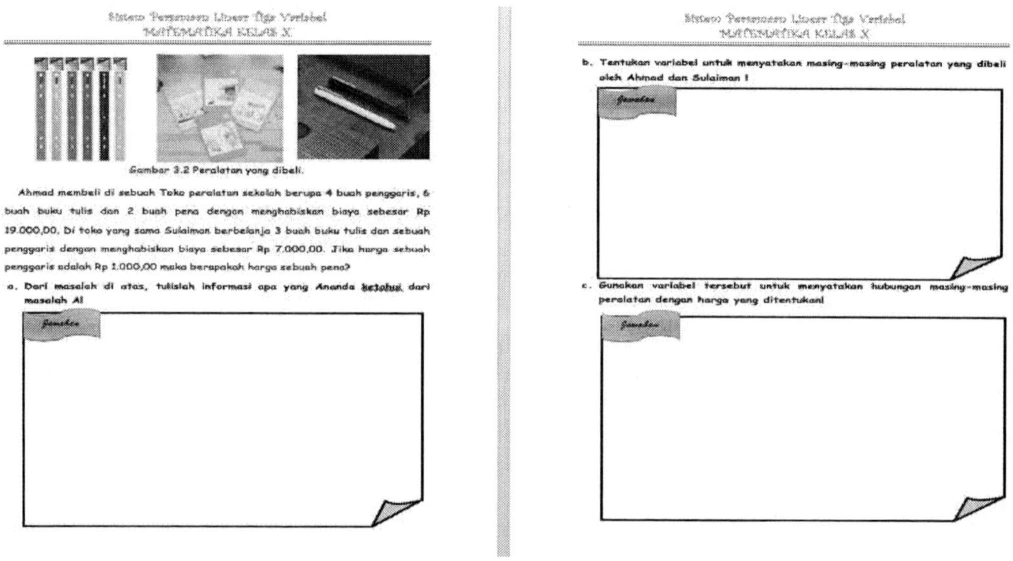

Figure 1. Example of Student's Worksheet.

The next phase was the assessment phase. In this phase, the testing of learning equipment to a large group of students of grade X-8 Senior High School 3 state of Padang was carried out. During the execution of the field test at the end of each session, the students were given 2 problem-solving problems. The assessment of this exercise was based on problem-solving rubric that was previously designed. The results of these exercises revealed that students' problem-solving skill increased from the beginning until the end of the meeting.

At the first meeting, the problem-solving skill of students for each indicator was still low because students were not accustomed to work on problem-solving problems. This can be seen from the percentage of problem-solving skills obtained for each indicator. The percentage of the indicator of understanding the problem was 18.1%. The percentage of the second indicator (devising a plan) was 33.6%. While the third indicator (carrying out the plan) obtained the percentage of 71.96%, the last indicator (looking back to the solution) obtained the percentage 14.65%. At the following meetings, the students were getting used to work on problem-solving problems, so that there had been an increase in problem-solving skill of students compared to the first meeting.

After the learning based on MEAs approach was completed, the questionnaires were given to teachers and students. The results of questionnaires filled by the teachers revealed that 86.1% of the convenience aspects were very practical, 75% of efficiency aspect was in the practical criteria, and

81.25% of appeal aspect also lied in the practical criteria. The equivalence aspect of student's worksheet was 91.7%, which lied in very practical criteria. Overall, the average of 83.5% was obtained. On the other hand, the results of questionnaires filled by students revealed that 80.2% of students were on practical criteria. Based on the questionnaires filled by the students, it was found that student's worksheet was easy to use; it was interesting and useful for the learning process. Additionally, based on the result of the anaysis of the observation sheet of lesson plan implementation based on MEAs approach filled by observer, it can be concluded that the implementation of lesson plan during learning was classified as practical with percentage of 82,78%.

After the learning was completed, then students were given a final test of problem-solving skill. Problems were given in accordance with the indicators of problem-solving skill. Students' answer sheets were assessed based on the assessment rubric. There were 36 students who participated in the final test. Based on the results of the final test conducted, there were 30 students who got the mark above the passing grade and there were 6 students whose mark was below the passing grade, with the percentage of classical completeness of 80%. From the results of the test conducted, there were more than 83.3% of the students whose mark was above the passing grade. This means that the learning equipment developed had been classified as effective to improve the problem-solving skills of students.

Finally, it can be said that the student's worksheet was developed, which started with the preliminary analysis phase, continued with the development phase, and the normative evaluation phase (self evaluation, one to ne evaluation, expert review, small group and field test) has contained principles of the MEAs approach. Student's worksheet that has been developed begins with the provision of contextual problems. Students are guided to present the model of the problem and ultimately result in a solution to the problem. So, it can be said that the student's worksheet has been in accordance with the MEAs approach.

4. Conclusion

Based on the results of this study, it was found that the designed learning material was in accordance with the data obtained in preliminary research. The developed equipment contained of the principles of the MEAs approach that was started with the contextual problems, then students constructed the models of the problems and lastly they socialised the models they created. Learning by using the equipment based on the MEAs approach had been able to improve problem-solving skill of students. This was because learning by using the MEAs approach made students get used to construct the model solution of the given problems. This happens because the students are accustomed to solving problems through the eliciting the model.

Learning materials that have been developed can be used as additional learning resources for teachers and students. Through learning material based on MEAs approach is expected to make the learning process more fun, make students to learn independently, accustomed to problem solving problems and more confident in solve a problem. Then, through this learning materials are expected to help teachers and learners in achieving the goals of learning. It is expected that the use of this learning material can also be used in other schools.

References

[1] Sumarmo U 2010 *Berpikir dan Disposisi matematik: Apa Mengapa, dan Bagaimana Dikembangkan pada Peserta Didik*. Bandung: UPI not published

[2] Kemendikbud 2014 *Peraturan Mentri Pendidikan Nasional Republik Indonesia Nomor 58 Tahun 2013 tentang Kurikulum 2013 Sekolah Menengah Pertama/madrasah Tsanawiyah* (Jakarta: Depdiknas)

[3] NCTM 2000 *Principles and Standards for School Mathematics* (Reston, VA: NCTM)

[4] Branca N A 1980 Problem Solving as A Goal, Proses, and Basic Skills in *Problem Solving in School Mathematics*: 1980 Yearbook edited by Krulik S and Reys R E (Reston, VA: NCTM)

[5] Ruseffendi 2006 *Pengantar Kepada Membantu Guru Mengembangkan Kompotensinya dalam Pengajaran Matematika* (Bandung: Tarsito)

[6] Polya G 1973 *How to Solve It: A New Aspect of Mathematical Method* Second Edition (New Jersey: Princeton Press University)

[7] NCTM 1989 *Principles and Standards for School Mathematics. Curriculum and evaluation standards for school mathematics* (Reston, VA: Author)

[8] Yee S P 2012 *Students' Metaphors for Mathematical Problem Solving* A dissertation submitted to the Kent State University College of Education, Health, and Human Services (Ohio: Kent State University)

[9] English L, Lesh R and Fennewald T 2008 *Future directions and perspectives for problem solving research and curriculum development* In: 11st International Congress on Mathematical Education, 6-13 July 2008 (Monterrey: International Congress on Mathematical Education)

[10] Kirkley J 2003 *Principles for Teaching Problem Solving* Technical Paper #4 Indiana University Copyright PLATO Learning, Inc.

[11] Lesh R and Doerr H 2003 *Beyond Constructivism: A Models & Modelling Perspective on Mathematics Teaching, Learning, and Problems Solving* Lawrence Erlbaum Mahwah NJ. Book Reviews ZDM Vol. **35** (6) Ernest von Glasersfeld, Amherst, MA (USA).

[12] Ekmekci, A and Krause G. 2011. *Model Eliciting Activities (MEAs).* 5th Annual UTeach Institute-NMSI Conference, May 26, 2011.

[13] Chamberlin, S.A dan Moon, S. M. 2008. "*How Does the Problem Based Learning Approach Compare to the Model-Eliciting Activity Approach in Mathematics*?" Diakses di http://www.cimt.plymouth.ac.uk/journal/chamberlin.pdf 26 September 2015

[14] Clark. R. M. Shuman, L.J. Sacre, M. B. & Yildirim T. P. (2008). *Use of Model Eliciting Activities to Improve Problem solving by Industrial Engineering Students*. Proceedings of the 2008 Industrial Engineering Research Conference. J. Fowler and S. Mason, eds. Pittsburgh, USA: Departement of Industrial Engineering, University of Pittsburgh.

[15] Yu, S & Chang, C. 2009. What did Taiwan mathematics Theacher Think of Model-Eliciting Activities and Modeling? *International Conference on the Teaching of Mathematical Modeling and Applications, ICTMA Vol 14*, University of Hamburg.

[16] Hanifah. 2015. Penerapan Pembelajaran Model Eliciting Activities (MEA) dengan Pendekatan Saintifik untuk Meningkatkan Kemampuan Representasi Matematis Siswa. Tersedia pada: http://-journal.unnes.ac.id/nju/index.php/kreano

[17] Dede A T, Hidiroğlu Ç N and Güze E B. 2017. Examining of Model Eliciting Activities Developed by Mathematics Student Teachers. *Journal on Mathematics Education.* (Electronic Material Vol 8 No. 2, July, pp. 223-242. DOI: http://dx.doi.-org/10.22342/jme.8.2.3997.223-242

[18] Lubis N H, Pulungan P S and Fauzi A KMS 2017 Model Eliciting Activities (MEA) Application in Online Group Discussion for Mathematics Learning. *International Journal of Science and Research* (Online) Volume 6 Issue 7.

[19] Plomp T dan Nieveen V 2007 *Educational Design Research* Enschede: Netherland Insti-tute for Curriculum Development (SLO)

The effectiveness of mathematics software aided learning tool with performance assessment on student independence and student learning outcomes

G A Mahayukti

Universitas Pendidikan Ganesha, Bali, Indonesia – 81116

E-mail: gustiayumahayukti@gmail.com

Abstract. This study aims to determine the effectiveness of the use of learning tools assisted mathematical software with performance assessment, on learning independence and student learning outcomes. The experimental design used was Posttest Only Control Group Design. The research was conducted in Mathematics Education Department of FMIPA Undiksha in Integral Calculus lecture involving 49 students as members of the study sample taken at random. Data were collected by questionnaires independent learning and achievement test Integral Calculus. Data were analyzed by U-Mann Whitney's Test for learning independence and t-test for learning outcomes. The results showed that the use of the device assisted learning mathematics software with performance assessment effectively improve learning independence but not significantly on student results.

1. Introduction

Integral calculus is a courses are offered in the second semester in the Department of Mathematics Education, Ganesha University of Education. When viewed from the level of the complexity of the material, Integral calculus is not too complex because it is a basic course and a prerequisite for other courses. Recognizing the important role of Integral Calculus to prepare students for lectures in the Mathematics Education Department, as a prospective educator, and face life in society Integral calculus learning is directed to make students able to think rationally, creatively, honestly, consistently and tough in facing problems. Therefore, Integral Calculus courses should be a favorite course, well studied by students.

But the learning of Calculus is often problematic. The low learning outcomes of students because the mastery of calculus concepts is very weak and procedural. This is because Calculus learning so far is still conventional, learning is still oriented towards achieving learning outcomes and often ignoring the process. In addition, the low result is caused by the lack of students' self-directed learning. The low ability of student problem solving is seen from the result of student's answer analysis on the final exam of semester, where many students fail to solve problems that demand analysis and problem-solving abilities. Independence of learning is very important owned by students, Considering in universities apply semester credit system, and education in universities should be able to establish student independence as stated in the Indonesian Law about Higher Education No. 12 Year 2012 and Permenristekdikti No. 44 Year 2015 which expressly states that the development of higher education should be able to handle the problem such as those related to independent human development.

To meet the needs of students about mathematical knowledge especially Integral Calculus, Its learning should take place interesting, fun, meaningful and insightful future. In order for learning to run optimally and meaningfully required learning tools which lecturers and students can use to describe the abstract concept to a more realistic situation [1]. But it seems that lecturers still rarely use instructional materials mathematics software, whereas technology helps to develop all kinds of thinking skills from the most basic level to high level of competence [2]. The technology in question here is mathematical software.

The role of this mathematical software provides an opportunity for students to explore their own abstract concepts learned, especially the concepts in Integral Calculus, this is due to the learning of mathematics in universities more emphasizing mathematics as "science." However, there is little mathematical difference as "science" with school mathematics, one of the differences is at the level of abstraction [3]. Object abstraction level mathematics in college is very high. Therefore the use of mathematics software in learning is very helpful for students. Mathematical software that is used as learning media in learning is Maple and GeoGebra. Some research results that use mathematical software in an effort to improve the quality of learning has been done as can be seen on [4], [5], [6], [7], and [8]. Similarly, the students' learning outcomes using Geogebra's help were higher than those that did not use Geogebra [9] and [10]. The same thing also revealed that students' problem solving skills improved significantly through the application of IDEAL problem solving model based on Maple. Therefore the use of technology helps develop all kinds of thinking skills ranging from the most fundamental level to the level of critical thinking skills including problem solving skills [11]. To explain the concept of integrals, the lecturer should provide a precise introduction of the integral concept. It is also suggested that students need visualization to help them understand the abstract integral concepts well [12] and [13]. This opinion is confirmed in different research which stated that in order to understand abstract mathematical concepts, students need a tool that is to motivate them like the use of mathematical software [14].

There are many factors that affect the independence of learning and student learning outcomes. Based on the study of research results, one dominant factor affecting independence and student learning outcomes is the form of assessment used. The form of assessment is thought to affect student learning outcomes, Because the form of assessment used by the lecturer in making the program plan to make instruction of learning process. The instruction in the learning device determines the interaction of the students in the class that participate in determining student learning outcomes.

Stiggins [15] stated that target performance assessment: knowledge, reasoning, skills, products, and affective. In performance assessment asks students to "complete complex and real tasks, by mobilizing initial knowledge, newly acquired learning, and relevant skills to solve realistic or authentic problems. Furthermore, its is known that adults are more interested in learning from problems than from the content of the course [16]. Acknowledgment, appreciation, and support for the adult learning process need to be created in a learning environment. These opinions basically emphasize that learning independence can be grown by designing learning as problem-solving activities through the use of appropriate forms of assessment.

Thus the use of learning media with performance assessment can reinforce each other's advantages. This is supposed to increase the independence of learning and student learning outcomes. To prove these allegations, the researcher conducting experimental research with the aim to know and analyze the effectiveness of the use of mathematical software-assisted learning tools with performance assessment on the independence of learning and student learning outcomes in Integral Calculus lectures.

2. Methods

This research uses Posttest Only Control Group Design. The effectiveness tested in this study is the use of mathematical software-assisted learning with performance assessment on the independence of learning and student learning outcomes on Integral Calculus lectures. The population of this study is all students of Mathematics Education Department who take Integral Calculus courses in Academic

Year 2015/2016 spread in 4 classes. Samples were taken randomly and randomization result was obtained by class B as experiment group which received treatment in the form of use of mathematical software as its learning media, combined with performance based assessment, while the class C got the conventional learning method.

Data were collected with a questionnaire of learning independence and test for Integral Calculus student learning outcomes. There are 6 indicators used in the preparation of learning independence instruments, namely: Independence of others, self-confidence, discipline, responsibility, self-initiation, and self-control [17] and [18].

All instruments have been validated to determine the validity and reliability of the instrument. Each instrument has a reliability of 0.89 for a learning independence questionnaire and 0.85 for Integral Calculus learning test results. Before the test the hypothesis, first tested the prerequisite.

Data were analyzed using t test for learning outcomes, while for learning independence was used U-Mann Whitney test with 5% significant level.

3. Findings

Summary of research results are presented in Table 1.

Table 1. Descriptive Data of Research Results

Statistics \ Learning	A_1Y_1	A_2Y_1	A_1Y_2	A_2Y_2
N	24	25	24	25
Mean	70	75,36	42,79	51,84
Median	70	75	40	55
Standard deviation	3,718	7,076	14,188	20,778
Varians	13,826	50,073	201,303	431,778
Span	13	25	56	70
Maximum	78	91	76	90
Minimum	65	66	20	20

Description: A1 = control, A2 = experiment, Y1 = learning independence, Y2 = learning outcomes

Before performing the hypothesis test, first tested the assumption, namely: Normal multivariate test.

Hypothesis 1:
H_0: learning independence data comes from normally distributed populations
H_1: learning absence data does not come from normally distributed populations
Hypothesis 2:
H_0: learning result data comes from normally distributed population
H_1: learning result data does not come from normally distributed population

Table 2. Tests of Normality

		Kolmogorov-Smirnov[a]			Shapiro-Wilk		
	Group	Statistic	df	Sig.	Statistic	df	Sig.
Independence learning	Control	.208	24	.008	.918	24	.054
	experiment	.224	25	.002	.895	25	.014
Learning outcomes	Control	.147	24	.193	.961	24	.458
	experiment	.080	25	.200*	.963	25	.488

Test the assumption of normality in this study using Kolmogorov-smirnov test, data processing using SPSS software assistance. From Table 2 output, it can be seen that the significant value of learning independence in both control and experimental groups is smaller than 5% significant level, Thus H_0 is rejected, this means that the data of learning independence does not come from a normally distributed population. In the dependent variable of learning outcomes, both in the control group and experimental group's scores significantly greater than 5% significant level, Thus H_0 is accepted, this means that the learning result data comes from a normally distributed population. The next step is to test the homogeneity of the variant.

Table 3. Levene's Test of Equality of Error Variancesa

	F	df1	df2	Sig.
Independence learning	4.663	1	47	.036
Learning outcomes	3.722	1	47	.060

Table 4. Test Statistics

	Independence learning
Mann-Whitney U	152.000
Wilcoxon W	452.000
Z	-2.975
Asymp. Sig. (2-tailed)	.003

The homogenity test of variance in this study used levene test, the results of data processing using SPSS. The results can be seen in Table 3. From these outputs, it can be seen that the significant value of learning independence is less than 5% significant level, So H_0 is rejected, thus the variance of learning independence in experimental and control groups is different. Because the dependent variable of learning independence does not meet assumptions (normality and homogeneity) then the test used U-Mann Whitney Test.

In the learning result data can be seen that the significance value is greater than the significant level of 5%, thus H_0 accepted. This means that the learning result variant in the experimental and control groups is different. Therefore, on the dependent variable, the learning outcomes meet the assumption of normality and homogeneity thus the test used is t-test.

Followed by testing the hypothesis as presented below.

Hypothesis I

H_0: $\mu_1 = \mu_2$, means that here is no difference in learning independence between the experimental group and the control group

H_1: $\mu_1 > \mu_2$, means that the learning independence group experiments higher than the control group

Hypothesis II

H_0: $\mu_1 = \mu_2$, means that there is no difference in learning outcomes between the experimental group and the control group

H_1:, $\mu_1 > \mu_2$, means that the experimental group learning outcomes are higher than the control group

From the output in Table 4 it can be seen that, the significance value is 0.003, this value is smaller than the level of significance applied. Thus H0 can be rejected, so the independence of student learning using mathematics software assisted device with performance assessment, higher than student learning independence using conventional learning device.

Table 5. Summary of Hypothesis Test Results Learning

Independent Samples Test

		Levene's Test for Equality of Variances		t-test for Equality of Means						
									95% Confidence Interval of the Difference	
		F	Sig.	t	df	Sig. (2-tailed)	Mean Difference	Std. Error Difference	Lower	Upper
hasil belajar	Equal variances assumed	3.722	.060	-1.773	47	.083	-9.048	5.104	-19.316	1.219
	Equal variances not assumed			-1.786	42.511	.081	-9.048	5.065	-19.267	1.170

From the output in Table 5. it can be seen that, the value of significance is 0.083, this value is greater than the significance level of 5%. Thus H0 is accepted, so the learning outcomes of students who use the Integral Calculus device assisted learning mathematics software with performance assessment does not differ significantly from the results of student learning that using conventional learning device.

4. Discussion and Conclusion

The result of the research shows that student's learning independence using mathematics software assisted device with performance assessment is higher than students using conventional learning tool. Thus the use of the device assisted learning mathematics software with performance assessment effectively improve learning independence. Complex tasks demanded in performance assessments have encouraged students to be creative in reading, writing, conveying ideas or arguing, which is believed to contribute significantly in fostering student self-reliance. This is in accordance with Adnyana et. al. [19] stated that the performance assessment is able to foster self-confidence, initiative and student motivation which is an important element for the growth of student independence.

While the used of mathematical software-assisted instructional tools are not effective in improving students' Integral Learning Calculus. Judging from the prominent advantage of the use of mathematical software in learning is the student may make visualization of abstract concepts and students get immediate feedback on their work simply by clicking an instruction, this should increase students' motivation and learning activities, but the convenience may lead less active students try different matter, proving the theorem or properties of integral or solving various problems of Integral Calculus. Another possibility that causes is not yet optimal student utilizing learning device assisted by mathematics software with assessment of performance in class.

In this study also elicited student opinions through interview after the experiment ended related to the use of mathematical software-assisted learning tools with performance assessment on Integral Calculus lectures. The results of the analysis of student comments show that 65%, expressing pleasure with the device used, 15% normal and 20% feel burdened. Furthermore, from the analysis of student answers during posttest for the experimental group obtained the results of 40% of students correctly answer the point item C2 level, And 10% correct answer point item C3 or C4,And 20% succeeded in solving point of C5 level. This may be due to student learning accustomed to being given an opportunity to pass arguments and to analyze the problem, and 30% of students failed to perform algebraic manipulations so that students were wrong in choosing the integration techniques used to solve the problem. While for the control group obtained the results of 50% of students correctly answer the point item C2 level, and 15% correct answer point items C3 or C4, And 15% succeeded in solving point item C5 level, This may be due to the fact that when students are accustomed to being given the opportunity to present arguments and analyze problems, But in doing algebraic manipulation only 20% of the control group failed. The number of students who fail in the selection of this integration technique may be influenced by 1) low student retention, this is related to the many existing integration techniques, and requires algebraic manipulation to be accomplished in accordance with applicable properties or theorems, thus making the students forget about the integration techniques. This is in accordance with the expression Deese [20] that someone who has many concepts to remember can cause low retention, while someone who has a little concept to remember causes high retention. 2) student ability to perform low algebraic manipulation, Because students are less practice to solve various types of problems or problems, It is also rarely done in the classroom based on the assumption of lecturers that the students are adults who certainly took the initiative themselves to do so. Therefore, the quiz should be implemented by the conceptualists.

These findings suggest that there are other factors that may affect such as factors derived from the students themselves such as the initial knowledge of the students, Student efficacy, and great mathematical logical intelligence of students as well as their contribution to student success. Because in studying the Integral Calculus whose material is abstract, In addition to reasoning required also needed a good understanding of the concept in order to have the skills to do algebraic manipulation. Visualization of the abstract concept would make it easier for students to understand a concept or problem, but without offsetting the practice of using the theorems and concepts will certainly result in less student retention. In addition to student factors, lecturer factors also have a big share in learning.

From several studies that have been done generally still partial, in the sense of the instructional model of mathematical software assisted with the conventional, As well as performance assessments compared to conventional assessments that are objective in nature. The combination of mathematical software-assisted learning tools with performance assessment so far has not been found, so it needs to be studied further fusion between the device assisted learning mathematics software with any suitable form of assessment, which can optimize student learning outcomes. Because practically should the use of mathematics-assisted instructional device promising increased interest, motivation, interest in student learning as well as increasing student participation by actively encouraging them to explore, and apply the knowledge they have acquired.

The use of mathematical software in learning actually has weaknesses also like two sides of the coin can have a positive impact and can have a negative impact. The use of inappropriate and poorly planned math software can have a negative impact, and vice versa. If the use of mathematical software only at the beginning or at the end of the course can lead to lazy students to determine the answers by manual, Students tend to choose a quick way to find answers simply by clicking on one of the commands. In this way it can cause students to have difficulties when faced with problems at a higher level. Such questions require higher order thinking. Of course this can not be allowed to last long because the impact on student learning outcomes. Therefore the use of mathematical software should be integrated with learning.

Based on the results of hypothesis testing and discussion of research results, the findings of this study are 1) student learning independence using mathematical software-assisted learning tools with higher performance assessment than students using conventional learning tools, While the students' Integral learning calculus did not differ significantly between the students using the mathematics-assisted instructional device with higher performance assessment than students using conventional learning devices. Referring to the findings of research results, it can be concluded that the use of mathematical software-assisted learning tools with performance assessment effectively improving and nurturing student self-reliance, but not effectively improving student Integral calculus learning outcomes.

Therefore, lecturers should apply learning-assisted mathematics software with performance assessment as an alternative that can foster student self-reliance. With appropriate media selection and assessment type, student self-reliance can also be developed.

Acknowledgments
The authors thank I Putu Wisna Ariawan and I Gusti Nyoman Yudi Hartawan from Mathematics Department, Universitas Pendidikan Ganesha for their help in data analysis.

References

[1] Mahayukti G A and Waluyo D 2015 *Pengembangan Model Pembelajaran Berbasis Masalah Berbantuan Pertanyaan-Pertanyaan Pembimbing dan Program Mathematica untuk*

Meningkatkan kemampuan Penalaran dan Komunikasi Mahasiswa dalam Perkuliahan Kalkulus Diferensial (Developing Problem-Based Learning Model Supported by Guided Questions and Mathematica Software to Increase the Students' Reasoning and Communication in Differential Calculus Course) Unpublished Research Report Singaraja: Lemlit Undiksha.

[2] Muderawan I W 2011 *Perkembangan Teknologi Informasi dan Komunikasi dan Aplikasinya dalam Pembelajaran* (The Development of Information, Technology and Communication and Its Aplication in Education) Paper presented on National Conference: Optimizing the Application of IT in Education Informational Technology Education Department Universitas Pendidikan Ganesha Singaraja

[3] Suparta I N 2014 *Kontribusi Pendidian Matematika Dalam Pengembangan Nilai Kemanusiaan* (The Contribution of Mathematics Education to Develop the Humanity Values) Paper presented on Inauguration of Professor at Mathematics Education Department, Universitas Pendidikan Ganesha Singaraja

[4] Suweken G and Mahayukti G A 2009 *Laporan Hibah Pengajaran: Eksplorasi Mathlet dan Kooperatif Think Pair Share dalam Perkuliahan MNSAB (Suatu Upaya Meningkatkan Kualitas Perkuliahan MNSAB)* (Singaraja: LP3)

[5] Zulnaidi H and Zakaria E 2012 The Effect of Using GeoGebra on Conceptual and Procedural Knowledge of High School Mathematics Students *Internasional Journal* **8** (11)

[6] Suweken G, Mahayukti G A and Wisna A 2013 *Pengaruh Model Pembelajaran SIOP Berbasis GeoGebra terhadap Hasil Belajar Matematika Siswa SMP Negeri di Kota Singaraja* (The Effect of SIOP Learning Model with Geogebra Towards Junior High School Students' Learning Outcome in Singaraja) Unpublished Research Report Singaraja: Lemlit Undiksha.

[7] Bozic R 2015 *The Impac of Dynamic Properties of the Software Packages Mathematica and GeoGebra to The Examining and Graphing of Functions with Parameters* European Union: Departemen of Mathematics and Informatics University of Novi Sad, Faculty Of Sciences,

[8] Hussain A, et al 2014 The Impact of Technologi in Mathematics Education: A Cass Study of Pakistan *Journal of Basic & Applied Sciences* **10** 67-72

[9] Arbain N and Shukor N A 2015 The Effects of GeoGebra on Students Achievement *Procedia-Social and Behavior Science* **172** 208-214

[10] Yilmaz Z , et al 2012 The Effect of Dinamyc mathematics Software Geogebra on Student Achievement in Teaching of Trigonometry *Procedia-Socialand Behavioral Sciences* **31** 183-187.

[11] Purwanto H, Indriani G and Dayanti E 2005 *Kalkulus* (Jakarta: Ercontara Rajawali)

[12] Sweden O and Yerushalmy M 2014 LearningThr Indefite Integral in Dynamic and Interactive Techonologycal Environtment *ZDM The International Journal on Mthematics education Springer* **46** (4) 531-517.

[13] Kauropatov A and Dreyfus T 2014 Learning the Integral Concept by Constructing Knowledge about Accumulatiion *ZDM The International Journal on Mathematics Education Springer* **46** (7) 533-548.

[14] Kocsis I 2007 Aplication of Maple ODE Analyzer in Investigation of Differential Equations in Higher Engineering Education *An International Journal for Engineering and Information Sciences* **2** 177-183

[15] Stiggins R J 1994 *Student Centered Classroom Assessment* (New york: Maxwell Macmillan International)

[16] Edmodson, et al 2012 Self-Directed Learning : a Meta-Analytic Review of Adult Learning Construction *International Journal of Education Research* **7** (1) 67 – 72.

[17] Zimmerman B J and Martines-Pons M 2001 Student's Differences in Self-Regulated Learning : Relating Grade, Sex, Giftedness to Self Efficacy and Strategy Use *Journal of Educational Psychology* **82** (1) 51-59

[18] Boekaerts M, Pintrich P and Zeinider M 2000 *Handbook of Self Regulation* (San Diego::

Academic Press)
[19] Adnyana D W, Candiasa, I M and Natajaya I N 2014 Pengaruh Implementasi Asesmen Kinerja terhadap Hasil Belajar Matematika Ditinjau dari Motivasi Berprestasi siswa kelas XI SMK Pariwisata Ubud *E-Journal Program Pasca-sarjana* **5** (1)
[20] Deese J 1959 *The Psychology of Learning* (London: McGraw Hill Book Company)

IOP Conf. Series: Journal of Physics: Conf. Series **1040** (2018) 012030 doi:10.1088/1742-6596/1040/1/012030

Developing blended learning environment to improve learning performance and self-reliance for junior high school students

I M G Sukawijaya and I G P Sudiarta

Universitas Pendidikan Ganesha

E-mail: imdgdsukawijaya@gmail.com

Abstract. Blended learning grows rapidly in recent decade. It is defined as a combination of online learning and face-to-face classroom learning. This research aimed at developing blended learning environment to improve the students' learning performance and self-reliance. This research is a part of a larger study which used development research as its method. The research subject were 38 (8 male and 30 female) grade 8 students of SMPN 3 Sukawati in the academic year 2016/2017. The data gathered from questionnaire, observation and students' online activities in the Learning Management System (LMS). The data were analyzed using content analysis techniques and coded based on type of students' responses. The result show that (a) the total of students' responses to the implementation of tutorial video in BLE were 67 consisting of 73,1% positive and 26,9% negative responses. (b) the total of students' responses to the implementation of online discussion in BLE were 68 consisting 61,7% positive and 38,3% negative responses. Based on the results of data analysis of students' responses to the implementation of tutorial video and online discussions in BLE, it can be inferred that the students were be helped to understand the content, active in learning, prepare themselves to be ready and confident while studying in the face-to-face classroom learning.

1. Introduction

Blended learning (BL) is a learning approach built on "blend" of electronic learning and face-to-face classroom learning [1]. Furthermore, BL also defined as a teaching strategy that combines ICT and traditional face-to-face learning in the classroom which can help students to convert from passive to active learning and overcome the problems of low motivation for course completion and low satisfaction [2]. BL has various model and variations of its "blend" to allow for different blended learning environments (BLE). In line with that, BL study with different "blend" of the learning process could have a different impact on students' learning performance and different research recommendation. For instance, a study conducted by Wu et al. [3] with a 'traditional' BL model where learning materials were given in a traditional face-to-face learning settings in the classroom and then student skills in problem-solving were enhanced through the provision of exercises in online learning settings. The results of the research were not satisfied. It was because the taught material was not flexible. It is merely given in the classroom and the students have no chance to ask when facing the problem while doing the exercises in the online learning session.

On the other hand, a research conducted by Mehmet and Arzu [4] has a better strategy whereby through the BLE used, learning materials were not only given in class but also given again at online meetings via text, images and video animation on the website. This was able to increase the students' understanding of the concept. However there still a negative response toward this study. It was given by the majority of students who feel they need an adequate discussion panel to discuss case studies that given at online learning session.

Moreover, there also BL with flipped classroom model where in this model lecture content is given to students at home through electronic means while face-to-face classroom learning is focused on practical application activities [5]. Research by Thuy, Bram, and Martin [6] showed that the flipped classroom model gives better results than other BL models, especially to students' self-efficacy and intrinsic motivation. This is also supported by the results of research from Montserra, Eliana, Jordi, Roger, and Javier [7] which showed that the flipped-classroom learning environment could give better impact than other models because it can promote self-learning, autonomous working, and time management concern in students and because it helps increase the classroom session effectiveness.

On the other hand, the advance of technology in recent decades have contributed to the development of BL models by integrating the innovation in the learning process [6]. Various ways can be used to present the learning material in a BLE, such as through e-books, e-text, images, animations, or video. Of the ways, the material learning that presented through the video tends to be more appropriate to use, especially at the elementary or middle level students. Certain concepts or skills could be presented and visualized in a more real and challenging manner so that it is easier to understand, especially on subjects that are considered difficult by most students, for example, mathematics. Furthermore, the various concepts that can be accessed through video can involve students in thinking and do learning activities more effectively, faster, and deeper [8].

In addition BLE that provides space for students to be able to inquire and discuss with other students or with online instructors (teachers) during online learning session is also important which is also as a recommendation from previous research [3] [4]. The availability of online discussion feature, students can ask about problems faced earlier and also enables to find the solution only by the online discussion process without having to wait for a face-to-face learning session in the classroom. It helps students prepare early before face-to-face meetings so that classroom learning activities become more effective by focusing on problem-solving exercises. In this case, the use of video tutorials and online discussions is seen as an important factor as a base in BLE practice. This particular research aimed to developing BLE based tutorial video and online discussion to improve students' learning performance and self-reliance. The formulations of the problems as follows *(1) How is the students' responses to the implementation of tutorial video in BLE? (2) How is students' responses to the implementation of online discussion in BLE?*

2. Methods

This research is part of a larger study which uses a development research approach, which consists of the following stages: Define, Design, Development and Dissemination [9]. In this current research have successfully defined and designed BLE prototype based on video tutorial and online discussion. The research subject were 38 (8 male, 30 female) grade 8 student of SMPN 1 Sukawati in the academic year 2016. Data about students' responses to tutorial videos and online discussion in BLE in this study obtained through in-depth interview and students' posts stored in online discussion board on the learning management system (LMS) which used as a tool to support the BLE in this study. Then data were collected and analyzed descriptively qualitative using content analysis technique and coded based on type of student responses. In total, 4 codes were created tutorial video and online discussion. Data about students' responses to tutorial video in BLE divided into groups in term of accessibility, content, effective learning, and motivation. While data about students' responses to online discussion in BLE divided in term of features, effective learning, accessibility and motivation.

3. Result and Discussions

3.1. Implementation of tutorial videos in BLE

Before face-to-face classroom learning is held, first, the students were given tutorial videos that present various concepts and mathematical skills on the website at online learning session. Expert judgment results for content validity of tutorial video indicated that the video in this study is valid and feasible to be used with attention to some suggestions for video improvement as seen in table 1 below.

Table 1. Expert advice.

No.	Advice
1	Make the duration of each video only 5 - 7 minutes
2	One video enough to discuss one indicator
3	Note the size of the video, make each video has maximum 5Mb
4	Adjust background music volume with vocals on video

In the following Table 2, the summary of students' responses toward the tutorial video in BL in online learning can be observed.

Table 2. The frequency of the positive and negative responses of students towards tutorial video.

Aspect	Positive		Negative		Total	
	F	%	*F*	%	*F*	%
Accessibility	16	61,5	10	38,5	26	38,8
Content	16	69,6	7	30,4	23	34,4
Effective Learning	7	87,5	1	12,5	8	11,9
Motivation	10	100	0	0	10	14,9
Total	*49*	*73,1*	*18*	*26,9*	*67*	*100*

From Table 2 it can be seen that around 61.5% of 38.8% students who responses to the accessibility of the video tutorial were expressed their positivity. In this aspect, the positive responses was indicated the advantages of tutorial videos in BLE, such as: it can be accessed anywhere and anytime so that the students can learn the material even though they cannot attend the class and it can also be watched repeatedly. This finding are supported by the results of Mehmet and Arzu [4] and Rosende, Garcia and Otero [10] studies in which the application of BL makes it easier for students to access learning materials whenever and wherever online. The asynchronous access to tutorial video can help the student who need time to review information or to understand the learning material better through pause, rewind and replay features [5]. Meanwhile, the negative responses indicated the students' difficulties in learning through the website, such as their limited access to internet at home, unstable internet connection which contribute to the video buffering and sometimes the video cannot be played through smartphones.

The previous study has predicted the possible obstacles to the application of BLE. Nielsen [11] explained inequities in students' home lives can also cause a barrier for the flipped blended-learning model. Technology needed for viewing or interacting with learning material may not be available in every student. In this study, the limitations of the Internet network became the main obstacle factor causing the number of students who did not have time to access and study tutorial videos before class began. Therefore, the teacher reminded the students to utilize the internet access that available in schools. Although it must be admitted that sometimes unstable connections also be complained by some students.

In addition, Montserrra et al. [7] in their study reported that unaccustomed students learning in new learning environments (BLE) is one of the important factors that could explain why student participation was low. Therefore, in the improvement of this study, the students were motivated by reward and punishment points to enhance students' activity.

Talking about the content, it was found that 69.6% students were responses positively. The students explained the benefits of learning using tutorial videos in BLE are as follows: 1) more easily understand the learning materials presented in video tutorials, 2) material presented on the video more complete and details than students textbooks, 3) animations video make the learning experience more memorable, 4) enthusiastic and less formal voice characters in explaining preferred material creating an interesting impression for students. While the negative responses are given by the students indicate some shortcoming such as students argue the background on the video should be more varied and the videos containing discussion of the problem should be more numerous.

Mehmet and Arzu [4] in their study, reported that students found the section of animation, video, and online assignment are more useful, but indicated that a discussion forum should be included in online learning session. The same is also mentioned by Wu et al. [3] where online forum discussion is needed in BLE especially when students faced problems when learning independently. While it has been addressed in this current study which in the online learning process, online discussion feature included, so it can be used to discuss the problems that faced by students. In addition, it can also be used to strengthen the concepts discussed in the video tutorial before face-to-face classroom learning that simultaneously also serves to further enhance the activity and understanding of students without requiring more resources [12].

Furthermore, students' responses indicated that the tutorial video was effective in helping them learning. It leads to a more meaningful learning activities where the students can easily build their cognitive understanding, not only memorize the formula and also tutorial video help students work on similar problems even more difficult. In addition, students also revealed by studying the material through the video early makes them more prepared when studying in the classroom.

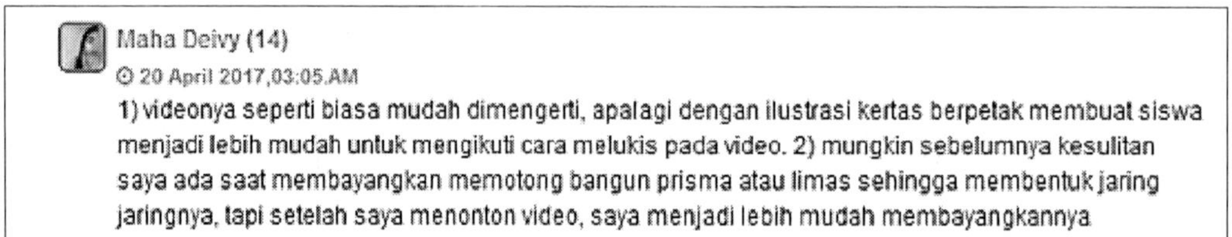
Maha Deivy (14)
20 April 2017,03:05.AM
1) videonya seperti biasa mudah dimengerti, apalagi dengan ilustrasi kertas berpetak membuat siswa menjadi lebih mudah untuk mengikuti cara melukis pada video. 2) mungkin sebelumnya kesulitan saya ada saat membayangkan memotong bangun prisma atau limas sehingga membentuk jaring jaringnya, tapi setelah saya menonton video, saya menjadi lebih mudah membayangkannya

Translation:
Maha Deivy (14): 1) the video as always easy to understand, especially with the illustration of grid paper makes me more easily understand how to draw 3D shape as exemplified in the video. 2) my previous difficulty was how to imagine cutting up the prism or pyramid to create its web. but after I learned the tutorial video, I became easy to imagine.

Figure 1. The example of positive response for video tutorial appeared in students' post.

The students expressed the benefit of the video help her to imagine the three dimensional shape that used to be hard for her. In conjunction with this result, Mehmet & Arzu [4] in their study identified that BLE is able to establish a positive learning environment where student' responses indicated learning activities became more interesting and easy to understand through 3D images and video. Moreover, Montserrra et al. [7] reported at their research the process of clarification given in the face-to-face classroom learning, based on students' performance in the self-learning activities before, improves the degree of understanding oh taught contents.

However, in certain circumstances, when tutorial videos explained what students already understand, there are some students who tend to feel bored watching the review and then prefer to skip that section. One of the sample of students' comment related to the video can be seen in the following Figure 2. After reflecting to the comment of the students, it was found that the cause of boredom of the video is because it contains many repetition of words.

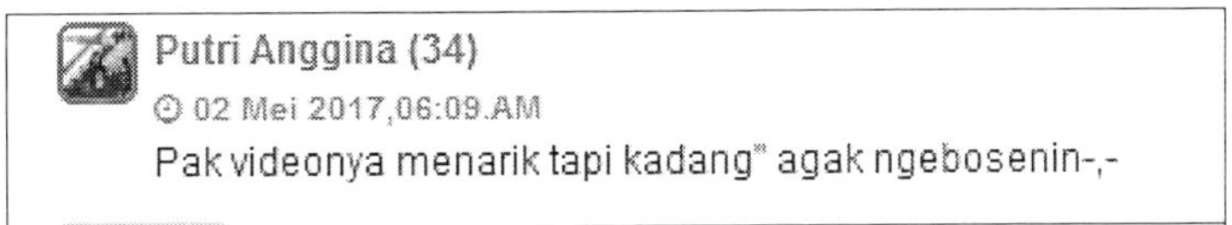

Translation:
Putri Anggina (34): Sir, the video is an interesting but sometimes little bit boring

Figure 2. The example of negative response for tutorial video appeared in students' post.

Lastly, all students' responses in the motivation aspect indicated that they feel more interested in having to study independently at home because the learning material presented in the video is quite easy to understand. In addition, students feel more active use their own leisure time to learn independently especially to use their gadgets and the internet to find positive things like other related learning resources.

3.2. *Implementation of online discussion in BLE*

Students' responses data to the implementation of online discussion during learning in BLE was obtained by using questionnaire. Various positive and negative responses are obtained from students' response for tutorial video and then all of answers are grouped into four aspects, including features, accessibility, effective learning, and motivation. The data is used to see how the quality and effectiveness of the implementation of online discussion in BLE corresponding to students' views as users.

Table 3. The frequency of the positive and negative responses of students towards online discussion

Konteks	Positive		Negative		Total	
	F	%	*F*	%	*F*	%
Features	9	37,5	15	62,5	24	29,6
Effective Learning	15	83,3	3	16,7	18	22,3
Accessibility	17	58,6	12	41,4	29	35,8
Motivation	9	90	1	10	10	12,3
Total	*50*	*61,7*	*18*	*38,3*	*67*	*100*

Table 3 shows that the students were not really welcome to the features given in the online discussion. This happened because of the difficulties in using the digital learning, such as not being able to upload images, malfunctioning the enter key on the keyboard to create new paragraphs, no equation for mathematical symbols, and there are no notifications appearing if someone replied to a students' post. However the rest of the students also stated that the discussion features help them to be more comfortable in conveying questions or opinions.

In term of effectiveness, 83.3% students express the benefits of the flexibility of online discussions in BLE. It could help them when studying independently at home where students can ask questions and discuss problems with other students and even with teachers or online instructors before attending face-to-face meeting. The example of how the students use the online discussion as a place for sharing can be seen in Figure 3.

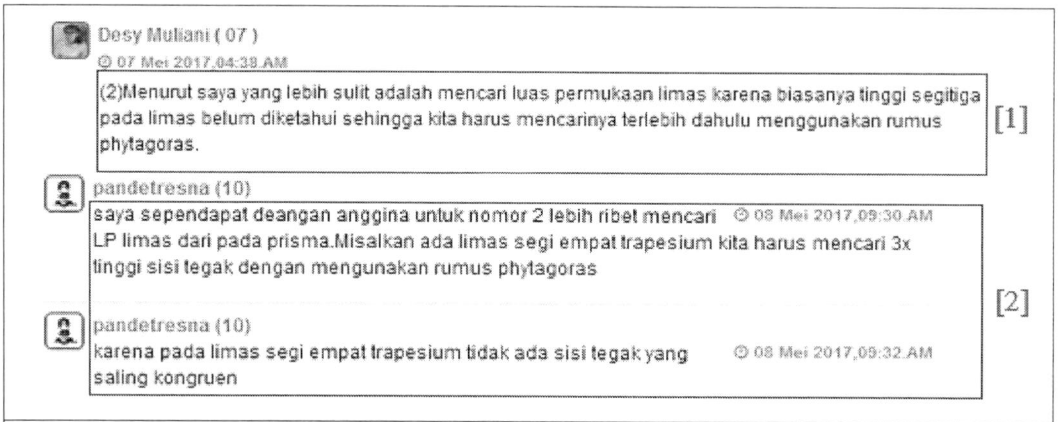

Translation:

Desy Muliani (07): I thought the more difficult is to find the surface area of the pyramid because usually the height of the plane of the pyramid is not yet known so we must search for it first using the formula Pythagoras

pandetresna (10): I agree with Anggina for question number 2 more difficult to find the surface area of pyramid than prism. Suppose there are pyramids that are trapezoidal-shaped then we should search for three times the height of the plane of pyramid using the formula Pythagoras

pandetresna (10): Because on the trapezoidal-pyramid there is no erect plane which is congruent

Figure 3. The sample of students' post contains reasoning competence on online discussion.

Figure 3 contains mathematical reasoning competency of students while discussiing the surface area of the prism and pyramid. The student with code (1) argued that the surface area of the pyramid is more difficult to determine because usually its height is unknown and must be searched first using the Pythagoras theorem. Next, the student with code (2) completed the previous answer by giving a specific example, that is determining the pyramidal surface area of a trapezoidal base where it is necessary to determine each height because there is no congruent side. Those opinions indicate that students have been able to analyze what might be faced when determining the surface area of the prism or pyramid and also show a good reasoning ability.

The various mathematical competencies that appear in online discussions such as reasoning, conceptual understanding, problem solving, connections and mathematical representations certainly could represent the extent to which students were able to construct their initial cognition by learning through tutorial video and online discussion. Furthermore, student's posts that contain various mathematical competencies significantly will be discussed more deeply in face-to-face classroom learning. Therefore, the teacher or online instructor has a responsibility to facilitate online discussions constantly conducive thus students' post that contains the competence of mathematics significantly always emerge in each learning activity.

Furthermore, the students also feel the accessibility of the online discussion is in fair level. It can be seen that 59.6% students stated it as positive while the rest negative. The positive point of accessibility is given to the availability of online discussion feature in BLE that able to students' self-reliance learning during online learning session. This is in accordance with the statement of Arnold and Garza [5] which mentioned the feature of flipped-blended learning can enhance the role of students who are more active and ultimate responsibility for their learning. However, the negative response is due to the limited network of internet owned by students in their homes. Moreover, the unstable internet connection sometimes causing students experiencing failed login and also could not post comments. This caused the degree of participation shown by students in following online discussion activity is quite low at the beginning of the stud but then slowly increase as the time being.

This finding is different from the results of research conducted by Geri, Gafni, and Winer [13] where student activities visit the website to access the material visually presented as a u-curve.

Finally, 9 out of 10 students were positively motivated to follow the online discussion. The students indicated that they feel challenged to be able to answer questions from online task. Moreover, they also interested and enthusiastic to discuss with other students and also teachers in online discussions. Meanwhile, the negative response on this aspect is more caused by feelings of reluctance or disappointment from students because sometimes students failed to login or added comments on online discussions when the internet connection is unstable.

At the beginning of the study, the utilization of online discussion by students did not go as expected. Students only employed online discussions to post the answers of online tasks and give corrections to other students' answers. But as time passes, students become accustomed and able to adapt well to customize their way of learning within BLE.

4. Conclusion

From the discussion of the findings, it can be concluded that the use of video tutorial in BLE is very helpful for students to more easily understand the concept or mathematics skills that tend to be boring if presented only through textbooks or explained by teacher traditionally. Furthermore, the advantages of tutorial video that can be accessed anywhere and anytime and can be viewed repeatedly makes learning activities more accessible and practical. Learning by videos that could present sound, images, text, and animation at once as well as help students more easily understand what they have learned with more meaningful. It also helps teachers in developing more effective learning activities and supports student self-reliance by reflecting on the perceived convenience of students when they have to learn independently through online learning sessions. However, it is important to note some weaknesses that emerge in this study, such as the limited internet connection on some students so that sometimes the video could not be accessed by students indirectly can decrease student motivation. In addition, the demands of students to get more interesting video and not so long in duration should be noted by the teacher before creating the learning materials.

Meanwhile, through the implementation of online discussion in BLE obtained various benefits for both students and teachers. For students, the availability of online discussion in the learning process is very helpful to overcome the difficulties often faced. Students feel facilitated where they can ask and discuss these difficulties more flexible, anytime and anywhere without having to wait for a face-to-face learning activities in the classroom

As for teachers, the implementation of online discussion in BLE is very helpful in developing independent learning attitudes for students. The availability of online discussions that can be a common forum to construct the initial cognition of students before face-to-face classroom learning is very potential to establish more effective learning and focus on improving students' learning performance. However, some weaknesses are also found from these online discussions such as students cannot post pictures, no math symbols are available, cannot create new paragraphs using enter key on the keyboard, there are none of the notifications when someone replies to a students' post became the thing most students complained about.

5. References

[1] I Gusti Putu Sudiarta 2008 *Jurnal Pendidikan dan Pengajaran* **41** 4
[2] Drysdale J S, Graham C R, Spring K J, & Halverson L R 2013 *The Internet and Higher Education* **17** 90-100
[3] Wu Y W, Wen M H, Chen C M and I-Ting Hsu 2016 *Eurasia J. Math. Sci. Technol. Edu.* **12**(5) 1347-1363
[4] Mehmet A O and Arzu D T 2014 *Eurasia J. Math. Sci. Technol. Edu.* **11** 647-683
[5] Arnold S and Garza 2014 *Communications in Information Literacy* **8** 1
[6] Thuy N T Thai, Bram De Wever, and Martin V 2017 *Computer & Education J.* doi: 10.1016/j.compedu.2017.01.003

[7] Montserrra I A, Eliana R R, Jordi H B, Roger B T and Javier T S 2016 *J. Technol. Sci. Edu.* **7** 119-135
[8] Jakaria, M H D, I Gusti Putu Sudiarta, & Suweken G, 2014 *Jurnal Penelitian Pascasarjana Undiksha* **3** 1
[9] Thiagarajan S, Semmel D S, and Semmel M L 1974 *Instruction development for training teacher of exeptional children* (Bloomington Indiana: Indiana University)
[10] Rosende S B, Garcia M J, and Otero J J 2011 *Higher Learning Research Communications* **1** 1
[11] Nielsen L 2012 *Technology and Learning Journal* **32**(10) 46-46
[12] Schullery N, Reck R, and Schullery S 2011 *International Journal of Business, Humanities and Technology* **1**(2)
[13] Geri N, Gafni R, and Winer A 2014 *Interdisciplinary Jurnal of E-Learning and Learning Objects* **10** 1-16

Investigation on students' mathematical online discussion: A case study in grade 8 SMPN 1 Denpasar

I G P Sudiarta, I N Sukajaya and I G P Suharta

Mathematics Department, Mathematics and Science Faculty
Universitas Pendidikan Ganesha, Bali – Indonesia, 81116

E-mail: gussudiarta@undiksha.ac.id

Abstract. Online discussions are now very important to build student thinking in a blended learning environment. This study aimed to explore the significance of mathematical abilities and communication skills that arose during online discussions. The research subjects were 40 (20 male and 20 female) grade 8 students of SMPN 1 Denpasar in the academic year 2016. The data were all students' posts stored in the learning management system used as a tool to support the blended learning process. Data were analyzed using content analysis techniques and coded using mathematical proficiency. The results showed that (a) there was significant mathematical proficiency and communication skills appeared in the online discussion dashboard, (b) The total students' posts were 556 consisting of 58.5% questions and 41.5% answers. The female students' posts were 39.7% while the male students' posts were 60.3%. (c) The posts of female students contained 75.6% mathematics skills, 17.2% communication skills and 7.2% unclear, while the male students' posts contained 89.3%, 9.0%, and 1.8 % for that category. It can be concluded that online discussion was able to create very significant mathematical proficiency (conceptual understanding, procedural fluency, strategic competence, adaptive reasoning, productive disposition) and communication skills (clarification, advice). Female students appeared less active in online discussions than male students did. Female students had given more contribution to the communication skill, while male students had more actively participated in building mathematical skills. Further research is needed to explain the relationship between student activities in online discussion and student performance.

1. Introduction

Internet technology has become more developed nowadays. The availability of bandwidth is getting bigger, making the internet access much faster and more affordable. It provides an opportunity for all sectors, education and learning sector is no exception. In Indonesian context, the lines of department and authorities of education, as well as teachers are trying to take advantage of internet availability to support education and learning. However, until this very second, it could be said that there is no measured and tested model on how school could truly make use of this internet technology. In schools, the use of internet still tends to help sporadically, which is sometimes useful and sometimes not. The use of internet has not been a continuous systematic program, so that the results and impacts could not be measured well. Most of the teachers are still confused too regarding how to use the internet technology to support their learning process. On the other hand, in the digital era of 21st century, mathematics education has also gone through a lot of changes and needs an ability in making use of internet technology.

The availability of advanced digital technology has changed the mind sets on mathematics and how effective mathematics learning is supposed to be. Several animated videos, which could be downloaded for free from the internet, are able to present and visualize mathematical problems in a real and challenging way, in order to improve the conceptual understanding, reasoning, problem-solving ability, as well as to increase curiosity and creativity. Either mathematical concept or skill that is presented in form of animated video could engage the students more into thinking and having the learning activities done more effectively, faster, and deeper compare to the conventional way, which could only be done by reading books, or face-to-face learning in class [1]. Moreover, if the animated video is presented with blended learning, which is a mixed learning process involving classroom face-to-face and online learning, in which the learning sources, such as the learning materials, assignments, or tests, which are in form of text, figures, audio, or video, could be put on the internet so that it could be accessed 24 hours by the students. YouTube is one of the richest source to contain animated video clips and short movies which would challenge the students to watch carefully and be involved in the thinking and learning process of mathematic. This rich source could be taken advantage of by the teachers as the source for learning process which utilized blended learning approach.

There are various understanding about blended learning. But in principle, it can be abstracted as follows. Blended learning is “a formal education program in which a student learns at least in part through online learning, with some element of student control over time, place, path, and/or pace” [2]. Meanwhile Lalima & Kiran Lata Dangwal [3] states that blended learning "is the concept that includes framing teaching learning process that incorporates both face to face teaching and teaching supported by ICT". The results of research on blended learning also varied. In the early days of blended learning experiments carried out with different features, depending on the availability of existing technological resources. Some authors also found different results, for example, [4] and [5] revealed that blended learning did not significantly improve the learning achievement. But in its development many authors also found that blended learning or online learning has good potential to improve students' performance. As Stenbom et.al. [6], Leong & Alexander [7] revealed that online inquiry model and web-based homework plays an important role in students' attitude possibly because of the immediate feedback that improves understanding. Osmanoglu et al. [8] also found that the use of online video case was effective to increase students' participation in discussion among students as well as between the students and their teacher, in connecting among mathematical ideas and in using what has been learned as a result of mathematics learning. The newest research had been conducted and revealed that the blended learning experience benefitted students [9]. It gave a positive effect not only on the learning outcomes, but also on students' attitudes toward mathematics. The male students and high-ability students were more motivated in the blended learning environment. Students also gave positive feedback on the use of blended learning platform.

About the implementation of blended learning is also strongly influenced by various things such as students' satisfaction. This was closely associated with clear guidelines on assignment, rubrics, and constructive feedback [10]. The non-availability, non-accessibility and inadequate students' ICT skills has also very significant role towards the result of blended learning in teaching and learning mathematics. There was significant difference on students' perceptions towards the challenges of blended e-learning tools [11]. The institution and teacher have to clearly identify the challenges and opportunities of blended learning and then resolve adequate practical support to the students. Besides that, Francis & Jacobsen [12] found that the nature and quality of the mathematics task impacted the quality and nature of the online interaction. Mathematics problems that incorporated easily drawn symbols and minimal text worked best in the online collaborative space. To support the blended learning environment, online collaboration exercises can be used as part of a diverse assessment. “Online collaboration served as a highly effective method for promoting and assessing generic capabilities such as writing in a context-relevant manner and the development of self-awareness with regard to mathematical strengths and limitations” [13]. Besides online collaboration exercises, online discussion has evolved into one of the most commonly used communication tools to support blended or online learning today. Online discussions, also known as online discourse or computer mediated communication, can be synchronous (e.g., chat rooms) or asynchronous (e.g., discussion boards). It is a commonly used to promote student understanding of a topic and to facilitate social interaction among students or between students and teacher. Some research concluded the promising benefits of online discussion in the blended learning environment.

A positive relationship was found between critical thinking and engagement, as well as peer interaction. This strategy can be applied to a wide range of educational environments to stimulate critical thinking and engagement [14]. Discussion forums are often utilized in the online classroom to build a sense of community, encourage collaboration and exchange, and measure time on task. There is evidence to support authentic knowledge transfer in the online discussion forums on a student-to-student basis but not from faculty to student [15]. Putman et al. [16] concerned about the lack of quality participation from students in online discussion. That is why the teacher needs to design a pre-structured form of discussion environment to scaffold the process of online discussion. The newest finding about the use of online discussion in blended learning environment indicates that using online discussion is likely to lead students to gain a better achievement [17]. In addition, it was revealed that there was a significant and positive relationships between student participation in online discussion and their final course mark. The social interaction and the collaborative nature in online discussion environments as well as the active learning in blended learning courses were likely to be the possible reasons for the increased achievement.

Based on the literature review above, it can be concluded that blended learning with its supporting features such as instructional videos, online assessment, online collaborative exercises, online discussions have great potential in improving students' performance and students' attitudes. However, a good and careful preparation for the implementation is required from students and teachers as well as schools. In short, well developed blended learning model in terms of content design and pedagogical approach for the teaching and learning of mathematics is a must before it is implemented. Blended learning needs high commitment, motivated teachers and students to guarantee its success.

However, in the context of implementing blended learning approach in Indonesia, several critical questions emerge which are very relevant, as follows (1) how is the best blended learning process which is also most suitable with the students' characters and schools in Indonesia, (2) Which blended learning model is the most effective and gives the most positive impacts towards essential mathematical competence such as (a) conceptual understanding, (b) reasoning and proof, (c) problem solving, (d) mathematical connection, and (e) mathematical representation and (3) which blended learning process is the most effective and gives the most positive impacts towards the students' essential characters, such as hard-working, honesty, curiosity, and even adoration for mathematics. This research is part of a larger study on the implementation of video-based blended learning in the mathematic learning process in the 8th grader. This particular research aimed to explore the significance of mathematical abilities and communication skills that arose during online discussions. The formulations of the problems are as follows (1) What types of meaningful students' posts do appear in online discussions? (2) How do the mathematical proficiency and communication skills significantly emerge in the online discussion? (3) How are the differences between male and female students' behavior in online discussion?

2. Method

This research is a part of the larger study. One graduate student and an experienced mathematics teacher collaborated to implement flipped blended learning model. They have trained for a week to create learning videos as the source for online learning, to practice the developed blended learning scenario and to moderate the students online discussion. The research subjects were 40 (20 male and 20 female) grade 8 students of SMPN 1 Denpasar in the academic year 2016. The data were all students' posts stored in the learning management system used as a tool to support the blended learning process. Data were analyzed using content analysis techniques and coded using mathematical proficiency [18].

3. Result and Discussion

Table 1: The Distribution of Students' Post on the Online Discussion

No.	Description	Students					
		Male		Female		TOTAL	
		VOL	%	VOL	%	VOL	%
1.	The number of students	20	50%	20	50%	40	100%
2.	Questions-Students' Post	187	56%	138	62%	325	58.5%
3.	Answers-Students' Post	148	44%	83	38%	231	41.5%
4.	Total number of students' post	335	100%	221	100%	556	100%
5.	Difference male and female students' post	60.3%		39.7%		100%	

Table 1 shows the distribution of students' post on the online discussion which took place for 4 weeks (8 meeting). The total students' posts were 556 consisting of 58.5% questions and 41.5% answers. The female students' posts were 39.7% while the male students' posts were 60.3%. It seems that the male students were more active in the online discussion. Furthermore, the male students proposed more answers and solutions than those of the female did (44%:38%), while the female students posted more questions than those the male did (62%: 56%).

Table 2: The Significance of Students' Posts on Online Discussion

No	The Significance of Students' Online Discussion	Students			
		Male		Female	
		Vol	%	Vol	%
A.	**Mathematical proficiency**				
1	Conceptual Understanding	108	*32%*	61,00	*28%*
2	Procedural Fluency	83	*25%*	65,00	*29%*
3	Strategic Competence	66	*20%*	14,00	*6%*
4	Adaptive Reasoning	35	*10%*	11,00	*5%*
5	Productive Disposition	7	*2%*	16,00	*7%*
	SUM	299	*89.3%*	167,00	*75.6%*
B	**Social Communication Skill**				
1	Clarification	19	*6%*	32,00	*14%*
2	Critic and Suggestion	11	*3%*	6,00	*3%*
	SUM	30	*9.0%*	38,00	*17.2%*
C	**Unclear**	6	*1.8%*	16,00	*7.2%*
	Grand Total	335	100%	221	100%

Table 2 shows the significance of students' posts on the online discussion in term of mathematical proficiency and social communication skill. The result revealed that students' post contained 89.3% mathematical proficiency, in which the conceptual understanding appeared predominantly, followed by procedural fluency, strategic competence, adaptive reasoning and productive disposition. There was also 9% of students' post contained social communication skill, and 1. 8% was unclear posts. It seems also that the male students posted more significant mathematical proficiency than those the female students did (89.3%:75.6%), while in term of communication skill the female students did better than the male students (9.0%: 17.2%). Finally, the unclear posts were delivered more by female students than those by the male students (7.2%:1.8%).

It can be seen in figure 1 several discussion made by the students and the teacher about the topic of slope.

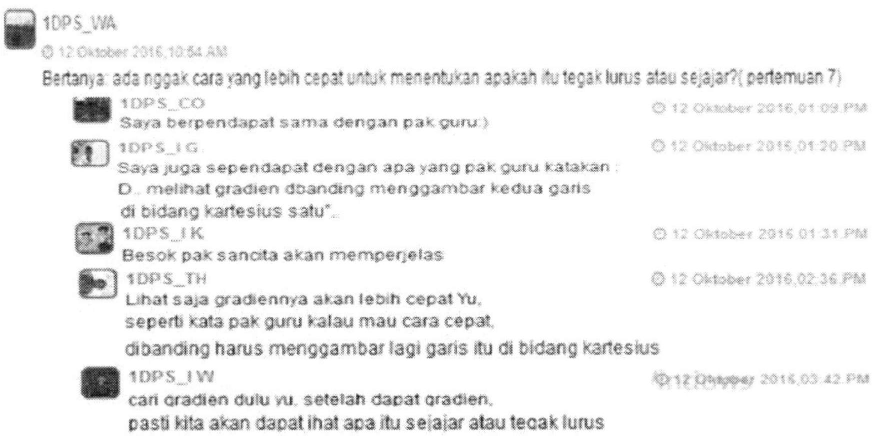

Figure 1: Example of students' post on online discussion

Translation:

1DPS_WA : I want to ask something. Is there any faster way to decide whether it is perpendicular or parallel?

1DPS_CO : My opinion is the same with our teacher

1DPS_IG : My opinion is also the same with what our teacher said (*smiling emoticon*). Seeing the gradient rather than picturing the line into the Cartesian field one by one is faster.

1DPS_I K : Mr. Sancita (*the teacher's name*) will elaborate more tomorrow.

1DPS_TH : Just take a look at the gradients Yu (*pointing to his friend*), that will be faster. It's just like what our teacher said, rather than drawing the lines again in the Cartesian field.

1DPS_I W : Find the gradient first, and after finding it we could see whether it is perpendicular or parallel.

The following are students' posts that emerged in the online discussion which can be viewed as significant representation of the mathematical proficiency [18] which consist of (1) conceptual understanding, (2) procedural fluency, (3) strategic competence, (4) adaptive reasoning, and (5) productive disposition

(1) Conceptual Understanding

This is an example of conceptual understanding appeared in students' post. It shows how students build their conceptual understanding of parallel and perpendicular line.

1DPS_THE : Parallel lines have the same gradients. Meanwhile, perpendicular lines have the opposite gradients.

1DPS_I WAY : Perpendicular lines gradients are opposing each other.

1DPS_I WAY : If the gradients are the same, it means the lines are parallel, and if the gradients are different, it means the position of the lines is different.

(2) Procedural Fluency

An example of procedural fluency appeared in students'post. The students discussed how the procedure to draw a graph of the implicit equation $ax + by = c$ is.

1DPS_DE : To find the coordinates, find the cutting point between x and y. For example $y = 3$ and $x = 4$ in the coordinates. So, you just have to cross-times it so that it becomes $3x + 4y = 12$ if we present it in form of $ax + by = c$
1DPS_MA : We find the coordinates first, then look for the lines equation.
1DPS_I G : Find the coordinates first. Then look for the lines equations.
1DPS_I P : In my opinion, we find the coordinates first, then substitute it to the formula that will be used.

(3) Strategic Competence

An example of strategic competence appeared in students' post. The students described how the most accurate strategy in determining the line gradient was.

1DPS_COK : It is better to use the formula of $ax + by = c$ or $y = mx + c$ because they are more accurate in defining the values of a and b.
1DPS_I G : Use the formula of $ax + by = c$ or $y = mx + c$ to get more accurate results in finding the value of a or b.
1DPS_TH : To make it more accurate, just use $ax + by = c$ or $y = mx + c$ to find a and b.
1DPS_I W : what's important here is that the value of x and y has been placed according to their places and order.

(4) Adaptive Reasoning

An example of adaptive reasoning appeared in students' post. The students build reasoning about slope and gradient.

1DPS_KO : There should be. Gradient is used to measure the slope of a function chart. If there is no function, then we could not measure the slope of that chart.
1DPS_TH : Satryadi said that $a = 4$, means that the gradient is -1/3 because $y2 - y1(a)$ is $3 - 4$ which is -1. If a is 1, it becomes $3 - 1$, the results we get is 2, Sir.
1DPS_MD : Maybe it was miscounted in $y - 2 = 2(x - 10)$ which was answered by our teacher with $-2x + 10$ which was supposed to be $-2x + 20$. Maybe there's a mistake there.

(5) Productive Disposition

An example of productive disposition appeared in students' post. It proves that students have been able to see the benefits of mathematics being studied for everyday life as well as other scientific fields.

1DPS_TH : Learning gradients could be implemented in daily life for example to help measuring the slope of objects such as ladder or others (if we want to be a designer, architect, or to make something, etc.)
1DPS_I W : Parallel and perpendicular lines are used to design a building. Parallel lines are found in the fame and the perpendicular lines in the foundation.

The results of this study indicate that concerns about the quality of online discussions as said by [16] did not occur. On the contrary, many empirical pieces of evidence showed that the online students' post contains significant and validated mathematical proficiency. This evidence was being mapped with 6 aspects of mathematical proficiency according to [18]. The male and female students produced different, important mathematical proficiency and communication skills in online discussion dashboard. The male students participated and delivered posts that contained conceptual understanding, adaptive reasoning, and strategic competence than those the female students did. Meanwhile, the female students contributed more on social communication skills as well as mathematical disposition. In short, it can be concluded that online discussion was able to create very significant mathematical proficiency (conceptual understanding, procedural fluency, strategic competence, adaptive reasoning, productive disposition) and communication skills (clarification, advice). This result is consistent with the previous research conducted by [15], [14] and [17]. But further research is needed to explain the relationship between student activities in online discussion and student performance.

4. Conclusion

(a) There was significant mathematical proficiency and communication skills appeared in the online discussion dashboard, (b) The total students' posts were 556 consisting of 58.5% questions and 41.5% answers. (c) The female students' posts were 39.7% while the male students' posts were 60.3%. (d) The posts of female students contained 75.6% mathematics skills, 17.2% communication skills and 7.2% unclear, while the male students' posts contained 89.3%, 9.0%, and 1.8 % for that category. (e) It can be concluded that online discussion was able to create very significant mathematical proficiency (Conceptual Understanding, Procedural Fluency, Strategic Competence, Adaptive Reasoning, Productive Disposition) and communication skills (Clarification, Advice). (f) Female students appeared less active in online discussions than male students did. Female students had given more contribution to the communication skill, while male students had more actively participated in building mathematical skills. (g) Further research is needed to explain the relationship between student activities in online discussion and student performance.

Acknowledgement

We gratefully thank the Directorate General of Research Enhancement and Development, Ministry of Higher Research, Technology, and Education of the Republic of Indonesia for the supports and funds of this research.

References

[1] Jakaria M H D, Sudiarta I G P and Gede Suweken 2014 'Pengembangan Media Pembelajaran IKRAR Berbasis Video Untuk Meningkatkan Prestasi Belajar Siswa Kelas VII (Developing Video Based IKRAR Learning Media to Improve Grade VII Students' Learning Achievement) ', *J. Pendidik. Mat* **3** 8-14

[2] Horn M B and Staker H 2014 *Blended: Using Disruptive Innovation to Improve Schools* (San Francisco: Jossey-Bass)

[3] Lalima and Dangwal K L 2017 Blended Learning: An Innovative Approach.' *Univers. J. Educ. Res.* **5** 129 - 136

[4] Joordens S, Le A, Grinnell R and Chrysostomou S 2009 Eating Your Lectures and Having Them Too: Is Online Lecture Availability Especially Helpful in "Skill-Based" Course? *Electron. J. E-Learn* **7** 281–288

[5] Orabuchi N 2013 *Effects of Online Visual and Interactive Technological Tool (OVITT) on Early Adolescent Students' Mathematics Performance, Math Anxiety and Attitudes toward Math* (Texas: Dissertation Texas Woman's University)

[6] Stenbom S, Hrastinski S, and Cleveland-Innes M 2012 Student-Student Online Coaching as a Relationship of Inquiry: An Exploratory Study from the Coach Perspective *J. Asynchronous Learn. Netw.* **16** 37–48

[7] Leong K. E and Alexander N 2013 Exploring attitudes and achievement of web-based homework in developmental algebra *Turk. Online J. Educ. Technol.* **12** 75–79

[8] Osmanoglu A, KOC Y and Isiksal M 2013 Investigation of Using Online Video Case Discussions in Teacher Education: Sources of Evidence of Mathematics Learning *Educ. Sci. Theory Pract.* **13** 1295–1303

[9] Lin Y W, Tseng C L and Chiang P J 2017 The Effect of Blended Learning in Mathematics Course *EURASIA J. Math. Sci. Technol. Educ.* **13** 741-770

[10] Lee J 2014 An exploratory study of effective online learning: Assessing satisfaction levels of graduate students of mathematics education associated with human and design factors of an online course *Int. Rev. Res. Open Distrib. Learn.* **15** 111-132

[11] Umoh J B and Akpan E T 2014 Challenges of Blended E-Learning Tools in Mathematics: Students' Perspectives University of Uyo *J. Educ. Learn.* **3** 60 - 70

[12] Francis K and Jacobsen M 2013 Synchronous online collaborative professional development for elementary mathematics teachers *Int. Rev. Res. Open Distrib. Learn* **14** 319–343

[13] Mallet D G 2008 Asynchronous Online Collaboration as a Flexible Learning Activity and an Authentic Assessment Method in an Undergraduate Mathematics Course *EURASIA J. Math. Sci. Technol. Educ.* **4** 143–151
[14] DeNoyelles A and Foster B R 2015 Using Word Clouds in Online Discussions to Support Critical Thinking and Engagement *Journal of Interactive Online Learn* **19** n(4)
[15] Tucker J P, YoungGonzaga S and Krause J 2014 A Proposed Model for Authenticating Knowledge Transfer in Online Discussion Forums *Int. J. High. Educ.* **3** 106 - 119
[16] Putman S M, Ford K and Tancock S 2012 Redefining online discussions: Using participant stances to promote collaboration and cognitive engagement *Int. J. Teach. Learn. High. Educ.* **24** 151–167
[17] Alzahrani M G 2017 The Effect of Using Online Discussion Forums on Students' Learning *Turk. Online J. Educ. Technol.* **16** 164-176
[18] Kilpatrick J, Swafford J and Findell B 2001 *Adding it up: helping children learn mathematics.* (Washington, DC: National Academy Press)

IOP Conf. Series: Journal of Physics: Conf. Series **1040** (2018) 012032 doi:10.1088/1742-6596/1040/1/012032

PISA-like problems using Indonesian contexts

B Murtiyasa, S Rejeki and R Setyaningsih

Universitas Muhammadiyah Surakarta, Indonesia

E-mail: sri.rejeki@ums.ac.id

Abstract. This study aims at developing Program for International Students Assessment (PISA)-like problems which are valid, practical, and have potential effect. The contexts used for this problems are Indonesian contexts involving traditional games, traditional clothes, tourist attractions, social lives. This is a development research which are analysed descriptively. Based on the analysis, it has been concluded that there have been eight PISA like problems consisting problems with quantity content category, problems with uncertainty and data category, problems with change and relationship content category, and problems with space and shape content category which are valid, practical, and have potential effect. The categories are also covering context and process categories in PISA framework. Those problems are valid according to judgment from experts based on the content, construct, and language aspects. Those problems are practical based on the results of the test to small groups of students. Moreover, those problems have potential effect on students' critical thinking skill.

1. Introduction

As the demand of globalization era, Indonesian people, including students, must be prepared to participate in international competition. Various kinds of competition are held to map the quality of education in many countries all over the words. However, the results suggested that Indonesia was still on the bottom of the ranking lists.

Based on the survey conducted by Trend in International Mathematics and Science Study (TIMSS) in 2015, the Indonesian students' achievement in mathematics was on the 45th position from 50 participating countries [1]. The TIMMS survey conducted every four year focusing on the domain of mathematics and science contents, particularly on students' cognitive skills. Besides that, Indonesian students' performance was on the 65t^{h} position out of 72 countries participated in PISA-Program for International Students Assessment [2]. PISA study conducted every three years is focus on assessing students' skills in identifying, understanding, and applying mathematics in daily life.

There might be many factors causing the results of the low of students' performances on both TIMMS and PISA. A possible reason might because Indonesian students rarely face and solve problems with the characteristic of TIMMS or PISA. This study will focus on the case of PISA.

There are three components examined by PISA. The components are content, context, and process. The content is divided into four categories namely change and relationship, space and shape, quantity,, and uncertainty and data [3]. The context is divided into three categories which are personal, occupational, public, and scientific. The process is divided into three categories namely formulating, employing, and interpreting. This study used those three components and its categories to design PISA-like problems using Indonesian contexts.

There have been several studies about Indonesian students' errors in solving PISA problems [4]–[6]. The studies found that Indonesian students mostly experienced difficulties in the early stages (comprehending and transforming) at solving context-based task. Moreover, [7] claimed that the low level of students' reasoning and creativity could be the main factors of their difficulties. Besides that, gender and mathematics textbook could be the factors influencing Indonesian students' performances in PISA [8]. Furthermore, a study about mathematics textbook for 8th graders in Indonesia showed that the percentage of mathematics problems in line with PISA problems is still low with about 33.40% [9].

As the mathematics textbooks could be the factors of the low level of students' skills in solving PISA problems, this study aims at developing PISA-like problems which are valid, practical, and have potential effect. The problems developed are expected to enrich learning resources used for 8th graders' classroom mathematics.

2. Research Method

This is a design research using the type of development study with formative evaluation consisting two stages; preliminary and prototyping stages. The preliminary stage is involving an analysis of students' background knowledge, PISA problems, and curriculum. The prototyping stage includes self-evaluation, expert review, one-to-one, small group and field test [10]. The subjects were 8th graders of a private school in Surakarta, Indonesia. The data collection was conducted using walk through, documentation, tests and interviews.

The preliminary stage was started by an analysis of students' background knowledge. This would be the consideration of the difficulty level of problems designed. Afterwards, PISA problems and curriculum analysis were conducted to map and match the problems designed with the content of mathematics for 8th graders based on the characteristics of PISA problems and the Indonesian curriculum. Lastly, the draft of the first prototype was designed based on the results of the analysis.

The prototyping stages begun with self-evaluation yielded the first prototype. Afterwards, walkthrough and one-to-one were conducted simultaneously to test the first prototype. In the walkthrough phase, the content, construct, and language of the first prototype are validated by experts. The researcher conducted one-to-one test to one student. The results of expert review and one-to-one produces the second prototype that will be tested at the small group phase. Six students from low, moderate, and high level of achievement in mathematics involved in this phase. Students are required to solve the problems and then provide suggestions or comments to know about the practicality of the problems that has been solved. The researcher carried out validity and reliability analysis of each items as well. The revision based on the result of small group phase produced the third prototype which will be tested in the field test. In the field test phase, thirty students worked on the third prototype.

3. Result and Discussion

This study was conducted for students who are at the first semester of 8th grade of junior high school. The preliminary stage involving an analysis of students' background knowledge, PISA problems, and curriculum and the prototyping stage including self-evaluation, expert review, one-to-one, small group and field test will be described in the following paragraphs.

3.1. Preliminary Stages

The preliminary stages was conducted by firstly analyzing students' preliminary knowledge. This study was conducted in the second half of the first semester of 8th graders. Therefore, the problems designed would be involving topics they have learned previously, either in the elementary school or in the 7th grade and the first half of the first semester of 8th grade in junior high school. Afterwards, an analysis of the characteristic of PISA problems and the Indonesian curriculum were conducted consecutively.

The analysis of the PISA problems was focus on the components and the categories of each component of the PISA frameworks. Whereas, the analysis of the Indonesian curriculum was focus on the basic competent of mathematics learned in the 7th and 8th grade of junior high school. Based on the analysis, the draft of the prototype was designed. It is involving four themes namely "Prambanan Temple" which is one of Indonesian Tourism Attractions, "Batik" which is one of Indonesian traditional clothes, "Ojek Online" which is one of Indonesian social life, and "Playing Kites" which is one of Indonesian traditional games. The problems of each theme have to fulfill the requirements of PISA-like problems. In other words, each problem was designed based on the categories of all components of PISA frameworks, namely content, context, and process. The map of the categories of the problems for each themes are described on Table 1. Whereas, the description about the map of problems of each theme based on the Indonesian curriculum will be described in the prototyping stage.

Table 1. Test Blue Print of The Prototype of PISA-like Problems

Theme	Unit	Content Categories	Context Categories	Process Categories	Item Type	Item No
Tourism Attraction	Prambanan Temple	Quantity	Scientific	Formulate	Essay	1
			Scientific	Interpret	Essay	2
Traditional Clothes	Batik	Uncertainty and Data	Societal	Interpret	Multiple Choice	3
			Societal	Employ	Essay	4
Social Life	Ojek Online	Change and Relationship	Occupation	Employ	Essay	5
			Societal	Interpret	Essay	6
Traditional Games	Playing Kites	Shape and Space	Scientific	Interpret	Essay	7
			Personal	Employ	Essay	8

3.2. Prototyping Stage

The draft of the prototype was evaluated by three researchers conducted this study and yielded the first prototype. Afterwards, in the walkthrough phase, an expert on PISA problems and an expert of Indonesian curriculum reviewed the prototype. The review related to the content, construct, and language of the problems designed. The one-to-one phase was conducted with one student to test the first prototype. The results of walkthrough and one-to-one phases were used to revise the first prototype and become the second prototype that will be tested at the small group phase.

In the small group phase, six students from low, moderate, and high level of achievement in mathematics were participated. Students are required to solve the problems and then provide suggestions or comments to know about the practicality of the problems that has been solved. The researcher carried out validity and reliability analysis of each items as well. To increase the level of validity and reliability, a focus group discussion involving mathematics educators (teachers and lecturers) was conducted. The revision based on the result of the small group phase and the focus group discussion produced the third prototype which will be tested in the field test. In the field test, thirty students worked on the third prototype. The prototyping stage will be described in detail based on each theme on these following paragraphs.

3.2.1. Prambanan Temple. The first theme is about one of Indonesian tourism attractions, Prambanan Temple. There are two problems designed for this theme which is related to numbers. As mentioned in Table 1, based on PISA frameworks, it is related to quantity. In the prototyping stage, there was no significant revision on the problems of this theme. One of the two figures is shown on Figure 1. Whereas, the description about the Indonesian curriculum is on Table 2.

Table 2. The Map of The First and The Second Problems

Based on Indonesian Curriculum

Indonesian Curriculum
Grade VIII

Basic Competence:

3.1 Making generalizations of the patterns of the sequence of numbers and of the sequence of object configurations.

4.1 Solving problems related to the patterns of the sequence of numbers and of the sequence of object configurations.

Figure 1. Prambanan Temple

In the field test, all students gave correct answers on the first and the second problems provided in this theme. The question is mainly about prediction about the number of all small temples (candi perwara) surrounded the main temples. However, students found difficulties to explain the steps to get the answers. Figure 2 shows one example of students' written works.

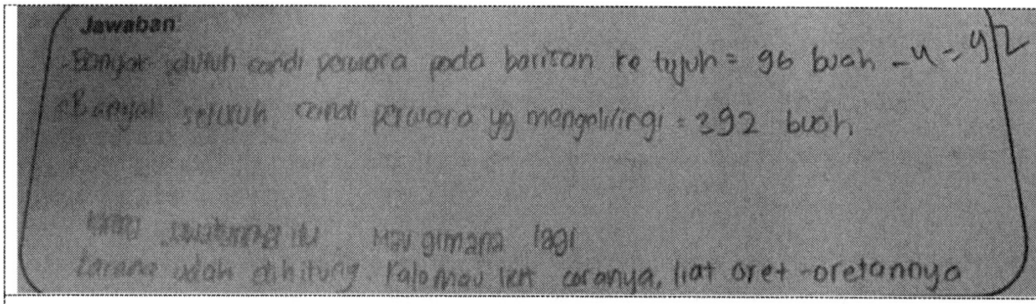

Figure 2. An Example of Students' Written Works on the Second Problem

The students said, "If you want to know the steps to get the answers, just see my calculation". It shows that the students found difficulty in communicate their thinking. However, by the mathematical discourse, students could step by step explain the way to get the answer. As [11] said that students had a better mathematical communication when they were asked more open-ended questions. However, to enhance students' performance, practicing writing about mathematics regularly and routinely are needed to gain experience with writing about mathematics [12].

3.2.2. Batik.. The second theme is about one of Indonesian traditional clothes, Batik. There are two problems designed for this theme which is related to statistics. As mentioned in Table 1, based on PISA frameworks, it is related to uncertainty and data. In the prototyping stage, there was significant revision on the problems of this theme. It was about the term "respondent" which was unfamiliar for students. One of the two figures is shown on Figure 3. Moreover, the description about the Indonesian curriculum is on Table 3.

Table 3. The Map of The Third and The Fourth Problems Based on Indonesian Curriculum

Indonesian Curriculum
Grade VII Basic Competence: 3.12 Analysing the relationship between data and the way to present the data (table, line diagram, bar chart, and pie chart. 4.12 Presenting and Interpreting data in the form of table, line diagram, bar chart, and pie chart.

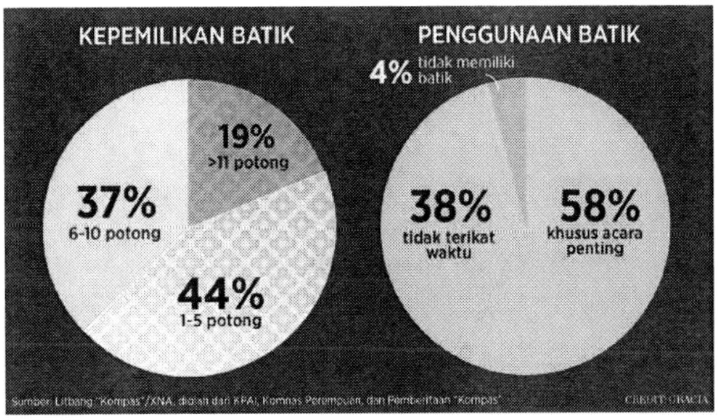

Figure 3. The Data of Batik Ownership of the Respondents.

In the field test, all students gave incorrect answers on the third and the fourth problems provided in this theme. The question is mainly about the number people having batik as shown in the pie chart. Based on the classroom discussion, students were careless in understanding the problems. Figure 4 shows one example of students' written works.

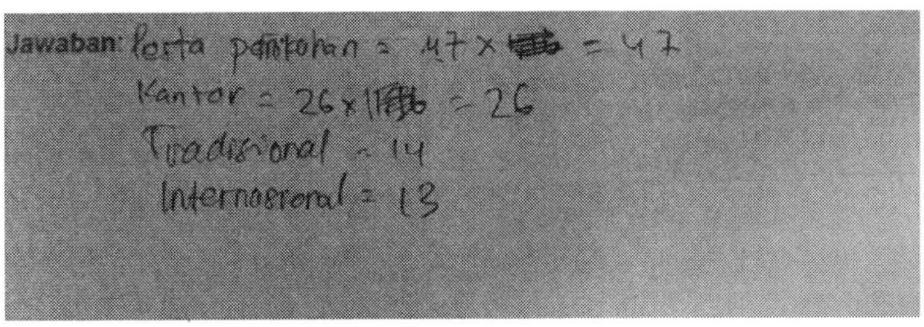

Figure 4. An Example of Students' Written Works on the Fourth Problem

The students answered that 47% means 47 people, 26% means 26 people, 14% means 14 people, and 13% means 13 people. Carelessness were usually become the main factor in translating mathematics word problems [7], [13], [14]. However, by the mathematical discourse, students could step by step understand the problem and find the correct answer.

3.2.3. Ojek Online.. The third theme is about one of Indonesian typical social life, Ojek Online, a mean of transportation commonly used by people in town. There are two problems designed for this theme which is related to algebra. As mentioned in Table 1, based on PISA frameworks, it is related to change and relationship. In the prototyping stage, there was no significant revision on the problems of this theme. One of the two figures is shown on Figure 5. Whereas, the description about the Indonesian curriculum is on Table 4.

Table 4. The Map of The Fifth and The Sixth Problems Based on Indonesian Curriculum

Indonesian Curriculum
Grade VII Basic Competence: 3.6 Explaining linier equation and linier inequalities with one variable and its solutions. 4.6 Solving problems related to linier equation and linier inequalities with one variable.

Figure 5. Ojek Online.

The question is mainly about choosing the cheapest fare of two different corporate of ojek online. In the field test, all students gave correct answers on the fifth and the sixth problems provided in this theme although they just found one possibilities of two possible answers. However, students found difficulties to explain the steps to get the answers. Figure 6 shows one example of students' written works.

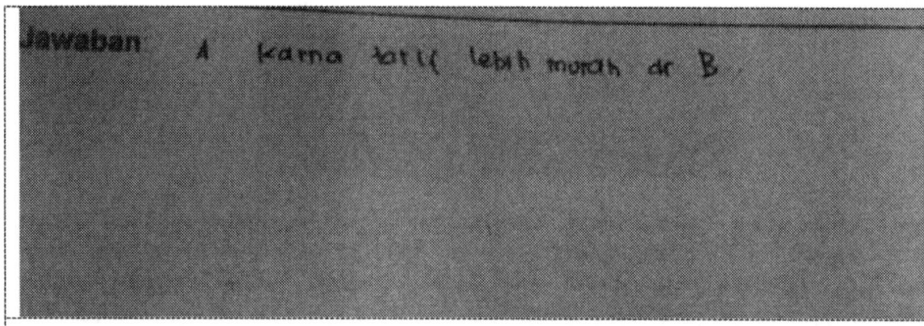

Figure 6. An Example of Students' Written Works on the Sixth Problem

The students just answering that A is cheaper than B without giving any explanation. Based on the classroom discourse, students did not really understand the problem and just guessing. As [15] stated that error in formulating mathematical model is the main difficulty of solving algebraic problems. Therefore, students tend to use trial and error in solving algebraic word problems [16].

3.2.4. Playing Kites. The fourth theme is about one of Indonesian traditional games, Playing Kites. There are two problems designed for this theme which is related to geometry. As mentioned in Table 1, based on PISA frameworks, it is related to space and shape. In the prototyping stage, there was significant revision on the problems of this theme. It was about the problem involving trigonometry which changed into problems about Pythagorean theorems. One of the two figures is shown on Figure 7. Whereas, the description about the Indonesian curriculum is on Table 5.

Table 5. The Map of The Fourth and The Eighth Problems Based on Indonesian Curriculum

Indonesian Curriculum
Grade VIII Basic Competent: 3.6 Explaining and proving Pythagorean theorem and triple Pythagorean numbers. 4.6 Solving problems related to Pythagorean theorem and triple Pythagorean numbers.

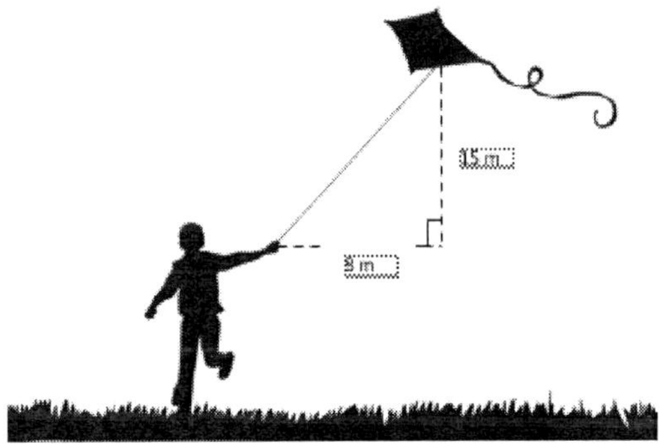

Figure 7. Playing Kites.

The question is mainly about the relations about the length of the thread, the height of the kite, and the distance of the child and the kite. In the field test, it was only about 25% students gave correct answers on the seventh and the eighth problems provided in this theme. In the classroom discussion, some students said that they forgot the Pythagorean theorems, while some others said that they were unfamiliar with the context. Figure 8 shows one example of students' written works.

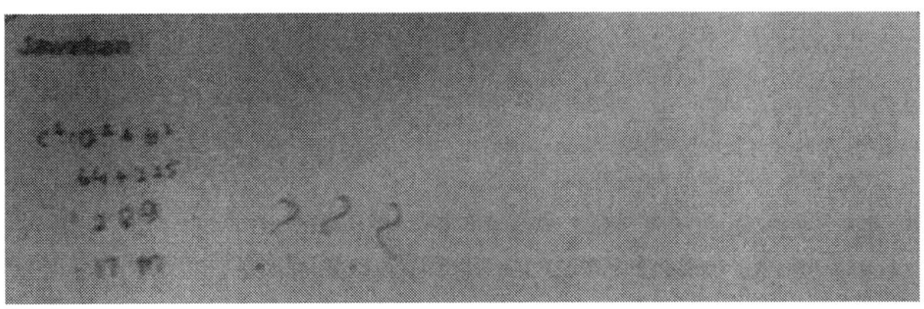

Figure 8. An Example of Students' Written Works on the Eight Problem

Students forgot the Pythagorean theorems could be because of the negative effect of classroom practices which mainly focus on remembering formulas [17]. Furthermore, realistic context might support students' understanding, but it must be choose carefully regarding students' knowledge of the contexts [18].

4. Conclusion and Suggestion

Based on the analysis, it has been concluded that there have been eight PISA like problems consisting problems with quantity content category, problems with uncertainty and data category, problems with change and relationship content category, and problems with space and shape content category which are valid, practical, and have potential effect. The categories are also covering context and process categories in PISA frameworks. Those problems are valid according to judgment from experts based on the content, construct, and language aspects. Those problems are practical based on the results of the test to small groups of students. Moreover, those problems have potential effect on students' critical thinking skill.

In addition, the use of PISA-like problems could stimulate students' critical thinking skills. However, classroom discourse and writing mathematics practices are needed to enhance students' critical thinking and mathematical communication skills. Furthermore, the use of appropriate contexts is important to avoid students' difficulties caused by unfamiliar contexts.

References

[1] Mullis I V S, Martin M O, Foy P, and Arora A 2012 *The TIMSS 2011 International Results in Mathematics*
[2] OECD 2016 Programme for International Student Assessment (PISA) Results from PISA 2015 Country Note: Indonesia pp. 1–8
[3] OECD 2013 PISA 2015 Draft Mathematics Framework
[4] Wijaya A., Van den Heuvel-Panhuizen M, Doorman M and Robitzsch A 2014 *Math. Enthus.* **11** (3) 541–554
[5] Wijaya A., Van Den Heuvel-panhuizen M and Doorman M 2013 *Identifying (Indonesian) Students ' Difficulties in Solving Context-Based (PISA) Mathematics Tasks*
[6] Sari Y M and Valentino E 2016 *An Analysis of Students Error In Solving PISA 2012 And Its Scaffolding* **1** (2) 90–98
[7] Wati E H, Murtiyasa B, and Surakarta U M 2016 Kesalahan Siswa SMP dalam Menyelesaikan Soal Matematika Berbasis PISA Pada Konten Change and Relationship i *KNPMP* **1** 199–209.
[8] P. Burckhardt 2014 *On the Factors Influencing Performance of Indonesian Students in PISA*
[9] Murtiyasa B, Rejeki Sand Murdaningsih S 2016 *An Analysis of Problems on Eight Grade of Mathematics Textbook Based on PIsa's* Framework *3rd International Conference on Research, Implementation, and Education of Mathematics and Science (3rd ICRIEMS)*, , 305–308.
[10] Tessmer M 1993. *Planning and conducting formative evaluations: Improving the quality of education and training* (London: Kogan Page Limited)
[11] Wichelt L 2009 *Communication : A Vital Skill of Mathematics* Lincoln
[12] Garner E and Duncan J Going from I can to I can and here ' s how !' Writing about math in a 3rd and 4th grade mathematics classroom Statement of Significance pp. 1–15.
[13] Dela J K B Cruz and Lapinid M R C*DLSU Res. Congr. 2014* 1–7
[14] Veloo A, Krishnasamy H N and Wan Abdullah W S 2015 *Asian Soc. Sci.* **11** (15) 324–334
[15] Jupri A and Drijvers P 2016 *Eurasia J. Math. Sci. Technol. Educ.* **12** (9) 2481–2502
[16] Egodawatte G 2011 *Secondary School Students' Misconceptions* (University of Toronto)
[17] Sembiring R K, Hadi S and Dolk M 2008 Reforming mathematics learning in Indonesian classrooms through RME *ZDM* **40** (6) 927–939
[18] Sullivan P, Zevenbergen R and Mousley J 2002 *Contexts in mathematics teaching: Snakes or ladders* pp. 649–656,.

IOP Conf. Series: Journal of Physics: Conf. Series **1040** (2018) 012033 doi:10.1088/1742-6596/1040/1/012033

Integrating ethnomathematics into open-ended problem based teaching materials

I P P Suryawan and Sariyasa

Mathematics Department, Mathematics and Science Faculty,
Universitas Pendidikan Ganesha, Bali – Indonesia, 81116

E-mail: pasek.suryawan@yahoo.co.id

Abstract. The idea of integrating ethnomathematics into open-ended problem-based presentation gives a new nuance to mathematics teaching at school. Ethnomathematics is mathematics that develops in a particular culture, and as an alternative in developing mathematics teaching materials which has so far tended to be conventional and less contextual. One way to realize it is by developing open-ended problem-based ethnomathematics which is valid, practical, and effective as an effort to enhance mathematics learning activities and achievement. The development of the teaching materials followed Plomp's product development procedure. The result of expert validation showed that the teaching materials that have been developed fell into the category of very valid with a few revisions. The student and teacher responses related to the use of the teaching materials showed that the teaching materials fell into the category of very practical with some suggestions for revision. The result of observation related to student activity fell into the category of active and the result of the mathematics learning achievement test of the students underwent an increase in every cycle, so that it showed that the teaching materials developed was effective in enhancing the students' activities and mathematics teaching achievement.

1. Introduction

An issue that motivated that development of Curriculum 2013 as the improvement of the school-based curriculum (KTSP) in Indonesia was the demand that mathematics teaching at school has to facilitate the students to be active, creative and critical (Abi [1]). This was based on the survey by International agencies such as TIMSS and PISA which still put Indonesia in the lower zone. In addition, the Ministry of Education and Culture explained that Indonesia is preparing for 2010-2035, the period when Indonesia enters the golden generation, since the number of population with the school ages is very high. The students are expected to survive when the condition is always changing , uncertain, and competitive. Furthermore, the Regulation of Minister of National Education No. 23 of 2006 on School Graduate Competence Standard stipulates that through mathematics, the students are expected to acquire logical, critical, creative and innovative thinking ability. Thus, the effort to enhance the quality of mathematics teaching at every level of education becomes a necessity to prepare an individual with critical and creative thinking ability who will be able to solve problems in his or her daily life.

At the elementary school level, the mathematics teaching which stresses on the importance of the students' rationality and integrates problem solving is more likely to be implemented (Muhsinin [2]). This is supported by the content of the curriculum that includes the mathematics competence standard that has to be attained by the students in learning. The intended competence standard is not the acquisition of mathematics as science, but as mathematical ability needed to understand the world, the ability to compete and succeed in life. Depdiknas [3] stated that the mathematics teaching activity is expected to start with the introduction of a problem based on the situation. By posing problems based on the situation, the students will gradually be guided to acquire mathematical concepts.

Based on the problem identification and the demand of the curriculum above, it is deemed necessary to make an effort to improve the teaching implementation that is student centered and is able to maximize reasoning ability and problem solving ability. The teaching that is capable of creating such learning condition is Problem-Based Learning. Yuliana [4] stated that by posing problems at the beginning of the lesson, the students are gradually guided to acquire mathematical concepts. However, teachers need to think of the appropriateness of the types of problems at the beginning of the lesson [5]. Selecting relevant types of problems to be given to the students in the problem-based learning is very important in order the students' logical, critical, creative and innovative thinking ability can be developed. The type of problems appropriate to facilitate student thinking creativity in PBL is open-ended problem. In the open-ended problem, the students are faced with a problem with many alternative solving procedures and one or many correct answers.

Sudiarta [5] states that open-ended problem is formulated in such a way that it has some or even many correct solutions, and can be solved by many ways. By providing the students with open-ended problems, a low ability student can also give an answer. Besides, the open-ended problems also require the students to explain their thinking patterns through reasoning which can become one source of information for the teacher in looking at the students' ability.

An important element to be considered in presenting open-ended problems to prevent meaningless and dry learning is the content of the problem itself. Problems presented should fit with the students' culture [6]. Some people say that students admit the importance of mathematics, but often have difficulty in learning it. This problem arises from a cultural conflict, the mismatch between the cultural tradition outside of the school , i.e., at home or in the community with what they meet at school [7]. Cultural difference causes differences in perspective and has an impact on differences in knowledge. Thus, a link is needed to bridge outside of the school mathematics and mathematics inside the school. Adam [8] stated that one of the ways is by making use of ethnomathematics as the beginning of formal mathematics learning which fits with the concrete operational stage of development of the students.

Gerdes [9] states that ethnomathematics is applied by a particular cultural group,labor/ farmer group, children in a particular social class, professional class, etc. This means that ethnomathematics is not a mere talk about a particular ethnic group. Mathematics teaching needs to bridge daily mathematics that is based on a local culture and school mathematics. Zaenuri explains that various forms of ethnomathematics in a community is related with the mathematical concepts such as the concepts of plane, space, set, etc. [1], so that they can be integrated into mathematics teaching both at elementary (elementary school and junior secondary school) and high school (senior high school/senior vocational school). The implementation of ethnomathematics as the means to motivate and stimulate students can overcome boredom and give a new nuance to mathematics teaching. Since the students are already familiar with ethnomathematics, it is easier to ask them to relate the parts of culture that they have known with the mathematics materials.

Based on the explanation above it can be interpreted that ethnomathematics is an integration of culture into mathematics teaching or, in other words , ethnomathematics has cultural elements. Cultural difference causes a difference in perpective which brings about an impact on differences in the students' knowledge, so that ethnomathematics nuanced open-ended problems are used as the bridge between outside of school mathematics and school mathematics. The culture introduced depends on where and to whom it is taught. In this paper the researchers take cultural elements and habits of Balinese community, especially in the district of Tembuku, Bangli Regency who work mostly as carpenters, artisans and unique traditional plays makers.

The examples of the elements that can be acommodated in the teaching are as follows. Women in Bali are probably not familiar with the definition of a circle as set of points with equal distances. They may not know how to draw a picture of a circle by using a compass as what the students do at schools. They may not know that the angle in a circle is 360°. But they can really draw a circle by using simple instruments, only by using *busung* (a fresh coconut leaf), *semat* (a sharp rib of coconut leaf that is used to stick parts of the *busung*), and a knife. The way how to cut the leaf into the same size is by folding the center and then sticking the edges, so that we get the shape as shown in Figure 1 below.

Figure 1. *Tamas* (Balinese Handicraft with a Circle Shape)

Another ethnomathematical shapc is the one that is used in buildings or ornaments that are found in Bali as shown in Figure 2 below.

Figure 2. Balinese Buildings Ornaments

The insertion of ethnomathematics into the presentation of open ended problems will give a new nuance in mathematics teaching at school based on the consideration that Indonesia consists of various ethnics and cultures , and every culture has its own way to solve the problem that it faces. Another consideration is that mathematics learned at school does not fit with the local community life that it is difficult for the students since there are two different schemata learned in the environment and at school. Hence, the presentation of an open-ended problem with ethnomathematic nuance is expected to facilitate the students' creativity which later will give a positive impact to the students' learning activities and achievements.

However, the lack of valid, practical and effective open-ended based teaching materials with ethnomathematical nuance is a constraint in the implementation of open-ended problem-based teaching model with ethnomathematical nuance at school. This is supported by the research by Sudiarta [5] that has developed an open ended-problem oriented mathematics teaching, but in its implementation the teacher still had difficulties in looking for ideal open-ended problems. In the same vein Tandidiling [6] in the study that has developed a school mathematics with local culture-based ethnomathematical approach states that no teaching materials that supported it.

Based on the problem above, it is necessary to do a research to obtain and develop mathematics teaching materials. Related to the aforementioned background, the question addressed in this study was "how the quality of the developed open-ended problem based learning materials with ethnomathematical nuance can be used to increase the seven grade students' learning activities and achievements in the triangle and rectangular topics?"

2. Methods

The development of teaching materials in this research followed the procedure of teaching materials development in Plomp [10]. The procedure was devided into four stages as follows.

Preliminary investigation. The activities at this stage were analyzing the situation and identifiying the problems faced by the students and the teachers in mathematics lesson in the classroom. The things done were: (1) reviewing the result of the 7th grade students learning achievement in the last two years and administering a pretest to the 7th grade students (class VII A) SMPN 3 Tembuku Bangli for a limited field try-out; (2) reviewing mathematical teaching process; (3) doing interviews with the teachers and the students about the problems in teaching mathematics; and (4) reviewing the teaching materials used. From the result of analysis a solution was made by analyzing the relevant theories and relevant research findings.

Design stage in which a probable solution to the problem identified at the preliminary investigation stage was designed by (1) reviewing the relevant theories to improve the learning quality, and planning the implementation of PBL model to be presented using open-ended problems and integrating ethnomathematical elements; and (2) designing teaching materials relevant to the characteristics of open ended PBL model that integrated ethnomathematics.

Realization stage in which the solution designed was realized to to produce Prototype 1. The prototype had open ended problem-based teaching materials integrating ethnomathematical values whose validity, practicality and effectivenes were tested.

Test, evaluation and revision stage with the following activities: (1) validating the teaching materials of prototype I which was done by two experts (validators). Based on the result of this validation, a revision was made to obtain teaching materials of prototype 2. After prototype 2 was obtained, a limited field try-out was carried out. (2) the try-out had the purpose of finding out the practicality and effectiveness of the teaching materials. The try-out activity was divided into three cycles. Every cycle consisted of planning, observation, and evaluation as well as reflection stages to see whether the teaching materials developed have met the criteria desired.

The data collected were then processed descriptively. The success indicator was that the teaching materials developed at least had to fall into the category of valid, practical, and effective according to Nieveen's criteria [11]. To attain the category of valid, the mean score of the validation sheet should at least reach $2.5 \leq Sr < 3.5$ (from validator 1 and validator 2) to make it possible to do a limited field try-out. The teaching materials developed can be called practical if at least the mean score of the sheet for observing the implementability of teaching materials given by the teacher and the researchers, the mean score of the questionnaire of the students' response, and the mean score of the teacher's response were in the interval of $2.5 \leq Sr < 3.5$. The teaching materials could be said effective if the test score of the students' mathematics learning achievement at least reached the standard set out by the school, i.e., 70, and the mean score from the sheet for observing the students' mathematics learning activities obtained was at least in the interval $2.5 \leq Sr < 3.5$.

3. Results and Discussion

3.1 Results

The results are explained based on the finding(s) of each stage in Plomp's development model. At the preliminary stage the causes of the low level of activities and achievement in mathematics learning achievement of the students were identified as follows: (1) the teaching did not give the opportunity to the students to find by themselves the concepts being learned. The teacher generally concentrated too much on the practices in solving problems which were more procedural and mechanistic than meaningful; (2) the teaching of mathematics seldom started with the introduction of a problem that fitted with the situation. By posing problems at the beginning, the students were gradually guided to acquire mathematical concepts. (3) the teaching tended to focus on the presentation of closed ended problems which did not give enough time for the students to be creative. (4) the students admitted that mathematics was important, but some of them often had difficulties in learning it. This was caused by cultural conflicts, the mismatch between the cultural tradition that the students met outside the school and what they met at school.

Based on the findings at the preliminary stage above, at the design stage, the researchers collaborated with the teacher in reviewing constructivist learning theory and relevant research findings to overcome the problem. The solution was by implementing open-ended PBL, and presenting open-ended problems which were integrated with ethnomathematic nuance. The first real attempt made was to design the teaching materials that supported the characteristics of the teaching, i.e., open-ended- based teaching materials with ethnomathematical nuance. Then, the researchers realized the design until having a draft or prototype 1 ready to be measured in terms of its validity, practicality, and effectiveness at the test, evaluation and revision stage.

At the test, evaluation, and revision stage, the prototype was validated by 2 validators from the Mathematics Education Department, Universitas Pendidikan Ganesha, validator 1 was chosen by considering his experience in doing a research about ethnomathematics, and validator II was selected based on his experience in doing a research about problem based teaching materials development. Thus, the two validators could be said to have met the criteria to evaluate the quality of the teaching materials developed in this research. The following is presented the recap of the results of the validators' evaluation of the open ended problem-based teaching materials with ethnomathematical nuance in Table 1 below.

Table 1. Recap of the Results of the Validation of Teaching Materials

No	Aspects evaluated	The mean score given by validators		Total	Mean
		I	II		
1	Content of the Teaching materials	3.50	3.50	7.00	3.50
2	Presentation technique	3.67	4.00	7.67	3.84
3	Physical form	4.00	4.00	8.00	4.00
	Total				11.34
	Mean				3.66

Table 1 shows that construct validity of the teaching materials meets the criteria of validity (very valid), since the mean score of validity was in the interval $3.5 \leq Sr \leq 4.0$. Then, the level of practicality of the teaching material was measured by looking at the students response questionnaire and teacher response questionnaire which were filled in by the teacher and the students after teaching using the materials. The following is presented the recap of the result of the validators' evaluation of the open ended PBL materials with ethnomathematical nuance in Table 2 below.

Table 2. Recap of the Results of the Practicality of Teaching Materials

No	Research Instruments	Practicality	
		Mean	Criteria
1	Student Response Questionnaire	3.36	Practical
2	Teacher Response Questionnaire	3.25	Practical

The level the effectivenes of the teaching materials was seen from two aspects, i.e.,: (1) the mean score from the observation of students learning activity by the teacher and the two researchers during the three cycles and (2) the result of the students' achievement in learning mathematics at every cycle after learning by using the materials developed. The recap of the results of the observation of the students learning activity and achievement a presented in Table 3 and Table 4 below.

Table 3. Result of the Observation of the Students Activity using the Teaching Materials

Observer	Cycle I			Cycle II			Cycle III		
	1	2	3	4	5	6	7	8	9
Observer 1	2.40	2.40	3.00	3.20	3.20	3.40	3.60	3.80	4.00
Observer 2	2.00	2.60	2.80	3.00	3.20	3.60	3.60	3.80	3.80
Observer 3	2.60	2.60	2.60	3.20	3.20	3.20	3.80	3.80	4.00
Mean		2.56			3.25			3.80	
Category		Active			Active			Very Active	

Table 4. Mathematics Learning Achievement in Every Cycle

Cycle-i	Criteria of Completeness	Number of Students	Percentage	Mean Score	Category
Cycle I	Score ≥70	19	63.33%	69.80	Not Yet Complete
	Score< 70	11	36.67%		
Cycle II	Score≥70	24	80%	73.56	Complete
	Score< 70	6	20%		
Cycle III	Score≥70	27	90%	76.43	Complete
	Score< 70	3	10%		

3.2 Discussion

This research has produced open ended PBL materials which integrated ethnomathematical content with the expected level of validity, practicality and effectivenes. There was an increase in students learning activity and achievement after using the materials. Thus, by presenting materials and integrating content closed to the students habit, the teaching will become more meaningful, interesting, and increase the students' learning activity and achievement in mathematics.

The increase was obtained through the stages in the open ended PBL model which facilitated the students to do problem solving activities and to use their reasoning in solving problems given. The most important characteristic in presenting open ended problems is the presence of the opportunity for the students to use a number of methods that they regard most suitable in solving the problems. It means the questions in the open ended problem are oriented to guide the development of understanding of the problem given. Sawada [12] states that open ended problem needs responses about the thinking process, ability to make a generalitation and to find relations between two concepts. Hence, the success in presentating open ended problem does not depend only on the format and the materials, but is very much dependent on the procedure, condition and the technique of presentation, so that the students are directed to use their critical and creative thinking ability to solve the problem given, which has the effect on the students' activity in the learning process. This is shown by the students' responses after using the materials, in which the students do not hesitate to use their abilities because they are free to use them in solving the problems presented in the teaching materials.

This problem arises from a cultural conflict, the mismatch between the cultural tradition outside of the school , i.e., at home or in the community with what they meet at school. Culture determine the students' view in responding to something, including in understanding mathematics teaching materials, when a materials is too far from the culture schema its students' they will have difficulty in understanding it. Cultural difference causes a difference in perpective which brings about an impact on differences in the students' knowledge, so that ethnomathematics nuanced open-ended problems are used as the bridge between outside of school mathematics and school mathematics. Francois [13] stated that ethnomathematical content is required in mathematics teaching which is able to relate mathematics and their culture. By integrating ethnomathematics into the teaching it makes it more possible a material learned to be understood by the students. The characteristics of ethnomathematics which is developed in this study is by integrating the mathematical topics of the area of triangle and rectangle that be implemented by the group of people, tradition, children, and professional classes in the Tembuku District, Bangli Regency, Bali.

IOP Conf. Series: Journal of Physics: Conf. Series **1040** (2018) 012033 doi:10.1088/1742-6596/1040/1/012033

To take an example, researcher implemented a traditional play of *kelas-kelasan* which is commonly played by children in Tembuku district, Bangli regency. *Kelas-kelasan* is a game that similar with *engklek* in Bahasa Indonesia. It can be done on the ground, field, or floors. The surface for the setting of the game usually in form of squares. In this study, the researcher modified the surface by using various plane, such as square, rectangle, trapezoid, and others. For drawing and finding the length and width the surface, the students only need a chalk and tape measure. In the following Figure 3, the example of students' worksheet that integrating the ethnomathematics in serving open-ended problem by using *kelas-kelasan* traditional game can be observed.

Have you ever played *kelas-kelasan*?
Remember that the area that you usually use in the game!
Look at the following example!

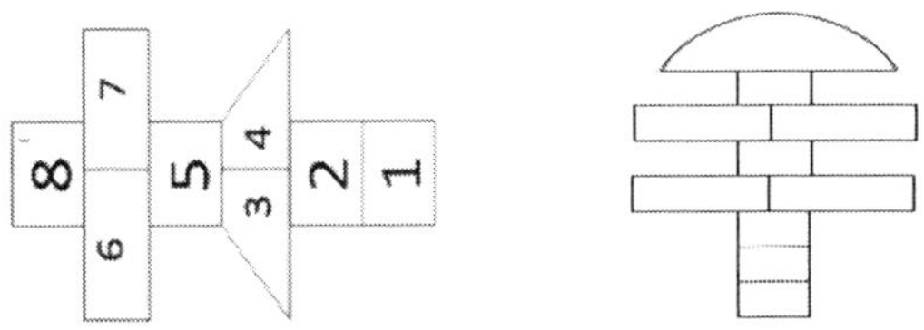

Make and draw two other possible area which you can use in the game!
Then, determine some characteristics of the picture in the *kelas-kelasan* area!

Figure 3. Integrating the *kelas-kelasan* game into the presentation an open-ended problem.

The integration of ethnomathematics in open-ended problem can also be seen in the technic used by traditional carpenters in Tembuku Regency. The majority occupation of the people in Tembuku is as a carpenter and most of the students help their parents to work after school or during the school holiday. An interesting fact of the carpenters in Tembuku Regency is they have no formal school background but able to implement mathematics in proper way, because it was taught by their ancestors as a culture. For instance, to put the lamp right in the center of ceiling which has the form as rectangle. Without knowing the characteristics of a rectangle they are certainly put two robes from the opposite angles and put the lamp in the intersection of the robes. The example of students' worksheet which is integrating the traditional carpentry techniques to the characteristics of rectangle can be seen in Figure 4.

Have you help or see a carpenter in putting the lamp at the ceiling of your house which has a form as a rectangle? Please draw, describe and explain his working procedures!

Then, solve the following case.
Mr. Putu told us that he divides a rectangle into 4 regions with the same area. Mr. Made doesn't agree that the area of four pieces of the rectangle are same. Who is correct among them? Why?

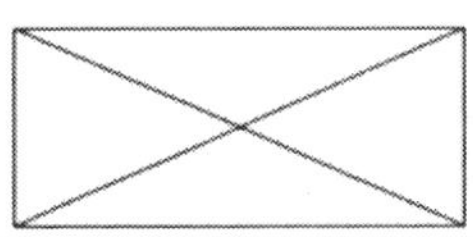

Figure 4. Integrating the carpenter's technic in an open-ended problem

Besides the aforementioned two ethnomathematical contents, there also other cases that can be used as a mathematical problem. It can be seen in the Balinese traditional crafting or ornaments in Tembuku Regency and Balinese traditional hand-made offering that mostly done by Hinduism women. The skills are informally come from one generation to the next generation and become the occupation for the Balinese people.

Therefore, the role of ethnomathematics in an open-ended mathematical problems can be classified into three major aim, which are: (1) provides a rich teaching and learning environment, not only inside classroom but also outside classroom, this create a positive motivation for the students to learn mathematics in meaningful way, (2) gives the students chance to construct their own concept and understanding based on their real experiences, and (3) develops a respect feelings toward the nation and its tradition, art and culture.

The result of the study comforms to the finding by Richardo [14] about the role of ethnomathematics in implementing mathematics teaching in Curriculum 2013, who also concludes that by integrating ethnomathematics the students are facilitated and able to construct mathematical concepts with their prior knowledge through their environment. In adition, the result of the study is similar to the result of the research by Eka Mahendra [15] about Project Based Learning with ethnomathematics content in mathematics teaching, that the implementation of Project- Based Learning with ethnomathematic content has a significant effect on the motivation and learning achievement in mathematics of the junior high school student.

4. Conclusion

Based on the discussion at can be concluded that through the stages in Plomp's product development model open ended problem based teaching materials with ethnomathematics content which are valid, practical, and effective to enhance the activity and mathematics learning achievement of the students of junior high school in plane geometry. The integration of ethnomathematical content into the presentation of open-ended problem, implied to the concept and knowledge that are constructed by students become real based on their on environment so that the teaching becomes more meaningful. The different culture of the students affect to the different point of view which create different understanding. Hence, it is important for the teacher to serve the open-ended mathematical problem with ethnomatematics context in the beginning of lesson which can be used as the bridge for the practical mathematics outside classroom and what they learn in the formal lesson. Therefore, the previous problem about the conventional and not contextual text book can be solved.

Acknowledgement

Our thanks go to all of the people who have supported the implementation of this research which has produced results that can be recommended as an alternative in innovative mathematics teaching. Our thanks is especially conveyed to people at SMP No. 3 Tembuku (headmaster, teachers and students) as the place were the limited field try out was conducted.

Refferences

[1] Abi A M 2016 *Integrasi Etnomatematika dalam Kurikulum Matematika Sekolah* Jurnal Pendidikan Matematika Indonesia **1** 1-6

[2] Muhsinin U 2013 *Pendekatan Open Ended pada Pembelajaran Matematika*. Edu-Math **4** 46 - 59

[3] Depdiknas 2004 *Kurikulum Mata Pelajaran Matematika SMP* (Jakarta: Depdiknas)

[4] Yuliana E 2016 *Pengembangan Soal Open Ended pada Pembelajaran Matematika untuk Mengidentifikasi Kemampuan Berpikir Kreatif Siswa* Prosiding Seminar Nasional

[5] Sudiarta I G P 2008 *Paradigma Baru Pembelajaran Matematika: Membangun Kompetensi Matematis Berpikir Kritis Melalui Pendekatan Open-Ended* (Singaraja: Undiksha Press)

[6] Tandililing E 2013 *Pengembangan Pembelajaran Matematika Sekolah dengan Pendekatan Etnomatematika Berbasis Budaya Lokal Sebagai Upaya Untuk Meningkatkan Kualitas Pembelajaran Matematika di Sekolah* Makalah disajikan dalam Seminar Nasional Matematika dan Pendidikan Matematika dengan tema "*Penguatan Peran Matematika dan Pendidikan Matematika untuk Indonesia yang Lebih Baik*" on 9th November 2013 at UNY

[7] Fatimah 2012 *Implementasi Etnomatematika dalam Pembelajaran Matematika Pada Jenjang Pendidikan Sekolah Dasar* Jurnal Lentera Pendidikan **15** 41-54

[8] Adam S 2004 *Ethnomatematical Ideas in the Curriculum* Mathematics Education Journal **16** 49-68
[9] Gerdes P 1994 *Reflection on Ethnomatematics* For the Learning of Mathematiccs **14** 19-21
[10] Plomp T 1997 *Educational And Training System Design* (Enschede: University of Twente, Faculty of Educational Science and Technology)
[11] Nieveen N 1999 *Prototyping to Reach Product Quality* (London: Kluwer Academic Publishers)
[12] Sawada T 1997 *Developing Lesson Plan* in J. P. Becker & S. Shimada (Ed.).The Open-Ended Approach: A New Proposal for Teaching Mathematics (Virginia: National Council of Teachers of Mathematics)
[13] Francois K 2010 *The Role of Ethnomathematics within Mathematics Education* Proceeding of CERME, January 8th February 1st 2009 Lyon Franc
[14] Richardo R 2016 *Peran Etnomatematika dalam Pembelajaran Matematika pada Kuriulum 2013* LITERASI **7** 118 - 125
[15] Eka Mahendra I W 2017 *Project Based Learning Bermuatan Etnomatematika dalam Pembelajaran Matematika* Jurnal Pendidikan Indonesia Undiksha **6** 106 - 114

IOP Conf. Series: Journal of Physics: Conf. Series **1040** (2018) 012034 doi:10.1088/1742-6596/1040/1/012034

On the implementation of e-learning with *mathlet* GeoGebra in Analytic Geometry course to improve students' engagement and achievement

G Suweken

Mathematics Department, Ganesha University of Education

Email: gede.suweken@undiksha.ac.id

Abstract. This study has been carried out to investigate the effect of the implementation e-learning with GeoGebra *mathlet* accompanied by its assessments to improve students' engagement and achievement. The problems of the study arose from our experiences that students' engagement during the course of Analytic Geometry course were very low. Because of that, an e-learning with GeoGebra *mathlet* complete with its assessments was set up using IMathAS, software especially developed for mathematics or mathematics related course. The choose of IMathAS was due to its capability to display mathematical notations, to randomize items order as well as items data, and especially to display explorative *mathlets*. After setting up the e-learning, a classroom action research was then carried out to investigate whether or not the implementation of this kind of e-learning was capable in improving students' engagement and achievement. Results showed that the implementation of this kind of e-learning was indeed capable of improving students engagements and achievement. This was clear from the progress they made during the learning process as evidenced by data obtained and its analysis.

1. Introduction

Analytic geometry is one course that usually taught in Mathematics Teacher Education and especially at the Department of Mathematics Education, Ganesha University of Education. This course is a *harmonious marriage* between Geometry and Algebra and it is a very good example to show to students that various mathematical concepts are in fact interconnected. This course is offered to students in semester III of Department of Mathematics Education, Ganesha University of Education and it is a continuation of high school mathematics subject. But, contradictive for being a continuation of high school mathematics, it is surprising that this course causes troubles to many students. This is reflected on the following table of students' achievement.

Table 1: Students achievement on Analytic Geometry on Academic year 2011/2012.

Class	Pass	Not pass	Total
3A	22	33	55
3B	17	22	39
3C	8	32	40

As a lecturer of this course for some years, various methods have been done to increase students engagement, namely (1) copying the textbook used, (2) involving students during lecture, by asking them to solve problems written on the book, and (3) asking students to present their answer to the board. However, as we mentioned earlier, students' achievement on this course is still far from our expectation. Based on our crude observation, it seems that students' engagement in this course is very low. If they are asked to do the problems on the book, then most of them are just wait and see their smart friend to come to the board, if the problems are used as a homework, then their answer will be uniform, indicating that only some of them are really do the work. Problems that appear on the book that has been copied is not interesting enough to attract students engagement.

Based on this finding, some innovative learning approaches were done to increase students' engagement. However, treatments being tried were failed to increase students engagement because there is no effort done to monitor students participation and then to value their participation, for example by considering this participations in determining their final marks. Solving the aforementioned problems will require much time, and conventional approach will not be able to overcome them. In academic year of 2015/2016, a new innovative approach have been devised to overcome this problems of engagement, that is by implementing an e-learning to students of semester III of Department of Mathematics Education, Ganesha University of Education. The e-learning was populated with (i) text material about Analytic Geometry, (ii) learning media (*mathlet*), and of course (iii) test that should be solved by students which is accompanied by feedback and grades that students achieve after doing the problems.

The e-learning have been built using a free open source software IMathAS [9]. The use of IMathAS was based on its capability to display mathematical symbols, graphics, and explorative GeoGebra *mathlets*, which is a very important tool to make students engage.

So, the research was aimed at (i) building e-learning for Analytic Geometry, (ii) increasing students' engagement, and (iii) investigating the effect of the e-learning to students achievement.

2. The importance of explorative *mathlet* in learning mathematics

Learning mathematics is learning about abstract ideas, their structure and their concepts inter-relationships. Since 'creatures' studied in mathematics are mostly abstract, then students, firstly need to be acquainted with them, study their behavior, and finally try to reconstruct them in their own way. "*Knowledge must be constructed by learners; it cannot be supplied by teachers. We all responsible for our own learning; no one can learn for us.*" [6]. Thus, students have to engage actively in the process of understanding concepts being learnt. According to Ausuble [7], even college students, during their preliminary steps in understanding some mathematical concepts will use a concrete approach, manipulate and explore it, before finally understand it abstractly. Indonesia Department of Education long ago have endorsed this approach in learning mathematics, by saying that mathematics can be defined as an activity of searching and discovering regularity and pattern, whether in number, geometry, or algebra. So, the learning of mathematics should be realized by: (1) Giving students opportunity to discover pattern and relationship, (2) giving students opportunity to explore and experiment, (3) encourage students to discover order, difference, comparison, (4) encourage students to draw conclusion, and (5) helping students in finding and understanding relationship among various concepts. [4].

The problem we then face is, how to make students, especially college students, to experiment or to explore a mathematical concept in order to understand it. This is not an easy task. To explore the concepts of addition, subtraction, or multiplication in elementary school, concrete materials can be used. But how to explore gradient, a focus and directrix of a parabola, locus of points satisfying a certain condition? [5] The only approach to overcome this obstacle is by using technology [9], that is computer with dynamic mathematics softwares and one of them is GeoGebra [8]. In this relation, a *mathlet*, a small program which is used to explore a certain mathematical concept, is very well known. *Mathlet* example in visualizing the concept of parabola -- a locus of points which are equidistant from a point (called focus) and a line (called a directrix)-- is as follows. [9]

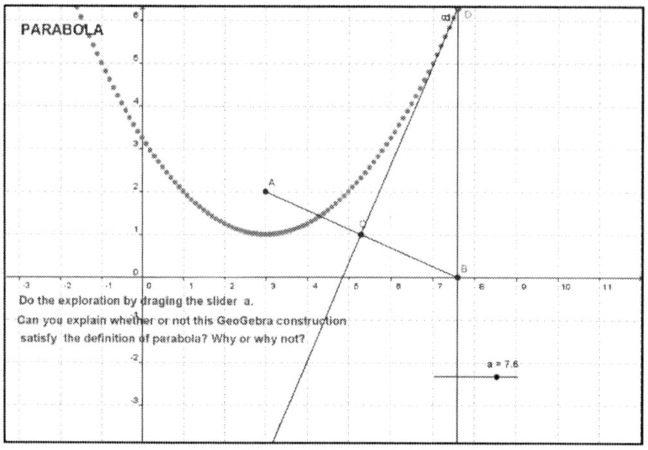

Figure 1: *Mathlet* to explore parabola.

With *mathlet* like the previous one, students will explore the concept of parabola such as whether or not the definition is satisfied. *Mathlet*, is also a very efficient tool to give students a sense of surprise, and from this surprise, hopefully, they will be interested in constructing proof of the concepts behind it. Here is an example.

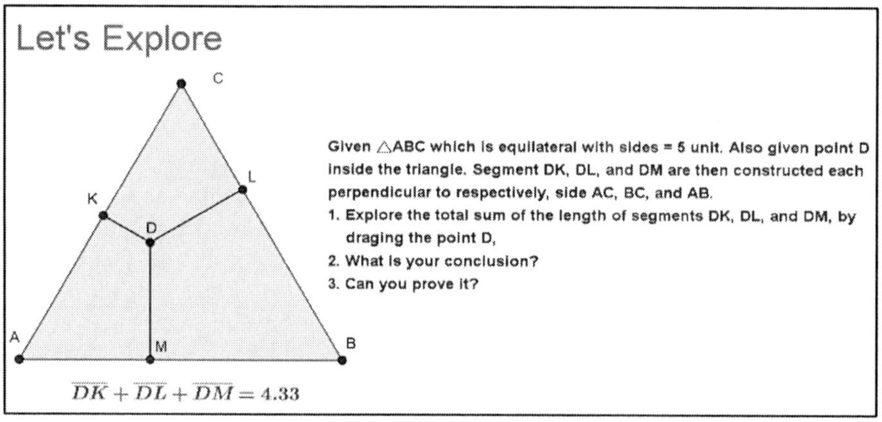

Figure 2: The exploration of The Viviani Theorem.

Mathlet visualization like this will give students another kind of representation of the mathematical concepts behind it. According to the Theory of Multiple Intelligence, humans have many kinds of intelligence, and one of them is a visual intelligence. Surprisingly, visualizations are rarely used in learning mathematics, in spite of its abundant appearance in students life all the day long. Albert A. Cuoco [3] states that: *"By helping people visualize and experiment with mathematical phenomena, modern computer technology have changed all people learn and work. In school, they can influence how mathematics is learnt and taught."*

Other factors making GeoGebra *mathlet* so powerful is its ability to give instant feedback and multiple representation of mathematical concepts. Giving instant feedback is very important feature of *mathlet* since students will instantly know whether their responds are correct or wrong, and accordingly make improvements. Meanwhile, displaying various representations will make students realize that mathematical concepts are in fact interrelated, that is a concept can be represented algebraically, numerically, or visually [5]. Especially visual representation, will help visual students a lot since they can understand ideas more easily using visualization. So far, in the teaching-learning mathematics, these visual students are rarely being given attention and helps. Teachers usually treat all students similarly, and deliver their teaching-learning analytically and rarely visually.[10,11]

3. Mathematics assessment software IMathAS

Another free and open source educational software, which is very useful in making qualified mathematics web, is IMathAS. This software is especially build to mathematics and many other subjects using mathematics. The interesting features of this software is its ability to display mathematical symbols and graphics perfectly, including dynamic graphics (Figure 4) produced by other software such as GeoGebra, as well as randomize items whether their order or their data. These features are very distinctive features of IMathAS compared to others software available. For example, if we want to ask about the gradient of a line which pass through two certain points, IMathAS is not only capable of randomizing the occurrence of this item, but also randomizing those two points (that is the item data). Therefore, theoretically each students can be planned to work out different problem.

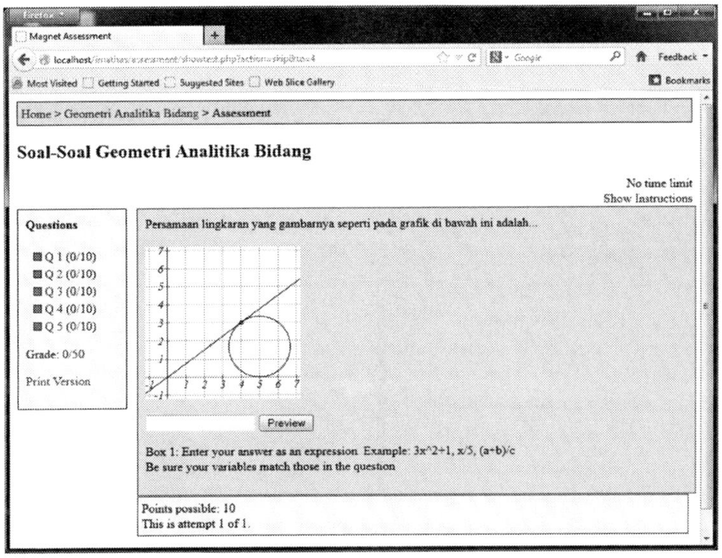

Figure 3. IMathAS display for a certain mathematical

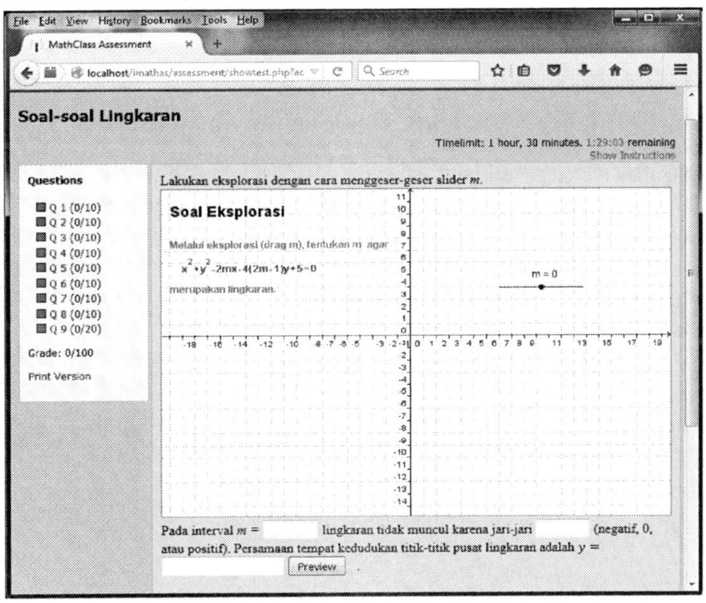

Figure 4. Explorative problem in IMathAS.

4. Results and discussion

To increase students' engagement in learning Analytic Geometry and to monitor their progress during learning process, a Mathematics e-learning called *MathClass* was built. This web set up was then followed by a Classroom Action Research (CAR) [11]. This research have been carried out using 2 classes consisting of 65 students of semester III of Department of Mathematics Education in academic year 2015/2016. In the starting period of the research, many obstacles should be overcome, e.g. many students still do not know how to input symbolic answers to the test being asked. Also, there is some instability with the web, that is, some correct answer is marked as incorrect by the web. To overcome this problems, we asked students to redo the test and submit their answer as a hard copy.

At the second period of the CAR, we thought that the web should be used as a place to learn. Therefore, we then set up a test which can be answered by students repeatedly until they get the right answer. Because the capability of the software to randomized the test data, students will always have different problems to work out. And because their progress will always be monitored, then they will be forced to do the problems. This is really a good way to make students learn. Bellow are results obtained during this period and notice that some students were able to achieve his/her maximum score (i.e. 80).

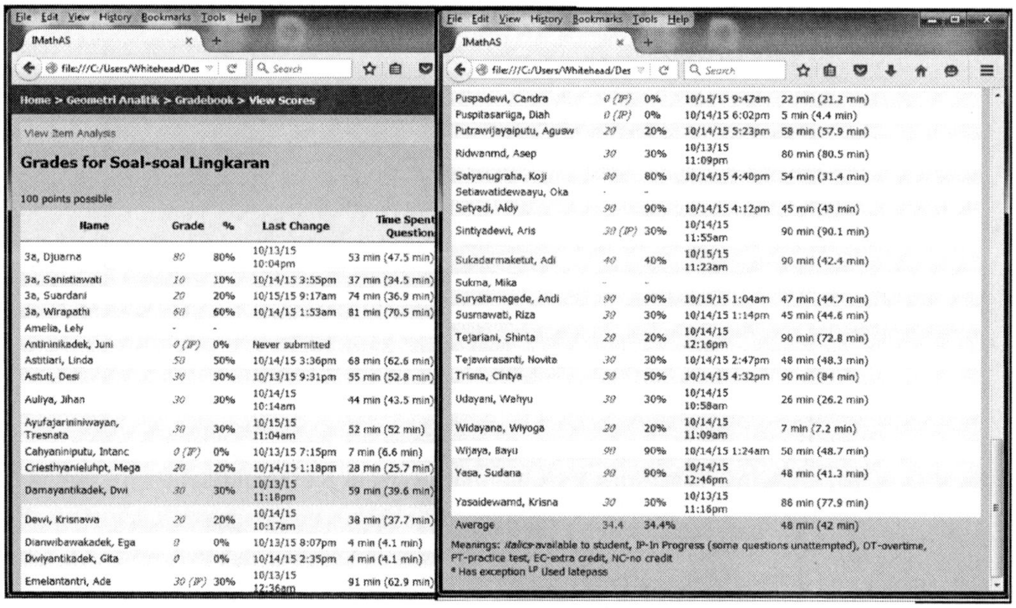

Figure 5. Students' result on test about a circle

So, in this cycle the average students score is 34.4 out of 100. This results, of course, is not very encouraging. This poor results is partly caused by uncommon way students achievement is assessed, that is by using web. In other words, students are still in transition state, from traditional assessment system to online assessment system. From the report is also clear that problem number 9 is rather difficult for students, while problem number 1, 2, and 5 are quite easy to students.

In cycle III, where students had been used to method used in learning, and which used Parabola and Ellipse as materials for assessment, it was obtained that the average students score has increased to become 65.6. This result was encouraging although was still not very satisfying.

Results obtained from the final test showed that the average of students score is 58. Compared to the previous results, i.e. 36, this average increased as high as 22, which is quite significant. In addition, based on the analysis of students responds, it was found that the e-learning was very useful to students and they have learnt a lot because their engagement were always being monitored and used to determined their final marks. In conclusion the implementation of the e-learning was successful in improving students' engagement and accordingly their achievement.

5. Conclusion

This study has been carried out to investigate the effect of e-learning in improving students engagement and achievement. The problem of the study arose from experiences that students' engagement in this course were very low. Therefore the intention of this CAR was to try to improve students' engagement for the course by the implementation of an e-learning complete with its GeoGebra *mathlet* and its online assessments. Through this e-learning, students engagement and progress were monitored, so they are forced to get involve in the process of learning. From the results obtained, it can be said that the use of e-learning with its *mathlet* and online assessments have been quite successful in improving students' engagement and achievement.

References

[1] Cuoco A A, Goldenberg P E and Jane M 1995 *Technology in Perspective* **87** 450 – 452
[2] Depdiknas 2004 *Kurikulum 2004 Sekolah Menengah Pertama (SMP): Pedoman Khusus Pengembangan Silabus Berbasis Kompetensi SMP Mata Pelajaran Mata Pelajaran Matematika* (Jakarta: Direktorat Jenderal Pendidikan Dasar dan Menengah)
[3] Gadanidis G 2000 *The Effect of Interactive Applet in Mathematics Teaching* Faculty of Education University of Western Ontario Accessed 5 November 2008
[4] Holzer S M. and Andruet R H 2000 *International Journal of Engineering Education* **16** (5) 372-384
[5] Hudojo H 2003 *Pengembangan Kurikulum dan Pembelajaran Matematika* (Malang: UM Press)
[6] https://www.geogebra.org
[7] imathas.sourceforge.net
[8] Klotz E A 1991 Visualization in Geometry: A Case Study of Multimedia Mathematical Education Project in Walter Zimmerman and Steven Cunningham *Visualization in Teaching and Learning Mathematics* (USA: The Mathematical Association of America)
[9] Oldknow A, Taylor R and Tetlow L 2010 *Teaching Mathematics Using ICT* (New York: Continuum International Publishing Group)
[10] Silverman L K 1998 *Guidelines for Teaching Visual-Spatial Learners (VSL)* www.visualspatial.org Assessed 1 December 2006
[11] Somekh Bridget 2006 *Action Research, a methodology for change and development* (England: Open University Press)

IOP Conf. Series: Journal of Physics: Conf. Series **1040** (2018) 012035 doi:10.1088/1742-6596/1040/1/012035

The effect of problem based learning model and authentic assessment on mathematical problem solving ability by using numeric ability as the covariable

I K Darma [1,2], I M Candiasa [2], I W Sadia [2] and N Dantes [2]

[1] Politeknik Negeri Bali, Bali, Indonesia – 80364
[2] Universitas Pendidikan Ganesha, Bali, Indonesia – 81116

Abstract. The research was intended to 1) recognize differences the ability in solving mathematic problems among students facilitated with problem-based learning model and conventional learning one, 2) determine the influence of authentic assessment towards mathematical problem-solving abilities, and 3) determine the effect of interaction between learning model with assessment model towards mathematical problem-solving abilities, once numerical ability is controlled. The research utilized experiment design using two factor measurements in factorial 2 X 2. The sample is 110 students. Data were collected using tests that had been validated both theoretically and empirically, then analyzed using Anacova stitistics. The results of the analysis show that the group of students facilitated by problem-based learning model and authentic assessment get the highest average score of mathematical solving ability compared to other groups. Once the numerical ability is controlled, the results of hypothesis testing show significantly: 1) there is different problem solving ability of mathematic between student which facilitated model of problem based learning with conventional learning model, 2) There is a difference problem solving skills of mathematics between students who facilitated an authentic assessment model with conventional assessment model, 3) there is an interaction effect between the learning model and the assessment model on the ability mathematical problem solving. In an effort to improve the effectiveness of mathematics learning, the collaboration of problem-based learning models with authentic assessment models can be considered as one of the classroom learning models.

1. Introduction

The ability to solve problems is one of the important aspects in making the students literate in mathematics. This ability is needed very much by the society, and is the most important in mathematics, not only by those who will learn mathematics as their specialization, but also for those who will apply it in other disciplines and daily life. For Polytechnic students, especially in the field of engineering, the ability is a potential that is needed in their educational process. Since this ability is important, many efforts have been made intensively and continually by the Indonesian government to improve education. However, many research results show that the Indonesian students' ability to understand and to solve mathematical problems in cluding Polytechnic students is still low. Among the results, Research and Development Agency in 2011 reported that the result of a survey conducted by Trends International Mathematics and Science Study (TIMSS) in 2011 shows that the learning achievement of the eighth grade students in Indonesia is in the 36th rank of 49 countries. The learning achievement of Indonesian students as reported by TIMSS 2015 is even worse, since Indonesia is in the 46th rank of 51 countries [1].

Similar to the result of TIMSS report, Progamme for International Student Assesment (PISA) shows that the learning achievement of the Indonesian students aged about 13 is still low. In PISA in 2011, Indonesia only gets the 64th rank of 65 countries, with the mean score of 397.

The PISA result in the literacy assesment in reading, mathematics and science in 2015 places Indonesia in the 67th rank of 76 countries. This data shows that the quality of education in Indonesia is still low [1].

The low quality of education in Indonesia as shown by TIMSS and PISA is obviously caused by many factors. One of them is the fact that Indonesian students are generally not well trained in solving mathematical problems (non-routine problems) which is the main characteristic of TIMSS and PISA problems. This can occur because the activity of problem solving in mathematics teaching has not been made the main activity. Another factor is the teaching model used by the teacher [2]. Teaching model plays an important role in creating succes in the teaching and learning process.

Improvement in problem solving ability in the students can be done through an innovative and effective teaching.The teaching process is done systematically and structurally using an effective model and method according to the charateristics of the course. Problem-based Learning (PBL) is a teaching model which makes problems as the basis of the students' learning [3]. Barrows states same characteristics of PBL, i.e., 1)The teaching process is student centered; 2) The teaching process occurs in small groups; 3) The teacher plays the role as facilitators or guide; 4) the problem presented becomes the stimulus for learning; 5) New information is obtained by self directed learning and 6) problems become the means to develope the ability to solve problems [4]. Indirectly, the attainment of the ability to solve problems can be developed by the PBL model.

The implementation of PBL in teaching gives some benefits : (1) Preparing the students better for applying what they learn in the real world: (2) Enabling the students to become the producers of knowledge rather than only the consumers ; and (3) Helping the students to develop communication, reasoning and critical thinking skill [5].

PBL compared to other teaching models, is a complex constructivist model of teaching and gives a great opportunity for the development of learning autonomy. PBL is very good for developing a high order thinking skill, i.e., critical and creative thinking. Complex, contextual and ill-structured problems will give the opportunity to the students to develop analytic, evaluative, and reflective thinking and to develop their creativity in finding various kinds of information, in developing a possible solution, and creating various resources to solve the problem that has to be solved [4].

A significant difference in the post test results in advance mathematics learning achievement between the students who were taught by using PBL and those who were taught by using traditional method, the advanced mathematical learning achievement of the students who were taught by using PBL was significantly better than that of those who were taught by traditional method [6]. Padmavathy, that PBL has an effect on mathematics teaching and improves the students' ability, the ability to use concepts in the real life, PBL is more efective for learning mathematics [7].

Assessment is an integral part of teaching process. Assesment is often regarded as one of three very important pillars in teaching. The three pillars are planning, implementation, and assessment.If the three pillars are synergetic and continuous, then they will determine the quality of teaching.Assessment has to be designed and implemented according to the planning and implementation of teaching. The assessment system has to be developed in line with the development of the model and the strategy of teaching.

The implementation of the teaching model in the classroom to develop the students' ability to solve mathematical problems will obviously become more optimal by the presence of an appropriate assessment. Wolf states that a good teaching quality has to be followed by a good assessment [8]. Assessment is an important part of teaching and a good teaching will not be succesful without a good assessment. Good assessment is generally present with good teaching and learning activities. The teaching and learning process will occur effectively if it is supported by an effective assessment. An assessment activity has to be made a means for improving the effectiveness of the teaching and learning process [9].

Authentic assessment is a process of collecting, reporting and using information about the students' learning achievement by applying assessment principles, continous implementation, authentic, accurate, and consistent evidence as public accountability [9]. Authentic assessment is appropriate for assessing problem based learning process [10].

PBL which is integrated with authentic assessment, the students are expected to be able to solve the problems given as the process to master the existing mathematical concepts. For example, teaching the application of the concept of comparison of trigonometric functions through the problem of determining the width of a scout tent as follows.

Problem

Figure 1

Figure 1 is a prism-shaped tent that is shaped prism built by students during the camp activity. The height of the tent pole is 2 m and is visible from the front entrance of the same triangle foot tent. The student wants to measure the width of the tendon, but does not carry a length measurement tool. The student only carries the degree arc in his bag. When they measure the angle between the foot of the tent and the ground they get 45^0. Can you help, how wide is the tent?

Illustration of the above learning, PBL with authentic assessment in the application of the concept of comparison of trigonometric functions. Students learn in small groups, lecturers play a role as facilitator to accompany students to discuss in their group to find problem solving which will then be presented. Assessment carried out during the learning process includes aspects of attitude and skills using checklist equipped rubric. While the knowledge aspect is implemented at the end of the learning using essay test which is completed with assessment rubric.

PBL situations encourage students to make connections between their knowledge and application in real life.The result of the discussion is then presented in the form of a simple report and is presented through a presentation activity which is one of the forms of authentic assessment. The authentic assessment which is integratedwith PBL model can increase the students' activities and motivation in learning and give the opportunities to the students to always assess themselves so that they can know their level of mastery. With the increase in their activities and motivation to learn and with the awareness of the relation between mathematics at school and their real life, it will be easier for them to understand the existing concepts. As the consequence, the students' mathematics learning achievement will be better.

The teaching model and the technique of assessment used by the teacher in the classroom are actually determining factors for the students' learning achievement, including for the ability to solve mathematical problems. In the attainment of this ability, every student has different characteristics and abilities. As an effort to improve mathematics learning achievement, in addition to considering the factor of teaching approach the teacher also has to consider the students' internal factor which is a basic factor.

One of the basic abilities that the students have to possess in relation to the ability to solve mathematical problems is numeric ability.Numeric ability is the ability to operate numbers, in the form of the ability to count which covers addition, substruction, multiplication, and division of numbers with speed and precision which facilitates the solution of mathematical problems.

Howard Gardner calls numeric ability numeric apptitude, i.e., logico mathematical intelligence in using numbers and reasoning. A person who has numeric intelligence, in general, has an organized way of thinking and solving problems.Those who are good at logical mathematics tend to learn mathematics quickly [11].

The students' numeric ability needs to be considered in mathematics teaching. Mathematics involves arithmatical operations such as substruction, addition, multiplication, and division. The higher a student's numeric ability the more chance he or she has to develop and achieve in mathematics. The students who have a high numeric ability tend to be active in learning, have an ability to solve problems, classify an categorize information, and do complex mathematical computation. Badru and Ademola also show a significant effect of numeric ability on mathematics learning achievement [12]. Singh and Kumar show a positive and significant correlation between numeric apptitude and academic achievement in mathematics among high school students [13].

Based on the explanation above, naturally, mathematical problems solving needs much mathematics computation.To find out the effect of PBL model and authentic assessment on the ability to solve mathematical problems, there is a need to control numeric ability scores so that the effect that occurs can be believed to be purely caused by PBL model and assessment.

The key problems investigated in this study, i.e., after numeric ability is controlled:1) is there any difference in the ability to solve mathematical problems between the students who are facilitated by PBL and those who are facilitated by conventional teaching model?; 2) is there any difference in the ability to solve mathematical problems between the students who are facilitated by authentic assessment model and those who are facilitated by conventional assessment model?; and 3) is there any effect of interaction between teaching model and assessment on the ability to solve mathematical problems?

2. Research Method

This study was conducted at Mechanical Engineering Department in Politeknik Negeri Bali, by using quasi-experiment design by measuring two factors in 2x2 factorial version.The sample taken consisted of 110 students, distributed in four classes.

The data were collected by using mathematical problem solving ability test and numeric ability test.Each instrument has been validated both theoretically and empirically. Theoretically, content validity was tested by using Aiken's statistical index approach, while empirically, it was tested by using classical test approach

The data were analyzed descriptively and by using 2x2 factorial ANACOVA. Before it was analyzed by anacova, prerequisite analysis testing was done, i.e; data distribution normality test, intergroup variance homogeniety, linearity test and multicoliniarity test.The results showed that the assumptions were met.

3. Results And Discussion

After treatment, PBL-AA group showed the highest mean in mathematical problem solving ability. The group of students who were facilitated by conventional teaching model and authentic assessment were better than those who were facilitated by PBL and conventional assessment. While the group of students facilitated by conventional model and conventional assessment were lower than other groups.

The testing of the hypotheses was done by using ANACOVA. The testing of all hypotheses was done consecutively, the recap of the testing can be seen in table 3.1 as follows.

Table 3.1 Results of 2x2 Factorial ANACOVA of Mathematics Problem Solving Ability Data

Source	Type III Sum of Squares	Df	Mean Square	F	Sig.
Corrected Model	2758,335[a]	4	689,584	35,996	,000
Intercept	20371,015	1	20371,015	1063,373	,000
X	1013,613	1	1013,613	52,911	,000
A	1266,843	1	1266,843	66,130	,000
N	300,847	1	300,847	15,704	,000
X * A	303,062	1	303,062	15,820	,000
Error	2011,483	105	19,157		
Total	611578,000	110			
Corrected Total	4769,818	109			

Notes:
Nu = Covariate (Numeric Ability)
X = Teaching Model
A = Assessment

Before the effect of the variable of numeric ability was controlled, the analysis shows that the effect of the interaction between teaching model and assessment on mathematical problem solving ability was significant. This was evidenced by the value of $F = 39.846$ ($p < 0.05$). It means there is an effect of interaction between teaching model and assessment on the ability to solve mathematical problems. The effect of teaching model on the ability to solve mathematical problems depends on the assessment used.

The Anacova analysis with numeric ability covarible shows that the interaction between teaching model and assessment on ability to solve mathematical problems is still significant as shown by the value of harga $F = 15.820$ ($p < 0.05$). Also significantly, numeric ability has an effect on the ability to solve mathematical problems ($F = 15.704$; $p < 0.05$).

The ability to solve mathematical problems is one of the form of the result of learning mathematics, and is the highest capability in thinking skill and intellectual skill. As one of the forms of the result of learning it is of course influenced by various abilities that are basic for the students, i.e. numeric ability. Numeric ability is the ability to operate numbers, in the form of arithmatic which covers addition, substraction, multiplication, and division of numbers with precision so that it facilitates the solving of mathematical problems. The students who have a high numeric ability tend to be active in learning, show the ability to solve problems, classify and categorize information and do a complex mathematical computation.

Fatoke, Ogunlade, and Ibidirin state that one of their findings shows that the problem solving learning strategy, numeric ability of the students and their performance improve in chemistry [14] Thus, after numeric ability was controlled, the mean of the ability to solve mathematical problems was caused by the effect of interaction between teaching model and assessment used,rather than solely caused by numeric ability.

PBL pedagogy helps to show and clarify the way of thinking and the richness of cognitive structure and process involved in it.In addition, in PBL the students can become self directed learners who have the desire to understand and learn, formulate learning needs, the ability to select and used the best learning resources. A self directed learner is an individual who directs himself or herself in the learning process. Sadia firmly states that in PBL situation, the students obtain learning experience to develop a high order thinking skill, analytic skill and problem solving skill as well as to develop autonomy and competitiveness. In such a situation the students learn how to use an interactive process and evaluate what they know, identify what they know, collect information, and collaborate in evaluating a hypothesis based on the data that have been collected [4] . This is similar to what is stated by Ibrahin and Nur, in PBL situation the students can be involved in a high order thinking and problem solving oriented to the real world problems, including learning how to learn. The students are expected to have a wholistic understanding of a material that is formulated in the problem, a positive attitude and gradual and continous skill. The students' mental activity is required to understand a concept, a principle, and skill through a situation or problems presented in the begining of the lesson. The students understand concepts and principles of a material from working and learning in a situation or problem given through investigation, inquiry, and problem solving.The students develop concepts and principles through their own ability which integrates the skills and knowledge that they have understood before [14].

While conventional teaching model is teacher centered. Coleman states that conventional teaching is an assimilation of information or classroom teaching, with the following characteristics: 1) acquisition of information through symbolic sources such as through teachers and reading; 2) assimilation or organization of information so that general information is understood, 3) use of a general principle in cases of specific nature; and 4) application of a general principle in a new condition [15]. Thus, in the conventional teaching model, the teacher's authority tends to dominate learning. The teacher has to demonstrate knowledge or skill to be trained to the students step by step. The teacher is required to play the role as an attractive model for the students, the material developed should meet students' interest. The students are passively receptive, their reasoning and knowledge develop only as far as the limit of thc teacher's. The students' activities become limited and this causes them unable to learn optimally. It is clear that it is better to implement PBL than conventional teaching model for the students since in PBL all of the student's senses of perception are involved in the teaching process. Therefore, the teaching outcome in the form of ability to solve mathematical problems of the students who learned through PBL is better than that of those who learn through conventional teaching model.

Assessment is a process of collecting various data which can give a description of the students' learning development [9]. This assessment stresses the importance of the ability of the students to demonstrate the knowledge that they have meaningfully and realistically. The activity of assessment does not only consist of asking or tapping the knowledge that the students have known, but of asking them to show the real performance about knowledge that has been understood and mastered. In addition, authentic assessment requires the students to demonstrate knowledge, skills and strategies by creating answers or products. The students are not only required to respond like in the conventional test, but are asked to create and produce answers based on theoretical knowledge.

The applications of authentic assessment in teaching that happens in the classroom can give a good psychological impact on the students. The involvment of the students in the assessment process can increase social interactions and mutual trust among themselves and with the teacher. The perceptions of the students that are involved in assessment activity generally improve their self confidence, responsibility and motivation. Social interactions can make the students and the teacher control the result of assessment made by each party, so that the assessment becomes valid, objective, and can be accepted by both students and teacher.

The conventional assessment is often called paper and pencil assessment. Such an assessment is separated from teaching. Paper and pencil assessment cannot be regarded as an integral part of the teaching process.The giving of feedback for the students will be delayed and even almost cannot be done, so that the objectives of correcting the teaching as the result of assessment cannot be done inmediately and even it is not if it will be done or not.This assessment model doesent provide information to the students during the learning process. Based on discussion, it seems that authentic assessment model tends to be better than the conventional assessment model in the attainment of the ability to understand concepts in mathematical problem solving.

There was an effect of interaction between teaching model and assessment on the ability to solve mathematical problems. The students facilitated by PBL and authentic assessment had the highest mean score for the ability to solve mathematical problems compared to other groups. The students facilitated by PBL and authentic assessment discussed in groups intensively, so that they would ask each others, answer, criticize, correct, and clarify every concept or mathematical argument that occurred in the discussion.In such a discussion will develop the students' ability to make, refine, and explore conjectures, so that it will strengthen their understanding of the mathematical concept that they are learning, or mathematical problem that they are solving. In the authentic assessment situation they have very broad opportunity to construct learning outcome. Since the teacher does not only ask the students to repeat what they have learned.The students are asked to construct what they have got when they face a concrete situation. In this way the students select and organize answers based on the knowledge that they have and analyze trhe situation in order their answers are relevant and meaningful.Such mathematical teaching will be able to improve the students' ability to understand concepts and solve mathematical problems.

The students facilitated by conventional teaching and authentic assessment also have a higher mean score for the ability to solve mathematical problems than that of the group of the students facilitated by PBL model and conventional assessment and the group of students facilitated by conventional teaching and conventional assessment.

The students in the class facilitated by conventional teaching model and authentic assessment model experienced teaching process that tend to be teacher centered. In the design of the teaching program, the teacher does not consider prior knowledge that the students have. The teaching process occurs one way, the role of the teacher is no longer as motivator, facilitator and mediator, but as a teacher with an authority in teaching. In the teaching activities, the teacher tends to used lecturing method with a few questions and answers. The teacher tries to transfer or transmit knowledge that he or she has to the students. This condition tends to make the students passive in receiving the lesson.Such teaching activities do not create active learning and even the students become very passive. But authentic assessment requires the students to demonstrate all skills, knowledge and competences needed to solve problems given, passivity of the students can be lessened and even they can become active. The students are given the opportunity to construct their learning outcome since the assessment does not only ask the students to repeat what they have learned. The students are asked to construct what they have got when they are facing a concrete situation. In this way the students will select and organize answers based on the knowledge that they have and the result of analyzing the situation in order the answers are relevant and meaningful..Such a mathematical teaching situation will facilitate the students in improving their ability in understanding concepts and solving mathematical problems.

From the above discussion each teaching model and assessment has the same direction, i.e., the attainment of the learning objectives. The learning objectives will be achieved when the teachers and students feel the learning process to be meaningful. This occurs if the students follow correct and appropriate steps of learning according to the characteristics /syntax of the teaching model implemented. Thus the optimization of the attainment of the learning objectives can be achieved. Based on the above discussion, we can see the strengths and the weaknesses of each teaching model, depending on the assessment model used.Thus in the teaching process the teacher should consider the assessment model used.The teaching process will occur effectively if it is supported by and effective assessment, assessment activitiers should be viewed as the means to improve the teaching process.

The implications that can be drawn from the results of this study is that the colaboration of teaching model and assessment needs to be considered in teaching mathematical problem solving.The application of PBL is very effective if it is collaborated with authentic assessment. Conventional teaching becomes effective if it is collaborated with authentic assessment. PBL is less effective if collaborated with conventional assessment in improving concept understanding mathematical problem solving ability. The implementation of each model obviously needs to consider the students, numeric ability.

4. Conclusion, Implication and Suggestion

In conclusion, once the numerical ability is controlled: there is different problem solving ability of mathematic between student which facilitated model of problem based learning with conventional learning model; there is a difference problem solving skills of mathematics between students who facilitated an authentic assessment model with conventional assessment model; there is an interaction effect between the learning model and the assessment model on the ability mathematical problem solvingAuthentic assessment strongly supports the effectiveness of problem-based learning in the classroom compared with conventional assessment. Authentic assessment can also improve the effectiveness of conventional learning

The implication is that improving students' mathematical problem-solving abilities can be pursued through the collaboration of problem-based learning models with authentic assessment. Implementation of learning models and assessments in the classroom is necessary to consider the numerical ability of each student

Suggestion. Before the problem-based learning model and authentic assessment are implemented in the classroom, the lecturers desperately need to make well-prepared preparations. Lecturers really need to provide a detailed explanation so that students understand the problems faced well. The motivational plants learn the students so that they have the confidence to succeed, so they become more active, creative and interested in solving the problems given. More time is maximized to lead students in discussion. Facilitate student discussions only when absolutely necessary. In a state of discussion of deadlock, lecturers can provoke students' ideas with challenging questions, or give key clues without turning off creativity

Acknowledgement
The author would like to thank Directorate Research and Social Service, Ministry of Research, Technology, and Higher Education for their financial support that this research was successfully undertaken.

References

[1] Badan Penelitian dan Pengembangan (Balitbang) 2016 *Laporan hasil TIMSS 2015* (Kementrian Pendidikan dan Kebudayaan)
[2] Barrows H 1996 *Problem-based learning in medicine and beyond: a brief overview: New direction for teaching and learning* (Jossey: Bass Publisher)
[3] Sadia I W 2014 *Model Pembelajaran Sains Konstruktivistik* (Yogyakarta: Graha Ilmu)
[4] Widjajanti D B 2011 *Makalah: Problem Based Learning dan Contoh Implementasinya* (Yogyakarta: FMIPA UNJ)
[5] Alfred O F, David M and Abayomi A 2013 Effek Problem-Based Learning on Senior Secondary School Student Achievements in Further Mathematics *Journal Acta Didactica Napocencia* **6** (3) 27-43
[6] Padmavathy R D and Mareesh K 2013 Effectiveness of Problem Based Learning In Mathematics *International Multidisciplinary e-Journal* **2** (1) 47-51
[7] Badmus G A 2007 *Changing Nature of Technical and Vocational Education and Students' Assessment Methods* www.iaea.info (Retrieved on Januari 2010)
[8] Sani R A 2016 *Penilaian Autentik* (Jakarta: Bumi Akasara)
[9] Eggen P and Kauchak D 2012 *Strategi dan Model Pembelajaran Menegajarkan Konten dan Keterampilan Berpikir* (Jakarta: PT Indek)
[10] Lucy B 2010 *Mendidik Sesuai Dengan Minat & Bakat Anak* (Jakarta Selatan: PT. Tangga Pustaka)
[11] Bandru and Ademola K 2016 Problem – Based Instructional Strategy and Numerical Ability as Determinants of Senior Secondary Achievement in Mathematics *Journal of Education and Practice* **7** (13) 89
[12] Singh P and Kumar N 2015 Achievement in Mathematics of Secondary School Students in Relation to their Numerical Aptitude *IJMSSR* **4** (5) 102
[13] Fatoke A O, Ogunlade T O and Ibidirin V O 2013 The Effects of Problem Solving Innstructional Strategy and Numerical Ability on Students' Learning Outcome *IJES* **2** (10) 97-102

[14] Ibrahim M and Nur M 2000 *Pengajaran Berdasarkan Masalah* (Surabaya: Universitas Negeri Surabaya)
[15] Santyasa I W 2014 *Asesmen dan Evaluasi Pembelajaran Fisika* (Yogyakarta: Graha Ilmu)

IOP Conf. Series: Journal of Physics: Conf. Series **1040** (2018) 012036 doi:10.1088/1742-6596/1040/1/012036

Exploration of table and graph literacy of statistics student at Universitas Negeri Makassar

M A Tiro, M K Aidid and A S Ahmar

Department of Statistics, Universitas Negeri Makassar, Makassar, Indonesia

E-mail: arif_tiro@unm.ac.id

Abstract. This study aim to explore and review the literacy of tables and graphs for statistics students at Universitas Negeri Makassar. To obtain this aim, descriptive analysis conducted on the students results of table and graph literacy test. This study is importance for leading statistical learning tendency which emphasis on technical procedures with much mathematical formula. From this exploration, we can identify weaknesses that must be looked for solutions, particularly would improve learning of tables and graphs literacy to improve student's visualization and communication skills. Indicators of tables and graphs literacy have four domains, namely: (1) understanding tables and graphs concepts, (2) the ability of tables and graphs application, (3) skills of making table and graph, and (4) the accuracy of tables and graphs interpretations. The four domains are mapped according to grade of student. Mapping of table and graph literacy will be the basis of statistical learning strategies. Conclusion of the data analysis that the level of achievement scores of students in understanding the concept of the tables and graphs is still low (≤43%), the ability of tables application is still too low (41%), but for the ability of graphics applications is slightly higher (58%). Furthermore, the skills to make the table is still very low (25%), but for the skills to make the graph is slightly higher (59%). The general conclusion that the achievement score of statistics students of students batches 2013, 2014 and 2015 was still below the expectations. Thus, learning strategies need to be formulated to improve every aspect of learning in statistics courses.

1. Introduction

As described Schield that the core curriculum should be based on a theory about discipline of statistics [1]. In statistical theory, data are not just numerical figures as studied by mathematicians, but data are numbers in context. Some changes can pervasive and important happening in the form and type of data which available to the general public. Data from large-scale studies are increasingly available. For larger sample sizes, small differences will be statistically significant. For example, an outcome of the *National Longitudinal Survey of Youth* (NLSY), involving 12,000 subjects, found a 0.4 point difference in IQ test scores between men and women that statistically significant at the significance level of 5% [2]. Someone who has a statistical literacy is equipped to recognize the effect of sample size on the statistical significance. Thus, in the large-sized data, assign people who have statistical literacy should anticipate that the statistical significance of almost meaningless because the relationship between the two variables are expected to be statistically significant. We can anticipate that understanding the context and the influence of confounding variables is much more relevant than understanding the statistical random variation and its statistical significance. The knowledge and skills are becoming increasingly important if the data population (ie, studies based on large samples), giving the effect became more common in decision-making, both for government and for business.

The results of the study are generally presented in the form of tables and graphs. A good table is an integral part of a publication, both for analytical articles as well as for research papers. Using tables effectively reduces the amount of data in text. The table also eliminate the purposes of discussion variables are less significant, are not important in the story line. Two important things to note in the presentation of the data in the table: (1) create table so that easy for the reader to understand the numbers in the table; (2) layout design and titles (label) of table in the form of direct and simple, so that the reader's attention focusing on the substance delivered by the data, rather than the structure of the table. Presentation is striking and disturbing contents from the observed important to avoid. Remember that technology is only waitress. We should not add useless notes and obscure elements on the tables. Keep the message simple for the reader. Thus, the purpose of the presentation of data can to be retained in mind when designing and creating of graph. Based on this description, the statistical literacy needs to be an emphasis on the learning process [3]. Education is a conscious and deliberate effort to create an atmosphere of learning and the learning process [4] and in the process of learning, a person have specific characteristics in learning. The characteristics of each person is different [5]. The primary basis for explaining the data and results of statistical analysis are the table and graph. Thus, literacy of table and graph as a strong foundation for the application of statistical literacy [6–8]. From this reasons, the researchers designed a study with the aim to: (1) obtain map of the quality of students' understanding of the consept and application of table and graph for explaining the data and results of statistical analysis; (2) obtain map of the quality of the students' ability using table and graph in real-life contexts; (3) obtain map of the quality of students' ability to create table and graph of the data and results of given statistical analysis; (4) obtain map of the quality of students' for interpreting the data and results of data analysis presented in tables and graphs.

2. Research Method

This study is a descriptive analysis conducted on students of Statistics Department of Universitas Negeri Makassar about literacy of table and graph. The instrument of this study was developed referring to Tiro & Ahsan [8] and Tiro & Sukarna [9]. Students were given the test and the results are analyzed, explored and described. Students are seen in the category of batch levels, namely the batches 2013, 2014, and 2015. Two research variables are defined as (1) table literacy is the ability to communicate statistical information's by tables and (2) graph literacy is the ability to visualize statistical information's by graphs.

Collecting Data using instrument (1) test of basic table literacy and (2) test of basic graph literacy both are in the form of multiple choice questions. Both of these tests are developed to test that simultaneously collecting data to perform analysis: (1) the level of difficulty of each test item; and (2) discrimination index for each item test.

The approach used to analyze the survey data is quantitative descriptive and classification. Classification analysis is used to determine the classification of the problems faced by students in literacy of tables and graphs. Furthermore, factor analysis was conducted to see the conformation of construct validity of operational definition variables. Thus, construct validity of the test, and the relationship between the two variables were analyzed with path analysis.

3. Result and Discussion

3.1. Result

In the descriptive analysis are shown in the median, mean, standard deviation, coefficient of kurtosis and skewers of the literacy of table and graph. Furthermore, the boxplot diagram of each indicator is presented to see the distribution of score of basic literacy of table and graph. Boxplot is selected because it showed a mean value, to describe the size of the central tendency; interquartile range (3rd quartile - 1st quartile) to describe the variability of 50% of the data in the middle, outliers that describes the data is extremely different from most other data. The boxplots can be seen in the Figure 1 and Figure 2. The are same outliers for application, skills of making and interpreting tables (Figure1), but only one outlier for understanding concept of graph (Figure 2).

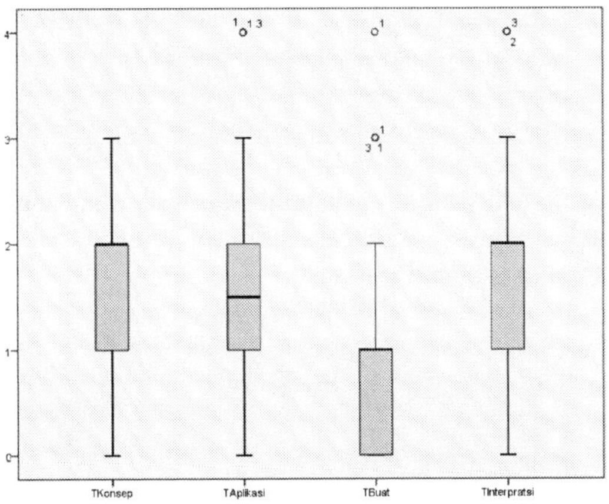

Figure 1. Boxplot four indicators literacy table

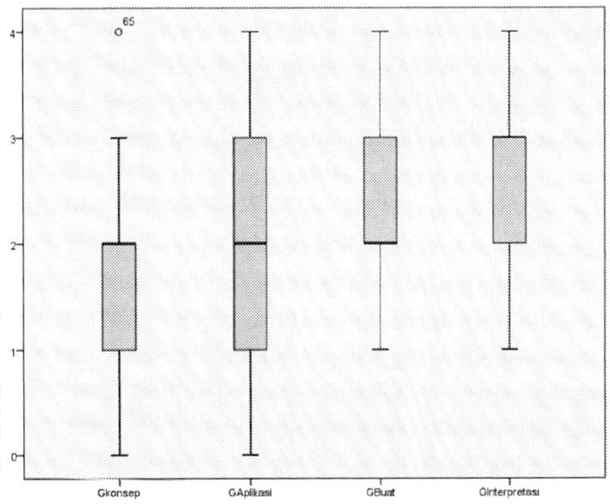

Figure 2. Boxplot four indicators literacy graph

3.2. Discussion

Basic literacy test of tables and graph were done to determine the level of difficulty of the questions. From this test shows that the number of items that are difficult on table literacy is 4 questions (25%) as much on graph literacy. The highest of difficult level in table literacy is an understanding consept (question number 9) ($p = 0.12$), it means that only 12% of the students were correct for this question. This questions measure knowledge of student about the type of tables, namely demonstration table of the smaller ones are eligible to be placed in the text and the greater of reference tables should be placed in the appendix of a paper.

Furthermore, difficult question of graph literacy is also on indicators of understanding the concept (question number 9) ($p = 0.06$), it means that only 6% of the students are correct in this question. This question measure respondents' understanding of the use of relevant graph to understand the characteristics of the data distribution. In this case, boxplot show the model of the data distribution, symmetrical, with a long tail at the bottom or the up. So, the topic of types of graphs are relevant with specific types of data still need attention in learning statistics. Then, the highest achievement of students in table literacy is an understanding of concepts with a score: 1.68 of 4.00 (42%). In addition, the lowest achievement of students in table literacy is create a table with score: 1.01 of 4.00 (25%). Similarly, the highest achievement in graph literacy is the accuracy of interpretation with score: 2.39 of 4.00 (56%). On the other hand, the lowest achievement in graph literacy is the understanding of the concept with score: 1.70 of 4.00 (42%). Thus, the main of information presented by tables and graphs need to be well understood by the students, in addition to the types of tables and graphs that are relevant for specific types of data.

From the discussion above, the strategy needs to be applied in teaching tables and graphs literacy are numerical. The table use to see the real of value in a comparative approach, and the graphs use to see trend of the data in the correlative approach.

The achievement of students in tables and graphs literacy can be seen in Table 1. This table shows that the ranges of level of achievement in table literacy range from 25% (skills to create a table) to 42% (understanding of the concept and application capabilities of table). Next, the range of achievement level of graph literacy from 43% (understanding the concept of the graph) to 60% (the precision of graph interpretation). So, student achievement in table literacy is still very low, but for graph literacy is slightly higher (though still needs to be improved).

Table 1. Level of achieved in tables and graphs literacy of students

Indicator	Table Literacy		Graph Literacy	
	average	%	average	%
Understanding of concept	1.68	42	1.70	43
Ability of application	1.66	42	2.32	58
Skills of making table and graph	1.01	25	2.35	59
accuracy of interpretation	1.61	40	2.39	60

Association between two variables (literacy of table and graph) were analyzed with path analysis. This relationship is presented in Figure 3, this image shows confirmatory factor analysis for each variable, literacy of table and graph, and correlation analysis of both.

Figure 3. Confirmatory path analysis and relationship between literacy of table and graph

From the path analysis, we have result:

Chi-square = 21, 860
Degrees of freedom = 21
Probability level = 0,408

Model	GFI	AGFI
default models	0.946	0.907
Independent models	0.895	0.866

From the results above, the value chi-square: 21.86 with $p = 0.408$ and GFI, AGFI exceed 0.85 or the default model is better than the independent models. This results show that the model supported by empirical facts. However, the correlation coefficient between literacy of table and graph is 0.003 with $p = 0.691$. Results of inferential analysis show that there is no significant correlation between scores of table and graph literacies. This might be due to differences in the orientation of the information presented in the table and graph. The table present information with the real scale and comparative, but the graph present information about scale approach and associative or tendencies.

This results of this study almost the same with the results of previous results by Tiro, Nusrang and Ahsan [10] about descriptive and inferential statistics where the level students achievement of an understanding of concepts (60%), the ability to apply the concept (60%), and the ability to interpret the results (60%). Another previous Sudarmin, Tiro and Irwan [11] about probability and random events where the level students achievement on understanding of concepts (46%), the ability to apply the concept (56%), and the ability to interpret the results (66%).

4. Conclusions

Based on the results of the discussion can be concluded that: (1) the level of student achievement on understanding the concept of the table is still relatively low (41%), as well as understanding of the concept of graph is still low (43%); (2) the level of student achievement on ability application of table is still too low (41%), but the ability application of graph is slightly higher (58%); (3) the level of students achievement in the skills to create a table is still very low (25%), but creativity to make the graph is slightly higher (59%) and (4) the level of student achievement in the accuracy of interpretation of table is still very low (25%), however accuracy of interpretation of graph is already higher (59%). The general conclusion that the student achievement based on batches 2013, 2014 and 2015 was still below expectations. Thus, learning strategy needs to be improved on any subject of learning statistics. The results of this study

Acknowledgments

The authors would like to express their thanks to Faculty of Mathematics and Natural Sciences, Universitas Negeri Makassar for the opportunity provided to carry out this research and supported fund, to the management of the Research Institute of Universitas Negeri Makassar for facilitation of this research, and the committee of IConMNS-2017 for giving opportunity to present the results of the research.

References

[1] Schield M 2004 Statistical literacy curriculum design *Curric. Dev. Stat. Educ* 54–74
[2] Bureau of Labor Statistics 1979 *https://www.bls.gov/nls/nlsy79.htm*
[3] Schield M 1999 Statistical literacy: Thinking critically about statistics *Journal of Significance* **1** 0–7
[4] Rahman A, Ahmar A S and Rusli 2016 The influence of cooperative learning models on learning outcomes based on students' learning styles *World Transactions on Engineering and Technology Education* **14** 425–30
[5] Rahman A and Ahmar A S 2017 Relationship of learning styles and learning achievement of mathematics based on genders *World Transactions on Engineering and Technology Education* **15**
[6] Tiro M A 2008 *Statistik Sebaran Bebas* (Makassar: Andira Publisher)
[7] Arif M 2010 *Penelitian Skripsi, Tesis, dan Disertasi* (Makasar: Andira Publisher)
[8] Tiro M A and Ahsan M 2015 *Penyajian Informatif: Tabel, Grafik, dan Statistik* (Makassar: Andira Publisher)
[9] Tiro M A and Sukarna 2012 *Pengembangan Instrumen Pengumpulan Data Penelitian* (Makassar: Andira Publisher)
[10] Tiro M A, Nusrang M and Ahsan M 2015 *Eksplorasi Pembelajaran Literasi Statistika dalam Paradigma Konstruktivisme Mahasiswa Program Magister PPs Universitas Negeri Makassar* (Makassar: Lembaga Penelitian Universitas Negeri Makassar)
[11] Sudarmin, Tiro M A and Irwan 2015 *Deskripsi Literasi Peluang dan Peristiwa Acak Mahasiswa Program Studi Statistika FMIPA UNM* (Makassar: Lembaga Penelitian Universitas Negeri Makassar)

IOP Conf. Series: Journal of Physics: Conf. Series **1040** (2018) 012037 doi:10.1088/1742-6596/1040/1/012037

Didactical design of integers: an elementary school teachers creation viewed from didactical situation perspective

S Prabawanto[1], D Suryadi[1], E Mulyana[1], D Ratnasari[2] and F Dewi[2]
[1] Departemen Pendidikan Matematika, Universitas Pendidikan Indonesia, Jl. Dr. Setiabudhi No. 229 Bandung 40154, Indonesia
[2]SD GagasCeria Bandung

E-mail: sufyani@upi.edu

Abstract. Integers was first introduced in the fourth grade of primary school. According to teachers, the main difficulty is how to introduce 0 and negative numbers for the first time. The use of recreational beach contexts that involve activities such as swimming, diving, and parasailing greatly help children understand the position of something refers to sea level. Prediction of teachers about the response as well as the alleged learning trajectory (HLT), proved to be an invaluable reference for developing both didactical and pedagogical interventions. Appropriate intervention would help students to create adaptation and acculturation processes, so the child's learning process takes place in good ways.

1. Introduction

Teaching integers in the fourth grade of primary school is an interesting event to be analyzed, at least based on three reasons. First, Students already know the positive integers since grade one, where they begin to learn counting (sound of numbers, one, two, three and so on), mapping the sound of numbers with the number of objects, compare numbers based on amount of objects, and perform arithmetic operations with simple addition and subtraction. Second, when the integers will be introduced in their entirety by adding elements 0 and negative integers, teachers and of course the children have difficulties because it is not easy to represent negative integers with concrete approach. In fact, the approach used by most teachers today as the debt to illustrate the negative numbers or forward and backward to represent positive and negative numbers, still leaves a problem in children. Third, didactical design model that exists today in textbooks issued both by government and the books that circulated widely, generally use a number line their explanation. With this direct approach, students do not have the opportunity to experience the thought process that leads to sense making for the meaning of zero and negative integers.

Teachers of an Elementary School in Bandung, Indonesia have been trying to address these issues, using the context of marine sports such as diving, swimming, and parasailing in the hope of encouraging good meaning of integers, including 0 and negative integers. The main objective of teaching was for the students to compare two integers and to sort the integers on a number line. This paper presents information about this information based on observations from the author in the teaching process and discussion with teachers after the lesson. Thus, the main description of this paper covers the theoretical framework as well as the analysis and discussion of the data of the observations. The theoretical framework used as the basis for in-depth look at each event in the classroom, especially from the perspective of the theory of didactical situations, learning obstacle, learning trajectory, and the didactical contract.

2. Theoretical Framework

Review about the theory of didactical situations that include a variety of didactical situations, learning obstacle, learning trajectory, and the didactical contract is described briefly as a frame of reference to analyze the implementation of the didactic design of the introduction of integers for fourth grade students.

Learning is always initiated action situations [2]. This situation is very important to make space for the student to use the experience and prior knowledge, so the perception of environment and action on the environment can occur well and the encapsulation process in the formation of new mental object can be facilitated [6]. When new mental objects began to form, the situation as a formulation situation, which substantially is the abstraction result of mental acts that occurred [2]. The process of interaction among students and between teacher and students will allow the emergence of negotiation of meaning so as to encourage the development of arguments, statements, or representations that drive the process of both internal and external validation. The learning process does not always go smoothly as expected. The situation and the plot developed to learn there are times when the cause of difficulties for children. Likewise, the intervention of both the didactic and pedagogical teacher, can came difficulties or obstacles to the learning process. There are three types of difficulties or barriers to learning (learning abstacles) ie ontogenic, epistemological and didactical obstacles that may occur in the learning process [2]. Ontogenic obstacle is the kind of difficulties related to the child's readiness to learn. Based on the author's experience observing the learning process for 3 years, there are at least three types of learning difficulties; psychological, instrumental, and conceptual. Ontogenic psychological obstacle is the lack of students' concentration in learning due to the psychological aspects such as low motivation and interest of the material being studied. Ontogenic instrumental obstacle is the technical difficulties that caused the student cannot fully follow the situation in learning as a result he does not understand the technical issues are key to the learning process. Meanwhile ontogenic conceptual obstacle is the kind of conceptual difficulties with regard to the level contained in the design less familiar corresponds to the situation of children seen from previous learning experience. Thinking or conceptual demands are too high can cause a child to lose their orientation to learn that frustration. Conversely, too low a conceptual challenge can be the cause of an underachiever in learning.

The second obstacle is epistemological obstacle. According to Duroux, this type of learning difficulties is more due to the limited context that is used when the concept was first studied [2]. For example, when the triangle concept was introduced to elementary school students, the teacher tends to use the default image as shown below (Figure 1). Segment AB is usually introduced as the base of the triangle and point C as a peak. As a result, students often have difficulty when faced with another triangle, especially when students are asked to determine the area of a triangle with the base and the height are not standard. The difficulty is not caused by not understanding the concept, but rather the lack of understanding related solely to the specific context (in this case a triangle in the standard form).

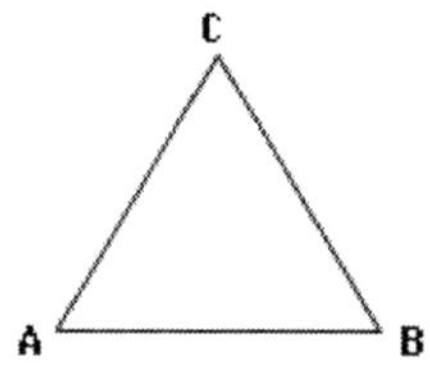

Figure 1: Picture a triangle that is often used

To achieve the learning goals that have been assigned by a teacher, every student has a trajectory of learning (learning trajectory) which is unique. Differences in the learning experience as well as a variety of mental objects formed from each of these experiences have a major impact on students' learning trajectory variations. The teacher can facilitate a variety of trajectories of learning that appear at the time of teaching, he/she should be able to predict the various possible students responses, so that the path of learning (Hypothetical Learning Trajectory or HLT) may be the main reference of facilitation process in the form of didactical or pedagogical interventions. A HLT was developed based on the objectives to be achieved, as well as the stages of learning in form of a series of continuous mutual didactical situation towards the achievement of the objectives [3], as shown below (Figure 2).

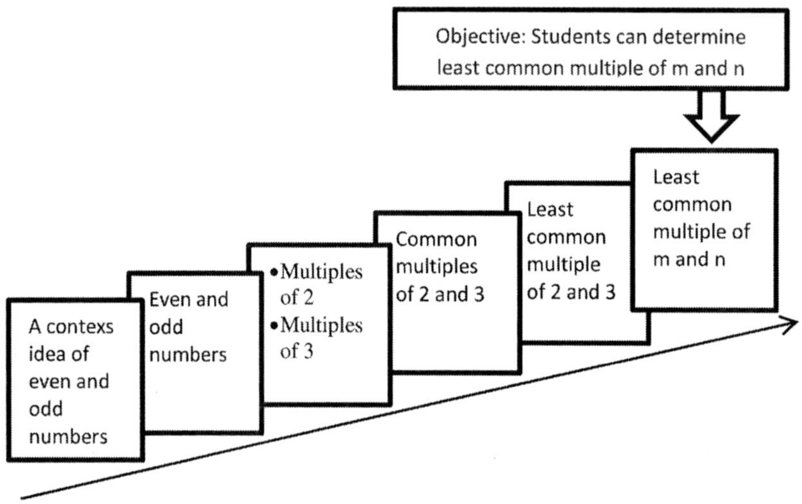

Figure 2: Hypothetical Learning Trajectory of Teaching LCM

In a process of teaching, teachers and students each have different ideas and expectations relating to the didactical situation that develop in the classroom. It needs the number of rules that determine a teacher and students responsibility related to didactical situation or a didactical contract [2]. The division of roles and responsibilities can span from teacher as the center to the students as center. There are three types of didactical contracts, namely ostention contract, mayeutic Socratic contract, and potential a-didactical contract [1].

The first types of contracts represent teaching process, where the teacher is a main actor in explaining the concept, demonstrating application of the concept, providing examples, as well as preparing the exercises. The second type, represent a teaching process where the teacher does not completely dominate, but to help students through the submission of the key questions in accordance with didactic situation that develop. The third type is a didactic contract is where the teacher provides space for the child as much as possible in accordance with their capacity, to encourage students' thinking independency through the teacher's didactical design. In practice, these three types of didactical contracts can occur alternately as needed and the design model. In addition, in the implementation process, termination of the didactical contract can occur (contract breaking) when there is a mismatch between the expectations of the teacher and the response shown by students during teaching process. For teachers who want to continue to grow, the situation of termination of the contract may be a trigger for improvements of design and didactical intervention model that are more appropriate for students.

A brief description of the theoretical study of the above can be used to see, interpret, and weave the didactic situations. The information can be used to make a better didactical design. Figure 3 below illustrates the thinking process of teachers using the theoretical perspective, prior learning (Reflection for action), during the learning process (reflection in action), and after learning (reflection of action). At this stage of reflection for action, teachers develop conjectures about the students' response that may occur on the didactical situation in the classroom. Based on these allegations, it could developed general possible students' learning track also called hypothetical learning trajectory (HLT). It can also developed both didactical and pedagogical anticipations based on the allegations about the various possible responses and groves of students learning.

The process of thinking when teaching occurs is based on a didactical situation that developed as well as allegations regarding about the students' response to the situation. When the teacher noticed the action by a student together with the learning environment, the milieu is formed so that the teacher can consider the process of interaction and negotiation of meaning that develop in the milieu, to provide a didactical or pedagogical intervention.

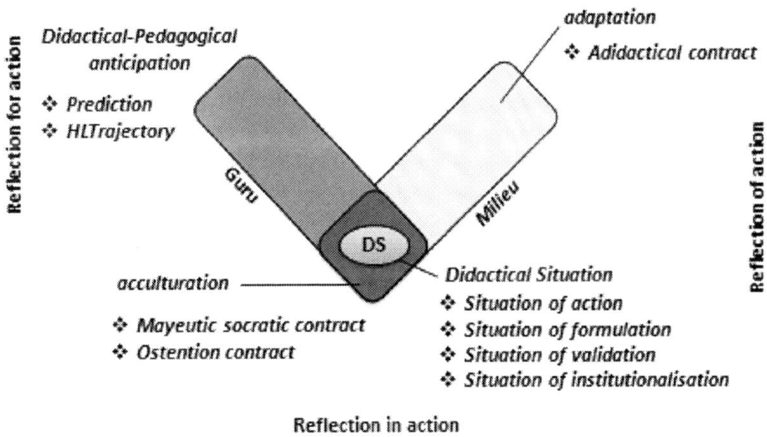

Figure 3: Thinking Process of Teacher

Decision of the teacher on the results of these considerations may represent the type of contract didactic, which can be ostention, mayeutic socratic, or a-didactical potential contract. In the event of termination of the contract (contract breaking), the teacher can decide to give a didactical intervention, so that the acculturation of the learning process can occur. However, if only a pedagogical intervention that be done, then there is a process of adaptation. Teachers also need to stay focused so that a series of situations and thinking processes that develop can shape a student's learning groove integrity of both structural (relation between concepts), and functional (linkages between the thought process) [4], [5]. In other words, teachers get to ensure that the aspect of coherence on a series of circumstances and the student's response in this situation. Given that the experience and previous knowledge students are diverse, the teacher should also consider the aspect of flexibility in choosing the type and level of intervention according to the evolving circumstances.

3. Analysis and Discussion

Didactical design on the introduction of integers developed by the teacher, in outline consists of four phases, the respective aim of each is to: (1) determine the position of objects with reference to sea level, (2) complete a vertical number line, and (3) compare the two positions object on a vertical number line. All of the stages form a groove of learning, so that its implementation in the classroom would be illustrated in the variety of didactical situation, learning obstacle, learning trajectory, and didactical contract.

In the first stage, the teacher asks the students to determine the position of the swimmer (R), diver (P), and parasailer (L) in a diagram similar to the one in figure 4. As the teacher predicted, not all students can put position R, P, and L correctly to be. This is evident from two student responses that be chosen by the teacher, where they put parasailer position is in incorrect position. From the third student response that be chosen by the teacher, where described positioning of the diver as below sea level, the swimmer at sea level, and the parasailer at above sea level. From the teacher's to the third student, obtained the answer that the parasailer should be placed on the surface of the ocean because of parasailer usually fly in the air. During the interviewing process, other students agreed with the correct student and made changes to the result of their work.

Students' difficulty relating to the placement of parasailing in the right position shows that there is a learning obstacle that can be indicated as an ontogenic obstacle. For a student who does not know the meaning of parasailing, the term of "parasailing" would prevent him/her be a source of trouble from understanding assignment of teacher. Teacher strategy featuring the first two responses was effective in revealing the source of the difficulty. Selection of a third child who was able to put parasailer in the correct position and give the right his reason was helped other students to reach an understanding. Teacher Strategy to facilitate the students' thinking through the discussion about three cases that were be selected, can created a situation in which the adaptation depicts the a-didactical contract. In this process, the students were able to reach an understanding as expected due to the pedagogical intervention by exploiting the growing response from the students themselves.

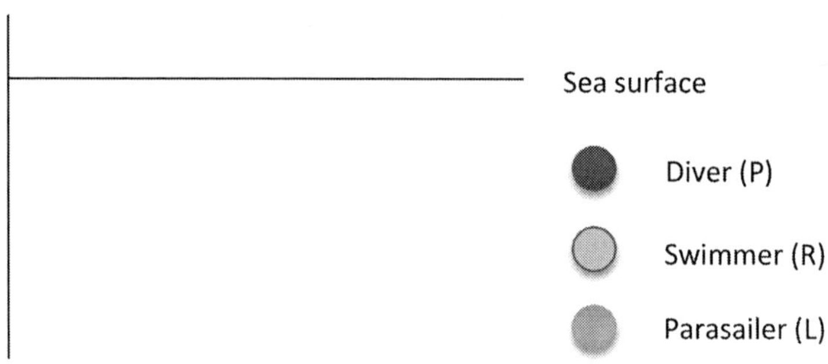

Figure 4: Placing of positions R, P, L in the picture

After the students were able to put the P, R, and L in the correct position, the teacher continued to the next didactic intervention that ask the students to determine at which position P, R, and L were placed based on the scale provided (distance between two points is considered as one meter). A Variety of responses were given by students on this task. For example, a student was using sea level as a reference for determining the position of the diver, so that by his calculations, the diver was at position 7 meters from sea level. However, to determine the position of the parasailer placed above sea level, he continued to count from sea level up by counting 8, 9, 10 and so on until he find a right number for the parasailer. Another response shown that the difficulty was the use of the lowest point as a reference, the diver was considered to be in a position 1 meter from the lowest point. From that point, he continued to count upwards so that the swimmer and parasailer positions were found. Another interesting response was the use of the sea level as a reference for determining the P and L position so that P in the position of 6 meters and L are also in position 6 meters. Being aware of these

IOP Conf. Series: Journal of Physics: Conf. Series **1040** (2018) 012037 doi:10.1088/1742-6596/1040/1/012037

responses is very important, the teacher then asked the question: "what is the difference of 6 meters at the position of divers and 6 meters at the position parasailer?" Most students responded that the 6 meters on the diver is at below of sea level, and the 6 meters for the parasailer is above sea level. The change in students' understanding as a result of the key questions asked by the teacher, describe the process of acculturation. In this process, teacher's didactical intervention (asking a question) has become a trigger of a change in students' understanding. The process can be interpreted as a Socratic mayeutic contract.

Based on the experience learned in the first stage, the teacher begins the second phase by providing a new task that asked students to label a vertical line as seen in Figure 5 by putting the appropriate numbers. In general there were three categories of response of the students, namely: (1) students had difficulty to interpret the point at the sea surface, (2) students who were trying to use the previous learning experience to describe each point on the line, and (3) students who used experience learned elsewhere so that they do not refer to the thinking sequence developed by the teacher.

In the first case, the student seems hesitant to put a number below the parasailer for the position at sea level, and also for the position of the diver. This indicates that the selection of the proper context does not necessarily help all the children to come to the understanding expected by the teacher. In this case there was a potential didactical obstacle, particularly related to the meaning of a point on the position of the swimmer. For students who had a case like this, still need to be helped by didactical intervention that can help him to interpret the point as a reference to determine the position above the sea level or below of the sea level.

In the second case, the students can use the context of the surface to be interpreted as a reference point and use to determine the point positions that were located above and below the sea level. Figure 5 illustrates one of the cases of the student response. The response shows that the student was able to represent the differences of numbers based on their position referring to the sea level. The position of the surface of the sea itself became a point of reference, represented by the statement "fit at the sea surface".

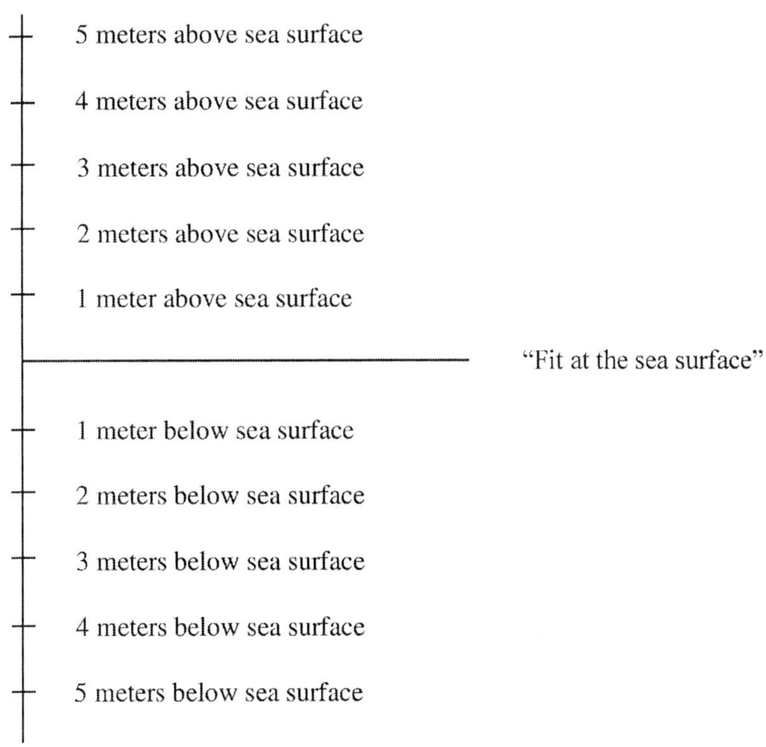

Figure 5: Completing the vertical number line

In the third case, a student used his experience from other places. He put 0 at the point position sea level, a positive integer at points above it, and negative integers at the points below. This response seems to have nothing to do with the thinking sequence that developed in previous activities. This situation seems to be a dilemma for the teacher, because on the one hand there were many students who had difficulty completing the points on the vertical number line correctly, on the other hand there was a student (a person) who had familiar with integers even using the positive and negative symbols. The three types of responses are in fact very important to create the adaptation process. The second response was used by the teacher to intermediate for students' understanding to the third response, by taking a deal so that the representation of the second type can be simplified by using a positive sign for above the sea level and a negative sign for below the sea level.

In the third stage, the teacher presents a problem to compare the positive numbers with a positive (A and B), a positive number with a negative (C and D), and negative numbers with the negative (E and F) based on the position of objects on the number line vertically along with reasons (Figure 6). Based on the responses given by students, generally they were able to determine the greater number based on its position. From these activities, students were able to express a conclusion, that the higher the position, the greater the value and the shorter the position, the smaller the value. This represents good adaptation, because through the pedagogical interventions of by using a particular student's response, other students were able make a good conclusion. From the standpoint of didactical contract, these events were included in the a-didactical contract.

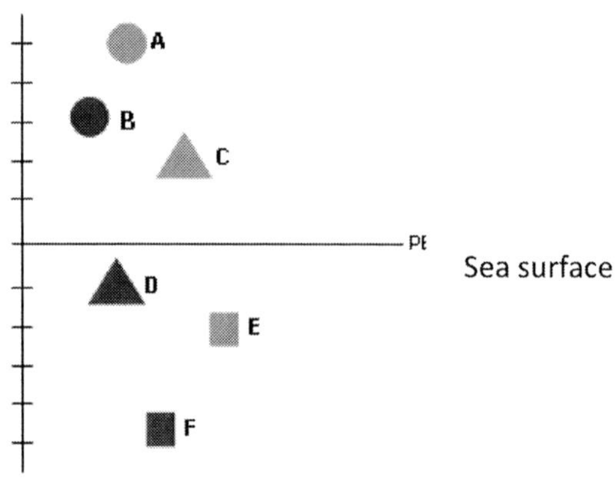

Figure 6. Comparing the two numbers

To ensure that students understood the conclusion, the teacher asked few questions regarding the comparison of two numbers. The teacher realized that the students had a good understanding using a vertical number line. The teacher gave the next challenge, if the number line is rotated clockwise so that a horizontal number line is obtained, can students complete it by putting integer in corresponding position?

4. Conclusion

Students' learning process was not always smooth in line with expectations. Situation and path of learning developed by a teacher can be the cause of obstacles to student learning. Similarly, the teacher improper intervention may also be a cause of students learning. The learning obstacles can be categorized into three types, namely ontogenic, epistemological and didactical obstacles. Ontogenic obstacle is the kind of difficulty related with student's readiness to learn; epistemological obstacle is the kind of difficulty related with limited context that was used when the concept was learnt the first time, and didactical obstacle is the kind of difficulty related to didactical design developed by a teacher or an improper didactical intervention in the classroom.

In the case of students difficult to put the parasailer in the right position indicated that students had an ontogenic obstacle, because of the parasailing was unknown to him. In the case of students difficult to interpret a point at the right position of the sea surface, there was a potential didactical obstacle, particularly relating to the meaning of a point at the swimmer position. One student who had knowledge of integers and even used the positive and negative symbols. The fact that many students had difficulty to complete the points on the vertical number line could indicate a potential pedagogical obstacle.

References

[1] Arias F and Araya A 2009 Analysis of Didactical Contracts in 10th Grade Math Clases. *Quaderni di Ricerca in Didattica (Matematica)* Suplemento n4, al n. 19. 155 - 163

[2] Brosseau G 1997 *Theory of Didactical Situations in Mathematics* (New York: Kluwer Academic Publishers)

[3] Clements D H and Sarama J 2009 *Learning and Teaching Early Math: The Learning Trajectories Approah* (New York: Routledge 270 Madison Ave)

[4] Suryadi D 2009 *Metapedadidaktik dalam Pembelajaran Matematika: Suatu Strategi Pengembangan Diri menuju Guru Profesional* Pidato Pengukuhan Guru Besar Universitas Pendidikan Indonesia Bandung

[5] Suryadi D 2010 Metapedadidaktik dan Didactical Design Research (DDR): Sintesis hasil pemikiran berdasarkan lesson study dalam T. Hidayat, I. Kaniawati, I. Suwarma, A. Setiabudi, and Suhendra (Eds.): *Teori, paradigma, prinsip dan pendekatan pembelajaran MIPA dalam konteks Indonesia* 55-75 Bandung: FPMIPA UPI.

[6] Tall D 1999 *Reflections on APOS theory in Elementary and Advanced Mathematical Thinking* Haifa: PME23

The identification of the 11th grade students' prior knowledge of electricity concepts

K Suma, I W Sadia and N M Pujani
Ganesha University of Education, Jln Udayana No 11 Singaraja, Bali, Indonesia

E-mail: ketut.suma@undiksha.ac.id

Abstract. According to constructivism, prior knowledge plays an important role in building students' scientific knowledge. Prior knowledge influences how students assimilate and accommodate new information. Prior knowledge affects how students perceive, organize, and create new information relationships. Therefore, in order to improve meaningful learning, students' prior knowledge needs to be identified. The purpose of this study was to identify the 11th grade-students' prior knowledge of electricity concepts. Students' prior knowledge was measured using a Three-Tier Electricity Multiple Choice Test, which consisted of 20 items. This test was given to seventy-five 11th grade students, who were 16-17 years of age. The data were analyzed descriptively. The results showed that students' prior knowledge about electricity concepts was very diverse; only 22.40% of the students had a scientific concept; 36.73% of students experienced misconception; 31.20% of students do not know the concept, and 9.67% of students experience the error. Students' misconceptions about electricity concepts that were found in this study are in mutual accord with the literature of misconception. The implication of the research results for teaching about electricity is that appropriate conceptual change strategies are needed.

1. Introduction

Studies showed that students come to formal instruction with their prior knowledge of electricity that they have constructed [1] [2] [3] [4]. There are some terms used for prior knowledge, such as preconception [5]; children science [6][1]; alternative conception [7] and misconception [8].

Some prior knowledge often takes the form of misconception [9] [5]. Misconception influences the way students learn about new knowledge and plays an important role in learning [10]. Misconceptions that students bring conflict with scientific explanation [11]. Facts show that misconception is the most significant factor that contributes negatively to students' academic success [12]. Wrong prior knowledge can cause difficulties in learning [13].

Electricity is an important topic in physics. In addition to being related to the materials in physics itself, electricity also plays an important role in other sciences such as chemistry and biology. To teach electricity meaningfully the teacher has to identify students' prior knowledge about electricity. This study was conducted to answer the following questions: what does the students' prior knowledge of electricity look like?

2. Theoretical Framework

Prior knowledge define as an individual's entire knowledge that is (1) dynamic, (2) that existed before instruction, (3) is structured, (4) can exist in various forms (declarative, procedural, and conditional knowledge), (5) is explicit and implicit , and (6) contains conceptual and metacognitive knowledge components [14]. Students' prior knowledge can be as an alternative conception or scientific conceptions possessed by the students [15].

According to constructivism, prior knowledge plays an important role in developing students' knowledge. In constructivist literature, in learning physics, there are three roles of prior knowledge i.e.: the knowledge-as theory, the knowledge-as-elements and the knowledge-as-system accounts [16]. Constructivism views learning as the construction and acceptance of new ideas or reconstruction of the existing ideas [6]. Prior knowledge influences how students accommodate and assimilate new information. Prior knowledge influences how students perceive, organize information and make new information connections [17].

In relation to prior knowledge in physic instruction, prior knowledge is one of the factors that influence students' learning [18]. Prior knowledge has a positive effect on students' learning achievement [19] [20]. However, on the other hand, the students' prior knowledge can also become the hindrance to learn because some prior knowledge may contradict with information that will be learned [21].

Prior knowledge or alternative conception is acquired by students through interactions with the environment through experiences. Every student has the different experience, hence their mental structure about a concept that they acquire is also different, some of which can cheat prior thinking [22]. The Alternative conceptions that students develop can hinder their conceptual understanding [23]. The explanation above shows that it is important for science teachers to know their students' prior knowledge before teaching new concepts. Students' prior knowledge can be used as a catalyst for learning and foster to understanding scientific concepts

3. Method

3.1 Subject of Study

This study was aimed at identifying the 11th grade students' prior knowledge of electricity concepts. This study was conducted at SMA Negeri 1 Singaraja, one of the favorite schools in Singaraja City. The students of the school generally also originated from the best graduates of the junior high school in Singaraja. They are the students who were accepted through national junior high school examination result and those who were accepted through academic potential test and achievement. There were 75 students aged 15-16 years old who participated in this study. Before the test was given to them they had not studied electricity material from senior high school curriculum, although they had learned electricity at the lower education levels before, at the elementary school and at junior high school.

3.2 Instrument

The instrument used to identify the students' prior knowledge f electricity concept was Three Tier Electricity Diagnostic Test (TTEDT) which was modified from [5] [24] [25]. This test was a multiple choice test with three levels. The first level, multiple choice with answer alternatives that should be chosen by the students based on the prior knowledge that they had. The second level contained reason choices from the answers in the first level. If in [5] all reasons were provided but in this test in addition to being provided with the reasons, the students were also given the opportunity to write down their own reasons if the reasons provided do not fit with their minds. The third level contained belief choices toward the students' options in the first level which contained two options, that is, sure and not sure. The used TTEDT had a reliability index r= 0.728.

3.3 Data Analysis

The data of the students' prior knowledge were analyzed descriptively. The TTEDT instrument would produce three kinds of information, that is, answers to the first level multiple choice, answers to the choice of reasons to answers to the first level, and answers to belief choices in the third level. The three types of data were then combined to make students' prior knowledge categories. The combination looks like that in table 1.

Table 1: Categories of the types of students' answers

Answer level1	Answer level 2	answer level 3	Prior knowledge category
True	True	Sure	Scientific Knowledge (SK)
True	True	Not sure	Lack Knowledge (LK)
Wrong	True	Not sure	Lack Knowledge (LK)
True	Wrong	Note sure	Lack Knowledge (LK)
Wrong	Wrong	Wrong	Lack Knowledge (LK)
Wrong	True	Sure	Error (E)
True	Wrong	Sure	Misconception (M)
Wrong	Wrong	Sure	Misconception (M))

Adapted from [26]

Based on the categoryze above, the percentages of the students who had scientific concept, lack concept, and misconception were then calculated. The students' conceptions of electricity were described qualitatively.

4. Result and Discussion

Table 2 shows the frequency of students with prior knowledge on electricity concept categorized into Scientific Knowledge (SK), Misconception (M), Lack Knowledge (LK) and Error (E), for each test item.

Table 2: Frequency and persentage of students with prior knowledge categorized into SK, M, LK, and E for each test item

Physics Prior Knowledge	Item No	SK		M		LK		E	
		f	%	f	%	f	%	f	%
How to light the bulb lamp	1	56	74.67	8	10.67	11	14.67	0	0.00
The direction of electric current in a simple circuit	2	26	34.67	21	28.00	19	25.33	9	12.00
Electric current definition	3	32	42.67	13	17.33	28	37.33	2	2.67
The magnitude of electric currents in single loop circuits	4	8	10.67	39	52.00	22	29.33	6	8.00
Ohm law.	5	40	53.33	11	14.67	21	28.00	3	4.00
I-V graph	6	19	25.33	25	33.33	23	30.67	8	10.67
Determinant factors of the resistance of the conductor	7	8	10.67	24	32.00	33	44.00	10	13.33
Conductor	8	26	34.67	20	26.67	25	33.33	4	5.33
Bulb lamp in parallel	9	43	57.33	14	18.67	8	10.67	10	13.33
Total resistance of resistor parallel	10	4	5.33	39	52.00	18	24.00	14	18.67
The electric current of bulb lamp in parallel	11	14	18.67	37	49.33	23	30.67	1	1.33
The electric current of	12	6	8.00	39	52.00	27	36.00	3	4.00

Physics Prior Knowledge	**Item No**	**SK**		**M**		**LK**		**E**	
		f	**%**	**f**	**%**	**f**	**%**	**f**	**%**
bulb lamp in series									
Kirchhoff first law	13	4	5.33	42	56.00	23	30.67	6	8.00
The electric current of bulb lamp in the combined circuit	14	12	16.00	29	38.67	29	38.67	5	6.67
Emf and potential difference across a real battery	15	1	1.33	35	46.67	34	45.33	5	6.67
The power resulted from battery series and parallel	16	2	2.67	35	46.67	30	40.00	8	10.67
	17	3	4.00	44	58.67	23	30.67	5	6.67
The power resulted battery series and parallel	18	10	13.33	28	37.33	26	34.67	11	14.67
How the fuse works	19	6	8.00	28	37.33	16	21.33	25	33.33
Voltmeter and ammeter	20	16	21.33	20	26.67	29	38.67	10	13.33
	Mean		22.40		36.73		31.20		9.67

From table 2 it is apparent in general that out of 75 students, there was only 22.20% of them had prior knowledge which fell into scientific knowledge category on the other hand, 36.73% of the students had prior knowledge which fell into misconception category, 31.20% fell into lack concept and 9.67% into error category. This result was similar to the result of research by [27] in Cimahi, Bandung, that shows that 13.9% of the students understood the concept, 39.9% had the misconception, 44.5% lack concept, 2.195 had errors.

The highest percentage of students who had scientific knowledge was in the test item number 1, that is, how to light the bulb lamp. While the highest percentage of the students had misconception in the test item number 17 on power specification of the electrical device. The highest percentage of students who lacked concepts occurred in the test item number 15, that is, emf and the potential difference across a real battery. The highest frequency of the students who had errors occurred in how the fuse works.

There were 22 types of misconception that could be identified in this study. Some of the misconceptions were similar to the ones found in some misconception literature. The following are the types of students' misconception on electricity identified through TTEDT.

(1) Bulb lamp can only light if the metal box part is touched to the positive pole of the battery and the hidden part is connected to the negative pole of the battery, otherwise, it does not light. This misconception is also found in [24]

(2) If a lamp is connected to a battery, electric current will flow from positive pole to the lamp and from negative pole to the lamp, meeting at the lamp so that the lamp lights. This misconception is similar to the one found by [3], in which the current flows to the lamp from both battery terminals with the same magnitude.

(3) Electric current flows from the low potential to the high potential. A similar misconception was found by [5], that is, the direction of electric current is from negative pole toward positive pole.

(4) The magnitude of electric current is the charge that flows multiplied by the duration of the flowing charge the longer the conductor is connected to the source of voltage, the greater the electric current.of electric current is the charge that flows multiplied by the duration of the

flowing charge the longer the conductor is connected to the source of voltage, the greater the electric current.

(5) Electric current at the point closer to the source of voltage is bigger than the electric current at the farther point from the source of voltage. A similar misconception is also found in [24].

(6) The closer to the positive pole of battery the greater the current. A similar misconception is also found in [3].

(7) The electricity current at the point close to the positive pole of the battery is greater than the electric current closer to the negative pole of the battery. This is similar to the students' misconception found in [5], that is, the current in the positive pole of the battery is always bigger than the current in the negative pole of the battery. This result is also similar to the one in [3]. The electric current on the positive terminal side of the battery is always bigger than the current on the negative terminal side.

(8) The steeper the slope of curve I-V of a conductor, the smaller the electrical resistance of the conductor.

(9) There is no relation between the steepness of curve I-V of a conductor and the electrical resistance of the conductor.

(10) The magnitude of the current that flows in a conductor does not depend on the dimension of the conductor.

(11) The magnitude of thc current that flows in a conductor does not depend on the dimension of the conductor.

(12) There is no relation between the conductor dimension and its conductor.

(13) The wider the area of conductor cross-section the greater the electrical resistance. This misconception is similar to the one found in [28].

(14) The students cannot determine the total electric resistance from two electrical resistances that are parallel arranged.

(15) The magnitude of the electric current that flows from the battery with electric motion force (emf) ε connected to lamp with electrical resistance greater than that connected to 2 identical lamps arranged in parallel.

(16) The magnitude of the electric current that flows from the battery with emf ε connected to a lamp with electrical resistance r smaller than that connected with an identical lamp serially arranged. This misconception is also found in [24].

(17) Electrical resistance in the battery does not have any effect on the result of measurement with a potential difference. The measured potential difference is the same as the emf of the battery.

(18) Electric resistance in battery causes the terminal potential difference is read bigger than its emf.

(19) Power produced by one battery is smaller than by two batteries parallelly arranged and is smaller than two batteries serially arranged. This is also found in [24].

(20) Fuse prevents short circuit that occurs because of a very big electric resistance.

(21) It is better to make a fuse from bigger wire to make it's strong.

(22) Fuse with a big current specification is safer.

(23) To measure the strength of current and potential difference in conductor with closed circuit, ampere meter and voltmeter are installed series with its conductor.

The students' prior knowledge on electricity identified in this study shows that before learning about electricity at a certain level, the students had varied prior knowledge some of them were in the form of scientific knowledge, misconception, lack knowledge and error. These findings further reinforce the view that it is important for the teacher to identify the initial range of students' prior knowledge before beginning the lesson. The types of prior knowledge shown are very important material for selecting teaching strategies, especially conceptual changes strategy. Learning should be done in order to reconstruct the student's prior knowledge.

5. Conclusion and Suggestion

5.1 Conclusion

The students come to a formal instruction bringing with them a variety of prior knowledge of electricity concepts. By using Three Tier Electricity Diagnostic Test, the students' prior knowledge can be categorized into four categories, that is the scientific concept, misconception, lack knowledge, and error. The results of the study showed that 22.20% of the students had prior knowledge in the form of scientific knowledge, 36.73% of the students had misconception, 31.20% lack knowledge and 9.67% had errors. Some of the students' misconceptions were similar to the ones in the literature on misconceptions. The types of the students' prior knowledge identified in this study can be used as the source by the teachers to plan meaningful electricity instruction. Hence, the identification of the types of students' prior knowledge of electricity is a necessity.

5.2 Suggestion

To physics teachers, it is suggested to identify students' prior knowledge of electricity concepts before beginning the lesson. This variety of students' prior knowledge should be used as a reference in the development of teaching materials and conceptual change strategies to build scientific knowledge and reduce misconceptions.

Acknowledgement

The writers would like to thank The Director of Research and Community Service, The Ministry of Research and Technology and Higher Education for funding this research through Post-Graduate Research Grant in 2017. Our gratitude also goes to the Principal of SMA Negeri 1 Singaraja for the permission to do the research at the school, and the tenth grade students who participated in this research. Without their cooperation, this research could not be carried out.

References

[1] Osborne R, Freyberg P, Bell B, Tasker R, Cosgrove M and Schollum B 1985 *Learning in Science The Implication of Children's Sience* (Auckland: Heinemann) p 5-27
[2] Maloney D P, O'Kuma T I, Hieggelke C J and Heuvelen A V 2001 *Am. J. Phys. Suppl.* **69** S12-S23
[3] Küçüközer H and Kocakülah 2007 *Journal of Turkish Science Education.* **4** 101-115
[4] O'Dwyer A 2009 *AISHE. Conf.* NUI Maynooth
[5] Turgut U, Gürbüz F and Turgut G 2011 *Procedia Social and Behavioral Sciences.* **15** 2011 1965–1971.
[6] Bell B F 1993 *Children's Science, Constructivism, and Learning in Science* (Victoria Deakin University) p 12
[7] Petersson G 2002 3rd *European Sym. on Conceptual Change*, June 26-28 2002, Turku, Finland
[8] Brown D E and Clement J *Instructional Science.* **18** 237-261
[9] Demerci N 2005 *The Turkish Online Journal of Educational Technology.* **4** 40-48
[10] Ozmen H 2007 *Asia Pasific Education Review.* **8** 413-425
[11] Broughton S H, Sinatra G M and Reynold R E 2010 *Journal Of Educational Research.* **103** 407-423.
[12] Ozkan G, Selcuk G S 2013 *Asia Pasific Forum on Science Learning and Teaching.* **14** 1-17
[13] Gürefe N, Yarar S H, Pazarbasi B N and Es H 2014 *International Journal of Educational Studies in Mathematics.* **1** 58-68.
[14] Dochy F J R C and Alexander P A 1995 *European Journal of Psychology of Education.* **X** 225-242.
[15] Hewson M G and Hewson P W 2003 *Journal of Research in Science Teaching.* **40** S86-S98
[16] Esanu A and Hatu C 2015 *Int. Scientific Conf. eLearning and software for Education Bucharest* 25-26

[17] Svinicki M 1993-94 *Essays on Teaching Excellence. Toward the Best in the Academy.* **5**
[18] Baser M 2006 *Journal of Maltase Education Research.* **4** 64-79
[19] Calisir F and Gurel Z 2003 *Computers in Human Behavior.* **19** 135–145
[20] Calisir F, Eryazici M and Lehto M R 2008 *Computers in Human Behavior.* **24** 439-450
[21] Cordova J R, Sinarta G M, Jones S H, Taasoobshirazi G and Lombardi D 2014 *Contemporary Educational Psychology.* **39** 164-174
[22] Akpınar M and Tan M 2011 *Western Anatolia Journal of Educational Sciences.* 139-144.
[23] Salame I, Sarowar S, Begum S and Krauss D 2011 *Chem. Educator.* **16** 190–194
[24] Engelhardta P V and Beichner R J 2004 *Am. J. Phys.* **72** 98-115
[25] Pesman H and Eryilmaz A 2010 *The Journal of Educational Research.* **103** 208–222
[26] Kaltakçi D and Nilüfer D 2007 *6th Int. Conf. of the Balkan Physical Union* (American Institute of Physics) pp 499-500
[27] Ismail I I, Samsudin A, Suhensi and Kaniawati I 2015 *Prosiding Simposium Nasional Inovasi dan Pembelajaran Sains 2015:* 8 -9 Juni 2015, Bandung, Indonesia.
[28] Bilal E and Erol M 2009 *Lat. Am. J. Phys. Educ.* **3** 193-201

IOP Conf. Series: Journal of Physics: Conf. Series **1040** (2018) 012039 doi:10.1088/1742-6596/1040/1/012039

Types of junior high school students errors in science problem solving

I M Mariawan[1] and N N Parwati[2]

[1] Department of Physics Education, Universitas Pendidikan Ganesha, 11 Udayana Street, Singaraja, Indonesia
[2] Department of Mathematics Educations, Universitas Pendidikan Ganesha, 11 Udayana Street, Singaraja, Indonesia

E-mail: made.mariawan@undiksha.ac.id, nyoman.parwati@undiksha.ac.id

Abstract. This research aims to identify the errors of junior high school students in Science problem solving. This study is a descriptive research with quantitative and qualitative approach. The subject of the research is 30 students of class 8C SMP Negeri 2 Singaraja that made mistakes in a test of Science problem solving. The sampling was conducted with a test of problem solving and interview. The research data is a description of the source, type, and causative factor of the errors done by the students in solving Science problem on temperature and heat material. The data was analyzed descriptively and qualitatively. The result of the analysis shows that a lot of students made errors in performing problem solving strategy, which are: (1) understanding problem 58, 12%; (b) planning problem solving 55, 15%; (c) executing plan 70, 18%; and reviewing 85, 7%. The types of errors done by the students are: (1) concept/misconception error, (2) linking between concepts, (3) principle error, and (4) error in writing the corresponding unit. Furthermore the causative factors of the errors done by the students are: (1) poorly trained in solving problem; (2) habitually doing routine tasks; (3) being incomplete/incorrect in writing what is known and asked; (4) did not understand with the requirements of the questions; (5) the use of incorrect formula; (6) being inaccurate; (5) forgot the concept; (6) unable to give reasons; and (7) did not do a review toward the problem solving result.

1. Introduction

Science Education in 2013 curriculum has an intention to make students understand about the concepts and the relation between concepts, being able to communicate, and applying the concepts to solve problems, whether academic problems or problems in daily life [1]. The purpose of science education beckons problem solving is based on the concept comprehension and its linkages, communication, and uses strategy in problem solving. The learning in problem solving nuance is urgently done, considering the quality and quantity of students' ability to understand concepts, solve problems, and communicate are the keys of science learning and parts of the main component in supporting science learning in 21th century [2].

Problem and problem solving are important aspects in science learning, because problem and problem solving are used to teach students in applying knowledge and skill they obtain in learning process [3][4]. Problem is different from task or routine question. If a problem given to a student and the student knows how to correctly solve it right away, then the question cannot be said as a problem.

Someone deals with a problem when there is a gap between current condition with the goal intended to be achieved, while it is unknown what must be done to achieve the goal [5][6][7]. Problem is a situation experienced by someone so that what he has experienced is different from what he wants [8][9]. Based on the points of view, it can be described that problem emerges because there is a gap between expectation with reality, what is owned with what is needed, what has been known with what people want to know.

The fundamental difference between doing of routine questions/exercises with problem solving in learning physics located at the solving process. In the completion of routine tasks/exercises, students are only required to immediately get the answers, for example counting by inserting the numbers into formulas, the operation of sum and multiplication of vectors, and so on. Problem solving in physics is when a student is unable to find out the solution right away, but the student needs to reason, guess or predict, and look for the simple concept or formula and then prove it. According to [3][10], the characteristics of a problem is it needs the power of thinking/reasoning, challenges students to figure out the concept and its relation, guesses/predicts the solution, how to get the solution, and it must be proved that the achieved solution is correct/right.

Problem can be solved with various steps in accordance with the context of the problem. According to [3][10] there are five steps in problem solving strategy, which are recognize the problem, describe the problem in terms of the field, plan a solution, execute the plan, and evaluate the solution. The recognition of a problem can be developed with qualitative description in form of pictures or words that can help students in finding the subject matter. In describing a problem, students can simplify the issue in its physical form and submit useful contexts. In planning a solution, students can make an equation structure based on the contexts submitted in the previous step. In executing the plan, students can manipulate the equations, insert discovered variables. Lastly, students have to evaluate the answers, which is by examining the errors and assuring that the answers have been met.

Problem solving requires completion steps or rules. Several problem solving steps are, [11] suggested problem solving steps through understanding the problem, qualitative analyzing of the problem, solution plan for the problem, applying the solution plan, and checking. [10] proposed problem solving steps through define the problem carefully, explore possible solutions, narrow your choices, and test your solution. [12][13] suggested problem solving steps through understanding the problem, devising a plan, carrying out the plan, and looking back. Taking a close look toward the problem solving steps, it seems that the arrengements of problem solving steps in science (physics) are: (a) understand problems which include concept comprehension and its linkages, (b) plan a solution that includes identifying and relating between variables, writing corresponding scale and unit system, describing solution sketch (picture, diagram, graphic), and establishing solving strategy (concept, formula, mathematical algorithm approach), (c) executing solving plan, (d) reviewing problem solving results and steps which include the use of alternative problem solving strategy, concept relevance and validity, concept/formula understanding, corresponding scale and unit, and mathematical algorithm. Based on the problem background, then the problem studied in this study is how are the types of errors done by junior high school students in solving science problems?

2. Method

This study is descriptive research type with quantitative and qualitative approach. The subject of the research is 30 students of class 8C SMP Negeri 2 Singaraja which got selected randomly by using intact group.

The instrument used in this research is written test which are questions about problem solving and interview guideline. The instrument in this research consists of 5 essay questions about problem solving on teaching materials of temperature and heat. The research instrument was validated by three validators, which are two lecturers of Universitas Pendidikan Ganesha of physics education study program and one science Junior High School teacher with the validation result that the applied instrument met the validation content and construction. Based on the instrument test result, information is achieved that the level of arranged problem reliability is classified to be reliable with reliability coefficient 0, 73.

The data collection was conducted with a test of problem solving and interview. The resulting data is a description of the source, type, and causative factor of students made errors in solving science problems in temperature and heat material. The data was analyzed by using error percentage descriptively and using Miles and Huberman's analysis stages qualitatively, included: data collection, data reduction, and verification.

3. Result and Discussion

3.1. Result

The analysis result of students' errors in science problem solving (physics) is located at the problem solving strategy stages, which are: errors in understanding problems, planning a solution, executing the plan, and errors in reviewing the result and steps of problem solving. The error percentage of students on each temperature and heat solving stage is presented on Table 1.

Table 1. The Error Percentage of Students on Each Stage of Science Problem Solving (Temperature and Heat)

Problem Solving Stage	No. Problem and Percentage of Problem Solving Errors					Average
	1	2	3	4	5	
Understanding Problem	68,76	77,36	82,14	80,18	85,10	78,71
Solution Plan	57,15	60,35	72,11	70,68	82,22	68,50
Plan Execution	71,12	75,20	80,17	76,15	78,47	76,22
Review	80,25	78,83	81,15	79,80	82,53	80,51

The average trend of students error percentage on each stage of problem solving is described in graphic form as shown in Picture 1.

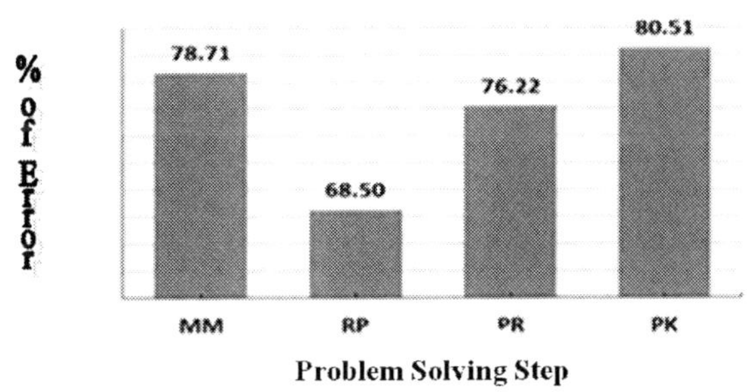

Picture 1. Percentage Graphic of Students Errors in Problem Solving

Based on the percentage of students errors in problem solving shown in Table 1 and graphic on Picture 1, it can be interpreted that most students errors in science problem solving (teaching materials of temperature and heat) are located on the review (PK), followed by mistakes in understanding problems (MM), executing plan (PR), and then making problem solving plan (RP)

The analysis result of the interview with 15 students reveals a type of students' errors in problem solving stages as shown in Table 2.

Table 2. Type and Percentage of Students Errors on Science Problem Solving Stage (Temperature and Heat)

Error Type on Problem Solving Stage	No. Problem and Percentage of Error Type in Problem Solving					Average
	1	2	3	4	5	

1. Understanding Problem						
(a) Identify concept	53,33	73,33	60,00	66,67	60,00	62,67
(b) Link between concepts	60,00	66,67	73,33	86,67	66,67	70,67
(c) Misconception	66,67	73,33	80,00	60,00	80,00	72,00
2. Solution Plan						
(a) Identify variable	46,67	60,00	46,67	66,67	60,00	56,00
(b) Determine the relation between variables	73,33	66,67	80,00	73,33	66,67	72,00
(c) Write the suitable scale and unit system	53,33	66,67	60,00	53,33	46,67	56,00
(d) Describe solving sketch (picture, diagram, graphic)	80,00	73,33	60,00	66,67	86,67	73,33
(e) Establish solving strategy (concept, formula, mathematical algorithm approach)	66,67	60,00	73,33	66,67	80,00	69,33
3. Plan Execution						
(a) Implement the plan in problem solving	73,33	60,00	80,00	66,67	53,33	66,67
4. Review						
(a) Use alternative strategy in problem solving	86,67	73,33	80,00	66,67	93,33	80,00
(b) Concept relevance, linkages between concept/formula, corresponding scale and unit, and matemathical algorithm	93,33	80,00	86,67	73,33	80,00	82,67
(c) Not done	86,67	93,33	80,00	73,33	80,00	82,67

The analysis result presented on Table 2 can be interpreted that the types of students' errors on each stage of problem solving are. (1) The type of students' errors in understanding problem stage, most students experienced misconception, followed by mistakes in linking between concepts, and identifying related concept with problem. (2) The type of students errors in planning problem solving stage, most students did mistakes in describing solving sketch (picture, diagram, graphic), followed by mistakes in determining the relation between variables, mistakes in establishing solving strategy (concept, formula, and mathematical algorithm approach), mistakes in identifying variables, and mistakes in writing the corresponding scale and unit system. (3) The type of students' errors in plan execution stage, most students were incorrect in implementing plan. Problem solving plan that has been appointed does not fit with the implemented one; and (4) The type of students errors in review stage, most students did not do it and were incorrect in applying relevant concept, linkages between concepts/formulas, suitable scale and unit system, and mathematical algorithm, then followed by mistakes in using alternative problem solving strategy.

3.2 Discussion

The analysis result of student's errors in science problem solving is located at the problem solving stages with different types of errors. This anaylsis result is suitable with the research result of [12][11] which found out that the struggles of students in solving motion problems are located at the problem solving stages, which are: stage of understanding concept, planning a solution, and executing the solving plan. [13] also figured out the struggles of students in solving rotation and torque problems are located at the problem understanding, problem solving planning, and the review of problem solving steps and result.

The students errors on the problem solving stages are caused by several factors, which are: (a) students are lack of training and not get used to with solving problem, (b) the problem representation presented in the learning process is limited to routine questions which use formulas to solve them, (c) mathematical misconception and operation. The following three factors affect the other errors, which were done by the students in problem solving stages. In the effect of training factor, [12] said a child who is given a lot of problem solving exercises, has higher grade in problem solving test compared to those who are lack of trainings. [11][14] also said the more problems that can be solved by a student, the more skills he will have that can help him to solve problems in daily life. In the effect of problem representation, [3] said, learning problem solving is essentially learning how to think or reason to apply the knowledge that have been obtained before, in solving new problems that never been encountered. In problem representation, [11] said, students who are studying science and wonted with non-routine questions or problems, these students will be skilled by applying various concepts in new situation, so they will be able to use various concepts of science that they have learned to solve problem in daily life. In the effect of misconception, [5] proposed that misconception is a form of delusion, which is a supposition of the true one, is false and the false one is true. This delusion can have the shape of a wrong understanding, a mistake, or cognitive disorder of students. [5][15] explained, students who experience misconception can influence the concept understanding, between-variable linkage, and concept application in problem solving.

4. Conclusion

Students' errors in science problem solving (physics) are located at the stages of executing problem solving strategy, which are: mistakes in understanding problem, planning a solution, executing solving plan, and mistakes in reviewing the result and steps of problem solving. The type of students errors on each problem solving stage are: (1) the type of students errors in understanding problem stage, most students experienced misconception, followed by mistakes in linking between concepts, and identifying related concept with problem; (2) the type of students errors in planning problem solving stage, most students were incorrect in describing solving sketch (picture, diagram, and graphic), followed by mistakes in determining the relation between variables, setting solving strategy (concept, formula, and mathematical algorithm approach), identifying variables, and writing the corresponding scale and unit system; (3) the type of students errors in plan execution stage, most students were wrong in implementing plan. The established problem solving plan are not corresponding with the implemented one; (4) the type of students errors in reviewing, most students did not do it and were mistaken in applying relevant concept, linkages between concept/formula, suitable scale and unit system, and mathematical algorithm, then followed by mistakes in using alternative problem solving strategy. Based on the result of this research, many students did various mistakes in solving science problems presented in learning. Therefore, Math and Science teachers are expected to perform learnings by emphasizing problem solving training. Problem solving skill is very important to be mastered by students when they plunge into society.

Acknowledgments

Acknowledgments are delivered to Universitas Pendidikan Ganesha (Undiksha) Institution that has provided the research funding. Head of Buleleng District Youth and Sport Education Office for the permission given to conduct this research. Rector and Chairman of LPPM Undiksha who have approved the execution of this research. All teachers and students in SMPN 2 Singaraja who were involved, and all parties who cannot be mentioned one by one in suceeding the execution of this research.

References

[1] Nugraheni A S 2015 *J. Educ. Pract.* **6** (2) 53–61
[2] Saavedra A R and Opfer V D *Phi Delta Kappan*, vol. 94, no. 2, pp. 8–13, 2012.
[3] NCTM 2010 Why is Teaching with Problem Solving Important to Student Learning?
[4] Kim M C and Hannafin M J 2011 *Instr. Sci.* **39** (3) 255–282
[5] Gilbert J K and Watts D M 1983 *Stud. Sci. Educ.* **10** (1) 61–98
[6] Parwati N N 2011 The effect of instructional strategies and problem types towards the problem

IOP Conf. Series: Journal of Physics: Conf. Series **1040** (2018) 012039 doi:10.1088/1742-6596/1040/1/012039

solving ability of elementary school students with different attitudes towards mathematics. Thesis (Malang: Universitas Negeri Malang)

[7] Parwati N N, Sudiarta I G P and Mariawan I M "Pengintegrasian Nilai Kearifan Lokal Masyarakat Bali dalam Model Pembelajaran Pemecahan Masalah Matematika untuk Membangun Karakter Positif Siswa SD Di Kabupaten Buleleng," in *Stranas Research Report, unpublished: Universitas Pendidikan Ganesha*, 2014.

[8] Wang M, Wu B, Kinshuk, Chen N S and Spector J M 2013 *Comput. Educ.* **68** 293–306

[9] Schoenfeld A H 1992 "Learning to think mathematically: Problem solving, metacognition, and sense making in mathematics," *Handbook for Research on Mathematics Teaching and Learning*. pp. 334–370

[10] Hentenryck P V and Coffrin C 2015 *Proceedings of the 45th ACM technical symposium on Computer science education - SIGCSE '14* 677–682

[11] Selcuk G, Çalişkan S and Erol M 2008 *Lat. Am. J. Phys. Educ.* **2** (3) 161–166

[12] Hmelo-Silver C E 2004 *Educational Psychology Review* **16** (3) 235–266

[13] Polya G 1996 *How To Solve It Qucosa.De*

[14] Parwati N N 2012 *J. Ilmu Pendidik* **18** (1) 65–70

[15] Docktor J L, Strand N E, Mestre J P and Ross B H 2015 *Phys. Rev. Spec. Top. - Phys. Educ. Res.* **11** (2)

The Effectiveness of problem-based hybrid learning model in physics teaching to enhance critical thinking of the students of SMAN

R Sujanem[1], S Poedjiastuti[2] and B Jatmiko[2]

[1]Universitas Pendidikan Ganesha, Indonesia
[2]Universitas Negeri Surabaya, Indonesia

E-mail: raisujanem@yahoo.com

Abstract. This research aimed at analysing the effectiveness of Problem Based-Hybrid Learning (Pro-BHL) to enhance the critical thinking skill (cts) of the students of SMA in Physics teaching. Pro-BHL model is a model of teaching which combines face – to – face PBL and online learning. Pro-BHL model is effective to enhance cts if: 1) there is a significant improvement in the students' cts at α=5%; and 2) The minimal N-gain average falls into medium category or consistent. This study involved 86 students who spread into three groups of grade X. The design of this study was one group pre-test and post-test group design. Before the teaching using Pro-BHL model, the students in three groups were given a cts test (pre-test). And at the end of the teaching, the students were given the same test (post-test). The collected data were analyzed by Paired-Test, normalized gain (N-gain T-Test) and ANOVA. The result showed that: (1) the teaching with Pro-BHL model can enhance the students' cts (α =5%); (2) the N-gain average for the three groups falling into high category; and (3) do not differ from one another. Thus, it this can be concluded that the Pro-BHL model is effective to enhance the students' cts in physics teaching.

1. Introduction

One of the paradigms of the 21th century is the expectation that the students master critical thinking (cts). Cts is a systematic process when a student makes a decision about what he or she believes and what he or she does [1, 2]. The indicators of cts include: to formulate a problem, to give an argument, to make a deduction, to make an induction, and to make a decision. Cts is very important to be trained since this critical thinking is carried from birth [3].

The importance of cts is in accordance with what is stipulated in curriculum 2013, which is the reference in teaching process to achieve the expected learning outcome [4]. According to Semerci [5], students with higher cts obtain higher learning achievement if they are compared to students with lower cts. Cts is the key in education to solve a problem. Most of jobs require workers with cts [6]. By examining closely how important is cts, teaching process should stress the importance of students' cts. However, the fact shows that cts in schools in Indonesia, especially at senior high schools has not been well trained so that the cts of senior high school graduates is still relatively low [7, 8]. The low cts of the graduate of elementary schools up to university level in Indonesia is often complained [9]. One of the factors which is assumed to be the cause of the low cts of the students, among others things is the type of test item given to the students uses Bloom taxonomy at a low level, so that it does not train the students thinking skill [2, 10]. The quality of physics teaching as part of science education up to now is still low and experiences a decrease as seen in the results of the study done by PISA (Program for International Student Assessment), i.e., a study that focuses on reading, mathematics and science literacies, that showed that the ranking of science in Indonesia is at the 62th from 69 countries [11]. This shows that the quality of science education in Indonesia, including Bali Province tends to be low. This is supported by the result of research done by Sadia [12] in some regencies in Bali that shows that the cts of the students of senior high school students at tenth grade fall into low qualification with the mean score of 49.38 in 0-100 scale. In adition, teachers still find difficulty in teaching cts to the students and there are still a few teaching models that teach cts to the students. The result of preliminary research on cts about temperature and heat at SMAN 1 Singaraja (Bali), showed that mean score was in the range of 35.30 – 41.10; falling into insufficient category [13, 14]. Based on the exposure to problems and examples above, it is identified by preliminary research that the students' cts is still low, the teacher knows less about how to improve the cts and we do not have adequate models of teaching to train cts. Teachers still teach the traditional teaching-learning. Traditional teaching strategies emphasize most of the materials by lecture, whereas the PBL method relies on the problem as a vehicle to guide learners to relevant content information [15, 16].

Based on the explanation above, the problem that arises is how effective is Pro-BHL teaching model to increase the cts of the students of SMAN ? This problem is formulated in more detailed as follows.(1) Is there any increase in students' cts after being given a teaching with Pro-BHL?; (2) What is the extent of the average level of the increase (average gain score) for the three groups; and (3) Is the average N-gain consistent for the three groups.

Regarding the still low cts of the students of senior high schools in Bali, an alternative solution needs to be found to train the cts of the students in accordance with curriculum 2013 [4]. One of the alternative models which assumed to be able to train the cts with Problem-Based Hybrid Learning model (Pro-BHL). Based on hybrid learning, learning is basically a social process which will be compromised if all the modules that take place in the virtual world are far from human interaction [17, 18]. Driscoll [19] shows that blended/hybrid learning can mean differently for different people. Pro-BHL model is teaching model which combines face to face Problem Based Learning (PBL) in the classroom and online learning. In the Pro-BHL model, the students are faced with the problem of ill-structure, then the students work in groups to formulate the problem. The teacher facilitates the students in learning, organizes the learning tasks, mediates problem formulation, and formulates hypotheses. Students in groups to solve problems that have been formulated earlier. Traditional teaching strategies emphasize the broad coverage of content areas through lecture, giving the example problems, and practice questions [20]. Students conduct investigations, collect data, analyze inductively and deductively, and make conclusions, to seek information through printed books available on the internet, have a discussion, are engaged in question and answer activities and present their work through a series of activities on the model Pro-BHL done carefully so it could be expected to train CTS.

Pro-BHL model in Senior high school Physics teaching contains ill-structured problems, physical phenomena, strategic and essential concept, contextual concept, animation/simulation, video, examples and contextual exercises; while learning tools such as lesson plans, worksheets, and teaching materials for Pro-BHL model can be accessed onlinely with address http://Probhl.com. Pro-BHL model in Physics teaching refers to Blended Problem-Based Learning model (BPBL) [18, 21] and Problem Based Blended Learning model (PBBL) Wannapiroon [22]. Donnelly [16] combines face to face PBL teaching in the class room with e-learning which is known as blended PBL (BPBL). Wannapiroon [22] combines blended/hybrid learning model and PBL which is known as PBBL.

2. Method

2.1. Research Design

This study emphasized the analysis of the effect of Pro-BHL model in Physics teaching on students' cts, therefore in this study used pre-experimental with pre-test and post-test design as follows:
O1 X O2 [23, 24]. Before teaching by using Pro-BHL model, the three groups where given pre-test (O1), and after teaching ws done by using Pro-BHL model (X), the three groups were given the same test again(post-test) (O2).

This research was done with three groups (A, B, and C) of the students of tenth grade in the academic year 2015-2016 at SMA Negeri 1 Singaraja with the total of 86 students, each group consisting of 29, 29, and 28 students respectively.

2.2. Method data collecting and Analysis of data

This research use test method for collecting the data. The test was used to know the improvement in cts of the students before and after the implementation of Pro-BHL model. The test consisted of pre-test and post-test according to cts indicators. The cts test instrument was in the form of essay test, consisting of the following indicators: (1) formulating problems, (2) giving arguments, (3) making and deduction, (4) making induction, and (5) making decision.

The data from the result of the cts test collected were analyzed by using paired t-test or non pharametic analysis of Wilcoxon test. Paired t-test was done if the criteria of population data had normal distribution [23, 25, 26, 27]. If the data did not meet the criteria of normal distribution, then Wilcoxon test was used. While, the computation of N-gain value was done to analyze the level of the students cts which was caused by Pro-BHL model implementation in Physics teaching. The mean of the cts improvement was determined by the value of the normalization of gain (N-gain). The category of N-gain according to Hake [28] are: (1) teaching with "high improvement average", if N-gain $\geq$ 0.7; (2) teaching with "medium improvement average", if 0.7 > N-gain $\geq$ 0.3; and (3) teaching with "low improvement average", if N-gain < 0.3. Statistical analysis was done using IBM SPSS Statistics 20 software [27]. After that, analysis of variance (ANOVA) was done to analyze consistence (no difference) in the average of cts improvement using Pro-BHL model among classes A, B, and C. ANOVA was done when the requirement of N-gain average for the three groups, each coming from population with a normal distribution and the three groups were homogeneous was met.

3. Result and Discussion

3.1. Result

The description of data of every group: A, B, and C is shown in diagram consisting of: mean score for pre-test, mean score for post-test, and N-gain average for cts. The description is shown in Figures 1 and 2 .

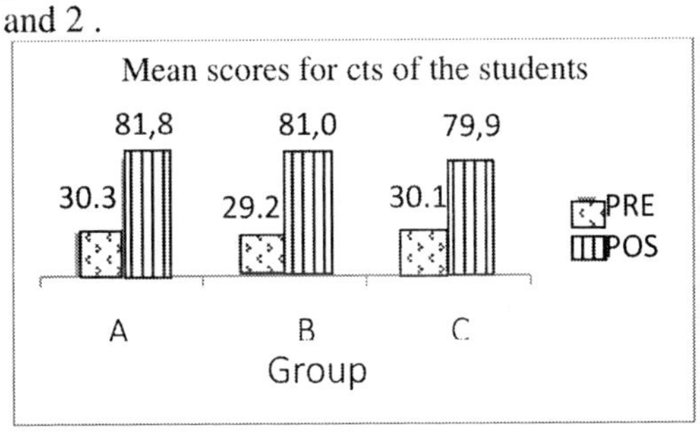

Figure 1. Mean scores for cts *pre-test, post-test*

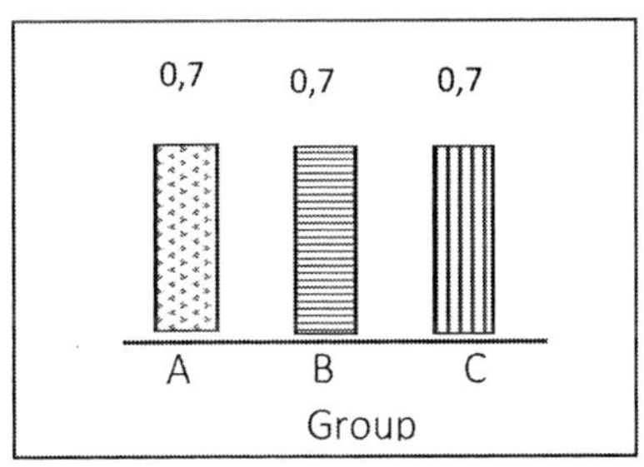

Figure 2. Mean scores improvement <g> cts

Figure 1 shows improvement in cts in three groups: A, B, and C. Improvement in cts of the three groups were categorized into high category shown by N-gain value as in Figure 2 . The result of paired t-test between pre-test and post-test of cts for the three groups had met the criteria of normality and homoginety as shown in Table 1. While, the ANOVA result about the improvement of cts of student for the three groups had met the criteria of normality and homoginety as shown in Table 2.

Table 1. Result of Paired T-Test between The Scores for Pre-Test and Post-Test

No	Data	Average Test (t)	Df	p (2-tailed)	Remark
1	Group A	-32,339	28.00	< 0.0001	Ho is rejected
2	Group B	-49,938	28.00	< 0.0001	Ho is rejected
3	Group C	-35,788	27.00	< 0.0001	Ho is rejected

$^{*}p < 0.05$ (2-tailed)

Table 1 shows that the value of p for the result paired t-test of the scores of pre-test and post-test in cts in the three group is < 0.05 and on the whole have negative values. This means that there is a significant difference (statistically) in the cts of the students between before and after the implementation of Pro-BHL model in teaching. The cts of the students after the implementation of Pro-BHL model was higher then before the implementation of Pro-BHL model.

Table 2. Results of ANOVA of the Improvement of the Students' cts

No	Data	F	Df	P	Remark
1.	Average N-gain of the cts of Groups A, and C	0,496	83	0.611	Ho is accepted

$^{*}p < 0.05$

Table 2 shows that the value of p in the result of ANOVA for groups A, B , and C is 0.611. Since the value of p > 0.05, then statistically, there is no difference in the average of improvement in the students' cts (consistent), siginificant at $\alpha = 5\%$ for the three groups A, B, and C.

3.2. Discussion

Pro-BHL model can be implemented in Physics teaching. This model requires the identification of unstructured problems, formulating problems, collecting information from the internet to solve the problems, processing information, presenting the result of problems solving, doing a discussion, asking and answering questions, so that the students are able to master cts.

The achievement of cts by the students is shown in Figure 1. The mean scores in cts for the pre-test in the three groups (A, B, and C) before Pro-BHL model was implemented were low, i.e., 30.3; 29.9; and 30.1 from 0 – 100 scale. After Pro-BHL model was implemented, the means score in cts of the students in the three groups were higher, i.e., 81.8; 81.0; and 79.9. The low score and cts of the students in the condition before Pro-BHL model was implemented was probably cause by the students who had not been cts after Pro-BHL model was implemented, their scores were higher. They improvemnet of the mean score in cts of the students is shown in Table 1, statistically significant $\alpha = 5\%$. In Table 2 it is shown that the average N-gain for every groups: A, B, and C, all have high categories acccording to Hake's criteria [28] that is 0.7.

In Table 2 it is shown the improvement of cts of the students, N-gains in the three groups: A, B, and C which are not different one and another or consistent. This shows that the implementation of Pro-BHL model could consistenly improve the students cts. This is in line with the schema theory that state when individual reconstructs information, he or she addapted which the knowledge that he or she had in mind [29]. Pro-BHL model is a model for integrating ICT in education which can give opportunity for the achievement higher level of cts. The effect of Pro-BHL model only improvement of cts is supported by the finding Wannapiroon [20] who state that higher level cts was achieve after teaching by using PBL model. Similarly this is also supported by the result of the research done by [30,31] who state that the students cts improve in Physics through the implementation PBL online; Elnetthra and Sulaiman [30, 31] who conclude that PBL online can improve cts; and [32] who shows that PBL significantly improve the students competence in planning and organizing teaching. The ability Pro-BHL model that provide hyperlinks familiarize the students to see the flexibility of the teaching material.

4. Conclusion

Based on the discussion of the result of the research it can be concluded that Physics teaching with Pro-BHL teaching model can effectively improve the cts of the students of SMAN 1 Singaraja, as shown by: (1) the improvement of the students cts significantly at $\alpha = 5\%$, (2) the average of cts gain (N-gain average) for the three groups = 0.7 categorize into high category, and (3) the N-gain averages are not different for the three groups of research or consistence.

Acknowledgement

The authors thank the Headmaster of State Senior High School-1 Singaraja who had been willing to provide a place for this research.

References

[1] Ennis R H 2012 *The Nature of critical thinking: Outlines of critical thinking dispositions and abilities* Retrieved from http://www.criticalthinking.net/longdefinition. html
[2] Ennis R H 1996 *Critical thinking* (Upper Saddle River, NJ: Prentice-Hall)
[3] Redhana I W and Liliasari 2008 Program pembelajaran keterampilan berpikir kritis pada topik laju reaksi untuk siswa SMA [Critical thinking skills learning program on reaction rate topics for High School Students] *FORUM KEPENDIDIKAN* **27** 103-112
[4] Kemdikbud 2016 *Salinan Lampiran Peraturan Menteri Pendidikan dan Kebudayaan Nomor 21 Tahun 2016 tentang Standar Isi Pendidikan Dasar dan Menengah* [Copy of Appendix of the Regulation of the Minister of Education and Culture No. 21 Year 2016 about Standard of Contents of Elementary and High School Education] (Jakarta: Kemdikbud)
[5] Semerci C 2005 The influence of the critical thinking skills on the students' achievement. *Pakistan Journal of Social Sciences* **3** 598-602
[6] Azami Z, Yuzainee M Y, Mohd Z O, Azah M, Norhamidi M and Ramli, M 2009 Perceptions and epectation toward engineering graduates by employers: A Malaysian study case. *WSEAS TRANSACTIONS on ADVANCES in ENGINEERING EDUCATION* **6** 296-305
[7] Ananiadou K and Claro M 2009 *21st Century Skill And Competency for New Millennium Learners in OECD Countries* (OECD Education Working Papers, No.41: OECD Publishing)
[8] Partnership For 21 St Century Skills 2009 P21 Framework Definition. Retrieved from http://www.p21. org.storage/documents/docs/P21_Framework_Definitions_New_Logo_2015.pdf
[9] Reta I K 2012 Pengaruh model pembelajaran berbasis masalah terhadap keterampilan berpikir kritis ditinjau dari gaya kognitif siswa [The influence of problem-based learning model to critical thinking skill in terms of student's cognitive style] *Jurnal Penelitian Pascasarjana Undiksha*, **2**(1), 1-17.
[10] Pursitasari I D and Permanasari A **2012** Model integrated problem solving based learning pada perkuliahan dasar-dasar kimia analitik [Integrated problem solving based learning model on lectures of analytical chemistry basics] *Jurnal Ilmu Pendidikan* **18** 172-178
[11] Gurría A 2016 *Programme for international student assessment 2 PISA 2012 results in focus* (OECD Secretary-General)
[12] Sadia I W 2008 Model pembelajaran yang efektif untuk meningkatkan keterampilan berpikir kritis [Effective teaching model for improving critical thinking skill] *Jurnal pendidikan dan Pengajaran Undiksha* **41** 219-237

[13] Sujanem R 2015 Profil Keterampilan berpikir kritis dan kemampuan pemecahan masalah siswa SMAN di kota Singaraja [Profile of Students' critical thinking skill and problem solving ability of SMAN students in Singaraja city] *Prosiding Seminar Nasional Sains UKSW 2015*, 19 September 2015

[14] Sujanem R 2015 Analisis kebutuhan model pro-bhl untuk meningkatkan keterampilan berpikir kritis dan kemampuan pemecahan masalah siswa dalam pembelajaran fisika SMA [Need analysis of Pro-BHL model to improve critical thinking skills and problem-solving abilities of the students of the senior high school in physics teaching] *Proceedings Seminar Nasional FMIPA UNDIKSHA V Tahun 2015* "Optimalisasi Peran MIPA dalam Membangun Sumber Daya Manusia Indonesia yang Kompetitif", 07 Desember 2015

[15] Barbara J D1995 *Problem-based learning in physics: the power of sudents teaching students.* Center For Teaching Effectiveness.

[16] Gijselaers W H 1996 Connecting problem-based practices with educational theory. *Ne Diection for Teaching and Learning* **68** 13-21.

[17] Graham C R 2004 *Blended Learning System: Definition, Current Trensd, and Future Directions.* In Bonk C J and Graham C R (Eds) (in press). Handbook of blended learning: Global Perspectives, local designs (San Francisco, CA: Pfeiffer Publishing)

[18] Donnelly R 2006 Blended problem-based learning for teacher education: Lessons learnt *Journal of Learning, Media and Technology* **31** 93-116

[19] Driscoll M 2002 Blended Learning: let's get beyond the hype, *e-Learning,* http://elearningmag.com/ltimagazine, March 1, 2002.

[20] Slavin R E 2009 *Educational psycology theory and practice* Eight edition (Bostond: Pearson

[21] Moeller S Spitzer K and Spreckelasen C 2010 How to configure blended problem based learning–Results of a randomized trial *Medical Teacher,* 32, 328–346.

[22] Wannapiroon P 2008 Development of problem-based blended learning in developing undergraduate students' critical thinking. *Journal of ICT to Improve Learning* **1** 1-7.

[23] Sugiyono 2012 *Metode penelitian pendidikan: Pendekatan kuantitatif, kualitatif, dan R&D* [Educational research method: quantitative, qualitative approach, and R&D] (Bandung: Alfabeta)

[24] Fraenkel J R and Wallen N E 2009 *How to design and evaluate research in education* (7th ed.) (New York: McGraw-Hill)

[25] Ary D Jacob L C and Sorenen C 2010 *Introduction to research in education* (Canada: Cengage Learning)

[26] Arikunto S 2010 *Dasar-dasar evaluasi pendidikan* [The Basics of educational evaluation] (Jakarta: Bumi Aksara)

[27] Priyatno D 2012 *Belajar Praktis Analisis Parametrik dan Non Parametrik dengan SPSSl & Prediksi Pertanyaan Pendadaran Skripsi dan* [Practical learning about parametric and non parametric analysis with spss & the prediction of questions and discussion of thesis] (Yogyakarta: Penerbit Gava Media)

[28] Hake R R 1998 Interactive-engagement versus traditional methods: a six-thousand student survey of mechanics test data *American Journal of Physics* **66** 64–74

[29] Santrock J W 2011 *Educational psychology 5th_edition* (New York: McGraw-Hill)

[30] Elnetthra F E and Sulaiman F 2013 The Role of PBL in improving physics students' creative thinking and its imprint on gender. *International Journal of Education and Research* **1** 1-10

[31] Sulaiman F 2013 The Effectiveness of PBL online on physics students'creativity and critical thinking: a case study at Universiti Malaysia Sabah *International Journal of Educational and Research.* 1 (3), 1-18

[32] Vidic A D 2010 The Impact of problem-based learning on statistical thinking of engineering and technical high school students.*ICOTS8 (2010) Invited Paper Refereed.* International Association of Statistical Education (IASE)

IOP Conf. Series: Journal of Physics: Conf. Series **1040** (2018) 012041 doi:10.1088/1742-6596/1040/1/012041

The contribution of school climate, achievement motivation, and self-concept to science learning achievement

Surayanah[1] and L Karma[2]

[1]Sekolah Tinggi Agama Islam Istiqlal, Buleleng, Bali – Indonesia
[2]STIKES Buleleng, Bali – Indonesia

E-mail: surayanah90@gmail.com, lewakarma@yahoo.co.id

Abstract. This research was aimed at investigating the contribution of school climate, achievement motivation and self-concept to science learning achievement in the seventh graders of MTs Al-Khairiyah Tegallinggah in the academic year 2015/2016. The research used the *expos facto* design. There were 46 seven grade students chosen as the sample. The data on school climate, achievement motivation and self-concept were collected by using questionnaire and the data on learning achievement by using a multiple-choice test. The hypotheses were tested by using simple linear regression and multiple linear regression. The results of the research showed that: 1) there is a positive and significant contribution of school climate to the students' learning science achievement (F = 39,88 and sig.5%); 2) there is a positive and significant contribution of achievement motivation to students' learning science achievement (F = 31,01 and sig.5%); 3) there is a positive and significant contribution of self-concept to students' learning science achievement (F = 15,91 and sig.5%); 4) simultaneously there is a positive and significant contribution of school climate, achievement motivation and self-concept to students' science learning achievement (F =15,27 and sig.5%). So, school climate, achievement motivation and self-concept can be predictors of students' science learning achievement.

1. Introduction

Science education has the same aim and essence as the aim and essence of the national education in general, i.e., (1) to understand ideas produced by science (concepts, models and theories), (2) to understand important issues in philosophy, history and methodology of science, and (3) to use science, in order that the students are capable of doing leadership activities and applying scientific knowledge in their lives [1].

As an effort to achieve the aim of education, the government has set up competence standards that a student should meet. The effort to meet the competence standards of education in Indonesia has been supported by the Regulation of the National Education Minister of the Republic of Indonesia No. 41 of 2007 on Process Standard. The process standard consists of the planning the instructional process, the instructional process implementation, the learning achievement assessment process, and the instructional process supervision. Education is run as the process of enculturation and empowerment of students which take place in the entire course of one's life. However, in reality, the aim of education has not been achieved optimally.

This can be seen from the result of a research done by PISA-OECD (*Programme for International Student Assessment-Organization for Economic Cooperation and Development*) in 2009 in which the reading literacy achievement of Indonesian students ranks the 57th, the mathematics, the 61st rank and the science literacy the 60th from 65 countries [2]. The results above show that the achievements of Indonesian children are still low.

One of the factors which cause the low learning achievement is a bad instructional management, viewed from planning, implementation and assessment done by the teachers. The instructional management done by the teachers is not yet prepared maximally. The instructional management is one of the factors that support the development of a good school climate. A bad instructional management causes the students not to pay attention, misbehave and not to do the tasks given by the teacher [3]. If that happens, it is just proper for the teacher to warn those students. The teaching will become meaningful if it is supported by a conducive, constructive and innovative learning condition [4]. This condition will affect and improve achievement and indirectly will create a good social relation. Othman *et al.,* [5] states that the mutualistic social relation between students and teacher is an important element in the school life. Teacher who is interested, caring, fair, democratic, and respectful toward the students is characterized as capable enhancing students' achievement.

Students will feel that the teacher pays attention to them if the latter is regarded as fair in giving attention to all students. The teaching process will raise the students' motivation to achieve [3]. Students' achievement motivation will develop if the teacher gives challenging tasks in the environment which support the acquisition of the material. The teacher has to give emotional and cognitive support, and meaningful and interesting materials to be learned and acquired, and gives enough support to develop students' autonomy and initiative. According to Kala, *et al.*, [6], achievement motivation is an effort to reach success by competing based on a standard of excellence. The standard of excellence can be in the form of the level of completeness of the result in task performance (related to tasks), in comparison to self-achievement (related to oneself), and in comparison to others (related to others).

Besides being positive in its effect on the students' academic achievement, school climate contributes positively to nonacademic achievements, such as the development of self-concept, self-belief and aspiration. Othman *et al.,* [5] states that self-concept development is influenced by other people who are close in the environment. There are two factors which influence self-concept, i.e., actor factor (parents, peers and the society) and substance factor (learning, association and motivation).

The end result of the management of school climate, achievement motivation and students' self-concept is the achievement of an optimal learning process and learning achievement. An optimal learning achievement is the ideal expectation that has to be achieved by a person after undergoing a learning process. In an effort to achieve an optimal learning achievement, according to the planning and expectation in the learning process, the school climate has to meet an adequate condition. In addition, school climate has to enhance achievement motivation and positive self-concept. This is a requirement that has to be met in a school. Based on the explanation above, a research on "the contribution of school climate, achievement motivation and self-concept to science learning achievement of the students of MTs Al-Khairiyah Tegallinggah" was conducted to see the relation between the three variables and learning achievement.

2. Method

This study belonged to an *ex post facto* research. The population consisted of 64 seventh grade students of MTs Al-Khairiyah Tegallinggah. The sample also consisted of 64 students.

There were two types of variables in this study, i.e., independent variables (school climate, achievement motivation and self-concept) and a dependent variable (science learning achievement). The topic of the lesson was matter and its changes. The study used questionnaire to obtain data on school climate. With the reliability index of 0.909 in the very high classification, another questionnaire to obtain data on achievement motivation with the reliability index of 0.926 in the very high classification, and another questionnaire to obtain data on self-concept with the reliability index was 0.870 in the very high classification. While to obtain data on learning achievement the study used a multiple choice test a with the reliability index 0.650 in high classification.

The data of this study were analyzed step by step. The steps were data description, assumptions testing, and hypothesis testing. The assumptions testing done consisted of (a) normality testing, (b) linearity testing and significance of regression direction testing, (c) multicollinearity testing, (d) auto correlation testing, and (e) heterocedasticity testing. If the prerequisites had been met, then it was continued with hypothesis testing using simple linear regression and multiple regression aided by *SPSS 16.0 for windows at significance level* $\alpha = 0.05$.

3. Results and Discussion

3.1 Description of Data on School Climate, Achievement Motivation, Self-Concept, and Science Learning Achievement

The description of the result of the study on school climate, learning motivation, self-concept and science achievement are presented in table 1 below.

Table 1. Description of Scores for School Climate (X_1), Achievement Motivation (X_2), Self-Concept (X_3), and Science Learning Achievement (Y)

No	Statistic	(X_1)	(X_2)	(X_3)	(Y)
1	Mean	155	146.17	156.24	77
2	Median	155	149	160	77
3	Mode	143	150	163	70
4	Standard Deviation	11.78	12.46	10.33	13.69
5	Variance	138.97	155.25	106.80	187.47
6	Minimum	130	123	133	47
7	Maximum	184	172	173	100
8	Range	54	49	40	53
9	Total Number of data scores	7130	6724	7187	3547
10	Number of Subjects	46	46	46	46

The result of descriptive analysis shows that the mean for school climate is 155 in high qualification and the mean for achievement motivation is 146.17 in high qualification. The score for students' self-concept is 173 with the mean = 156.24 in high qualification. The score for science learning achievement of the students of seventh grade of MTs Al-Khairiyah Tegallinggah ranges from 47 to 100 with the mean score of 77 in the high qualification.

3.2 Hypothesis Testing

Table 2. Recap of the Result of Analysis of the Data on the Inter-variable Contribution

No	Inter-variable Correlation	Regression Line Equation	Correlation Coefficient	Contribution (%)	Effective Contribution (%)
1	$X_1$and Y	$\bar{Y}$=36,74 + 0,34 X_1	0,58	40,17	34,27
2	$X_2$and Y	$\bar{Y}$=27,58 + 0,28 X_2	0,53	32,78	27,96
3	$X_3$and Y	$\bar{Y}$= 34,43 + 0,43 X_3	0,48	27,04	23,01
4	X_1, X_2, $X_3$and Y	$\bar{Y}$=12,44 + 0,26 X_1 + 0,16 X_2 + 0,14X_3	0,92	99,99	85,24
	Note:	Significant and Linear	Significant		

From the result of the testing of the first hypothesis was found the linear regression equation $\bar{Y}$=36.74 + 0.34 X_1 in which $F_{obs.}$= 39.88 (sig.=0.000) so that F is significant, and it can be concluded that the variable of school climate gives a positive effect on the students' science learning achievement. Thus, the null hypothesis which states that there is "no positive and significant contribution of school climate to science learning achievement" is rejected. This shows that there is a positive and significant contribution of school climate to science learning achievement.

A good school is characterized by a condition of the people in the school who create together a harmonious and conducive atmosphere, obeying the rules consistently, and give appreciation to the people of the school who have a good achievement, both in the academic and nonacademic fields. Due to the significant enough contribution of school climate to science learning achievement, it is a must and a proper thing for the school to always pay attention and create a school condition which enables the students to learn peacefully, and comfortably that in the end the students can enhance their learning achievement in general and in science in particular.

Magen-Nagar *et al.*, [7] states that one of the ways to improve the quality of education that has been mostly studied and practiced is through improving school climate. They have proven that students can reach a better learning achievement when they are in a comfortable, secured, conducive and pleasant school climate. The teachers too can present a good performance, committed to make the school make a better progress and enhance the students' learning achievement.

From the result of of the testing of the second hypothesis was found the linear regression equation $\bar{Y}$=27.58 + 0.28 X_2 in which $F_{obs.i}$=31.01 (sig.=0.000) so that F is significant, and it can be concluded that the variable of achievement motivation gives a positive effect on the variable of students' science learning achievement. Thus, the null hypothesis which states that there is no positive and significant contribution of achievement motivation to science learning achievement is rejected. This shows that there is a positive and significant contribution of achievement motivation to science learning achievement is rejected. This shows that there is a positive and significant contribution of achievement motivation to science learning achievement.

This study finding is also in line with the study done by Kala et al., [6] that the students with a high achievement motivation tend to take a mild measured risk, like the work situation that requires personal responsibility, want to increase knowledge about good work procedure, investigate the environment, and use the existing resources and learn in a unique and innovative way. Each student shows an effort to be able to pursue a better learning achievement than their friends in the class.

Winkel [8] states that achievement motivation is the power in an individual to reach the highest possible learning achievement to satisfy his or her drive. Kokila *et al.,* [9] states that achievement motivation is reflected in the orientation toward the goal and dedication in order the goal can be achieved well. Furthermore, it is stated that an individual with a high achievement motivation likes a job which challenges his or her expertise and his or her ability to solve problems. The power of one's achievement motivation depends on the amount of success expectation. Achievement motivation in this case is defined as a drive to get the result of an activity as well as possible in an effort to reach a goal, an effort to get superiority, and an effort to avoid failure.

From the result of of the testing of the third hypothesis was found the linear regression equation $\bar{Y}$= 34.43 + 0.43 X_3 in which $F_{obs.}$= 15.91 (sig.=001) so that F is significant, and it can be concluded that the variable self-concept gives a positive effect on students' science learning achievement. Thus, the null hypothesis that states that there is no positive and significant contribution to science learning achievement is rejected. This shows that there is a positive and significant contribution of self-concept to science learning achievement.

Human conceptions about him self or herself influences his or her choice of behavior and expectation in his or her life. Djaali [10] believes that self-concept is an individual's whole perspective about the physical dimension, characteristics, personality, motivation, weaknesses, intelligence, and failure Suarni [11] defines self-concept as a set of expectations, and behavioral assessment that refer to the expectations.

Othman *et al.*, [5] states that the difference in self-concept between high achieving students and low achieving students causes the difference in looking at the achievements they have achieved. The students who look upon themselves negatively take the success that they got just as something that occurs by chance.While, the students who look upon themselves positively will take the success as an indication of their ability. In addition, the students who have a positive self-concept will recognize themselves from the point of view of their weaknesses and strengths so that they can design goals that are suitable for them and in keeping with the reality, thus will be wise in determining their careers toward the future. On the other hand, the students with a negative self-concept will have a pessimistic attitude toward competition so that will hesitate to use their strengths and weaknesses

From the result of of the testing of the third hypothesis was found the linear regression equation $\bar{Y}=12.44 + 0.26\ X_1 + 0.16\ X_2 + 0.14\ X_3$in which $F_{obs.} = 15.27$ (sig. = 0.000) so that F is significant, thus it can be concluded that simultaneously, the variables of school climate, achievement motivation, and self-concept give a positive effect on students' science learning achievement. Thus, the null hypothesis which states that there is no positive and significant contribution simultaneously of school climate, achievement motivation, and self-concept to science learning achievement is rejected. This shows that there is a positive and significant contribution simultaneously of school climate, achievement motivation, and self-concept to science learning achievement.

Santrock [3] states that the supporting factors in enhancing learning achievement are school climate, achievement motivation, and self-concept. Suastra [12] states that science naturally, has two dimensions, i.e., science as a process and science as a product. The series of activities or mental processes have to be activated to produce knowledge. To activate it, it needs a strategy that contains procedural meanings known as scientific method,that consists of various mental actions, work patterns, procedures and technical ways to obtain and develop the gain of knowledge.

Student's success in achieving the goal of learning such as in achieving learning achievement is determined by internal and external factors. The internal factors, i.e., achievement motivation, aptitude, and other intelligence entities. The external factors, i.e., environment, school climate, school system, and the teachers', parents' and peers' motivational powers. One of the factors with the most support to science learning achievement is instructional approach or method implemented by the teacher in the instructional process. Othman *et al.,* [5] states that instructional method is one way which is used by the teacher based on various considerations such as the instructional goal, learner characteristics, facilities and environment that are meant to create conducive learning condition, so that the students can learn joyfully.

An important element which has to be considered in the instructional process is the aspects of school situation and condition.

These aspects and condition of school have play an effective role in giving extrinsic motivation to the students. A good school condition is needed very much since it gives a significant enough effect to the comfort, peace in learning and increases the students' enthusiasm to learn inside and outside the classroom. The students who learn in a pleasant situation can become more highly motivated to achieve. The orientation of achievement motivation is a drive which comes from oneself to reach success as a reward for achieving something. Besides, school climate actually has a direct contribution to the instructional process at school, in the form of the development of students' character, which includes among others, the students' self-concept. Self-concept is a comprehensive view of oneself, both about the personal characteristics, life values, life principles, life values, morality, the weaknesses and potentialities which are formed from experiences and interactions with others which can help the individual in self actualizing himself or herself freely and responsibly in attaining a particular goal as what is expectation.

4. Conclusion

First, there is a positive and significant effect of school climate on science learning achievement in the seventh grades at MTs Al-Khairiyah Tegallingah as shown in the linear regression equation $\bar{Y}=36.76 + 0.34\ X_1$ with $F_{obs.}= 39.88$ sig. 5%.

Second, there is a positive and significant effect of achievement motivation on science learning achievement in the seventh graders at MTs Al-Khairiyah Tegallingah as shown in regression linear equation $\bar{Y}$=27.58 + 0.28 X_2 with $F_{obs.}$= 31.01 sig. 5%.

Third, there is a positive and significant effect of school climate on science learning achievement in the seventh grades at MTs Al-Khairiyah Tegallingah as shown in the linear regression equation $\bar{Y}$=34.43 + 0.43 $X_{1\,X}$with $F_{obs.}$= 15.91 sig. 5%.

Fourth, there is a positive and significant effect of school climate, achievement motivation, and self-concept on science learning achievement in the seventh grades at MTs Al-KhairiyahTegallingah as shown in the linear regression equation $\bar{Y}$=12.44 + 0.26 + 0.16 X_2+ 0.14 X_{3X} with $F_{obs.}$= 15.27 sig. 5%.

In the light of these findings, and discussion, it can be concluded that school climate, achievement motivation, and self-concept give a positive and significant effect simultaneously to the science learning achievement of the students of the seventh grade at MTs Al-Khairiyah Tegallinggah both partially and simultaneously.

Based on the findings of this study, some suggestions can be made, among others, the results of this study shows that school climate, achievement motivation, and self-concept can be used as predictors in an effort to enhance the students' learning achievement.

For this purpose, schools should give support to the creation of a comfortable and conducive school climate through supervising the teachers and the administration staffs at the schools in order they perform their tasks and obligations optimal the role of various parties, i.e., the students, the parents and the teachers, are needed very much. The teachers and parents are expected to act as motivators for the students at school and at home, both in relation to the teaching and learning activities and the parents' guidance at home. For other researchers who would like to do a study of the same type as this study it is expected that they develop this study further by using a larger sample.

References

[1] Mubarrak L. 2009. The Web-Based Learning Model on Dynamic Fluid Concept to Improve Student's Science Generic Skills *Paper Presented in Proceeding International Seminar on Science Education* Universitas Pendidikan Indonesia.

[2] Balitbang 2012 Survei International TIMSS Available at http://litbangkemdiknas. Net. Accesed on 7th July 2017

[3] Santrock J W 2008 *Psikologi Pendidikan: Edisi Kedua* (Jakarta: Kencana Prenada Media Group)

[4] Trianto 2010 *Mendesain Model Pembelajaran Inovatif, Progresif, Konsep, Landasan, dan Implementasinya pada Kurikulum Tingkat Satuan Pendidikan (KTSP)* (Jakarta: Kencana Prenada Media Group)

[5] Othman N and Leng K B 2011 The Relationship between Self-Concept, Intrinsic Motivation, Self-Determination and Academic Achievement among Chinese Primary School Students *International Journal of Psychological Studies* **3** 90-98

[6] Kala P C and Shirlin P 2015 A Study on Achievement Motivation and Socio Economic Status of College Students in Tirunelveli District *International Journal of Research Granthaalayah* **5** 57 - 64

[7] Magen-Nagar N and Azuly D 2016 The Contribution of School Climate and Teaching Quality to the Improvement of Learning Achievements, According to an External Evaluation System *Creative Education* **7** 1773 - 1784

[8] Winkel 1999 *Psikologi Pendidikan dan Evaluasi Belajar* (Jakarta: Gramedia)

[9] Kokila K and Kalaiselvi 2017 A Study on Achievement Motivation among Higher Secondary School Students in Pudukkottai Educational District *International Journal of Informative & Futuristic Research* **4** 7588 - 7593

[10] Djaali 2011 Psikologi Pendidikan (Jakarta: Bumi Aksara)

[11] Suarni, Ni Ketut. 2004. Meningkatkan Motivasi Breprestasi Siswa Sekolah Menengah Umum di Bali dengan Strategi Pengelolaan Diri Model Yates (Studi Kasus Eksperimental pada Siswa Kelas I SMU di Bali). *Dissertation* (unpublished). Jogjakarta: UGM.
[12] Suastra, I W. 2013. *Pembelajaran Sains Terkini: Mendekatkan Siswa dengan Lingkungan Alamiah dan Sosial Budayanya*. Singaraja: Universitas Pendidikan Ganesha.

The development of *smoky glass box* as a physics instructional medium on light subject

S Hartini, M Syahrirani, and A Salam

Physics Education Program, Universitas Lambung Mangkurat, Jln. Brigjen H. Hasan Basry, Banjarmasin, Indonesia

E-mail: srihartini_pfis@unlam.ac.id

Abstract. We had conducted a research and development (R&D) about one of the physics instructional media namely *smoky glass box* on the light subject. This research aimed to produce proper *smoky glass box* instructional medium based on the validity, practicality, and effectiveness of the media.The type of research is research and development which refers to the ADDIE model. The subjects of the try out were the eight grade of SMPN 9 Banjarmasin. The instruments of this research were validation sheet, questionnaire, and learning outcome test. The results showed that: (1) the category of validity media is highly valid, (2) the category of practicability media is good, and (3) the category of effectiveness media is high. Based on the results of the research *smoky glass box* instructional medium is proper to use in learning physics.

1. Introduction

Education is a conscious and planned effort to create a learning atmosphere and process so that learner can actively develop his/her potential to have spiritual strength, self-control, personality, intelligence, character, and skills that needed by himself/ herself, nation and state society [1]. Education is the representation of a nation progress. Qualified education is supported by an effective learning system. An effective learning is a learning that allows learners to learn easy and fun and can achieve the learning objectives that have been set [2]. Effective learning cannot be separated from the role of teacher, learning condition, involvement of learner, facilities/media/resources/learning aids, classroom size/ condition/atmosphere and effective time [3].

Instructional media play an important role in the learning process. They can convey messages and stimulate students' thoughts, feelings, and willingness on the learning processes [4]. Instructional media can make students' attitude more positive, encourage their self-motivation, demonstrate associated factors and ideas, highlight specific topics and concepts, encourage relevance and credibility, and enhance understanding [5]. Through the instructional media, the obscurity of the material can be clarified, especially for sciences that have a high level of abstraction [6]. Props is one type of instructional media [7]. Props are instructional media that contain or bring the concepts studied [8]. They can change the abstract teaching materials become more concrete so that it can help students understand the concept of the material [9]. Therefore, the use props as instructional media is an important aspect of the learning process of science studies that have a high level of abstraction.

Based on the results of interviews at SMP Negeri 9 Banjarmasin, the instructional media used, especially the visual aids in the light-matter is less practical to use. The available props cannot properly display the light phenomena in experimental activities. This makes teachers struggling to use props.

One of many efforts to overcome the problem is through the development of instructional media in the form of props. Developed aids should be able to display the phenomenon of light well so that students can obtain the understanding of the material taught maximally. Props play an important role as a tool to help the teaching-learning process so that communication can work well and effectively [10]. This is in line with [11] the use of props will make it easier for students to understand a concept, in this case, instructional media of *smoky glass box* is developed to teach light teaching material.

Based on the background, a research and development of *smoky glass box* instructional medium on the Light subject was done for junior high school students. The purpose of this research is to produce proper *smoky glass box* instructional medium on the Light subject for junior high school students based on validity, practicality, and effectiveness of the instructional media developed.

2. Method

2.1. Type of Research

This research is a development research with ADDIE development model (Analysis, Design, Development, Implementation, and Evaluation) [12]. This *smoky glass box* instructional medium was tested to 35 students of the eight grade of SMP Negeri 9 Banjarmasin as the subjects of this research. The instruments used were validation sheet, questionnaire, and learning result test.

2.2. Data Analysis Technique

2.2.1. Validity. Validation was done by two media validators by using a validation sheet. The validity in this study is calculated based on equation (1) [13].

$$Validity = \frac{X}{X_{max}} \times 100\% \qquad (1)$$

Where are X = the total of the obtained scores and $X\,max$ = maximum score. The validity category of the instructional media is shown in Table 1 [13].

Table 1. Validity category for the instructional media

The score of validity (%)	Category
$86 < x \leq 100$	Highly Valid
$71 < x \leq 85$	Valid
$56 < x \leq 70$	Moderately Valid
$41 < x \leq 55$	Poorly valid
$25 < x \leq 40$	Not valid

The percentage of agreement from Borich on equation (2) was used to calculate the reliability value of the validation results of two validators [14].

$$PA = 1\text{-}\frac{A\text{-}B}{A+B} \times 100\% \qquad (2)$$

Where are A = larger validator score; B = smaller validator scores. The results of the calculation of the reliability are described in the form of categories in Table 2 [15].

Table 2. Reliability category

Reliability coefficient (r)	**Category**
$0.80 \leq r$	High
$0.40 < r < 0.80$	Moderate
$r < 0.40$	Low

2.2.2 Practicality. Instructional media practicality was counted by using equation (3).

$$P = \frac{\sum K}{\sum N} \times 100\% \quad (3)$$

Where are P = students response percentage, $\sum K$ = sum of obtained score, and $\sum N$ = sum of maximum score. The results of the instructional media practicality is described in the category of Table 3 [15].

Table 3. Practicality category of instructional media

The mean of the practicality (%)	**Criteria**
$20.00 < x \leq 36.00$	Poor
$36.01 < x \leq 52.00$	Not Good
$52.01 < x \leq 68.00$	Moderately Good
$68.01 < x \leq 84.00$	Good
$84.01 < x \leq 100.00$	Very Good

The reliability of practical instructional media is analyzed using equation (4) [15].

$$r_{11} = \left(\frac{k}{k-1}\right)\left(1 - \frac{\sum \sigma_b{}^2}{\sigma_t{}^2}\right) \quad (4)$$

Where are r_{11}= reliability ; k= number of item; $\sum \sigma_b{}^2$= number of item variance; $\sigma_t{}^2$= total variance. The variance value on equation (4) is given in equation (5).

$$\sigma = \frac{\sum X^2 - \frac{(\sum X)^2}{N}}{N} \quad (5)$$

Where are X= total score; N= number of respondents. The results of the calculation of realizability are described as in Table 2 [15].

2.2.3 Effectiveness. The effectiveness of the instructional medium was analyzed by using equation (6).

$$\langle g \rangle = \frac{S_f - S_i}{S_{max} - S_i} \quad (6)$$

Where are S_f = mean score of *posttest*, S_i = mean score of pretest, and S_{max} = maximum score that can be achieved.

Before analyzing using equation (6), the data obtained should meet the prerequisite test of Normality test. The data is normal if sig (p) > 0,05 [16]. The result of the instructional media effectiveness is described in the category of Table 4 [17].

Table 4. Category of the effectiveness

Effectiveness	**Category**
$\langle g \rangle > 0.7$	High
$0.3 \leq \langle g \rangle \leq 0.7$	Moderate
$\langle g \rangle < 0.3$	Low

3. Result and Discussion

3.1 Product of the research and development

The result of the development of *smoky glass box* instructional medium is shown in Figure 1. The developed instructional media can display fourth experimental activities: light travels in a straight line, the law of reflection, the law of refraction, and shadow formation on mirrors and lenses. *Smoky glass box* developed uses a glass box that serves as a smoke confinement. The confined smoke in the box is aimed to make the fired light through the smoke show a propagation line. This is because smoke is one of the colloidal particles that can cause Tyndall effects. If the light laser/flashlight used in the experiment can show the light propagation, then the light phenomenon can be displayed better than by using experimental devices in the school. The developed instructional media can display more real light phenomenon so that the Light subject that is delivered becomes more concrete. This is in accordance with [9] who stated that props can change the abstract teaching materials become more concrete so that it can help students understand the concept of the material that is being taught.

Figure 1. Result of the development of *smoky glass box* instructional medium

3.2 Validity

Validation of *smoky glass box* instructional medium consists of three aspects, namely suitability, quality and usability (ability). The suitability is categorized as highly valid. The quality aspect is categorized as highly valid. The usability aspect is categorized as highly valid. The overall validity of *smoky glass box* instructional medium is categorized as highly valid. The reliability coefficient is categorized as high. According to [18] the higher level validity of a teaching media, more authentic it becomes. Based on [19], the developed instructional media must meet the needs, and the components must be based on science and all components must be consistently connected to each other. If the product meets these requirements, then the product is valid.

Table 5. Validity result of learning media

Aspect	Mean (%)	Category
Suitability	92.00	Highly Valid
Quality	87.60	Highly Valid
Usability	93.75	Highly Valid
Validity	90.95	Highly Valid
Mean of the validity	91.08	Highly Valid
Reliability	0.99	High

3.3 Practicality
The practicality of this medium consists of three aspects namely: benefit, efficiency, and convenience. The aspect of benefit is categorized as good. This shows that benefits of the instructional media developed can be perceived by students. The aspect of efficiency is categorized as good. This shows the instructional media is efficient when used by the students. The aspect of convenience is categorized well. This shows the developed instructional media is easy to use by the students. The overall practicality of *smoky glass box* is categorized as good. The reliability coefficient is categorized as high. This shows that *smoky glass box* developed is practical. Students can use the instructional media that is developed practically when experimenting. According to [20], practicality is seen from whether or not the props can be applied/used in the classroom when the learning process takes place. According to [21], practicality means that the props that are produced are easy to use by the user, in this case, the student. Practicality refers to the level that users consider the props can easily be used for them in a way that is largely in accordance with what the researchers intended [22].

Table 6. Practicality result of instructional media

Aspect	Mean (%)	Category
Benefit	79.10	Good
Efficiency	70.94	Good
Convenience	78.12	Good
Mean of the practicality	76.05	Good
Realiability	0.84	High

3.4 The effectiveness
The effectiveness of the media was analyzed by using N-gain test. Before analyzing using N-gain, the normality test was done. The result of the normality test of the pretest data (0.123) and the posttest data (0.130) are significance. The results of this test show that the pretest and posttest data are normally distributed because (p)> 0.05. The result shows that the N-gain score is categorized as high. This shows that smoky glass box can help students achieve learning objectives and meet the feasibility of developing an instructional media on the aspect of effectiveness. According to [22], instructional media will be effective if it gives a positive impact on the learning process. The effectiveness of an education product that indicates the characteristic of the high-quality product is a product that can give the desired result [19].

Table 7. The effectiveness of instructional media

Score			$\langle g \rangle$	Category
Pretest	Posttest	Max		
3.09	76.89	100.00	0.76	High

The developed props can be functioned as instructional media that can support the learning process, give information, generate students' motivation, and improve students' understanding of the materials that are taught. Understanding that is achieved from directly try or do through practice is better than just listening and looking. In accordance with the opinion of [23], learning activities will be better if the learning subject experience directly or do so through practice to more effectively build attitude, skills, critical thinking and others when compared with learning memorization course that is verbalize. The *smoky glass box* instructional medium can have an effect on achieving the expected learning objective or competence because it can be directly perceived by the students. In line with the opinion of [24], visual aids can help students understand better the existing material and provide a real experience to the students. The result can help students achieve better learning achievement.

4. Conclusion

The product produced in this research is *smoky glass box* instructional medium on the Light subject for junior high school students. This media is proper to use in the learning process. This is supported by the experimental results of the *smoky glass box* instructional medium that is categorized as highly valid, the good category of practicality, and high category of effectiveness.

The researchers suggest for further development that the design of the *smoky glass box* props need to be developed so that the props are more interesting. It is also suggested that the material for the box of glass material can be developed from the fiber glass material so that it is not easily broken and it is portable.

Acknowledgment

The authors thank Faculty of Teacher Training and Education, Universitas Lambung Mangkurat due to the funding of DIPA-2017 for this research.

References

[1] *Undang-Undang Nomor 20 Tahun 2003 tentang Sistem Pendidikan Nasional* (Jakarta : Penerbit Sinar Grafika)
[2] Mansur H R 2015 *E-Buletin* **4** 1-12
[3] Syaodih E 2007 *Educare* **1** 1-25
[4] Anhar K and Zuhri S 2013 *Jurnal Pendidikan Teknik Elektro* **2** 575-582
[5] Onasanya S A 2004 *Institute Journal of Studies in Education* **2** 127-133
[6] Mediawati E 2011 *Jurnal Penelitian Pendidikan* **1** 61-68
[7] Sharon E S, Deborah L L and James D R 2014 *Instructional Technology & Media For Learning* (New Jersey: Pearson Education Inc)
[8] Prasetyarini A, Fatmaryanti S D and Akhdinirwanto R W 2013 *Radiasi* *2* 7-10
[9] Yulianti, Eni, Zulkardi, Siroj and Rusdy A 2010 *Jurnal Pendidikan Matematika* **4** 25-32
[10] Yensy N A 2012 *Exacta* **1** 24-35
[11] Wicaksoni H T, Kurniawan E S and Maftukhin H A 2013 *Radiasi* **3** 142-144
[12] Tegeh I M and Kirna I M 2013 *Jurnal IKA* **1** 12-26
[13] Akbar S 2016 *Instrumen Perangkat Pembelajaran* (Bandung: PT Remaja Rosdakarya Offset)
[14] Viana R V and Subroto 2016 *Jurnal Pendidikan Fisika* **5** 311-319
[15] Ratumanan T G and Laurens T 2006 *Evaluasi Hasil Belajar yang Relevan dengan Kurikulum Berbasis Kompetensi* (Surabaya: Unesa University Press)
[16] Nisfiannor M 2009 *Pendekatan Statistik Modern untuk Ilmu Sosial* (Jakarta: Salemba Humanika)
[17] Hake R R 1998 *Am. J. Phys* **66** 64-74
[18] Widodo O, An'nur S and Mahardika A I 2017 *Berkala Ilmiah Pendidikan Fisika* **1** 18-28
[19] Akker J V, Bannan B, Kelly A E, Nieveen N and Plomp T 2007 *Proc. Sem. Conducted at the East China Normal University* (Shanghai) (Netherlands: SLO/Netherlands Institute For Curriculum Development) p 37-50
[20] Semadiartha I K S 2012 *Jurnal Penelitian Pascasarjana Undiksha* **1** 1-16
[21] Widiyahti U N, Suprapto E and Adamura F 2015 *JIPM* **1** 1-12
[22] Mahyuddin R S, Wati M and Misbah 2017 *Berkala Ilmiah Pendidikan Fisika* **2** 229-240
[23] Rejanti A and Prabowo 2012 *Inovasi Pendidikan Fisika* **1** 94-99
[24] Sulistyarsi A 2016 *Premiere Educandum: Jurnal Pendidikan Dasar dan Pembelajaran* **2** 21-37

Character building in physics learning for Indonesia children

I M Astra

Universitas Negeri Jakarta, Rawamangun Muka, Jakarta Timur, 13220

E-mail: imadeastra@gmail.com

Abstract. This paper aims to develop the character of Indonesia children in physics learning. Indonesia now face moral crisis that is very dangerous and threatening for young generation. Violance, sexual abuse, and corruption occur more frequently. This condition needs solution to make Indonesia better. The most effective solution is character building reflected Indonesia character that is known as nobility, tolerance, and religious. Character education is important to grow since childhood to build the young generation representing living value of Indonesia. Method used in this study is literature study from various references Mardapi (2011) states that character education is part of affective domain. Unfortunately, affective domain gets the least concern if it is compared to cognitive and psychomotor domain. Therefore, affective domain needs more concern in every learning activity so education character can be applied properly. A physics teacher can build good character well if he understands the characteristic of subject and applied appropriate learning model that is suitable for students characteristic and school properties. Students' character that can be developed by applying problem possing and collaborative learning model are creative thinking, critical thingking, logical thinking, honest, good behavior, team work, and tolerance.

1. Introduction

In this globalization era, technological breakthrough is developing fast so it creates competitions in various aspects of our lives, where education is one of that. Education is a way to build qualified human resources. Therefore, it is an obligation to increase the education quality.

Increasing the quality of human resources needs a good learning strategy because it is expected to improve the education system. This is supported by government by introducing act no 20/2003 about education system. The objective of the act is stated on article 3 "The National Education functions to develop the capability, character, and civilization of the nation for enhancing its intellectual capacity, and is aimed at developing learners' potentials so that they become person imbued with human values who are faithful and pious to one and only god; who possess morals and noble character; who are healthy, knowledgeable, competent, creative, independent; and as citizens, are democratic and responsible". According to that statement, it is clear that education is not olny focus on cognitive and psychomotor domain but also on affective domain that concerns on character building. Bloom separates education aim into three domains, namely: cognitive, affective, and psychomotor domain [1]. Cognitive domain focus on activities related to intellectual capacity, such as: knowledge, understanding, and thinking skill. Thinking skill on cognitive domain involve remembering, understanding, apllying, analyzing, creating, and evaluating. The second domain, psychomotor, focus on physical ability involving human body. The last domain, affective, is related to interest and attitude. The objectives of all domain can be achieved by responsibility, teamwork, commitment, discipline, honesty, confidence, respect, and self-controlling.

Indonesia now face moral crisis that is very dangerous and threatening for young generation. Violance, sexual abuse, and corruption occur more frequently. This condition needs solution to make Indonesia better. The most effective solution is character building reflected Indonesia character that is known as nobility, tolerance, and religious. Character education is important to grow since childhood to build the young generation representing living value of Indonesia.

Mardapi states that character education is part of affective domain [2]. Unfortunately, affective domain gets the least concern if it is compared to cognitive and psychomotor domain. Therefore, affective domain needs more concern in every learning activities so education character can be applied properly.

Physics is branch of science concerned with the nature and properties of matter and energy [3]. In physics learning, there are a lot of topics that can be used for character building. For example, when students learn about heat transfer, they can also learn about justice, giving, and solidarity by observing the properties of heat and reflecting to the rich and the poor interaction [4].

Based on the problem above, the research problem is "How to build character in physics learning for Indonesia children?".

2. Literature review

2.1 Character

Character etymologically comes from greek language "kasairo", which means "blue print", "fundamental shape", "fingerprint". In this condition, Character is something given or something that being natural behavior of human in their groups. However, the terminology of character is ambiguous[5]. Mourier proposes two ways of interpretation for character. First, character is a set of condition that is taken for granted, or something that have been exist on human. In other words, this can be assumed as given. Second, character can be understood as level of ability that can be mastered by individuals [6]. That character is called a willed process. Another expert define character as shown below:

Character determines someone 's private thoughts and someone's actions done. Good character is the inward motivation to do what is right, according to the highest standard of behaviour, in every situation" (Hill, 2002).

Character education teaches how to think systematically and have a good behavior that help individuals to live and work together as family, society, and citizen and make responsible decisions. Character that have been standard is found in The Six Pillars of Character which is introduced by *Character Counts Coalition (a project of The Joseph Institute of Ethics).* The six characters mentioned above are:

a. *Trustworthiness,* a kind of character that makes individuals have: integrity, honesty, and loyality.
b. *Fairness,* a kind of character that makes individuals become open minded and not to exploit other people.
c. *Caring,* a kind of character that makes individuals have a care to other people and social condition.
d. *Respect,* a kind of character that makes individuals have admiration and appreciation to other people.
e. *Citizenship,* a kind of character that makes individuals have awareness to obey the law and norms.
f. *Responsibility,* a kind of character that makes individuals have discipline and awareness to make certain tha particular things are done well

Character is similar with nature that is composition of every permanent behaviour so it becomes special feature that make an individual have a different with others. The character can be interpreted from behavior when individuals interact each other. The interaction contains psychological and ethic meaning. Based on psychology, character is a visible behavior representing theirself. While based on ethic, character is something that shows human value and trusted attitude, so charactered individual can present firm standpoint, kindness, commend, and trust. Character has a meaning in moral value, where the behavior is responsible and firm. Ministry of National Education develops culture education and nation character through Pusat Kurikulum as shown on the table below:

IOP Conf. Series: Journal of Physics: Conf. Series **1040** (2018) 012043 doi:10.1088/1742-6596/1040/1/012043

Table 1. Value and Description of culture education and nation character

Value	Description
1. Religious	Attitude and behavior that obey the obligation of religion, tolerance with others, live together peacefully
2. Honest	Attitude based on efforts to make him as a person who is always trustworthy on words dan actions.
3. Tolerance	Attitudes and actions that respect religious differences, ethnic, opinion, attitude, and other persons' actions that is different from him.
4. Discipline	Measures that show orderly and obedient behavior on various terms and regulations.
5. Work Hard	Behavior that shows earnest effort in overcoming various barriers to learning and duty, and complete the task as well as possible.
6. Creative	Think and do something to make way or new results from something that already have.
7. Independent	Attitudes and behaviors that are not easily depend on other people in completing tasks.
8. Democratic	A way of thinking, and acting based on equality between rights and obligations of himself and others.
9. Curiousity	Attitudes and actions that always strive for knowing deeper and wider than something which he learned, viewed, and heard.
10. Spirit of nationality	The way of thinking, acting, and being insightful placing the interests of the nation and state aboveself-interest and group.
11. Patriotism	How to think, behave, and do the show loyalty, caring, and high appreciation to the language, physical environment, social, cultural, economy, and politics of the nation.
12. Meritocracy	Attitudes and actions that drive him to produce something useful for society, and recognize, and respect the success of others.
13. Friendly	Actions that demonstrate happy speaking, get along, and work with others.
14. Peacefull	Attitude, words, and actions that cause people others feel happy and secure for their presence.
15. love reading	The habit of providing time for various reading a reading that gives him the virtue.
16. Environmental care	Attitudes and actions that always try to prevent damage to the surrounding natural environment, and developing efforts to improve the natural damage that has occurred.
17. social care	Attitudes and actions that always want to help to others and to the needy.
18. Responsibility	Attitudes and behavior of a person to carry out the taskand obligations, which he should have done, against oneself, society, environment (nature, social and cultural), the state and God Almighty.

2.2 Builiding of character education

The steps in character building according to Najib Sulham (2010: 20) in Sofan Amri, et al (2011: 43) are as follows:

a. Incorporating the concept of character in each lesson by:
   1) Implanting the value of goodness in children (knowing the good) instilling the silf-concept to each child when enter the lesson.
   2) Using a way that makes a child have a reason or desire to do good (desiring the good)
   3) Giving some examples to the children about the character being fault. For example, through stories with figures that are easy to understand students
   4) Developing a lovinf attitude. Rewarding to a child doing the good and Treating the children doing something bad by giving educated punishment
   5) Performing good action (acting the good). Character application in learning process during school.

b. Creating a slogan that is able to grow the habit in all the behavior of the school community
c. Continous monitoring is a manifestation of character building implementation.
d. Assessment of parents has a great role to build the character of the child.

2.3 Learning definition

Learning is an activity undertaken by teachers programmed in instructional design that creates a process of interaction between fellow learners, teachers with learners and with learning resources. Nana Sudjana states that learning is a process marked by a change in a person. Learning by Morgan in Agus Suprijono is a permanent change in behavior as a result of experience. One of the signs that a person has learned can be observed by the change of behavior within him. The behavioral changes include the cognitive domain, the psychomotor domain, and the affective domain [7] [8].

Learning can be done face to face or without face-to-face. The face-to-face learning is carried out in the classroom where the interaction between learners and teachers takes place. In face-to-face learning, learning can be carried out in a classroom by not requiring direct teacher and student interaction. Open learning can be helped by using technology to enable the development of mobile learning (m-learning). Along with the development of technology, m-learning can take advantage of android applications in learning activities [9] [10] [11].

3. Method

Method used in this study is literature study. Steps in this method are collecting and analyzing data. On the first step, various references discussing about character building and physics learning are collected. After that, the references are analyzed to construct this study [12].

4. Discussion

From the litarature review of character education development studies, character education can be designed in the curriculum, promoted in the community, facilitated in the classroom, given examples and role models. Some ways of character development that physics teachers can do in learning, first a teacher understands the basic competencies that must be achieved by students such as:

Basic competence 3.6 (knowledge): Applying the concept of static electricity and direct current electrical law. Basic competence 4.6: solving technological and engineering problems related to static electricity and direct current electricity [13]. After that the teacher made a concept map about the material that will be developed to achieve basic competence.

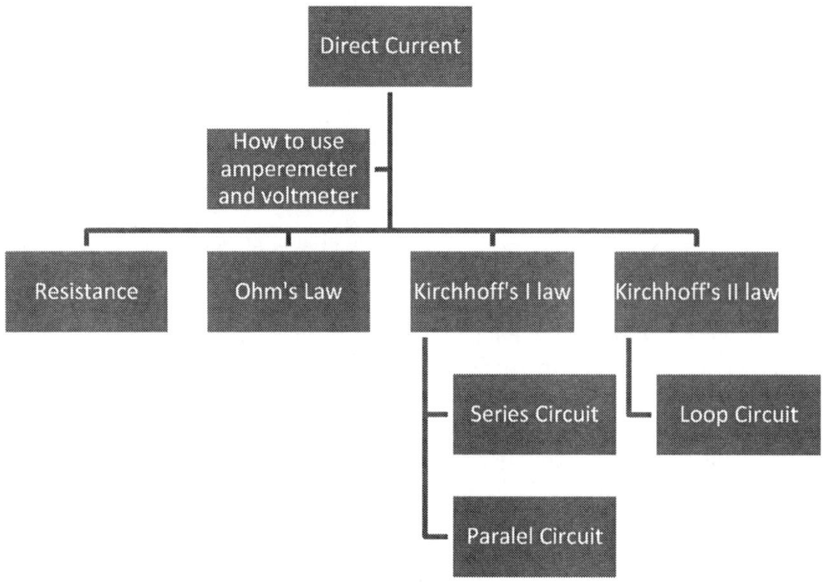

Figure 1. Mind Map

Then the teacher understands the characteristics of the students as well as the appropriate learning model to achieve Basic Comptetence (KD) as for basic competence above will be tried to be achieved with collaborative learning model of group investigation type, it is suitable for passive students and low learning outcomes.

Marjan and Ghodsi mentioned that collaborative learning is an educational approach to teaching and learning involving groups of learners working together to solve problems, complete tasks, or create a product [14]. Here the role of the teacher is to facilitate students in group discussions, as consultants when there is conflict within the group, as well as to observe progress processes in group activities.

In Group Investigation (Slavin, 2009), students in groups work out, execute, and report on problem solving outcomes. Students participate in discussing a problem or project with friends and teachers. Students also gain practical experience in giving or receiving criticism. It provides students with an intensive topic of study and gained specific knowledge on a topic. Through this technique it is hoped that their understanding of the importance of the invention will increase. The purpose of the study cited is actually to improve the process and the results of physics learning through the application of collaborative learning model of group investigation type. This model has been tested in SMA N 14 MIA class 10 obtained results shown only related to the characters that want to be developed as follows:

TABLE 2. Assesment Result of Student-Student Interaction

Aspect	**Cycle**		
	Cycle I	**Cycle II**	**Cycle III**
Team Work	61,11	65,97	76,39
Information exchange	45,14	64,58	75,69
Delivering opinion	45,14	64,58	75,69
Listening to others' opinion	45,14	64,58	75,08
Average	49,13	64,93	75,87

TABLE 3. Learning outcome of Affective Domain

Aspect	**Cycle**		
	Cycle I	**Cycle II**	**Cycle III**
Curiousity	61,11	65,97	76,39
Team work	45,14	64,58	75,69
Discipline	45,14	64,58	75,69
Responsibility	45,14	64,58	75,08
Care	49,13	64,93	75,87
Politeness	61,11	61,8	75
Activeness and	54,86	59,03	68,06
Responsiveness	61,8	61,8	75,69
Respect	57,55	61,63	70,225
Average			

TABLE 4. Learning Outcome of Psychomotor Domain

Aspect	**Cycle**		
	Cycle I	**Cycle II**	**Cycle III**
Observing	61,11	65,97	76,39
Questioning	45,14	64,58	75,69
Associating	45,14	64,58	75,69
Presenting	45,14	64,58	75,08
Average	49,13	64,93	75,87

From the results of research shows there are changes in the character of students during the classroom action research. Other examples that have been tried for active students as well as students who have moderate learning outcomes as well as schools that have a complete laboratory at SMA Labschool class II to:
Basic competence 3.7: Applying the laws of magnetism in everyday matters
Basic Competence 4.7: Demonstrate magnetism and electromagnet
concept map of material distribution made by teacher:

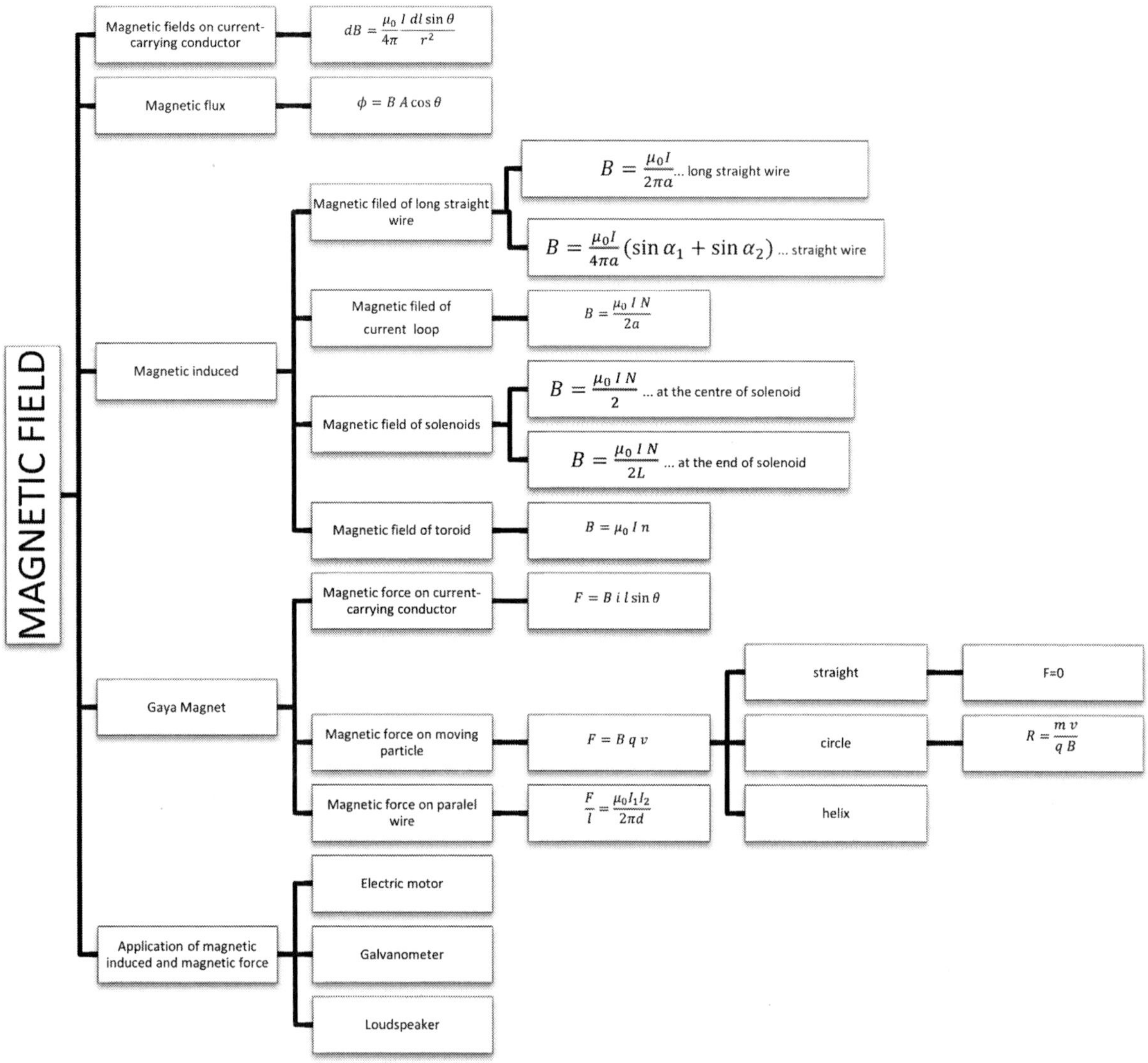

Figure 2. Mind Map

Basic competence will be achieved using Problem Posing model of Pre-Solution Posing type Implementation of Problem Posing Type Posing Pre-Solution Posing. Problem posing pre-solution posing type is one of the learning model that involves students actively in teaching learning process. This learning model requires students to make their own questions and answers based on the questions given by the teacher [15] [16]. Based on Silver's opinion [17], the application of learning model problem posing type of pre-solution posing is as follows:

a. Describing content
   The teacher explains the material to the student if necessary to clarify the concept of using, in this step the teacher provides the student with a code.
b. Illustrating the problems
   The teacher gives examples of problems, with problem posing model of pre-solution posing type that gives stimulus such as a picture, story, diagram, exposure and others, then students illustrate problem / describe problem given by identifying given stimulus.

c. Creating problems
Teachers provide practice with problem posing model of pre-solution posing type by relating problems related to their daily life.

d. Discussing problems
In this step, a teacher becomes a facilitator to guide the students to discuss to solve the problem. The facilitator or teacher only monitors and directs the course of teaching and learning activities, should not be involved in problem solving. It is important to cultivate students' confidence that they have the ability to solve problems themselves.

e. Discussing alternative solution
Teacher Discussed the assigned task with problem posing model of pre-solution posing type and teacher trains students to look for other possible questions from the stimulus provided. In this research, this model will be applied in physics learning.

According to the result of the research, it is found that there is influence of application of learning model of problem posing type pre-solution posing to physics learning outcome. In this study, the physics topic used is the magnetic field. In addition to applying this model in its entirety, to strengthen the concept that is taught also experiment, on this experiment students do experiment in group because of the limited tools. When students do experiment. Their character can be built. The character are carefully, honestly, responsible, not manipulate data, able to work together, mutual respect, polite, open minded, creative thinking, critical, logical and able to communicate the results of the experiment.

Based on the two results of classroom action research, it can be seen if a teacher wants to develop the character in the learning. The teacher must understand the characteristics of the subject and choose the appropriate model of learning, then determine the character that focuses to be developed. It is not apart from the characteristics of students and infrastructure facilities is in school.

5. Conclusion

From the results of the discussion above can be concluded that learning is an activity undertaken by teachers programmed in instructional design that creates a process of interaction between fellow learners, teachers with learners and with learning resources. In the implementation of the learning, teachers can use various learning models designed to the characteristics of the subject to instill character, including the application of problem posing model of pre-solution posing type, collaborative model of group investigation type, therefore this model can be used in physics learning in the classroom. Through this model of learning can also be developed character of students include creative thinking, critical and logical, work carefully, honest and polite behavior and social skills such as working together and mutual respect. The character of Indonesian children can be formed in physics learning.

References

[1] Bloom B S, Engelhart M D, Frost E J, Hill W H and Krathwohl D R 1956 *Taxonomy Educational objectives: Handbook I: Cognitive domain* (New York: Daved Mc Kay)
[2] Mardapi D 2011 *Penilaian Pendidikan Karakter, Pendidikan Karakter Dalam Perspektif Teori dan Praktik* (Yogyakarta: UNY Press)
[3] Giancoli D C 2001 *Fisika [Alih Bahasa : Yuhilza Hanum] Edisi Kelima Jilid 1* (Jakarta: Penerbit Erlangga)
[4] Martin M 1991 Science Education and Moral Education in *History, Philosophy, and Science Teaching*. Hal 102-11; ed. Michael Mathews (Toronto & NY: OISE Press, Teacher Collage Press)
[5] Doni K A 2007 *Pendidikan Karakter* (Gramedia Widisarana Indonesia)
[6] Mounier E 1956 *The Character of Man* Translate into English by Cynthia Rowland (New York: Harper dan Brothers)

[7] Dimyati and Mudjiono1999 *Belajar dan Pembelajaran* (Jakarta: PT. Rineka Cipta)
[8] Purwanto M N 2007 *Psikolog Pendidikan* (Bandung : PT Remaja Rosdakarya)
[9] Georgiev T, et al. 2004 M-Learning-A New Stage of E-Learning *International conference on computer system and technologies*
[10] Mahizer 2009 *Mobile Learning: New Era in Malaysia* Malaysia: Jurnal Pendidikan Dasar
[11] Samsudin A, et al. 2012 *Pengembangan "CELLS" Dalam Eksperimen Fisika Dasar Untuk Mengembangkan Performance Skills Dan Meningkatkan Motivasi Belajar Mahasiswa.* Semarang: Jurnal Pendidikan Fisika Indonesia
[12] Mulyatiningsih E 2011 *Metode Penelitian Terapan Bidang Pendidikan*. Bandung: Alfabeta
[13] Departemen Pendidikan Nasional, 2004, *Kurikulum Berbasis Kompetensi Standar Kompetensi Mata Pelajaran Fisika untuk SMA dan MA* (Jakarta: Departemen Pendidikan Nasional)
[14] Goodsell, et al. 2010 *Collaborative Learning A Sourcebook for Higher Education* (United States: National Center on Postsecondary Teaching, Learning, and Assesment)
[15] Dunker 2010 Problem posing-Adding a creative Increment to Technologi Problem Solving. journals/JTE.
[16] Lyn D E 1998 *Journal for research in mathematics Education* Vol.29. No.183.106
[17] Silver E A 1994 *On mathematical problem posing: For the Learning of Mathematicas* (Vancouver, British, Columbia, Canada: FLM Publishing Association)

IOP Conf. Series: Journal of Physics: Conf. Series **1040** (2018) 012044 doi:10.1088/1742-6596/1040/1/012044

Characterization the geometry of the peat soil of Pontianak using fractal method

J Sampurno[1], A Muid[1], Zulfian[1], F D E Latief[2]

[1]Department of Physics, Faculty of Mathematics and Natural Sciences, Tanjungpura University
[2]Department of Physics, Faculty of Mathematics and Natural Sciences, Institut Teknologi Bandung, Jalan Ganesa 10, Bandung 40132, Indonesia

E-mail: jokosampurno@physics.untan.ac.id

Abstract. Characterization of the microscopic pore structure of various types of peat soil of Pontianak has been done. The data consist of three types of peat (Sapric, Hemic, and Fibric), in which every type consists of three samples. The characterization was done in macro and micro level. At the macro level, samples were measured their physical properties namely bulk density, porosity and particle size distribution. In the micro level, samples were scanned using Micro-CT Scanner to reconstruct their 3D microstructure imagery. The images then analyzed using the fractal method to obtain their regularity level. The result shows that the regularity of pore distribution of each sample was varied. Yet, based on their types, Hemic has the most irregular pore size distribution.

1. Introduction

Pontianak is the capital of West Kalimantan Province of Indonesia, which has area 107.8 km^2 [1]. This city is located in the Kapuas river delta region. The land of the city is dominated by mud, gravel, sand, and plant debris [2]. The area of Pontianak is mostly covered by peat soil, which occupies 51.42% of the total area of the city [3].

Peat soil is a type of soil that rich in organic content [4,5,6]. The organic content of peat soil can reach more than 75% [5]. Peat soil can also be defined as fibrous soil resulting from decomposition and decay of microscopic and microscopic flakes of plants [7]. Because it consists of weathering fibers, this causes the porosity of the soil to be generally high in size [8].

Based on the level of organic maturity, peat soil can be classified into three types: Fibric, Hemic and Saprik [9]. The maturity level of peat soil is based on the content of organic fibers. Peat soil is classified as Fibric if the fiber content is greater than 67%, classified as Hemic if its fiber content between 33% - 67% and classified as Sapric if its fiber content less than 33% [9]. Fibric is a type of peat that most of its material has not decomposed and generally reddish brown. Hemic is a peat that about half its material has decomposed and generally dark and reddish. Sapric is peat which most of its material has been decomposed, normally has a darker color than the other types, and it contains the highest organic element than the other [10].

The previous works show that fractal geometry has been successfully used to interpret the 2D image of peat soil structure [11-13]. Fractal geometry is a geometry that constructed by repetition and sequence of simple geometric shapes that are replicated repeatedly and combined with each other on a variety of scales. Fractal geometry has self-similarity characteristic which shows that fractal consist of parts that are similar in shape to each other.

In this study, the physical properties of peat soil are observed based on its type. The objective of the present study is to compare the geometry of 3 types of peat soil of Pontianak using fractal method. For that purpose, analysis of peat samples was conducted in two levels, macroscopically and microscopically. At macroscopic level observation, the analysis was done on two parameters namely bulk density and porosity. While at the microscopic level, the analysis process is performed on the 3D image of peat soil that is reconstructed using Micro-CT Scanner. Furthermore, fractal analysis was conducted in order to compare the physical properties of peat soil with its pore structure.

2. Method

Peat soil samples were taken from three locations (L, P and S). The first location (L) is located at coordinates 00° 01' 29.2" N and 109° 22' 26.4" E. The second location (P) is located at coordinates 00° 04' 08.7"S and 109° 19' 19.5" E. The third location (S) is located at coordinates 00° 04' 26.2" S and 109° 20' 18.9" E (See Figure 1).

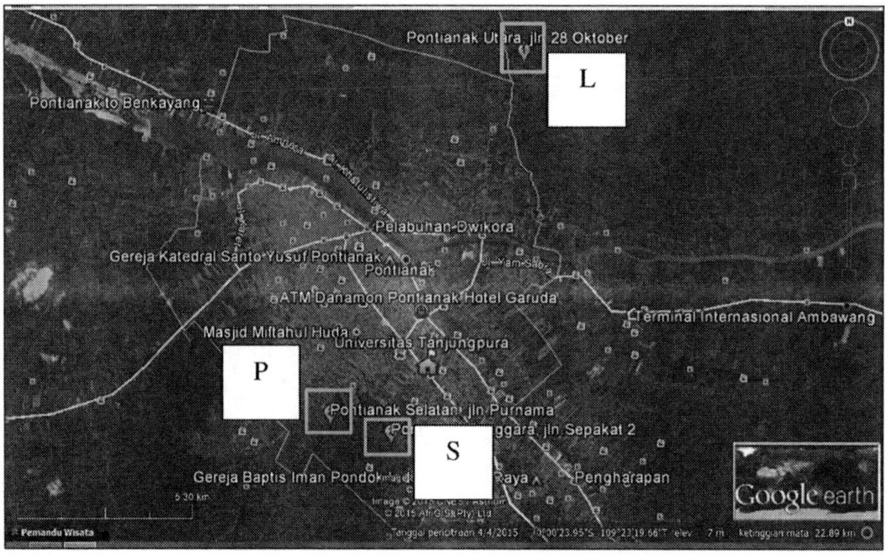

Figure 1. Location of peat soil samples[14]

For every location (L, P and S), three samples were taken from different depths. The number of peat soil samples were taken is 9 (3 samples from each location). Samples are labeled according to the location and the type of the peat soil. For example, sample from location L is labeled with L1 for Sapric, L2 for Hemic and L3 for Fibric. This procedure is also carried out on samples from P and S sites. The depth of all peat soil samples is presented in Table 1.

Table 1. The depth and the type of all samples.

Location	Sample code	Depth (cm)	Type
L	L1	0 – 10	Sapric
	L2	30 – 40	Hemic
	L3	70 – 80	Fibric
P	P1	10 – 20	Sapric
	P2	40 – 50	Hemic
	P3	90 – 100	Fibric
S	S1	10 – 20	Sapric

S2	40 – 50	Hemic
S3	90 – 100	Fibric

2.1. Macroscopic Level

At the macroscopic level, laboratory tests are performed to determine the porosity and the bulk density of the peat. To determine the bulk density of the soil, firstly the dry weight of the sample was measured. Then, the bulk density is determined from the ratio between dry weight and sample volume. The volume is calculated using the cylindrical volume equation $\pi r^2 t$, where r is the radius and t is the height of the sample tube.

Porosity (ϕ) is the ratio between empty space (including pore space, fractures, cracks, etc.) in the sample and total sample space. Thus, porosity (ϕ) is calculated based on the volume of pore space (V_p) and the total volume of sample (V_t) using the following equation:

$$\phi = \frac{V_p}{V_t} \times 100\% \qquad (1)$$

2.2. Microscopic Level

At the microscopic level, the sample is reconstructed into a digital image. Digital images are reconstructed using Micro-CT Scanner Skyccan 1173. First, peat soil samples are made in the cubes form (2cm x 2cm x 2cm). Then, the sample is scanned to produce three-dimensional images (3D) which has a pixel size 300x300x300 pixels^3. Figure 2 shows the result of peat soil reconstruction in RGB for each sample with image resolution 28 μm.

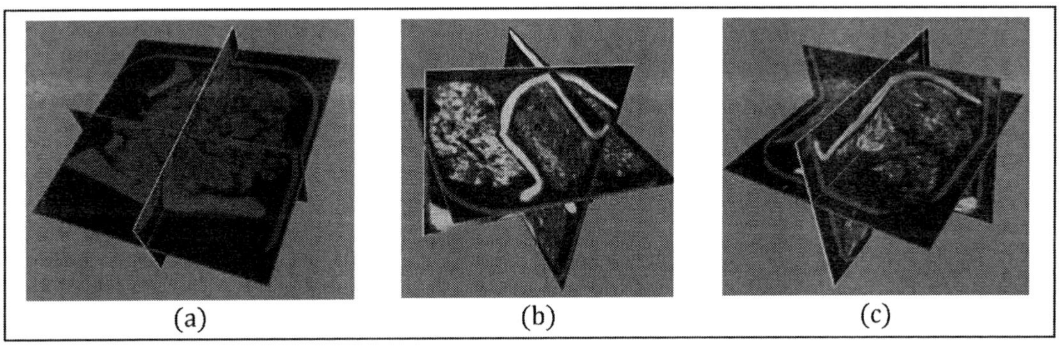

Figure 2. The Results of peat reconstruction in 28 μm resolution (a) Sapric; (b) Hemic and (c) Fibric.

Furthermore, the image is transformed into a binary image to separate between the pore and the matrix. Binary image is then analysed using software CTAn [15] to calculate the pore size distribution and analysed using fractal method to characterize the regularity level of soil pore structure. The calculation of pore size distribution of the peat, which using CTAn, has been done in 2 stages [16]. Firstly, a "skeletonisation" is performed to recognize the average axis of all structures. Next, a "sphere-fitting" estimation is made for all the voxels lying along the axis.

2.3 Fractal Analysis

Fractal dimension is calculated using Box-Counting method. In the Box-Counting method, the fractal dimension of a 3D image is calculated by [17]:

$$D = \lim_{r \to 0} \frac{\log(N(r))}{\log\left(\frac{1}{r}\right)} \qquad (2)$$

Where D is fractal dimension, N (r) is the number of grid boxes which contain pore, and *r* is the size of the box.

3. Result and Discussion

The result of the measurement of all samples shows that the bulk density of Pontianak peat is varied. At location L, the bulk density of the peat ranges from 0.10 g/cm^3 to 0.23 g/cm^3. At location P, the bulk density of peat ranges from 0.08 g/cm^3 to 0.21g /cm^3. At location S, the bulk density of the peat ranges from 0.06 g/cm^3 to 0.20 g/cm^3.

Yet, commonly the bulk density of the peat depends on its maturity and its compaction. It can be seen at the Table 2 that the bulk density of peat from the largest to the smallest are Sapric, Hemic and Fibric respectively. The more mature the peat soil, the greater the bulk density. However, in general the bulk density of peat soil is relatively low. This causes the bearing capacity of the soil is also low.

From the Table 2, it can be also seen that porosity is affected by maturity and compaction processes. The less mature of the peat, the greater of its porosity. On the other hand, the greater of the bulk density, the smaller of its porosity and vice versa. Peat porosity from the smallest to the largest is shown by Saprik, Hemik and Fibrik respectively. The porosity of all peat samples ranged from 77.23% to 95.18%.

Table 2. The bulk-density and porosity of the peat of Pontianak

Sample Code	Type	Bulk density (g/cm^3)	porosity (%)
L1	Sapric	0,23	77,23
L2	Hemic	0,13	88,85
L3	Fibric	0,10	92,12
P1	Sapric	0,21	79,38
P2	Hemic	0,11	90,08
P3	Fibric	0,08	93,62
S1	Sapric	0,20	80,20
S2	Hemic	0,11	90,21
S3	Fibric	0,06	95,18

Next, peat samples were reconstructed using Micro-CT Scanner SkyScan 1173. In the scanning process, the scanning parameters used were: current source: 50 μA, voltage source: 55 KV, illumination time: 500 ms, and resolution: 28 μm. The reconstruction result is displayed in 3D grayscale image, which has a size of 300x300x300 pixels^3. Figure 3-5 show the reconstruction results for all peat samples, with white as granules.

(a) (b) (c)

Figure 3. 3D Reconstruction of peat samples from location L, where: (a) Sapric; (b) Hemic; and (c) Fibric

(a) (b) (c)

Figure 4. 3D Reconstruction of peat samples from location P, where: (a) Sapric; (b) Hemic; and (c) Fibric

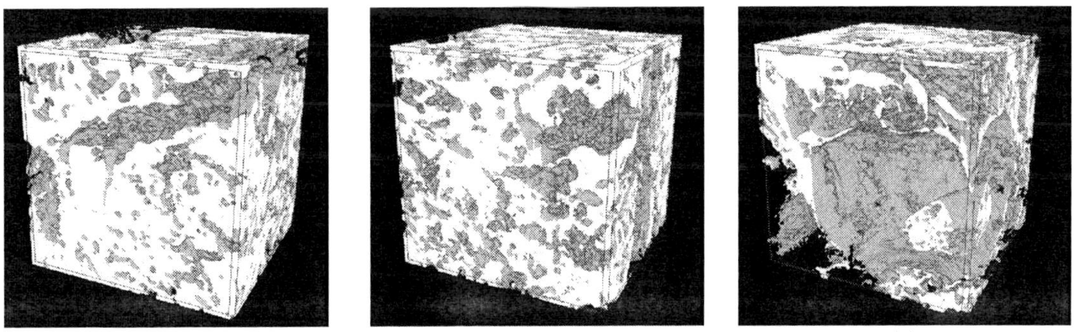

(a) (b) (c)

Figure 5. 3D Reconstruction of peat samples from location S, where: (a) Sapric; (b) Hemic; and (c) Fibric

Table 3. Fractal Dimension of all samples.

Sample Code	**Type**	**Pore size (pixel)**	**Fractal Dimension**
L1	Sapric	2 - 34	2,90
L2	Hemic	2 - 66	2,73
L3	Fibric	2 - 70	2,70
P1	Sapric	2 – 76	2,69
P2	Hemic	2-84	2,58
P3	Fibric	2 - 132	2,69
S1	Sapric	2 – 56	2,58
S2	Hemic	2 - 50	2,15
S3	Fibric	2 - 100	2,76

From Table 3, it can be seen that the fractal dimension of the 3D image of Pontianak peat ranges from 2.15 to 2.90. The fractal dimension closer to three indicates that the pore distribution is more regularly, and vice versa. According to the Table 3, Hemic has the smallest fractal dimension when compared to Sapric and Fibric. It indicates that the distribution of Hemic pore is the most irregular when compared with other peat types. This condition is because Hemic is a kind of semi-mature peat, which means some of its material has been decomposed and some has not. This causes the distribution of pore size is more varied.

The results of digital image reconstruction can also be used to calculate the peat pore size distribution. Figure 6 - 8 show the results of the calculation. From the figure 6-8, it can be seen that the largest pore size interval belongs to Fibric, followed by Hemic and Sapric. Sapric has the smallest pore size interval because it is the most decomposed and most mature type of peat. So that, most of its pore has almost similar size.

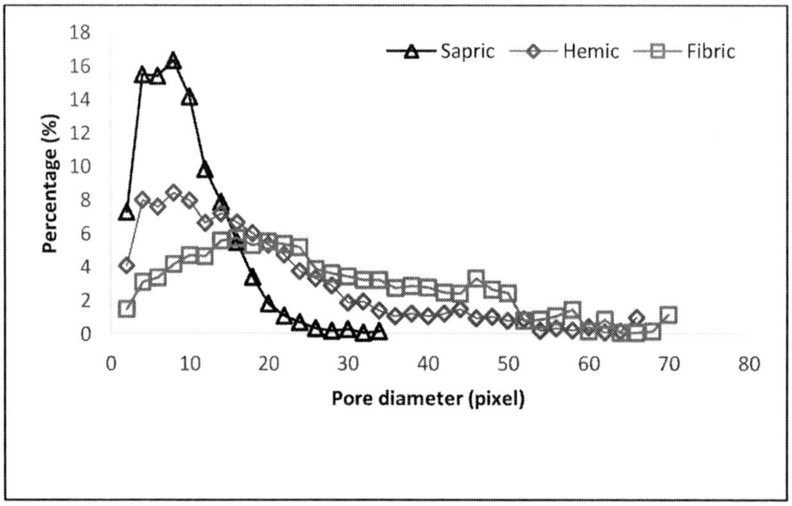

Figure 6. The distribution of peat pore size at location L

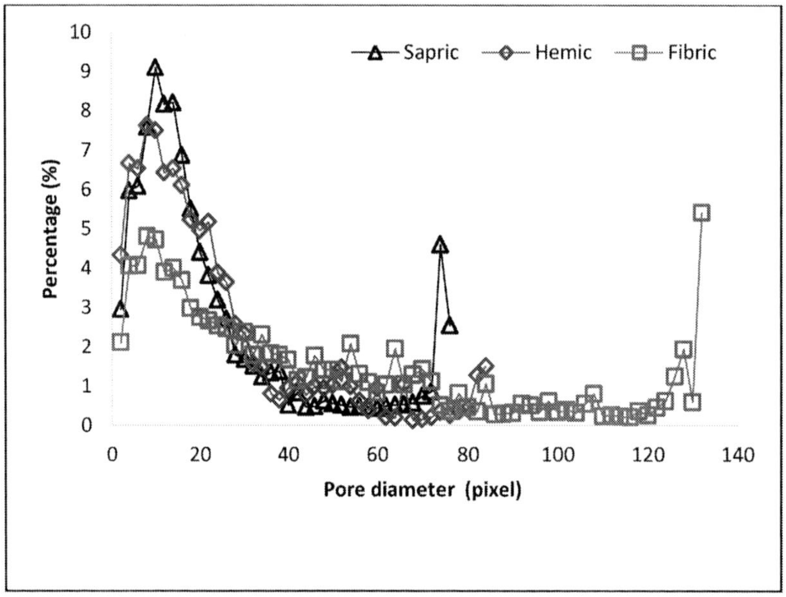

Figure 7. The distribution of peat pore size at location P

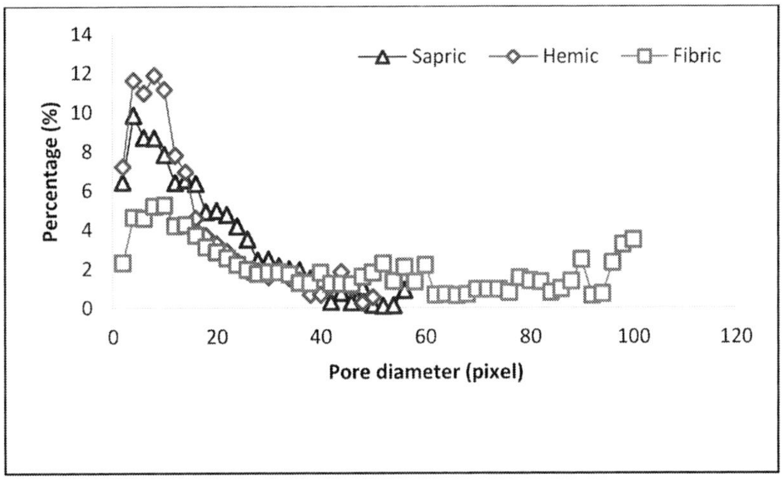

Figure 8. The distribution of peat pore size at location S

4. Conclusion

From the data above, it can be concluded that the porosity of all types of samples from the highest to the lowest belongs to Fibric, Hemic, and Sapric respectively. Next, the bulk density of all types of samples from the highest to the lowest belongs to Sapric, Hemic, and Fibric respectively. Similar to that, the range of pore size distribution of all types of samples from the widest to the narrowest also belongs to Sapric, Hemic, and Fibric respectively. Based on their fractal dimension, the regularity level of pore distribution of each type sample depends on the location. Yet, Hemic almost has the most irregular pore size distribution.

Acknowledgments

This research is supported by Directorate General of Higher Education, Ministry of Research, Technology and Higher Education (Indonesia), through DIPA of Faculty of Mathematics and Natural Sciences, Tanjungpura University.

References

[1] Gandasasmita K, Munibah K, Widjaja H, Darea, Anshari G Z, Muslihudin R, Achsani H, Hastuti S, Yanwijaya D and Soli T I 2012 *Masterplan Pengelolaan Ekosistem Gambut Berkelanjutan Provinsi Kalimantan Barat* (Indonesia: Kementerian Lingkungan Hidup)
[2] BPT 2011 *Pengelolaan Lahan Gambut Berkelanjutan* (Bogor: Balai Penelitian Tanah)
[3] Dinas Urusan Pangan Kota Pontianak 2001 *Profil Agribisnis Aloe Vera di Kota Pontianak. Provinsi Kalimantan Barat* (Pontianak: Dinas Urusan Pangan)
[4] Murtedza M, Padmanabhan E, Mei B L H and Siong W B 2002 *The peat soil of Serawak* (Sarawak: STRAPEAT status report University Malaysia Sarawak) pp 16-20
[5] Zulkifley M T M, Ng T F, Raj J K, Hashim R, Ghani A, Shuin M K and Ashraf M A 2013 *Bull. Eng. Geol. Environ.* **72** 547–553
[6] Agus F and Subika I G M 2008 *Lahan Gambut Potensi untuk Peretanian dan Aspek Lingkungan* (Bogor: World Agroforestry Center (ICRAF))
[7] Terzaghi K and Peck R B 1967 *Soil Mechanics in Engineering Practice* (New York: John Wiley & Sons)
[8] Dunn I S, Anderson L R and Kiefer F W 1980 *Fundamentals of geotechnical analysis* (New York: Wiley)
[9] ASTM 1989 D4427 *Annual Book of Standart: Soil and Rock: Building Stones: Peats* vol 4.08
[10] Huat B B K 2004 *Organic and Peat Soils Engineering* (Serdang: Universiti Putra Malaysia Press) pp 146–150

[11] Sampurno J and Srigutomo W 2015 *AIP Conference Proceedings* **1656** (1), 070004
[12] Sampurno J, Faryuni I D, Latief F D E and Srigutomo W 2016 *Journal of Physics: Conference Series* **739** (1) 012071
[13] Sampurno J, Azwar A, Latief F D E and Srigutomo W 2016 *Journal of Mathematical and Fundamental Sciences* **48** (2) 106-114
[14] Google earth V 6.2.2.6613. (April 4, 2015). Pontianak, Indonesia. 0° 00' 23.95" S, 109° 23' 19.66"E, Eye alt 22.89 km. DigitalGlobe 2015. http://www.earth.google.com [April 26, 2017].
[15] http://bruker-microct.com/products/downloads.htm [April 30, 2017]
[16] Bruker microCT *Introduction to porosity analysis* Retrieved from http://umanitoba.ca/faculties/health_sciences/medicine/units/cacs/sam/media/MN059_PorosityAnalysis.pdf [December 13, 2017]
[17] Silva D D, Boudon F, Godin C, Puech O, Smith C and Sinoquet H 2006 *International Symposium on Visual Computing (ISVC)* p 754

IOP Conf. Series: Journal of Physics: Conf. Series **1040** (2018) 012045 doi:10.1088/1742-6596/1040/1/012045

Dielectric properties and bioactivity of PVA/PEG/TiO_2 fibers for capacitive based body sensor

M Diantoro[1,2], T Istirohah[1], A Fuad[1,2], and I Ristanti [1]

[1]Department of Physics, Faculty of Mathematics and Natural Sciences, Universitas Negeri Malang, Jl. Semarang 5 Malang, Indonesia
[2]Minerals and Advanced Materials Science Laboratory, Universitas Negeri Malang, Jl. Semarang 5 Malang, Indonesia

E-mail: markus.diantoro.fmipa@um.ac.id

Abstract. TiO_2 nanoparticles are known as a material which widely used in optoelectronic, industrial, and health sectors application. This material has a controllable dielectric constant making it possible to be utilized as a capacitive based body sensor. Many efforts had been studied by means of synthesizing TiO_2 nanoparticles in the form of conventional film depositions. It is also obtained that rare information of comprehensive study related to dielectric properties of the materials. In this report, we propose to synthesize PVA/PEG/TiO_2 fibers using electrospinning method at various TiO_2 concentration. Dielectric properties and bioactivity of samples are a focus of study. Electro spinning process has been performed with a voltage of 20 kV, and flow rate of 100 for 3 minutes. We used ITO glass as substrates. The FTIR spectra showed a successful synthesis of PVA/PEG/TiO_2 fibers. This is indicated by the appearance of the corresponding characteristics of PVA, PEG, and TiO_2 bonds. The XRD patterns showed the main peaks of TiO_2 as well as the diffraction peaks of PVA. We obtain that the average size of PVA/PEG/TiO_2 fiber is 268.27 nm. The dielectricity of PVA/PEG/TiO_2 fibers increased with the addition of TiO_2 nanoparticles. In addition, the increase of frequency effect to the decreased of dielectric constant. The toxicity test of PVA/PEG/TiO_2 fibers showed that the samples were non-toxic. On the other hand, the result of the anti-bacterial analysis showed that the samples had no anti-bacterial properties against *S. aureus* bacteria.

1. Introduction

Titanium dioxide (TiO_2) has been widely developed in both industrial and medical fields. The characteristic of TiO_2 which has been extensively studied and utilized is the optoelectric properties, which is transmittance and semiconductive. The dielectric constant of nanoparticles of TiO_2 was previously reported at 85 [1]. It was reported that the dielectricity of TiO_2 could be affected by several factors, including the mass of nanoparticles and particle size. Higher percentages of TiO_2 nanoparticle particles result in higher dielectric constant values [2]. Small particle size affects the number of particles per unit volume increased, resulting a high dipole moment per unit volume and dielectric constant. In addition to optical properties, bioactivity of TiO_2 nanoparticles also widely studied in some applications.

Bioactivity of TiO_2 nanoparticles is also investigated regarding its applications in the health field. The bioactivity properties of TiO_2 nanoparticles that have been studied are the toxicity and antibacterial properties. TiO_2 nanoparticles have antibacterial activity against e-Coli bacteria. It is reported that the higher of TiO_2 nanoparticles induce, the higher bacterial mortality rate in agar medium [3]. The pristine TiO_2 anatase nanoparticles (without coating) shows a toxic level. This nanoparticles demonstrate great appeal to proteins, which can damage cell mitochondrial function and make the decrease of cell survival [4-9]. Surface modification of TiO_2 nanoparticles with polyethylene glycol (PEG) can affect its toxicity properties. PEG coatings can reduce the toxicity of TiO_2 nanoparticles [10]. On the other hand, a non-toxic combination material with PEG is required to increase the water solubility level and to increase mechanical flexibility. PEG may act as a surfactant on the dissolution of TiO_2 particles as well as to decrease the particle agglomeration rate.

So far there are many reports on the dielectricity of TiO_2 thin film. But, it has not been comprehensively studied about TiO_2 thin films with PEG combinations and their bioactivity properties. So it is essential to investigate PVA/PEG/TiO_2 fiber synthesis by analyzing the dielectricity and bioactivity properties. In this report, we synthesized fiber by electrospinning method using polyvinyl alcohol (PVA) as fiber matrix and TiO nanoparticles as filler. A complete analysis of fiber structure was performed using XRD, SEM and FTIR characterization. Dielectricity of fiber were analyzed as a function of the mass of TiO_2 nanoparticles and light intensity, while fiber bioactivity was expressed by analysis of toxicity as well as antibacterial properties.

2. Method

This research was focused on the synthesis and characterization of dielectric response and bioactivity properties of PVA/PEG/TiO_2 fiber at the various mass of TiO_2 nanoparticles. The research was performed at Nanomaterial Laboratory of Universitas Negeri Malang, while the characterization and testing process was observed in Central Laboratory and Microbiology Laboratory of Universitas Negeri Malang also in Biomedical Laboratory of Muhammadiyah University of Malang. The synthesis of PVA/PEG/TiO_2 fibers was performed in two stages. First, by preparing the fiber solution using stirring and sonication method and followed by electrospinning process. The fiber material consists of PVA polymer as a matrix and TiO_2 nanoparticles as a filler [21].

The synthesis of PVA/PEG/TiO_2 fiber solution begins by preparing a solution of TiO_2 nanoparticles trough mixing aquades (1.3 mL), TMAH (11 μL), and PEG-400 (99.6 μL) at 90 °C. Next, we add a 0.1125 grams of TiO_2 nanoparticles into the previous mixture for further stirred at 90°C for 15 minutes. After that, the TiO_2 nanoparticle solution was sonicated for one h to obtain a more homogeneous solution (A). The next step is dissolving PVA in aquades (3 mL) which was stirred at 120 °C with a 9% PVA concentration in the total solution. The solution was distilled for 1 hour at 120 °C to obtain a homogeneous PVA solution (B). The third step is to dissolve (solution A) and (solution B) with stirring 700 rpm at 120 °C for 1 hour, and then mixing the solution for 2 hours to obtained (solution C). The synthesis was carried out at different mass of TiO_2 nanoparticles (0.225, 0.15, 0.1125, 0.09, and 0.075 g).

Furthermore, the electrospinning process begins with stirring (solution C) at 90 °C under 360 rpm for 15 minutes, followed by preparing the equipment such as putting ITO substrate on a plate and filling (solution C) on stirring. The controlled variable of electrospinning process is voltage (20 kV), flow rate (100), and runtime (3 minutes). The structure characterization of fibers is represented by its crystallinity using XRD, morphology investigated by SEM, and functional groups obtained from FTIR. The dielectricity test was performed by employing different frequency, and light intensity in each sample analyzed using LCR meter. The bioactivity properties of PVA/PEG/TiO_2 fibers were observed by antibacterial test with agar diffusion method use *S. aureus* bacteria and toxicity test by isolated organ method.

3. Results and Discussion

3.1 FTIR Spectra of PVA/PEG/TiO_2 Fiber

The FTIR spectra of the bonding groups of TiO_2 powder, PVA, PEG-400, and PVA/PEG/TiO_2 fibers, as well as the IR spectra of sample 1 to sample 5 are shown in Figure 1.

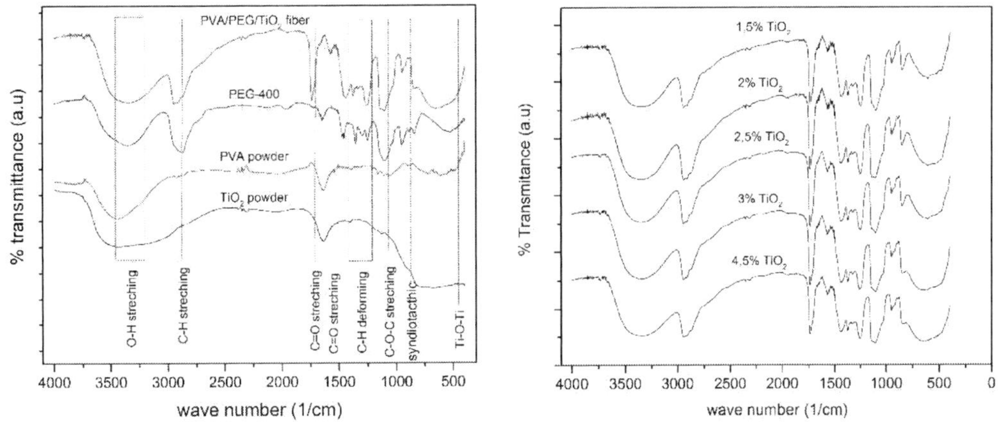

Figure 1. IR Spectra of PVA/PEG/TiO_2 Fiber

Figure 1a shows the IR spectra of TiO_2 nanoparticles that appearing in the range 450 cm^{-1} to 600 cm^{-1} indicating the presence of Ti-O bond vibrations in the TiO_2 [11-14]. Peak 1633,71 cm^{-1} and the peaks between 3100 cm^{-1} and 3600 cm^{-1} show the vibrations of the hydroxyl group (O-H) from the residual solvent water after synthesis. IR spectra of PVA/PEG/TiO_2 fiber in Fig. 1a is more similar with the peaks of PEG-400 and PVA. The IR spectra of PEG-400 at 845-900 cm^{-1} indicates the presence of stretching O-O [16], whereas the presence of 1109 cm^{-1} peaks shows symmetry of C-O-C in the aliphatic ether. Spectral peak at 2882 cm^{-1} indicates the presence of C-H alkyl is stretching, whereas 1352 cm^{-1} indicates the presence of deforming of C-H alkyl.

IR spectra of PVA showed peak about 3434 cm^{-1} that indicating the presence of polymer blend of free hydroxyl groups and vibrations of bound O-H band [16];[11]. PVA spectrum at ~ 1650 cm^{-1} of wave number indicates the presence of vinyl alcohol deformation, and this peak is the specific peak of PVA [16]. The peak between 3450-3230 cm^{-1} shows the strain band of the hydroxyl group O-H, whereas the peak of 2940 cm^{-1} and 2916 cm^{-1} is strains of the group bands CH_2 and CH_2. Peak 2836 cm^{-1} indicate the C-H stretching vibration, peaks at 1732 cm^{-1} and 1569 cm^{-1} show the stretching form of the remaining C=O carbonyl group of the acetate group [11];[16]. The peak of 1430 cm^{-1} shows the symmetry bond of CH_2, and the peak of 1375 cm^{-1} denotes a mixture of the C-H and O-H mode bonds and the combined alcohol. The 918 cm^{-1} peak shows the syndiotactic structure and exhibits strong vibration, while the 849 cm-^{1} & 608 cm^{-1} peaks indicate stretching of C-C vibration and OH [11].

The IR spectra of PVA/PEG/TiO_2 fiber almost all show the PEG-400 and PVA spectra with different shifts and intensities. At 3452 cm^{-1} there is a Ti-OH bond [11], resulting from the hydrogen bonding between OH and PVA molecules with Ti^+ titanium ions resulting in charge transfer. This complex charge transfer shows that with the increase of nano TiO_2 mass will increase the amount of charge. Figure 3b shows the variation of the mass of nanoparticles TiO_2 compared to pure PVA. A small shift is observed of OH, CH groups and Ti-O-Ti bonds. The OH group and CH group are observed at the broad peak of 3400 cm^{-1} and ~ 2800 cm^{-1} regions. The new peak formed at around 400-600 cm^{-1} represents the existing of Ti-O-Ti bond. These peaks indicate that the composite has created, and the peak at 514 cm^{-1} signifies the Ti-O-O bond [11]. The reduced mass of nanoparticles TiO_2 produces a lower %transmittance at 400-500 cm^{-1} region.

3.2 Diffraction Pattern and Crystal Size of PVA/PEG/TiO_2 Fiber

The crystal size of TiO_2 nanoparticles was analyzed using Scherrer Equation showing that the crystal size is 16.77 nm. While the results of Rietica calculations obtained the average size of 19.45 nm nanoparticles with (χ^2) is 1.175.

$$D = \frac{k\lambda}{\beta cos\theta} \quad (1)$$

The result of XRD analysis using Rietica software is shown in Table 1.

Table 1. The result of structure analysis of TiO_2 nanoparticles

Parameter	Result of analysis
a (Å)	3.7913
b (Å)	3.7913
c (Å)	9.5203
Crystal size (nm)	19.45
Rwp	25.44
Rp	17.83
χ^2	1.175

Figure 2. shown the diffraction pattern of TiO_2 powder, PVA fiber, and PVA/PEG/TiO_2 fiber.

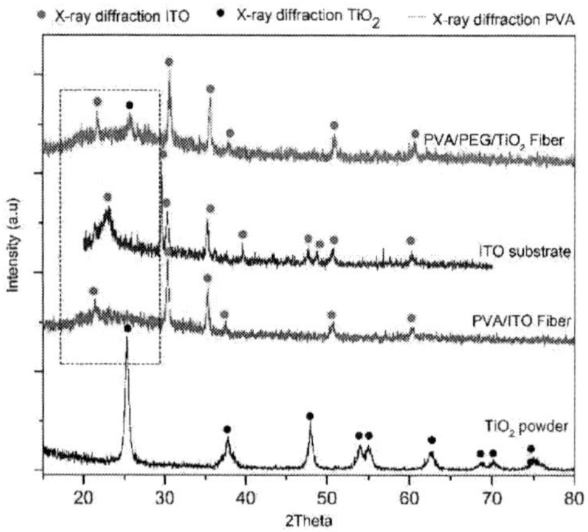

Figure 2. X-ray Diffraction Patterns

The diffraction peak of PVA appears at the 2θ from 15° to 30° that containing amorphous and crystalline phase [11]. The diffraction patterns of PVA/PEG/TiO_2 fibers exhibit a more dominant peak with PVA intensities, i.e., a broad peak in the 2θ region between 15° and 30°. The diffraction peaks of TiO_2 nanoparticles in PVA/PEG/TiO_2 fibers showed at 2θ = 25.307° which is the specific peak of the anatase TiO_2 nanoparticles. The diffraction peak of PVA/PEG/TiO_2 fibers indicates that TiO_2 nanoparticles have been successfully composed of fiber. The other peaks formed on PVA and PVA/PEG/TiO_2 diffraction patterns at 2θ = 20°, 30°, 35°, 50°, and 60° are the diffraction peaks of the ITO substrate that has been used [17].

3.3 Fiber Morphology Analysis using SEM-EDAX

The morphology of PVA/PEG/TiO_2 fiber can be analyzed using the result of SEM analysis (Figure 3a). ImageJ software analysis was implemented to obtain the average diameter of the fiber. We obtained that nanofiber is 268.2695 nm in diameter. The percentage element of TiO_2 nanoparticles in PVA/PEG/TiO_2 fibers can be determined from EDAX analysis (Figure 3b). The result of EDAX shows the presence of the TiO_2 peak and the conductive substrate used, i.e., ITO (Indium Tin Oxide). ITO peaks clearly appear because the arranged fiber in substrate still has large empty spaces, so the ITO substrate was also detected by electrons gun during characterization.

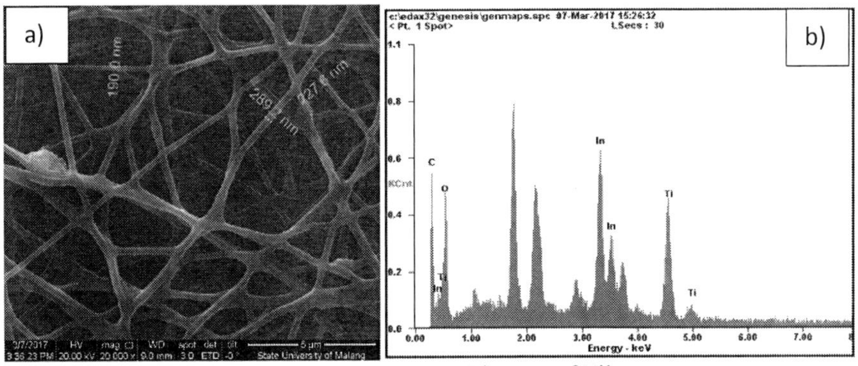

Figure 3. SEM-EDAX image of Fiber

The EDAX analysis is shown in Table 2. The percentage of TiO_2 nanoparticles by EDAX characterization for the total area obtained 0.47% of weight is much smaller than the %weight of EDAX of a single point. These results confirm that nanoparticles that have been composed of fiber are still in small quantities. While EDAX results for one selected point show the %weight of TiO_2 nanoparticles is 14.37%. We can assure that TiO_2 nanoparticles are entering the fiber.

Table 2. Chemical Element Percentage on the PVA/PEG/TiO_2 fiber

Element	Total area		One point	
	% Weight	% Atom	% Weight	% Atom
C	15.04	35.18	19.50	36.98
O	29.04	50.98	35.14	50.03
In	55.44	13.56	30.99	16.15
Ti	00.47	00.28	14.37	06.83

3.4 Toxicity of PVA/PEG/TiO_2 Fiber

The result of fiber toxicity testing of PVA/PEG/TiO_2 is shown in Figure 4. Toxicity testing was taken using the liquid sample. The observed data is the contraction curve of the marmot's small intestine (*ileum*) which may indicate the response toxic or nontoxic of the injected sample. The *ileum* with no treatment is shown in Figure 4a, indicating a constant contraction that means the *ileum* is alive, as well as when injected sample 1 to 5 (represented by sample 1) (Figure 4b). After washing, the *ileum* contractions can return as well as the initial condition. The ability of the *ileum* to contract again indicates that the *ileum* still alive and the sample is not toxic.

As a comparison, controlled toxicity testing was conducted on *ileum* marmot with an injection of Atrhopin drugs. Atropin drugs exhibit high toxicity when used in 100% concentrations. After the injection of Atropine, it appears that the *ileum* contraction rises drastically and then the curve cannot return to the initial condition (Figure 4c), this indicates that the *ileum* was death. This result then confirms by washing; the *ileum* cannot contract as in the initial condition before being treated.

From the toxicity test, we may conclude that the sample is non-toxic. From several reviews, it is known that the fundamental properties of nanoparticles TiO_2 are toxic, whereas the PEG-400 polymer serves as a surfactant and PVA that acts as a fiber matrix is non-toxic. In the liquid solution of PVA/PEG/TiO_2 composite fiber, the percentage of the TiO_2 nanoparticle is low, starting from 1.5% - 4.5% of the total weight. We cannot observe the toxic part from TiO_2 due to encapsulating of TiO_2 by PVA-PEG. The encapsulation using PVA-PEG causes the toxic effects of the TiO_2 nanoparticles cancelled.

IOP Conf. Series: Journal of Physics: Conf. Series **1040** (2018) 012045 doi:10.1088/1742-6596/1040/1/012045

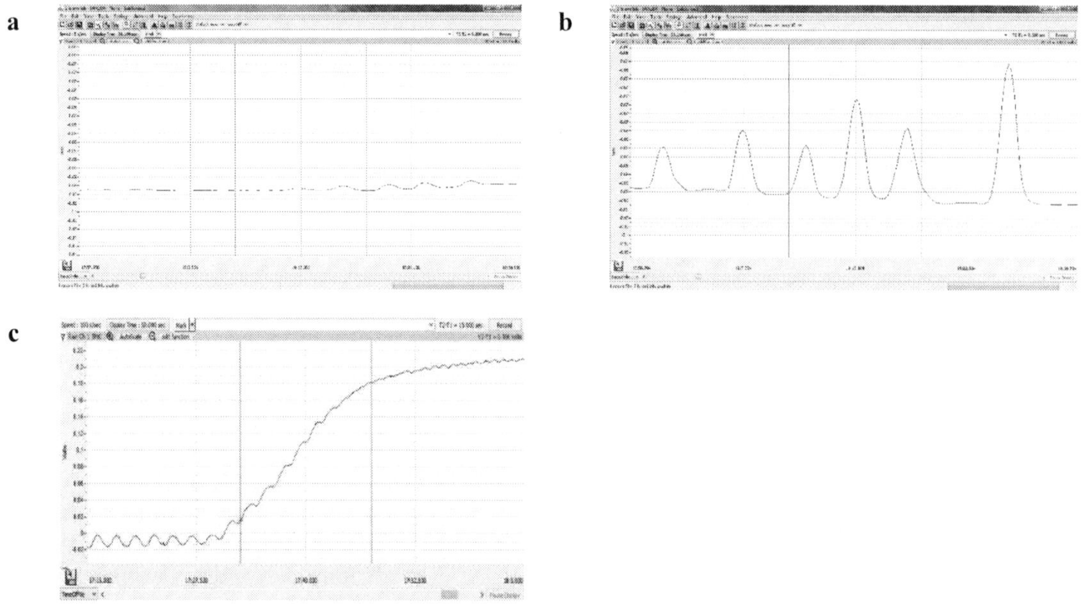

Figure 4. Toxicity Testing of PVA/PEG/TiO_2 Fiber

3.5 Antibacterial Testing of VA/PEG/TiO_2 Fiber

The antibacterial activity of material occurs from the interaction of the electric field between the material and the cell wall of the bacteria. In this study the analysis of anti-bacterial properties of PVA/PEG/TiO_2 fibers using *S. aureus* bacteria seen after 24 hours incubation. Selection the type of bacteria is adjusted with the application of fiber PVA/PEG/TiO_2 that applied to the skin, because of that used the kind of bacteria that often breed in human skin. In a previous study of TiO_2 nanoparticles anatase is to have antibacterial activity against E. coli bacteria.

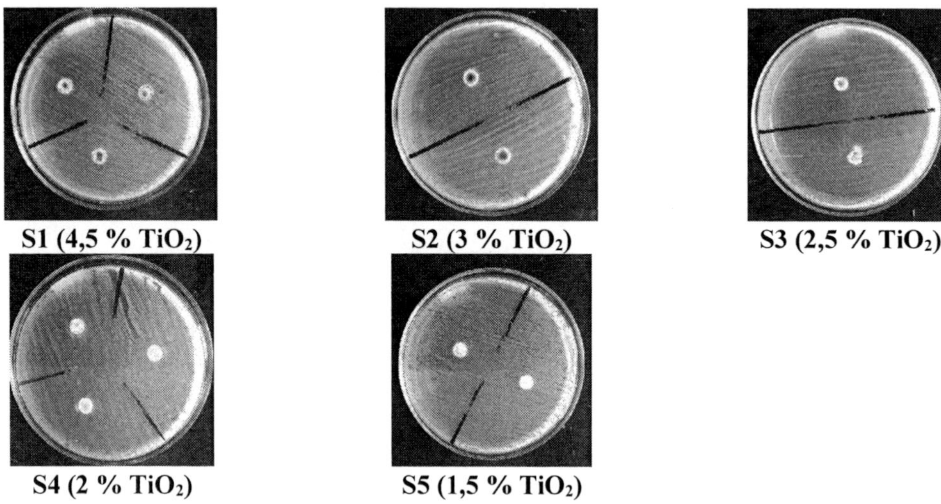

Figure 5. Antibacterial Testing

The test results in Figure 5. showed that an inhibited zone (transparent zone) is not formed around the sample. This result is caused by the type of Bacteria which is used in this research is S. aureus bacteria which are gram-positive bacteria with one layer of the thick cell wall with thickness (15-80 μm). While the bacteria E. coli is a gram-negative bacteria, which has a thin 3-layer cell wall (10-15 nm). So the results of this test prove that the nanoparticle TiO_2 is not strong enough to produce an electric field interaction with the cell wall of *S. aureus* bacteria that resulted in nanoparticles cannot penetrate the thick cell wall of bacteria.

3.6 Dielectric Constant of PVA/PEG/TiO_2 Fiber

The influence of frequency varies to the dielectric constant of PVA/PEG/TiO_2 fiber showed in Figure 6a. The graph in Figure 8b shows the effect of light intensity on dielectric constant PVA/PEG/TiO2 fiber. The value of the dielectric constant was obtained from the calculation of capacitance (C), with the sample area (A), and the sample thickness (d) using Equation 2.

$$\varepsilon = \frac{Cd}{\varepsilon_0 A} \tag{2}$$

Figure 6a describes the decreasing of dielectric constant with increasing frequency. The effect of frequency on the dielectric constant has been studied intensively. At low frequencies, the electrons can oscillate following the applied field, and the polarization mechanism can follow the applied field. However, at high frequencies, electron oscillations can not follow application field fluctuations and cause the polarization mechanisms cannot follow the applied field [18].

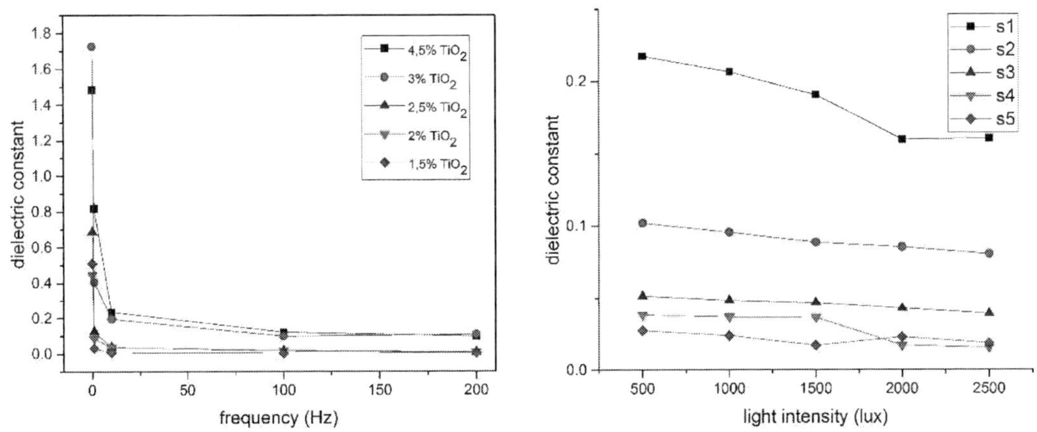

Figure 6. a) Frequency influence and **b)** Light intensity influence of PVA/PEG/TiO_2 Fiber

After some data treatment, we found the effect of light intensity on the dielectric constant of PVA/PEG/TiO_2 fibers as in Figure 6b. The higher light intensity decreases the dielectric constant. The dielectric constant of the sample is close to a semiconducting material. This is due to the interaction between light and sample. The intensity of light that exists in the sample gives the energy that causes in the excitation of electrons in the atom. This situation produces an atomic swelling that occurs as the radius of the atom gets bigger. As the atomic radius increases, the electron binding energy becomes weaker, allowing for the electrons to escape from their core.

Figure. 6a and 6b also explain the higher percentage of TiO_2 nanoparticle mass, have the higher the dielectric constant of PVA/PEG/TiO_2 fibers. The discussion in the previous section demonstrates that the dielectric constant of TiO_2 nanoparticles is 100 [19], 20-80 [20]. Because of the high dielectricity constant, the addition of a mass percentage of TiO_2 to PVA/PEG/TiO_2 fibers also increases the fiber dielectric constant.

4. Conclusion

We conclude that the addition of TiO_2 nanoparticles affects to PVA/PEG/TiO_2 fiber structure. The increase of TiO_2 nanoparticles induces the increase of IR spectra peak of Ti-O-Ti bond. We found that the sample shows a nontoxic material. It is obtained that the addition of TiO_2 nanoparticles did not improve the anti-bacterial properties of PVA/PEG/TiO_2 fibers for S. aureus bacteria. It is possible that the TiO_2 fraction is too small. In addition, this condition caused by a kind of bacteria is gram-positive which has a thick cell wall. We further obtain that the addition of TiO_2 nanoparticles increases the dielectric constant of PVA/PEG/TiO_2 fibers. The higher the intensity of light the lower the fiber's dielectricity. The range and the possible control of dielectric constant open to this material to be applied to capacitive based body sensor.

Acknowledgments
The authors would like to thank DRPM, Ministry of Research, Technology and Higher Education (Ristekdikti) Republic of Indonesia for supporting a research grant for MD.

References
[1] Hasan E S, Jahiding M, Hasrin I N, and Idris I Karakterisasi Lapisan Titanium Dioksida (TiO_2) Sebagai Bahan Dielektrik Sensor Kelembaban Jenis Kapasitif
[2] Kim Y J, Kim J, Kim Y S and Lee J K 2013 *Org. Electron.* **14** (12) 3406–3414
[3] Rahman Z and Singh V P 2014 *Int. Biodeterior. Biodegrad.* **91** 97–103
[4] Gerloff K, et al. 2012 *Chem. Res. Toxicol.* **25** (3) 646–655
[5] Jin C, et al. 2011 *Biol. Trace Elem. Res.* **141** 3–15
[6] Lin X, et al. 2014 *PloS One* **9** (10) p. e110247
[7] Liu H et al. 2009 *Biol. Trace Elem. Res.* **129** 170–180
[8] Numano T et al. 2014 *Asian Pac J Cancer Prev* **15** (2) 929–935
[9] Yu Q, Wang H, Peng Q, Li Y, Liu Z and Li M 2017 *J. Hazard. Mater* vol. 335 125–134
[10] Mano S S, Kanehira K, Sonezaki S and Taniguchi A 2012 *Int. J. Mol. Sci.* **13** (3) 3703–3717
[11] Shehap A M and Akil D S 2016 *Structural and optical properties of TiO_2 nanoparticles/PVA for different composites thin films*
[12] Luo Y L, Chen L L, Miao Y and Xu F 2013 *Ind. Eng. Chem. Res.* **52** (4) 1571–1580
[13] Ba-Abbad M M, Kadhum A A H, Mohamad A B, Takriff M S and Sopian K 2012 *Int J Electrochem Sci* **7** 4871–4888
[14] Hamadanian M, Reisi-Vanani A and Majedi A 2010 *J. Iran. Chem. Soc.* **7** (1) S52–S58
[15] Beranek R and Kisch H 2008 *Photochem. Photobiol. Sci.* **7** (1) 40–48
[16] Kuptsov A H and Zhizhin G N 1998 *Handbook of Fourier transform Raman and infrared spectra of polymers* **45** Elsevier
[17] Kim H et al. 2000 *J. Appl. Phys.* **88** (10) 6021–6025
[18] S. F. Mansour 2005 *Egypt J Solids* **28** (2) 263–273
[19] Halliday R 2001 *Fundam. Física* **2** p. 6
[20] Hubbard K J and Schlom D G 1996 *J. Mater. Res.* **11** (11) 2757–2776
[21] Nasikhudin, et al. 2016. *Preparation of PVA/TiO_2 Composites Nanofibers by using Electrospinning Method for Photocatalytic Degradation.* IOP Conference Series: Materials Science and Engineering.

IOP Conf. Series: Journal of Physics: Conf. Series **1040** (2018) 012046 doi:10.1088/1742-6596/1040/1/012046

New simple synthesis, crystal system and physical properties of Zn_5S_2 compound

A D Kusumawati[1], T Suprayogi[1], M Diantoro[1,3], I Dasna[2,3] and Subakti[1,3]

[1] Department of Physics, Faculty of Mathematics and Natural Sciences, Universitas Negeri Malang, Jl. Semarang 5 Malang 65145, Indonesia

[2] Department of Chemistry, Faculty of Mathematics and Natural Sciences, Universitas Negeri Malang, Jl. Semarang 5 Malang 65145, Indonesia

[3]Laboratory of Minerals and Advanced Materials 65145, Universitas Negeri Malang, Jalan Semarang 5 Malang, Indonesia

E-mail: subakti@um.ac.id

Abstract. We report a large-scale solid-state synthesis of non-stoichiometric Zn_5S_2 powders at considerably low (400 K) sintering temperatures. The DTA measurement has been carried out to reveal the temperature dependence phase transformation of the samples. Our best powder XRD refinement analysis so far shows that the studied compound belongs to monoclinic $P2_1m$ with a = 5.2538Å, b = 19,3137 Å, c = 3,4543 Å, β = 90.176°. The I-V measurement shows that our Zn_5S_2 samples to be potentially p-type semiconducting material. Thus, from our preliminary characterizations, we can conceive that these Zn_5S_2 could be potentially great p-type thermoelectric materials. The ongoing thermoelectric properties characterizations of these compounds are discussed in this contribution.

1. Introduction

Indonesia is the fifth most populous country in the world [1], the human population is proportional to energy demand [2]. Indonesia abounds with energy resources (steam coal, natural gas, oil). But, we face several challenges are increasing energy efficiency and using renewable energy. The gas sector should be further developed becomes renewable energy, especially geothermic [3]. Geothermic energy potential in Indonesia is estimated about 40% of the world geothermic energy potential. However, only approximately 4.5% is being utilized as energy. Another type of renewable-energy potentials such as hydro energy, wind energy, wave energy and biomass potential. But these sources depend on weather, climate and expensive [4].

Based on this reason, we need to increase quantity and quality of renewable energy. The thermoelectric generator is one of renewable energy alternative [5]. Thermoelectric can generate electric energy from thermal energy and vice versa. Therefore thermoelectric materials are crucial in renewable energy conversion technologies to solve the global energy crisis [6]. Recent studies about thermoelectric are Tellurium based, but Tellurium is rare, and toxic [7]. So, we need to synthesis non-Tellurium compound. We choice sulfide compound because of one of the most widespread on Earth's crust [8].

Thermoelectric technology is regarded as an alternative and environmentally friendly technology for harvesting and recovering heat which is directly converted into electrical energy [9]. In the case of thermoelectric materials, the thermoelectric figure of merit (ZT) can be defined as follows: $ZT = \alpha^2\sigma/\kappa$, which α = Seebeck coefficient, σ = electric conductivity, κ = thermal conductivity, and $\alpha^2\sigma$ = power factor. For ideal thermoelectric materials, ZT should be 1 to obtain a conversion efficiency of > 10% [6]

2. Research Method

Zn_5S_2 was prepared with high purity elements of Zinc (Zn) and Sulfur (S) as sources, which were weighed according to their respective stoichiometry. The reaction equation follows

$$5\ Zn_{(s)} + 2\ S_{(s)} \rightarrow Zn_5S_{2\,(s)}$$

Based on this equation, to s synthesis 7 grams of Zn_5S_2, we need 5.852 grams Zn, and 1.148 gram S. We synthesis this compound using solid-state sintering is given in Figure 1. After being uniformly mixed, the mixture was pressed into a pellet. The sample was heated at 130 °C for 14 hours, 140 °C for 19 hours and 38 hours. On heated process, we covered the sample by carbon active to pretend oxidation. The design of procedure of Zn_5S_2 synthesis is described in Figure 1. The scheme of sample structure during sintering process is illustrated in Figure 2.

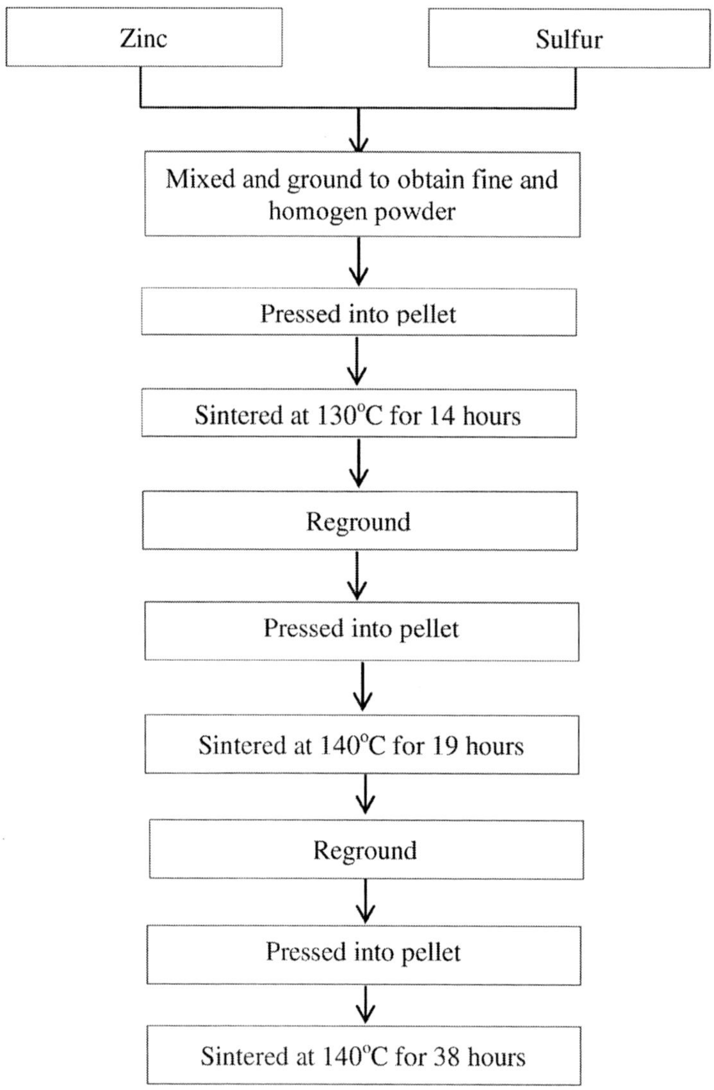

Figure 1 Procedure of Zn_5S_2 synthesis

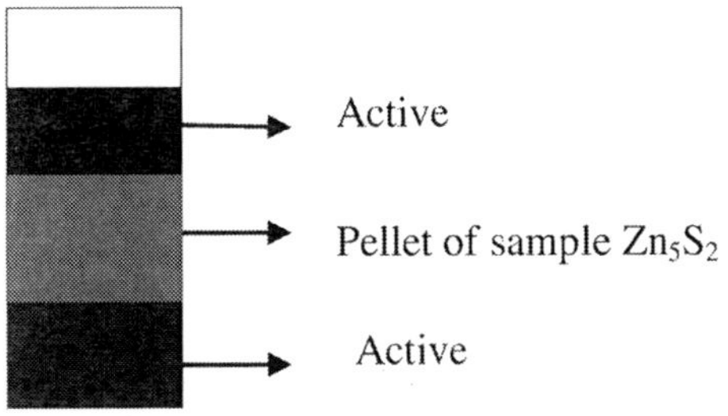

Figure 2. Schematic illustration of sample arrangement in the sintering process

We used Energy Dispersive X-Ray Spectrometer (EDX) to obtain the elemental composition, Differential Thermal Analysis (DTA) and Thermogravimetric (TGA) for thermal analysis, X-Ray Diffractometer (XRD) to analysis structure. We further measure electrical conductivity using I – V meter and characterize band gap using Ultraviolet-Visible Spectrometer (UV Vis). We further analyze X-RD pattern to search the possible crystal system using DICVOL under Fullprof software.

3. Results and Discussion

3.1 Composition Analysis

The composition analysis of this compound was measured by Energy Dispersive X-ray spectroscopy (EDX). The result of EDX is given in Table 1.

Table 1. EDX composition of Zn_5S_2

Compound	Composition (%)
Zn	49.98
S	19.51

Our analysis shows that the studied compound belongs to Zn_5S_2. This compound was suitable for the purpose of this research.

3.2 Thermal Analysis

The thermal analysis was investigated by Differential Thermal Analysis (DTA). There are two results of DTA measurement. First, measurement on 140°C with increment 1°C, the peak was found on 116°C. Second, measurement on 600°C with 5°C increment, the peaks were found on 119.21°C and 422.49°C. Both of these results are explained in Figure 3 and Figure 4. Based on Figure 3, the peak of 116°C is endotherm.

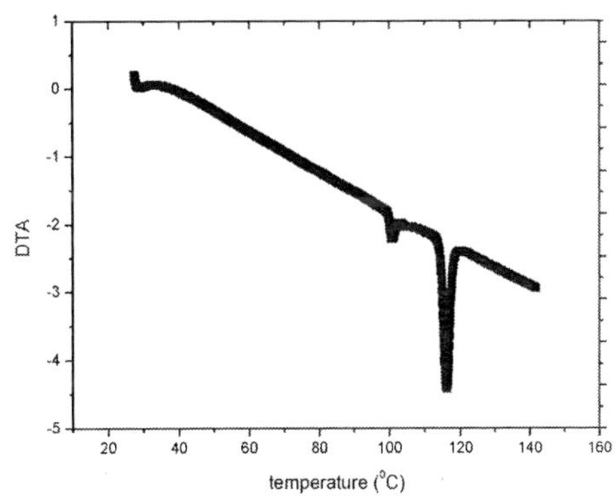

Figure 3. Graph of DTA Zn_5S_2 on 140°C with an increment of 1°C

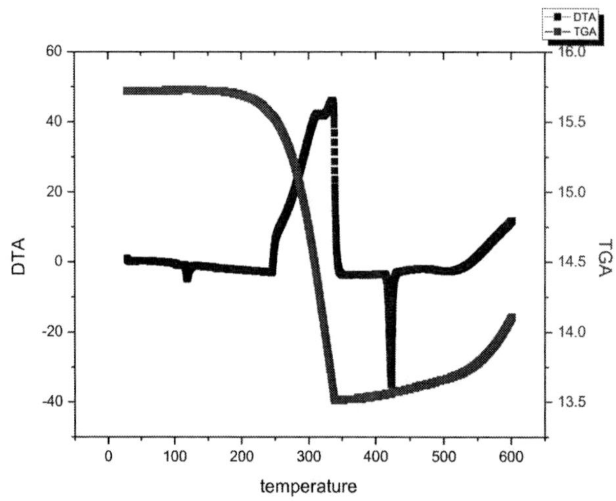

***Figure 4**. Graph of DTA Zn_5S_2 on 600°C with an increment of 5°C*

Figure 4 shows two measurements, i.e., Differential Thermal Analysis (DTA) and Thermogravimetric (TGA) measurements. The peak at 119.21 °C is showing exothermic, while the existing peak at 422.49 °C is indicating an endothermic. TGA is the measurement of the mass of a sample as the temperature increase. The result shows that the mass of Zn_5S_2 loss as the temperature increases because the reactant became product and gas.

3.3 Structure Analysis

The structures were investigated by X-Ray Diffraction (XRD). The structure pattern of this compound is given in Figure 5.

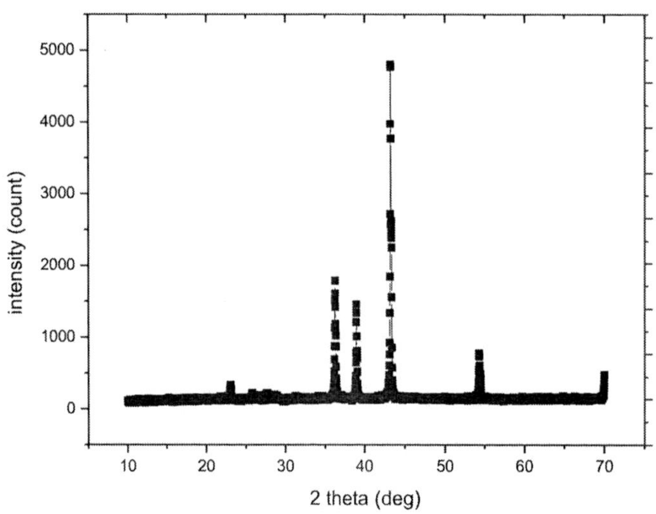

Figure 5. XRD patterns of Zn_5S_2

We have compared the observed pattern with the only ZnS sphalerite in the database. By observing the pattern, it should be a single phase and ZnS sphalerite phase. We then further analyzed this compound using DICVOL optimized under Fullprof program. By inputting the xy data from X-Ray Diffraction pattern, we firstly plot this data using winPLOTR. Then we select peaks by using automatic peak search button on point selection section. We choose all crystal system and press Run DICVOL button on the External application. We found several structures from this analysis which is displayed in Table 2.

Table 2. The lattice parameter of Zn_5S_2 from DICVOL analysis

System crystal	a (Å)	b (Å)	c (Å)	Volume ($Å^3$)	Figure of merit M	F
Tetragonal	13,9918	13,9918	78,2780	153,45	18.3	9,1
Orthorombic	15,4670	52,1114	49,5693	399,53	13,4	6,6
Monoclinic	5.2538	19.3137	3.4543	350,50	10,2	4,7

We choose the lattice parameter, stucture of this compund by the smallest error value, belongs to $P2_1/m$ a = 5.2538 Å, b = 19.3137 Å, c = 3.4543 Å, β = 90.176 °, volume= 350.50 $Å^3$, figure of merit M = 10,2 and F = 4,7. The volume of Zn_5S_2 crystal 2 times greater than volume of ZnS crystal from database of standard ZnS sphalerite.

3.4 Electrical conductivity

The electric conductivity of Zn_5S_2 was measured by 2 probe I-V meter. The measurement was repeated 75 times. The graph voltage and current Zn_5S_2 explained in Figure 6. The result of electric conductivity is 5.95 x 10^{-9} (Ω^{-1} m-1). The positive σ values imply a p-type conducting behaviour [7].

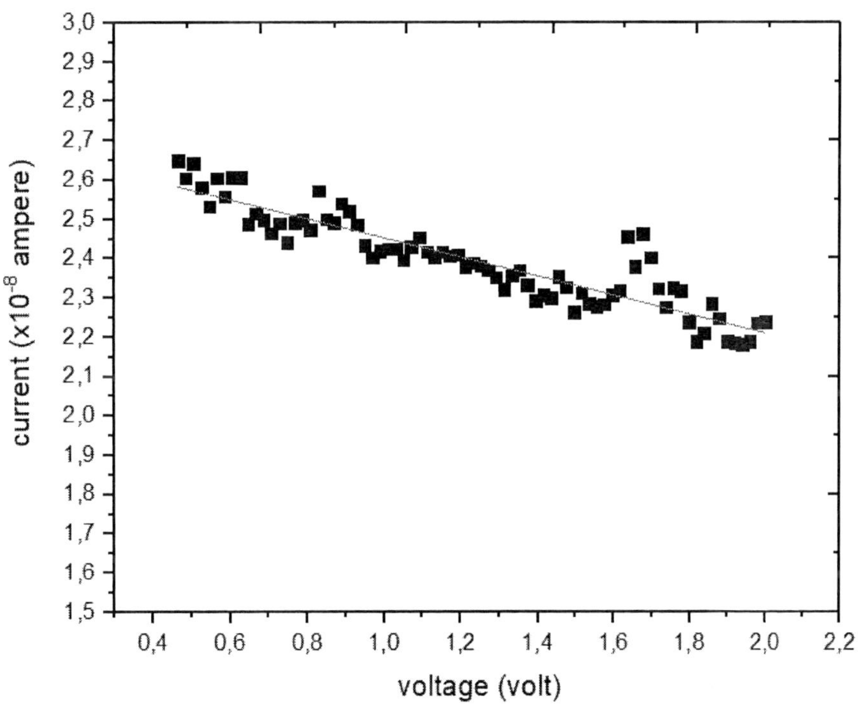

Figure 6. Graph of voltage and current Zn_5S_2 by I-V meter 2 probe measurement

3.5 Bandgap analysis

In a solid material, electrons exist at energy levels that combine to form energy bands. The region between the valence band and the conduction band is called the forbidden band, where no electrons exist. For a conductor, no forbidden band exists. For an insulator, this band is so broad that the electrons require high energy to move from the valence band to the conduction band. For semiconductors, the gap of the forbidden band is smaller than for an insulator [10].

The band gap of Zn_5S_2 was analyzed using absorbance of Ultraviolet-visible spectroscopy (UV Vis) characterization. The data is depicted in Figure 7. The calculation of band gap energy was using Tauc Plot method given in Figure 8. The result of band gap energy analysis is 5.98×10^{-19} J = 3.7 eV; this result has a suitable value of ZnS band gap energy because ZnS has wide energy band semiconductor with a range of 3.6 – 3.9 eV. [11].

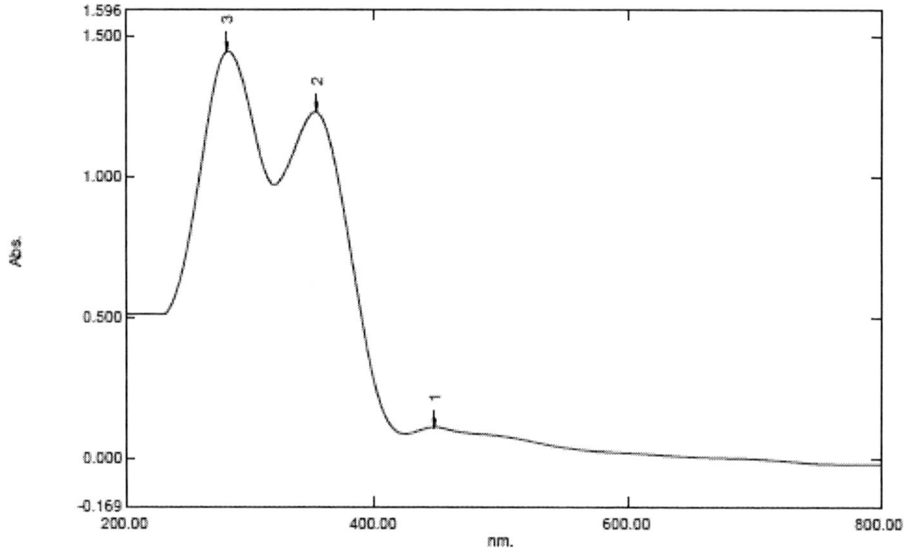

Figure 7. Graph of wavelength and absorption Zn_5S_2 by UV Vis measurement

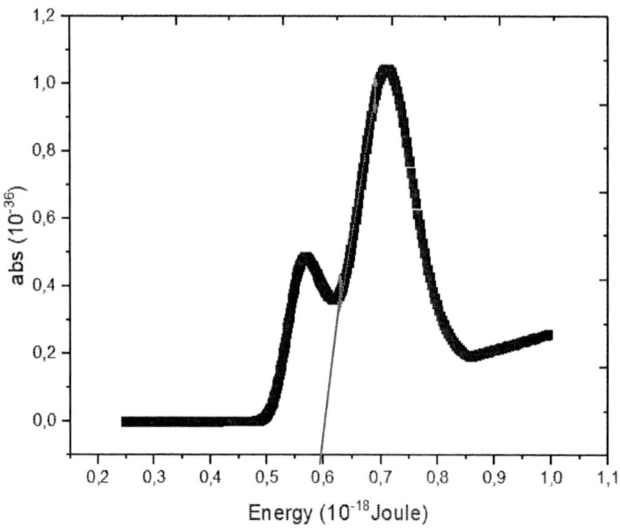

Figure 8. Graph of energy and absorption calculation Zn_5S_2 by Tauc Plot Method

4. Conclusion

The new compound Zn_5S_2 has been successfully synthesized by using a simple method solid state sintering utilizing sandwiched carbon active. The most possible crystal system of this compound fall to monoclinic under space group of $P2_1/m$, with lattice parameters of a = 5.2538Å, b = 19,3137 Å, c = 3,4543 Å, β = 90.176°. The thermal analysis we found that there were two phases, exothermic on peak 119.21 °C and endotherm on 422.49 °C. The electric conductivity of this compound is 5.95 x 10^{-9} (Ω^{-1} m $^{-1}$). The positive $\boldsymbol{\sigma}$ values imply a p-type conducting behavior. The energy band gap about 3,7 eV and has sa uitable value for sa emiconductor. So, we conclude that Zn_5S_2 compound has potential become thermoelectric or supercapacitor materials.

Acknowledgment

The work was conducted in part under Grant for University Applied Research on Smart Material Areas: ''Prototyping Design of Induction Furnace and their Application for Synthesis of Functional Intermetallic Compound M_xSb_y and M_xSe_y'' and "Supercapacitors of PUPT from Ministry of Applied Research, Technology, and Higher Education of Indonesia.

References

[1] Chen J J and Mark M Pitt 2017 Sources of change in the demand for energy by Indonesia households: 1980-2002 *Energy Economics* **61** 147-161

[2] Purba N P, Jaya K, Rona S, Syahrir G, Resti A I P, Fatimah M, Marine K M 2015 Suitable Locations of Ocean Renewable Energy (ORE) in Indonesia Region-GIS Approached *Energy Procedia* **65** 230-238

[3] Dutu R 2016 Challenges and policies in Indonesia's energy sector *Energy Policy* **98** 513-519

[4] Nasruddin, M Idrus A, Yunus D, Arief S, Agus S, HB Aditya, T M I Mahlia 2015 Potential of geothermal energy for electricity generation in Indonesia: A review *Renewable and Sustainable Energy Reviews* **53** 733-740

[5] Zhu Q, Hopper EM, Ingram BJ, Mason T O 2011 Combined Jonker and Ioffe analysis of oxide conductors and semiconductors *J. Am. Ceram. Soc.* **94** pp 187–193.

[6] Gayner C, Kamal K K 2016 Recent advances in thermoelectric materials *Progress in Materials Science* **83** 330-382

[7] Ge Z H, Peng Q, Dong S H, Xiaoyu C, Dan F, Yi H J, Jing F, Jiaqing H 2017 Highly Enhanced Thermoelectric Properties of Bi/Bi_2S_3 Nanocomposites *ACS Applied Materials & Interface* doi:10.1021/acsami.6b14803

[8] Fasolin, Stefano; Fiameni, Stefania; Fanciulli, Carlo; Battiston, Simone; Famengo, Alessia; Fabrizio, Monica. 2017. Nanostructured Tetrahedrite Synthesis for Thermoelectric Applications *Journal of Nanoscience and Nanotechnology* **17** 1645-1649

[9] Twaha, Ssennoga., Jie Zhu., Yuying Yan., Bo Li., 2016. A comprehensive review of thermoelectric technology: Materials, applications, modelling and performance improvement *Renewable and Sustainable Energy Reviews* **65** 698-726

[10] lsheikh, Mohamed Hamid; Dhafer Abdulameer Shnawah; Mohd Faizul Mohd Sabri; Suhana Binti Mohd Said; Masjuki Haji Hassan; Mohamed Bashir Ali Bashir; Mahazani Mohamad, (2014), A review on thermoelectric renewable energy: Principle parameters that affect their performance *Renewable and Sustainable Energy Reviews* **30** 337-355

[11] Infahsaeng, Yingyot and Sarute Ummartyotin 2017 Investigation on structural properties of Al-substituted ZnS particle prepared from the wet chemical synthetic route *Results in Physics* **7** 1245-1251

Development of digital viscometer based on sensor technology and microcontroller

Yulkifli, Yohandri and R Kurniati

Physics Department, Faculty of Mathematics and Natural Science, Universitas Negeri Padang, Jl. Prof. Hamka, Padang, Sumatera Barat, Indonesia 25131

E-mail: yulkifliamir@gmail.com

Abstract. The development of a digital Viscometer of fluid using photodiode and ultrasonic sensor has been done. The time taken of the ball to pass a distance in the fluid is controlled using two photodiode sensors. On the other hand, the ultrasonic sensor is used to measure the distance between the photodiode sensors. An Arduino microcontroller is used as data processing and interfacing with digital displays (LCD). The Viscometer system is validated by an expert and a valid criterion is obtained of 91.67%. Empirically, the measurement error of the system is obtained of 0.5%. Viscosity measurement is carried out on three types of fluid which are glycerine, oil and cooking oil. Based on the measured data, the viscosity of the glycerine, oil and cooking oil are 1.12 Pa s, 0.78 Pa s and 0.39 Pa s, respectively. The satisfy result show the digital Viscometer is promising to be used in industry and schools as a tool to improve the understanding of the viscosity concept of the student.

1. Introduction

Physics has provided a strong foundation of technological evolution. Advances in the physical sciences such as sensor systems, components, electronic circuits, and materials provide a considerable impact on the development of technology. Therefore, understanding the concepts of physics in teaching is very important. An experiment is an important instrument for understanding the concepts of physics. In general, some experimental instrument in laboratory is working manually, where less effective and efficient. Manual instrument is produced lack of proper data due to some mistakes made in the experiment, such as general and systematic error. In systematic error, the calibration and parallax error greatly affect to the quality of the measured data. Viscometer is one the experiment system that requires a accuracy and precision system

Viscometer is an instrument used to measure the viscosity of a fluid. Viscosity may be thought of as the internal friction of a fluid [1]. There are several methods used in the determination of the coefficient of viscosity of the fluid, one of which is the falling ball method. This method uses a ball that was dropped on a tube containing the fluid and travel time of the ball from one point to another point is recorded. Based on this travel time, the viscosity of a fluid can be calculated [2]. In manual instrument, the travel time is measured using a stopwatch and the viscosity of the fluid being calculated manually. In this manual experiment, a data error in measurement and calculation can be occurred. The automation of measurement using sensors and digital technology is one solution that can be applied [3][4]. In this work, the digital viscometer is developed for physics experiment based on sensor technology and microcontroller.

The digital viscometer is composed of a pair of photodiode sensors and ultrasonic sensors. Photodiode sensor is a semiconductor device with a p-n junction that designed to operate in reverse bias as a light detection [5]. This photodiode is operated as timer controller, where the ball passes the first sensor the timer is started and when the second sensor the timer is stopped. A pair of these sensors can be shifted upwards or downwards in order to vary the distance of the ball fall in the liquid. Ultrasonic sensors (Figure 1a) is a sensor works by emitting ultrasonic wave and receive a reflection of waves that can be used to interpret the distance of an object [6].

Arduino microcontroller (Figure 1b) is an electronic component that can be programmed and execute the program. Microcontroller equipped with supporting peripherals to as a complete computer on a chip level. A simple microcontroller is an IC that comprises RAM, ROM, parallel I/O, counter, and a clock circuit [7]. A microcontroller is a microprocessor system in which there already exists CPU, Read Only Memory (ROM), Random Access Memory (RAM), Input-Output, timers, interrupts, clocks, and internal equipment others already interconnected and well organized in a single chip that is ready to be used [8]. Arduino is a microcontroller board that has been integrated with a microcontroller ATmega328. Arduino data will be displayed in a Liquid Crystal Display (LCD) module. LCD is a flat-panel display or other electronically modulated optical device that uses the light-modulating properties of liquid crystals.

(a) (b)

Figure 1. Part of the system, (a) ultrasonic sensor and (b) Arduino

2. Experimental

The digital viscosmeter built by several circuits and electronic components. Ranging from the power supply, Arduino, photodiode and ultrasonic sensor as well as LCD displays this experiment, three kinds of liquid is used as a sample and the sensor distance will be varied. The diagram block of digital viscometer is shown in Figure 2.

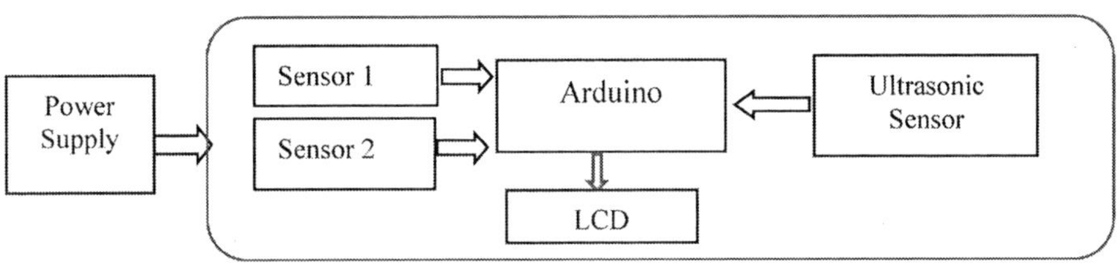

Figure 2. Block diagram of the digital Viscometer

In this design, Arduino acts as a timer which is controlled by a pair of photodiode sensor. On the other hand, Arduino also serves to receive data from the ultrasonic sensors to determine the distance between two sensors. The schematic of the program inside the Arduino is presented in flow chart as shown in Figure 3.

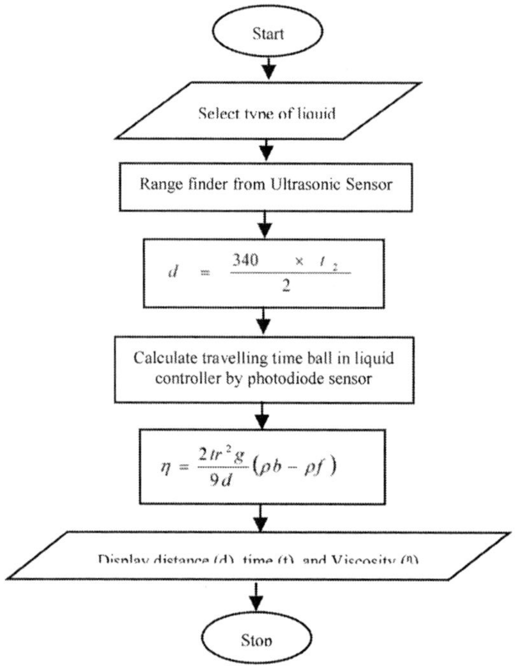

Figure 3. Flowchart of the software of the system

The accuracy of the system is determined based on measured results and know a coefficient of viscosity of the fluid. Each of the liquid samples are measured and compared with its standard coefficient of viscosity. Meanwhile, the determination of the precession of the system performed by take measurements repeated ten times for each sample fluid. Determine the average value, standard deviation, absolute error and relative error and report the results of measurements [9].

3. Results and Discussion

3.1. Results

3.1.1. Specification of the System

In general, the digital viscometer consists of sensor system, liquid container, and electronic system. Figure 4 are photographs of the sensor system and liquid containers. The sensor system is composed of a pair of photodiode for timer controller and ultrasonic sensors for a rangefinder. In addition, the glass tube is used as a container of fluid to be measured. Tubes with a height of 43 cm and a diameter of 8 cm, equipped with a hook ball. The circuit system installed in electronic case, which consists of an LCD display, push-button, power supply, Arduino, reset button, power on/off, sensor and AC ports. On the Arduino, 2, 3, 4, 5, 6, 7 port is used for the LCD display. Ports A0, A1, A2, and A3 are used for fluid and a reset button. Port 8.9 to sensor distance and port 5 Volt and the ground is used simultaneously as a voltage photodiode (Figure 5a). Box consists of two parts, front and rear. The front and back of the box can be seen in Figure 5b and 5c. The front part is where the LCD display, reset button, the on / off switch and three buttons for liquid selection. The rear consists of the AC port, fuse, and sensor ports. The port sensor has three pins including a proximity sensor pin, pin the start and stop of the sensor photodiode.

IOP Conf. Series: Journal of Physics: Conf. Series **1040** (2018) 012047 doi:10.1088/1742-6596/1040/1/012047

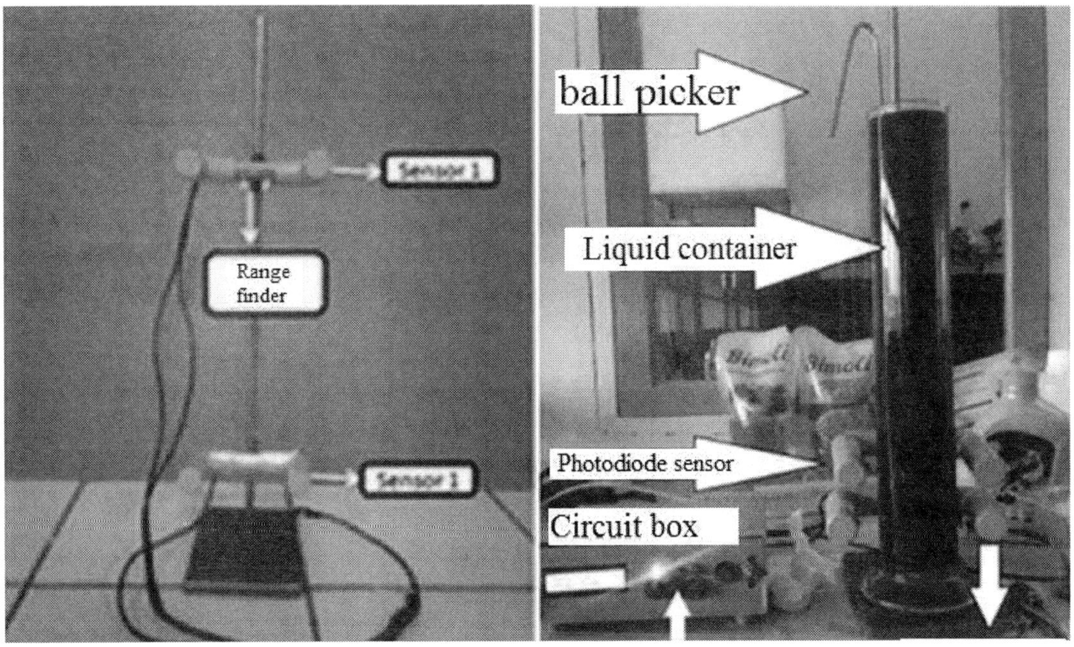

Figure 4. Photograph of the sensor system and liquid container

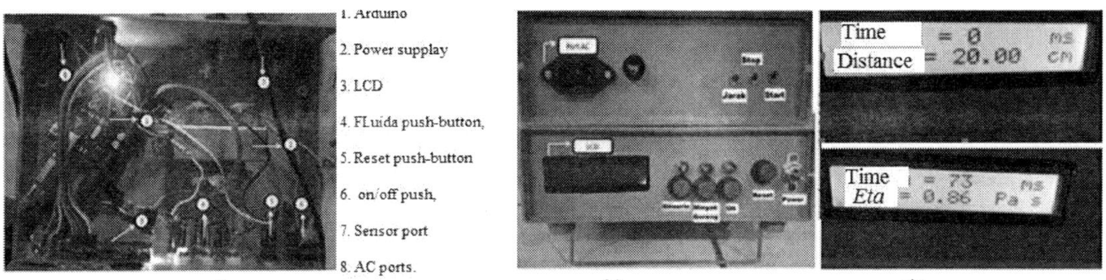

a) b) c)

Figure 5. Electronic system, a) electronic circuit, b) front and rear panel, and c) LCD display

3.1.2. Specification of the Design

The accuracy is determined by comparing the measured data with the known or standard viscosity of the liquid. Some parameters such as ball and the liquid density are entered to system before performing the viscosity measurement. The accuracy of the system is calculated for three kinds of the liquid which is oil, cooking oil and glycerine. Based on the relationship between distance and travel time the ball in liquid, the viscosity each liquid can be measured. Figure 6-8 show the relationship between the distance and the time for oil, cooking oil and glycerine, respectively.

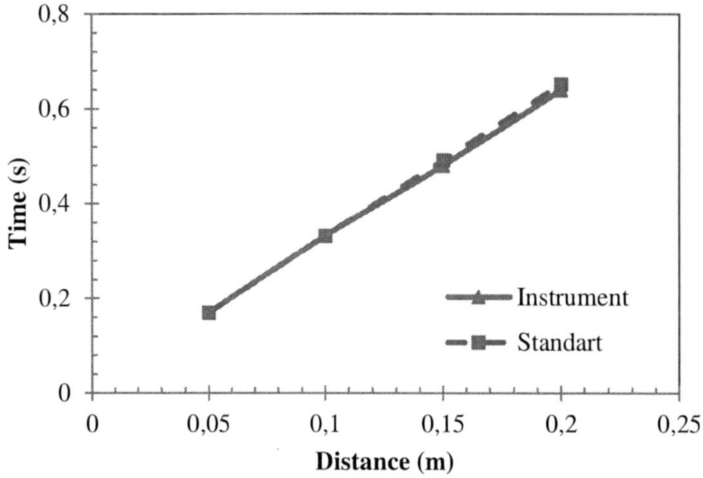

Figure 6. The relationship between the distance and the time for oil liquid

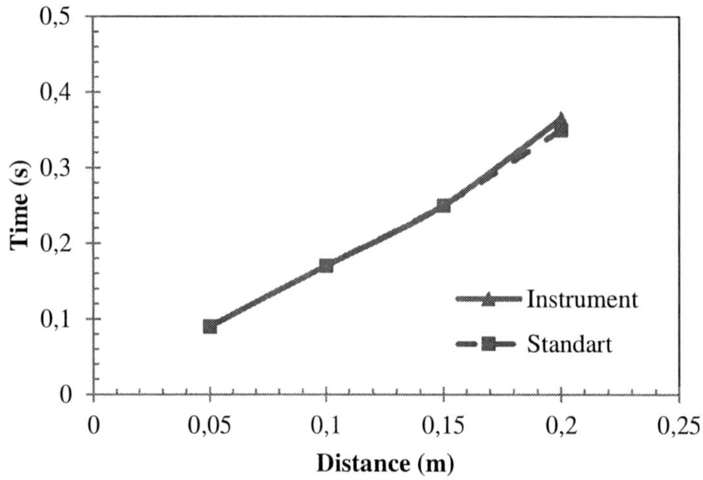

Figure 7. The relationship between the distance and the time for cooking oil liquid

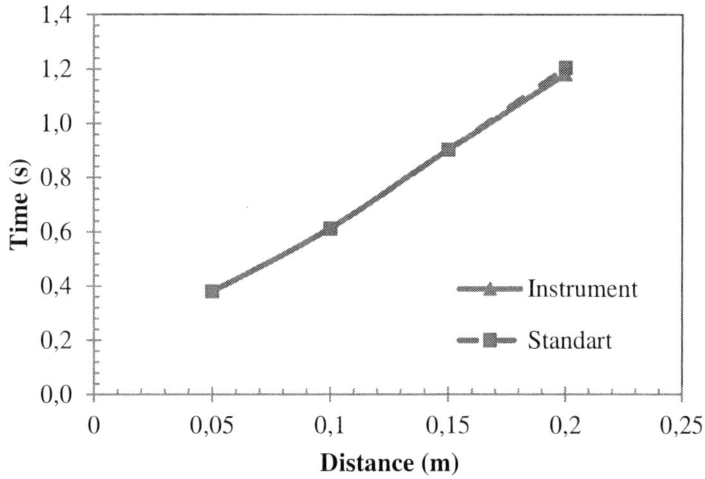

Figure 8. The relationship between the distance and the time for glycerine liquid

In Figure 6, comparison of measured and theoretical viscosity of the oil is presented. The measured data are approaching the theoretical value. The experiment proves that the value of the viscosity of 0.785 Pa s with accuracy percentage of 98%. In Figure7, the viscosity of cooking oil was 0.395 Pa s. As can be seen on the graph, the blue line is always close to the red line. So we can say with a theoretical approach viscosity measurement accuracy percentage of 98%. The glycerine has an average viscosity 1.112 Pa s and the percentage of accuracy of 99%. The relationship between the distance and the time for both measured and theoretical can be seen in Figure 8.

Different fluid types have different viscosity. Figure 9 shows the difference in viscosity fluid used in the experiment. The third type of fluid has a different coefficient value. Based on measured data, the viscosity coefficient of the oil, cooking oil and glycerine are 0.785 Pa s, 0.395 Pa s and 1.112 Pa s, respectively.

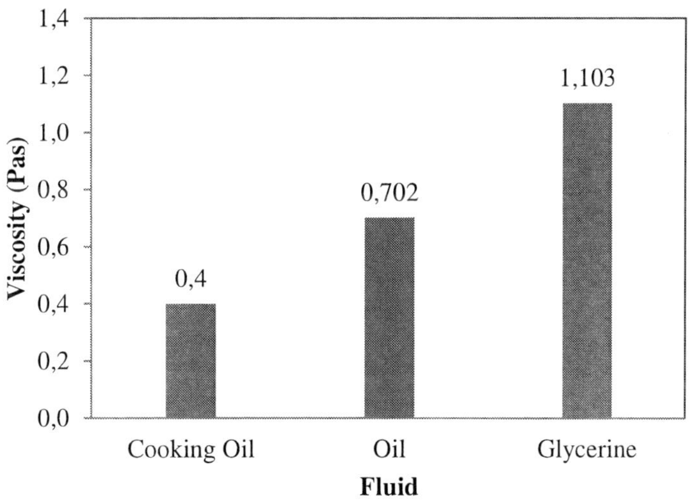

Figure 9. Comparison the viscosity coefficient of the liquids

The accuracy of the system is obtained based on the repeated measurement. In this experiment, the ten repeated measurements are performed. The accuracy is calculated based on the average value, standard deviation, and error percentage. The repeated measured data for accuracy calculation are listed in Table 1, Table 2 and Table 3. Based on the data in tables, the accuracy of the system of glycerine, oil and cooking oil are 99.085%, 98.101% and 97.938%, respectively. Measurement error derived from the digital viscometer is smaller than an instrument that has been developed previously [9].

Table 1. The measurement accuracy of viscosity coefficient for the glycerine

Distance (m)	η measured	η Standard	$\Delta\eta$	Relative Error (%)	Accuracy (%)
0.050	1.124	1.120	0.004	0.357	99.643
0.100	1.122	1.120	0.002	0.179	99.821
0.150	1.102	1.120	0.018	1.607	98.393
0.200	1.103	1.120	0.017	1.518	98.482
Average	1.113	1.120	0.010	0.915	99.085

Table 2. The measurement accuracy of viscosity coefficient for the oil

Distance (m)	η measured	η Standard	$\Delta\eta$	Relative Error (%)	Accuracy (%)
0.050	0.800	0.79	0.010	1.266	98.734
0.100	0.780	0.79	0.010	1.266	98.734
0.150	0.770	0.79	0.02	2.532	97.468
0.200	0.770	0.79	0.02	2.532	97.468
Average	0.780	0.790	0.015	1.899	98.101

Table 3. The measurement accuracy of viscosity coefficient for the cooking oil

Distance (m)	η measured	η Standard	$\Delta\eta$	Relative Error (%)	Accuracy (%)
0.050	0.420	0.400	0.020	5.000	95.000
0.100	0.402	0.400	0.002	0.500	99.500
0.150	0.411	0.400	0.011	2.750	97.250
0.200	0.400	0.400	0.000	0.000	100.00
Average	0.408	0.400	0.008	2.062	97.938

3.2. Discussion

The working principle of fluid viscosity measurement system based on digital are photodiode sensor will detect a ball that was dropped onto the surface of the fluid in a tube and an ultrasonic sensor to determine the distance from the sensor 1 to sensor 2. The sensor's position can be raised or lowered in order to vary the distance. When the ball is dropped, the sensor 1 detects the presence of the ball and starts the timer on the Arduino. Once the ball reaches terminal velocity and then through a sensor 2 which is located at the bottom, then the sensor 2 will stop the timer on the Arduino. Simultaneously, the ultrasonic sensors measure the distance from the sensor 1 to sensor 2. Based on the time and the distance, the Arduino circuit will automatically do the calculations to obtain the viscosity and displayed on the LCD. Digital viscometer can be operated for three kinds of the liquid that can be selected on the front panel. On the Arduino program, each button has a different type of mass values. Oil has a density of 859.5 kg/m3, cooking oil has a density of 884.8 kg/m3 and glycerine has a density of 1244.8 kg/m3. The value of the density of each fluid obtained from measuring the mass of fluid and divide by its volume.

4. Conclusion

The development of digital viscometer based on sensor technology and microcontroller has been presented. The digital viscometer is built consist of viscosity tube, photodiode and ultrasonic sensor, electronic circuit and instrument case. An electronic circuit is composed using Arduino kit, power supply circuit and LCD module. A pair of photodiode sensor acted as starter and stopper of the timer. In addition, the ultrasonic sensor is performed as a rangefinder between two photodiode sensors. The performance of the system in term of accuracy and practicality are satisfied for implementing in a physics experiment at school or laboratory. Based on the accuracy, precision and measured results, viscosity measurements can be performed more accurately and precisely.

Acknowledgment

The authors would like to thank Directorate General of Higher Education (DIKTI), Ministry of Research, Technology and Higher Education, Indonesia, for the Research Grant (Hibah TIM Pascasarjana 2017), No. 446/UN35.2/PG/2017.

References

[1] Scars Z 1954 *University Physics Mechanics, Heat, and Sound* (USA: Addison-Wesley Publishing Company)
[2] Setiadi 2012 *Pengukuran Viskositas Bola Jatuh Fluida Dengan Mikrokontroler Atmega 8535.* (Jawa Tengah: Universitas Kristen Satya Wacana)
[3] Yulkifli and Yohandri 2016 Pengembangan Teknologi Sensor Menjadi Alat-Alat Praktikum Fisika Dalam Mendukung Implementasi Kurikulum 2013 *Prosiding Semirata* 22-23 Mei 2016 Palembang ISBN 978-60271798-1-3
[4] Ihsan N, Yulkifli and Yohandri 2017 Development of speed measurement system for Pencak Silat kick based on sensor technology *IOP Conference Series: Materials Science and Engineering*, **180** (1) 012171.
[5] Pandiangan J 2007 *Perancangan Dan Penggunaan Photodiode Sebagai Sensor Penghindar Dinding Pada Robot Forklift* (Medan: USU)
[6] Yulkifli 2011 *Jurnal Otomasi, Kontrol & Instrumentasi* **3** (3) 1-10.
[7] Yohandri 2013 *Mikrokontroler dan Antar Muka* (Padang: Universitas Negeri Padang)
[8] Heri S et al. 2013 *Perancangan Sistem Telemetri Wireless Untuk Mengukur Suhu dan Kelembaban Berbasis Arduino Uno R3 Atmega328P dan XBEE PRO* (Riau: Universitas Maritim Ali Haji)
[9] Nurry P T, Yulkifli and Zulhendri K 2014 *Jurnal Sainstek* **6** (1) 71-83

IOP Conf. Series: Journal of Physics: Conf. Series **1040** (2018) 012048 doi:10.1088/1742-6596/1040/1/012048

Influence of high energy milling time on nano-quartz structure from West Sumatera

Ratnawulan, A Fauzi and Y Zahara

Department of Physics, Faculty of Mathematics and Natural Science, Universitas Negeri Padang, Indonesia

E-mail: ratnawulan@fmipa.unp.ac.id ; afz_id@yahoo.com

Abstract. Mineral silica is widely available in nature. Silica in nanometer size has a high economic value in the world steel industry. Methods that can be implemented in making nanomaterial one of them is High Energy Milling 3D Motion. The purpose of this study was to determine the effect of grinding time on nanostructures in the form of quartz crystal structure and particle size. The results showed that for grinding time 6, 12, 24 hours and 30 hours decreased the size of quartz particles. The size of the quartz particles was consecutively equal to 464 nm, 380 nm, 345 nm, and 293 nm with hexagonal structures.

1. Introduction

Silica is a mineral found in quartz sand mining. Quartz sand mining is commonly found in West Sumatra, Indonesia [1]. Silica is a type IV metal oxide which has good abrasion resistance, high insulators and high thermal stability. Silica has two structures: amorphous silica and crystals. The structure of quartz atoms is tetrahedral, a silicon atom surrounded by 4 oxygen atoms [2]. The content of silica minerals such as quartz, cristobalite, and trydmite, has many applications in technology especially if it is nanometer-sized (smaller than 100nm). The nanometer-sized materials have a number of chemical and physical properties that are superior to bulk materials. Nanomaterials have the advantage of being readily absorbed, larger touch surface areas [3], and anti-corrosion [4].

Many methods can be used to create nano-sized materials. Each method has its advantages and disadvantages [5][6][7][8]. One method for producing materials in nano sizes is the high energy milling method. High energy milling is a unique technique using the collision energy between the crushing ball and the wall of space that is rotated and driven in a certain way. The qualified nanoparticles depend on, among others, the time of milling [9][10][11]. In other words, knowledge of optimum milling time will provide information about the structure, size and characteristics of the surface and bulk of nanoquartz particles which impact on its application in the industry. In this paper, we focus on the effect of high energy milling time on the nanoquartz structure.

2. Research Method

2.1. Preparation of manganese Powder

Silica powders were prepared by high energy milling method using natural quartz sand from daerah Tanah Datar, Sumatera Barat, Indonesia (88.9% SiO) as starting material. The tool used for the preparation of silica is the Ball Mill High Energy Milling Ellipse 3D Motion (HEM-E 3D). Variations of milling time used are 6, 12, 24 and 30 hours.

2.2. Characterization

The Silica oxcide samples were characterized by an X-ray powder diffractometer (XRD, CubiX3Cement) using Cu-Kα radiation (λ = 1.5406 °A) at 45 kV and 40 mA and were used to determine the identity of any phase present and their crystallite size. The crystalline size was calculated using the Scherrer formula, $D = 0.9\ \lambda/\ \beta\cos\theta$, where λ is the wavelength of X-ray radiation, β is the full width at half maximum (FWHM) of the peaks at the diffracting angle θ [12][13]. The surface morphology ofthe quartz was obtained using scanning electron microscopy (SEM). The stages of the research are summarized in figure 1.

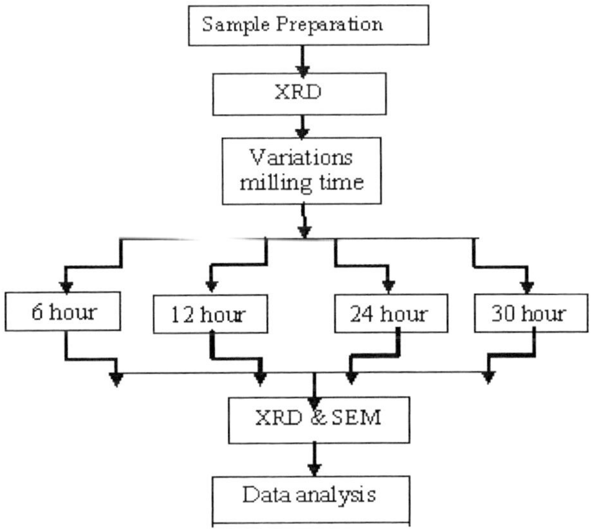

Figure 1. The stages of the research

3. Results and discussion

3.1. Quartz Structure

Figure 2 shows the XRD pattern of samples prepared with grinding time at 6 hours and room temperature (26^{0}C). The results of the diffraction pattern showed silica samples were dimilled with a time of 6 hours mineral quartz periclase minerals, cristobalite, and brucite. The peak on the Quartz XRD pattern is compared with the standard [11]. The XRD pattern of quartz formed at 20.89, 26.66, 36.56, 39.48, 40.30, 42.46, 45.81, 50.15, 54.89 and 59.97^0. Quartz Structure for hexagonal.

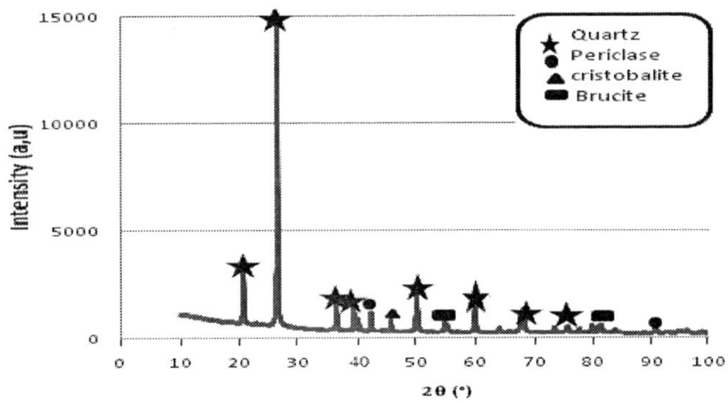

Figure 2. Spectra XRD of silica with milling time at 6 hour

Figure 3 shows an XRD pattern of samples prepared with milling time at 12 hours. The results of the diffraction patterns showed in the sample of silica, mineral periclase, cristobalite, and brucite, increasingly disappearing from the sample. Quartz Structure for hexagonal.

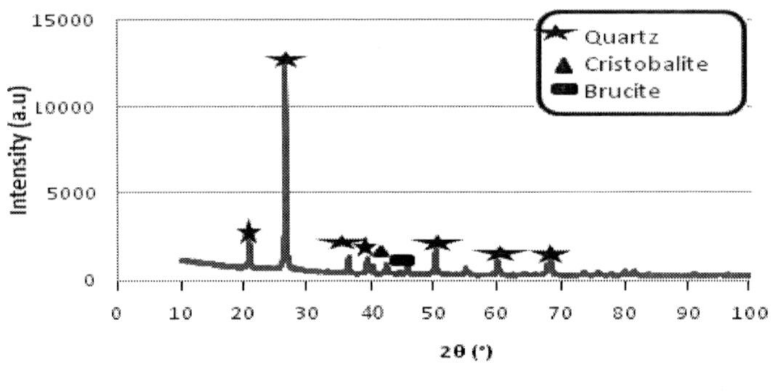

Figure 3. Spectra XRD of Silica with milling time at 12 hour

Figure 4 is a plot of intensity against the 2 angles of theta (2θ) of the X-ray diffraction pattern of the silica samples being dimilled for 24 hours. The results of diffraction patterns show that the highest peak is dominated by quartz minerals. While the other peak appears mineral periclase the longer milling time then the peak will be lost. Quartz Structure for hexagonal.

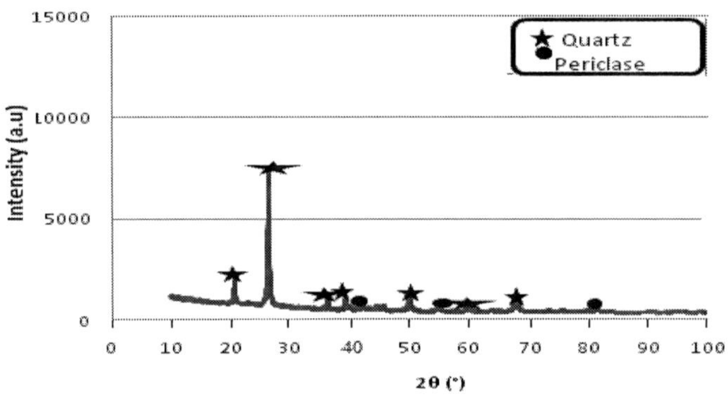

Figure 4. Spectra XRD of Silica with milling time at 24 hour

Figure 5 is a plot of intensity against the 2 angles of theta (2θ) of the quartz X-ray diffraction pattern of 30-milling results seen that the highest peak is dominated by quartz minerals in which no other mineral appears. Quartz Structure for hexagonal

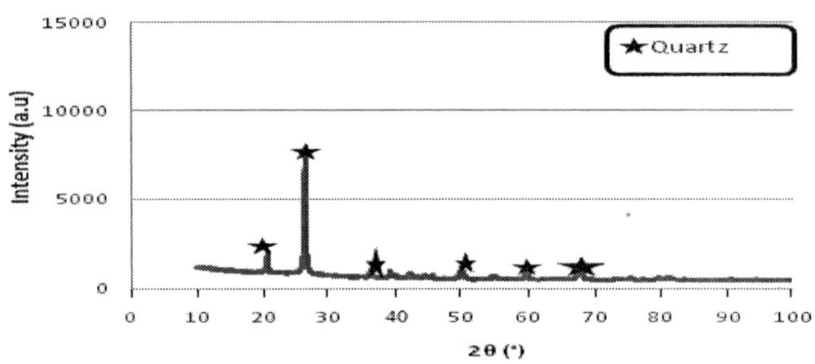

Figure 5. Spectra XRD of silica with milling time at 30 hour

Based on X-ray diffraction pattern on silica powder with milling time variation it can be seen that at milling time 30 hours obtained pure mineral quartz. Uncontaminated with other minerals. In addition, the increase in milling time will decrease the intensity of mineral quartz. The decrease in intensity due to the smaller particle size, but not affect the structure of mineral quartz.for hexagonal. In addition to the decrease in intensity, the widening of the curve also occurs at any time the milling is done. The decrease in intensity due to changes in crystal size, this change not results in the crystal structure of the shaping element changed [9]

3.2. Morphological characteristics

The surface morphologies of quartz with milling at different time were examined by SEM technique. Figure 6 shows a typical SEM image of quartz at milling time from 6, 12, 14, 30 hour with 33.000^x magnifications. The SEM images indicated nearly spherical quartz

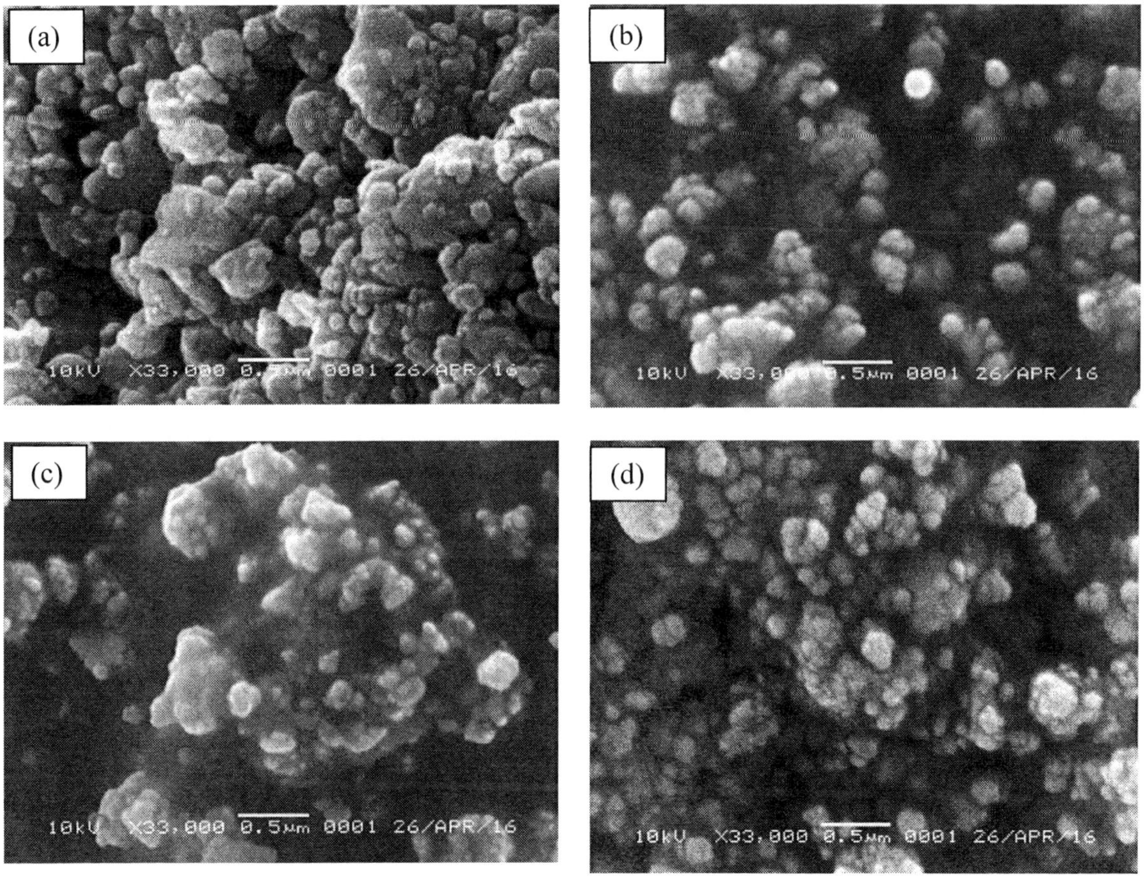

Figure 6. SEM images of nanoquartz at milling time(a). 6 hour ; (b).12 hour; (c).24 hour; (d).30 hour; with 33,000^x magnification

The SEM results also show that the particle size is getting smoother as the milling time increases (Table 1). Grinding time is 6 hours, the average particle size is 464 nm. At 12 hours of grinding, the average particle size was reduced to 380 nm. Furthermore, for a 24-hour grinding period, the particle size becomes 345 nm. At milling time 30 hours the average particle size is 293 nm. Based on the quartz mineral morphology form, for a 6-hour grinding time, the shape is round but not homogeneous. At the time of the 12-hour grinding, the shape begins to be slightly homogeneous. At the time of grinding 24 hours, there is little merging between particles. At 30 hours of milling, the quartz mineral is in the form of a homogenous round.

Table 1. SEM Data Analysis of Milling Results Against Particle Size

Samples	Milling Time (hours)	Avg. Particle Size (nm)
1	6	464
2	12	380
3	24	345
4	30	293

4. Conclusion

In short, silica powder from natural minerals is synthesized by using high energy milling. The quartz mineral crystal structure does not depend on milling time. The milling time variation produces the same hexagonal quartz crystal structure. Increased milling time causes morphological changes and quartz particle size to become smaller. Because the particle size of the quartz is still in the order of hundreds of nanometers, it is advisable to continue milling by increasing the milling time, to obtain a quartz particle size below 100 nm

Acknowlegment

The authors thank to Ministry of Research, Technology and Higher Education for financial support through MP3EI Grant Number 535/UN35.2/PG/2017 for this work.

References

[1] Ratnawulan, Fauzi A and Hayati A E S 2018 *Proc. The 2th Int.Conf. on Mathematics, Science, Education, and Technology* (Padang: IOP Conference Proceedings) **335(1)** 012008
[2] Horacio E B 2006 *Colloidal Silica Fundamentals And Applications* (CRC Press: New York)
[3] Rawat M 2006 *Biol.Pharm. Bull* **29(9)** 1790-1798
[4] Anwar K and Tausif A M 2014 *Mater. Sci* **38** 181-189
[5] Radu D, Yu I C, Yu H P, Pizzi N 2011 *Silica Nanomaterials (NSN) For Carbon Capture Application* (Department of Chemistry: Delaware State University Dover DE)
[6] Challa S, Kumar S R, Hermes J and Leuschner C 2005 *Nanofabrication Towards Biomedical Applications* (Germany: Wiley-Online Library)
[7] Tavangarian R and Emadi R 2009 *Journal of Alloys and Compounds* **485**(1-2) 648-652
[8] Suryanarayana C 2004 *Mechanical Alloying and Milling* (New York: Marcel Dekker)
[9] Sanskriti A, Parashar Y, Kale D, Yashesh H G and Dattu R J 2005 *International Journal of Luminescence and Applications* **5(1)** *69-72*
[10] Sarimai, Ratnawulan and Fauzi A 2018 *Proc. The 2th Int.Conf. on Mathematics, Science, Education, and Technology* (Padang: IOP Conference Proceedings) **335(1)** 012004
[11] Riazian M and Bahari A 2012 *Int. J.NanoDimens* **2** 127-139
[12] International Centre for Diffraction Data (ICDD) 2000 *Joint Committee on Powder Diffraction Standards, Diffraction Data File* **05-0661** 12-23
[13] Ratnawulan, Fauzi A and Sukma H A E 2017 *Proc. The 4th Int.Conf. on Research, Implementation, and Education for Developing Scientific Attitude in Sciences and Mathematics* (Yogjakarta: AIP Conference Proceedings) **186** 060009

Testing on porosity of composite material composed by ultrafine amorphous silica (UFAS) from rice husk using X-ray micro-computed tomography

P Widiarini[1] and G B Suparta[2]

[1]Universitas Pendidikan Ganesha, Indonesia
[2]Universitas Gadjah Mada, Indonesia

E-mail: putuwidiarini27@gmail.com

Abstract. In general, this study was aimed to test the porosity of composite materials. The test was using X-ray micro-computed tomography (micro-CT) system developed at the Department of Physics, Universitas Gadjah Mada, Indonesia. Testing was done for three composite materials comprised by ultra fine amorphous silica (UFAS) from rice husk. Samples were made in three variations of compositions M1, M2, M3 namely a) UFAS1 plus rice husk powder, resin and catalyst, b) UFAS2 plus rice husk powder, resins, and catalysts, and c) powder UFAS3 plus rice husk, resins, and catalysts. Rice husks powder has an average fineness of 150 meshes. The testing procedure included preparation, process of scanning, data processing, and data analysis. The results showed that the linier attenuation coefficient of materials M1, M2, and M3 were $(9.2 \pm 0.5) \times 10^{-3}$ cm^{-1}, $(9.7 \pm 0.1) \times 10^{-3}$ cm^{-1}, and $(10.0 \pm 0.1) \times 10^{-3}$ cm^{-1} respectively. The porosity of materials M1, M2, and M3 were 13.74%, 11.31%, and 7.22% respectively.

1. Introduction

Composite material is a mixture of two or more distinct constituents or phases [1]. One kind of particle composite was made from rice husk [2]. As a new material that will be used as a substitute material for the manufacture of wood carving craft, it is necessary to test the feasibility of it so that the quality of products made from these materials is assured. Therefore, it is necessary to test for internal quality composite material with nondestructive testing (NDT), should had no internal structure damage. One of NDT techniques that can be used to inspect the internal parts of a material is a computed tomography (CT) [3]. Tomography is an imaging technique that presented a cross-section of an object without going through surgery [4]. Tomography was first used to diagnose women who have brain tumors [5], then it used in variety test such as examine the structure of polymers, steel, and ceramics [6], capacitors [7], frozen biomaterial [8], pore structure of material [9].

A set of micro-CT system has been developed at the Department of Physics, Universitas Gadjah Mada, Indonesia. Micro-CT is operated using a computer device, from the process of scanning until the process of reconstructed images to produce cross-sectional images of objects. Testing of the internal structure of material would provide the results of the internal structure of materials that can be used as basic analysis of the quality of material. The quality of the composite material is determined by several things such as stiffness, strength of materials, density comprising homogeneity and porosity of materials [1]. Porosity is closely related to the mechanical properties of composite materials such as stiffness and strength of the material [10]. Porosity testing aims to determine the ratio between the total volumes of pores to the amount of the overall volume of materials.

The micro-CT system that has been developed in the Department of Physics, Universitas Gadjah Mada, Indonesia, was adopted the first generation CT. First generation CT is called a translation-rotation tomography that has a single X-ray source and a single detector [5]. There are interaction between X-ray and object when the X-ray passes through the object. According to the Beer-Lambert law, the intensity of X-rays that pass through the object to be diminishing exponentially can be expressed in Eq. (1)

$$I = I_o \exp\left(- \int_{l_\phi(x_r)} \mu(x_r, y_r) dy_r \right), \quad (1)$$

with I is X-ray intensity after its interaction with an object, I_o is initial intensity of X-ray, and $\mu(x_r,y_r)$ is linear attenuation coefficient of the object.

Projections from different angle inspection were arranged into a set of sinogram. Projection is a set of ray-sum in translation trajectory from $-R$ to R in Radon space. The ray-sum $p_\phi(x_r)$ is obtained by line integral of linear attenuation coefficient of ray-path across the object along the trajectory. In other to produce good quality of reconstructed image, so it used the Summation Convolved Back Projection (SCBP) as reconstruction method on the operation. Thus, the distribution of linear attenuation coefficient was expressed by

$$\mu(x, y) = \int_0^\pi p'(x_r, \phi) d\phi \quad (2)$$

with $p'(x_r,\phi)$ is convolved ray-sum. The reconstruction method gives an accurate reconstructed image. The reconstructed images that have been obtained from the reconstruction method represent the distribution of linear attenuation coefficient of the certain cross-sectional slice according to Eq. (2).

The basic compositions of composite material are UFAS, rice husk powder, resin and catalyst. These composite materials are categorized as particle-reinforced composites. There were three kinds of UFAS namely UFAS1, UFAS2, and UFAS3. The UFAS successively obtained from the combustion at a temperature of 400°C, 600°C and 800°C. Porosity associated with the permeability of material. Surely, porosity of material affects the quality of the material, especially the mechanical properties of composite materials. One method for measuring the porosity of a material is through tomography image analysis. The results will display the cross-sectional slices of the material, so that the internal part of the material is clearly visible between the pores and solid parts. Based on the image of the cross section can be calculated pore volume and the overall volume of material. Porosity measurements with CT method can be done by calculating the total black area of the entire slice on each material compared with the total volume of material. This statement can be formulated as

$$P\% = \frac{\sum_{n=1}^{n} A_{Bn}}{\sum_{n=1}^{n} A_{Tn}} \times 100\%, \quad (3)$$

with P is material porosity, n is the number of cross-sectional slice of each material, A_{Bn} is total black area of the n-sliced, and A_{Tn} is total area of the n-sliced. In addition, by calculating comparing total black area with total area of certain slice, we can measure the area porosity of each slice.

One method of testing porosity is measurement by CT method [9][11]. Each slice cross-sectional tomography shows the internal structure of a material that cannot be seen directly from the outside of the material. Pore structure in the material will affect the value of the material porosity. This paper presents our study of micro-CT system that has been developed expected to be used for testing composite materials. The results will be able to provide information on the quality of the internal structure of material that can be proposed to be a measure of quality composite material. Linear attenuation coefficient, porosity, and profile of reconstructed image were observed and compared for each composite material.

2. Methods

Figure 1 shows the procedure of experiment. The micro-CT system that has been developed at the Department of Physics, Universitas Gadjah Mada, Indonesia comprises an X-ray generator of a Molybdenum anode target, rotation system of object stage, detector with fluorescence screen, and computer unit. The fluorescence screen was coupled with CCD camera to capture the radiograph of object. The object stage is rotated with a controller motor. The X-ray generator was operated on 40 kV and current 30 mA. The objects of this study are three kinds of composite materials with diameter 1 cm and height 2 cm. By rotation of the object, radiographs of the composite material on different angle can be obtained along the scanning process.

A set of 2D projections was collected for each angle from 0° until 360° with interval 1°. A set of 3 radiographs was taken for every 1° angle projections. The 360 radiographs were collected from the system. Then, the data row from a certain height of radiograph was considered as a projection of cross-sectional image of the object at that height. This data row was called a sinogram that has 360 projections. Then, a set of sinogram can be reconstructed to a set of axial cross-sectional slice of the object. The cross-sectional slice of the object is called tomography image.

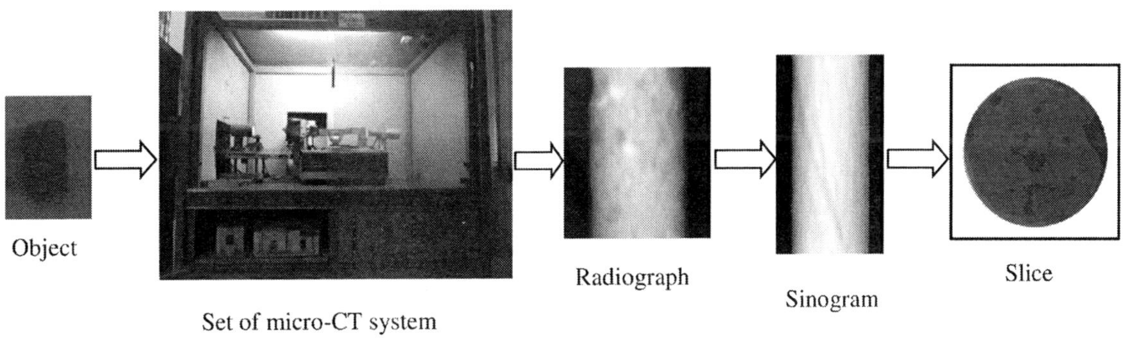

Figure 1. Scheme of experiment procedure consists of a composite material, set of micro-CT system, radiograph, sinogram, and cross-sectional slice or original CT image.

The internal structure of object can be observed from the cross-sectional slice for each composite material. The cross sectional slice represented linear attenuation coefficient distribution along the area of image. Thus, we can measure the linear attenuation coefficient of each material. Figure 2 shows the original cross-sectional slice that represented the pore structure of material. We assumed that the darkness area is pore of material. In other to make a simple analysis, the original CT image was changed into binary particle image by selecting the threshold value of it. Black area indicates the pores. Then, the analysis of porosity was done by calculating the total black area.

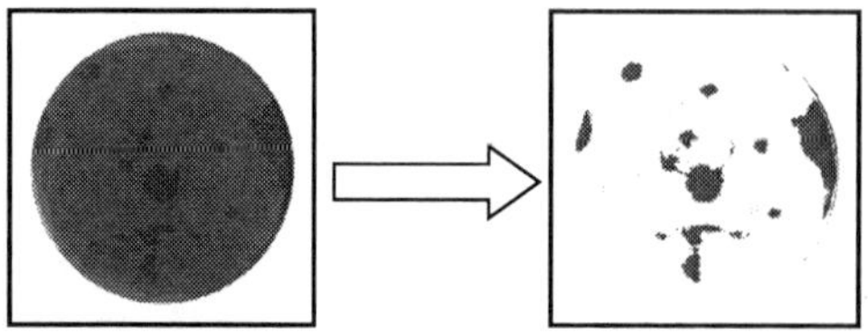

Figure 2. The binary particle image (pore is black) from original CT image in other to porosity analysis

3. Results and Discussion

The result of reconstruction process was produce 540 cross-sectional slices for each material. The cross-sectional slice or reconstructed image obtained has good quality. The images clearly showed the contrast of pores on the each slice area as is shown in Figure 3. The X-ray energy of 40 kV and 30 mA was able to provide good quality reconstructed image and to distinguish between pores and total area of each slice.

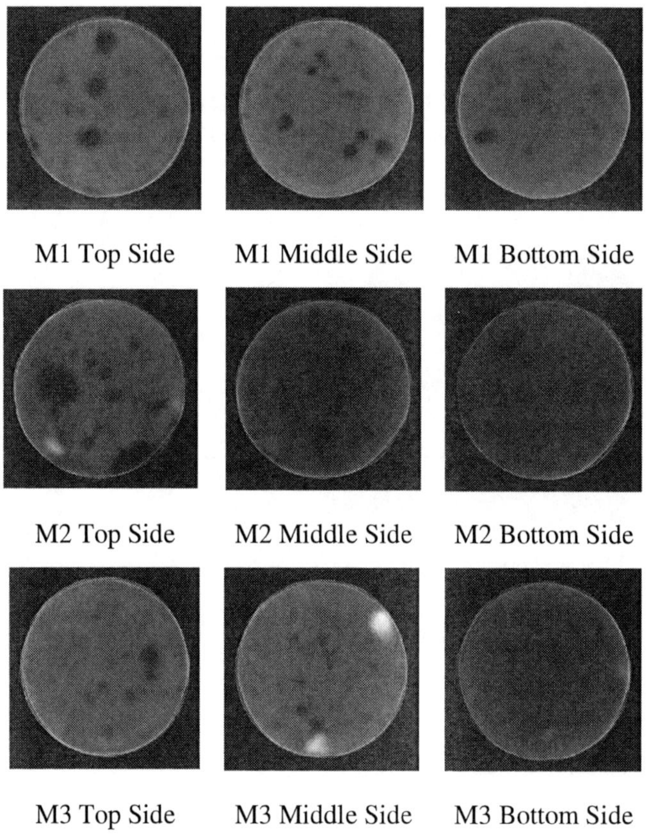

Figure 3. Result of reconstructed image of composite material from inspection with X-ray on 40 kV and 30 mA.

The result of each material linear attenuation coefficient (μ) and each material porosity analysis is shown in Table 1 and Table 2.

Table 1. Linear attenuation coefficient of each material

Materials	Linear attenuation coefficient (x 10^{-3} cm^{-1})
M1	9.2 ± 0.5
M2	9.7 ± 0.1
M3	10.0 ± 0.1

Table 2. Porosity of each material

Materials	Porosity (%)
M1	13.74
M2	11.31
M3	7.22

The result indicated that μ value of composite materials M1, M2, and M3 were respectively greater, but porosity value of composite materials M1, M2, and M3 were respectively smaller. The difference of material linear attenuation coefficient and porosity value on each composite material is influenced by the different temperature combustion and different composition of it.

The total pore structure in the material affects the density of material indirectly. If the material porosity is being greater, then the material density is being smaller. Material density has inverse effect to the linear attenuation coefficient. This theory is suitable to the experiment results that have been obtained. In addition, based on the result in Table 1 and Table 2, there is an inverse relationship between the values of material porosity with linear attenuation coefficient of material. The analysis result showed that the greater of material porosity, so the material linear attenuation coefficient gets smaller.

In general, the material is required to have porosity value between 10% until 20% for commercial needs. The result of porosity testing by X-ray micro-CT is consistent with the standard method. The advantages of using it for the porosity testing are 1) the testing is one kinds of non-destructive testing because the testing object is not damaging and changing its integrity, 2) the internal structure of object was displayed more clearly by the cross-sectional slices that have been obtained, 3) we can distinguish the pores and the total area of each slice clearly, and 4) we can see the internal condition of the object.

4. Conclusion

The micro-CT system that was developed at Department of Physics, Universitas Gadjah Mada, Indonesia can be used for porosity testing of composite material. The X-ray generator was operated on 40 kV and 30 mA. It was able to provide good cross-sectional slice of each composite material and to distinguish pores of each area slice. The results showed that the differences of linear attenuation coefficient and material porosity were affected by the different composition of each composite material. The higher temperature combustion of material gives effect the greater value of material porosity and the smaller value of linear attenuation coefficient. Besides that, we can estimate that M1 and M2 have a requisite porosity value to be a good material for commercial needs. Therefore, MI and M2 can be used as a substitute material for the manufacture of wood carving craft. We suggested testing other properties of the composite material in other to get a clearly quality of it.

Acknowledgement

The authors would like to express gratitude to Mr. I Wayan Karyasa for his sharing during the experiment. Thank you also to Gede Arya Wiguna, and I. B. G. Putra Pratama for their assistance during the experiment.

References

[1] Matthews F L, Davies G A O, Hitchings D and Soutis C 2000 *Finite Element Modeling of Composite Materials and Structures* (USA: CRC Press)

[2] Karyasa I W 2014 Producing of ultra fine amorphous silica (UFAS) from rice husk *J. of Sains and Technol.* **3(1)** 263-274

[3] Mix P E 2005 *Introduction to Nondestructive Testing* (A Training Guide) (USA: John Wiley & Sons, Inc)

[4] Buzug T M 2008 *Computed Tomography* (From Photon Statistics to Modern Cone-beam CT) (Germany: Springer)

[5] Cierniak R 2011 *X-ray Computed Tomography in Biomedical Engineering* (London: Springer)

[6] Petit C, Meille S and Maire E 2013 Cellular solids studied by X-ray tomography and finite element modeling-A review *J. Mater Res.* **28(17)** 2191-2201

[7] Pratama I B G and Suparta G B 2015 Quality inspection on electrolytic capacitors using X-ray micro-computed tomography *Proc. SPIE 9302 Int. Conf. on Experimental Mechanics*

[8] Bischof J C, Mahr B, Choi J H, Behling M and Mewes D 2006 Use of X-ray tomography to map crystalline and amorphous phases in frozen biomaterials J. *Annals of Biomedical Engineering* **35(2)** 292-304

[9] Ai-xiang W, Bao-hua Y, Yong X and Huai-chun J 2007 Pore structure of ore granular media by computerized tomography image processing *J. of Central South University of Technol.* **12(2)** 220-224

[10] Aqida S N, Ghazali M I and Hashim J 2004 Effect of porosity on mechanical properties of metal matrix composite (An overview) *J. of Technol.* **40** 17-32

[11] Lin C J and Miller J D 2000 Network analysis of filter cake pore structure by high resolution X-ray micro-tomography *J. of Chemical Engineering* **77** 79-86

Correlation of seismotectonic parameter and seismic quiescence z-value in West Sumatra Indonesia

Syafriani[1], Yulkifli[1], A Z Sabarani[2] and F D Raharjo[2]

[1]Departement of Physics, Faculty of Mathematics and Science, Universitas Negeri Padang

[2]Indonesian Meteorology, Climatology and Geophysics Agency

Email: syafri@fmipa.unp.ac.id

Abstract. West Sumatra is a tectonically active area where there have been several major earthquakes happened. Earthquakes can happen anytime and anyplace, so that need to know some precursor of earthquake before the earthquake occurred, such as local stress condition of region. Interrelationships between of the local stress condition a-value, b-value and z-value for the time period extending from 1981 to 2015 are examined in the West Sumatera region. For this purpose, a distribution relationship with the frequency-magnitude earthquake and temporal changes of this parameters were analyzed in order to reveal the potential of future earthquake in the study region. Beside that, standard deviation value of Z-value was estimated in the beginning of 2015 to identify of current seismicity rate changes. The result showed that the area had low spatial variation of b-value around 0.8. Low b-value indicated that the area had high stress condition and a-value had low spatial variation between 5.5 to 7 with low seismic activity. In contrast, several region had positif z-value that indicated that there were significant decreased in seismic activity. Based on this analysis it appears that these three parameters (a,b,z value) were related to each other. In some part of West Sumatra there were region that decreased earthquake activity and indicated that there was a large accumulation of energy which high local rock stress condition with significant decreased in seismic activity or seismic quiescence. This mean that the potentially devasting earthquake can be happened in West Sumatera.

1. Introduction

The Sumatra region is at a meeting of the Eurasian plate with the Indo-Australian plate. The Indo-Australian Plate moves under the silent Eurasian Plate at a rate of 50-70 mm per year [1]. Because of this shift, there is a fault system along the island of Sumatra, which is called Sumatran Fault System. This fault formed segments that stretched along the island of Sumatra.

In West Sumatra, stretches the SFS segment from South Solok to Rao area in Pasaman District. The fault segment is Sumpur, Sianok, Sumani, and Suliti segment. In addition, in the ocean area there is also a fault system called Mentawai Fault System (MFS). The existence of active faults around land and sea, causing West Sumatra is potentially the occurrence of earthquakes.

West Sumatra has experienced several significant earthquake shocks and damage due to the activity of the plate and the faults. Based on the Meteorology, Climatology and Geophysics Agency (BMKG) data, in West Sumatra has been shaken several times by earthquakes such as events in Singkarak (1926 and 1943), Pasaman (1977), Solok (2004), while in Gunuang Rajo (2007), two times earthquake with a strength of 6.4 SR and 6.3 SR. In addition, in the Sianok segment there was also a major earthquake on March 6, 2007 with coordinates at 0.48 ° LS, 100.37 ° East with a depth of 33 km and a magnitude of 6.4 SR [2]. While earthquakes originating from subduction zones occurred in 1779 (8.4 SR), 1883 (9.2 SR), 1861 (8.3 SR), 2007 (7.9 SR), and 2009 (7.6 SR) [3]. There are two significant earthquakes that occur in the ocean of West Sumatra was the earthquake of Pariaman, September 30, 2009 magnitude 7.6 and the earthquake of Pagai, Mentawai that occurred on October 25, 2010. Both earthquakes cause severe damage, loss of property and claimed many lives.

To understand the process of the occurrence of significant and destructive earthquake in West Sumatra required a strong understanding of the precursor shown by seismotectonic and earthquake activity. Seismotectonic parameters were obtained by calculating the b-values and a-values as well as their relation to the phenomena that preceded the significant earthquakes in the region. This parameter describes local tectonic stress conditions and earthquake activity in West Sumatera. While precursors that precede the significant earthquake is called the phenomenon of seismic quiescence is a phenomenon. This phenomenon is characterized by a decrease in seismic activity over a period

Characteristics of the seismotectonic parameters can be determined by looking at the spatial distribution of b-values and a-values in the region based on the Frequency-Magnitude Distribution relationship (FMD). The value a represents the seismic activity level of a region, whereas the b-value reflects the tectonic conditions associated with the region's stress conditions. Meanwhile, to observe the seismic quiescence phenomenon can be done using the z-value method. The z-values method measures the difference in the intentness level moving in the time window with the background seismicity using the standard deviation z [4].

The method of z-value has been done before in various places to observe seismic quiescence. Among them, by Katsumata [5] who examined the Precursory Seismic Quiescence Before the Mw = 8.3 Tokachi-Oki, Japan, Earthquake on September 26, 2003 Revealed by a re-examined Earthquake Catalog. The results of his research found that before the Tokachi-oki event there were two adjacent seismic quiescence anomalies. The anomaly appeared in 1999 and lasted for 5 years before the Tokachi-oki 2003 event [5]. In addition, Sunardi [6] also observed changes in earthquake level before Tasikmalaya earthquake, September 2, 2009. Based on observations made by Sunardi [6] obtained spatial distribution of z-value in the period that has been determined. A positive value of z-value indicates a decrease in seismicity. The decrease in seismicity occurred around the mainshock beginning three years before a strong earthquake occurred. The decline in the mean seismicity level of the years leading up to strong earthquake events is suspected to be an earthquake precursor [6].

Based on the spatial distribution of low-a values means low seismic activity and indicates the presence of energy accumulation (asperity) [7], on the contrary for high-a values. Whereas spatial high-b values are associated with low rock stress conditions [8] [9]. High b-values have high heterogeneity medium conditions, but low-b values correlate with high stress rock conditions and have low heterogeneity medium [10]. A region with high stress rock conditions has a b-value of 0.4-0.9, while under low stress conditions the b-value is about 1.2 [11].

This study focuses on an understanding of the relationship of seismotectonic parameters to the seismic quiescence phenomenon based on their spatial distribution. It aims to information the potential of seismicity and its relationship with earthquake disaster mitigation.

2. Methods

The relationship of the frequency distribution and earthquake magnitude is a linier equation that is influenced by the constanta of seismotectonic parameter. The relationship of the frequency-magnitude distributions satisfies in the following equation [12]:

$$\text{Log}\ N(M) = a - b\,M \tag{1}$$

where N (M) is the cumulative number of earthquakes having Richer magnitude larger than M, a is a measure of seismic activity and the b-value is the slope on the log N~M and is a cosntant parameter. The seismotectonic parameter (a-values and b-values) is determined using the maximum likelihood method. This method is a part of the seismological statistics. The b-values are obtained by using the equation:

$$b = \frac{\log e}{M - M_o} \tag{2}$$

Where log e = 0.434, M is the sampling average of magnitudes and M_0 is the minimum magnitude. Then a-value can be determined according to the equation:

$$a = \log N\,(M \geq M_0) + \log(b \ln 10) + M_0 b \tag{3}$$

The z-values method measures the difference in the seismicity level moving in the time window with the background seismicity using the standard deviation, Z [4]. The aim is to detect a possible period of low seismic anomalies before the mainshock near the epicenter, and to evaluate statistically the ratio of significance of seismic decrease with all possible seismic drops in random locations and times [5]. Measurement z-value is performed using LTA (Long Term Average) function with the aim that the anomaly obtained is not statistically biased [6]. Figure 1 illustrates the observed changes in the mean seismic rate with the LTA function.

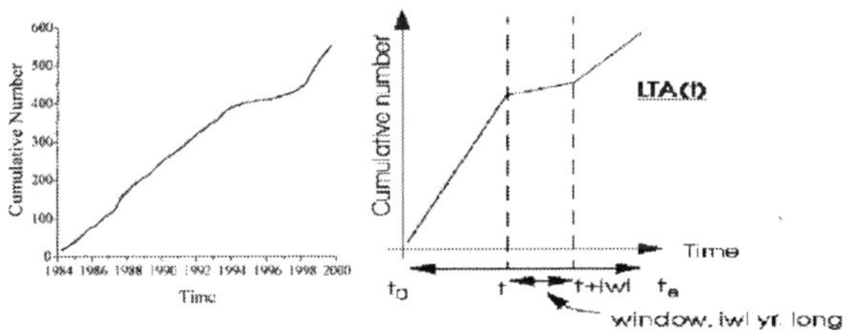

Figure 1. The Curve of Average Rate Changes Curve with Long Term Average (LTA) Function (Wiemer, 2001).

Based on Figure 1, a curve showing cumulative number changes the seismic rate using the Long Term Average (LTA) function. The curve is plotted for each node. The curve starts from the time (the initial period) and ends at the time (the final period).

The standard deviation Z (z-value) is defined in equation 1.

$$z(t) = \frac{(R_{bg} - R_w)}{\sqrt{\frac{S_{bg}}{n_{bg}} + \frac{S_w}{n_w}}} \tag{4}$$

Where: R_{bg} is the mean rate in period 1 (from t_0 to t) and R_w is the mean rate in period 2 (from t to t_0), where t is the "current" time ($t_0 < t < t_a$). The term S_{bg} and S_w are the standard deviation in these periods, and n_{bg} and n_w are the number of samples.

The data used in this research was taken from BMKG and USGS catalog for West Sumatera and its surroundings. The study area is limited by coordinates 2°LU - 4°LS and 96°BT - 104°BT with depth of earthquake less than 350 Km. The observation time period was 1980 - 2015 with earthquake magnitude bigger than 4.0 SR.

The data processing is done with several stages, 1) classify the data from earthquake catalogs, 2) perform dekluster of earthquake, 3) plotting of cumulative number curve and earthquake frequency distribution in order to see the completeness of earthquake data, 4) calculates the-a, b-values, and z-values by using ZMAP version 6.0 software, and than, 5) the spatial distribution of seismotectonic parameters (values-a and b-values) and z-value was analized by using gridding technique with a 0.1° x 0.1° grid spacing and for every grid the earthquake maximum only 50.

3. Results and Discussion

Data of the earthquake that has been used in this study amounted to 3794 events after the process of decluster and can be seen on figure 2.

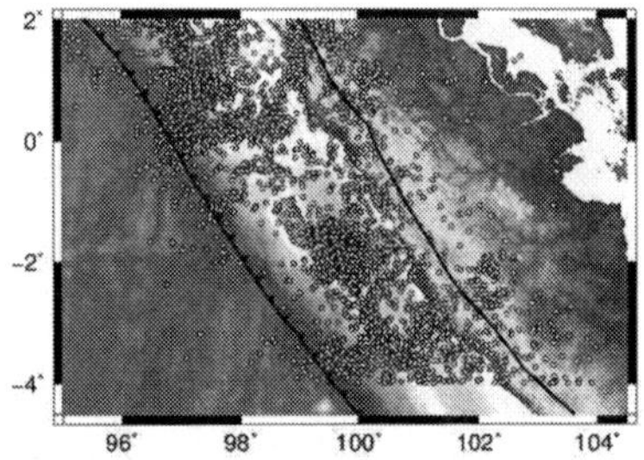

Figure 2. Seismicity Map of West Sumatra Region with the Period from 1980 - 2015. (the source of data from BMKG and USGS / NEIC)

High seismicity in West Sumatera region is shown in figure 2. The density of seismicity of West Sumatera is more visible around the island of Nias, to the south of Pagai Island, around the Mentawai Islands and along the active fault zone of Sumatra as well as the subduction pathway. It is suspected that this high seismic density is affected by tectonic activity in West Sumatera. The tectonic activity is a subduction zone located in the western part of Sumatra Island along the coastline, the presence of underwater active faults such as the Mentawai fault and the activity of the Sumatran Fault. Most of the earthquakes caused by the subduction zone and the active fault zone of Sumatra are potentially large and cause damage.

The number of earthquakes that occurred and the time of the incident is shown in figure 3. This curve shows the tendency of earthquake occurrence in West Sumatra region. The tendency of earthquake events to increase from year to year, as shown in figure 4, that between 1980 and 2005 there was little change in seismic activity, whereas after 2005 there was a significant change in seismic activity.

The frequency-magnitude distributions in West Sumatera region is shown in figure 4. On this curve obtained the general value b is about 0.928, while the value of an obtained is about 7.43. The value of Magnitude Completeness (Mc) in West Sumatra region according to the catalog used is 4.5. this indicates that the earthquake catalog used to record well the earthquake with the smallest magnitude of 4.5 in West Sumatra.

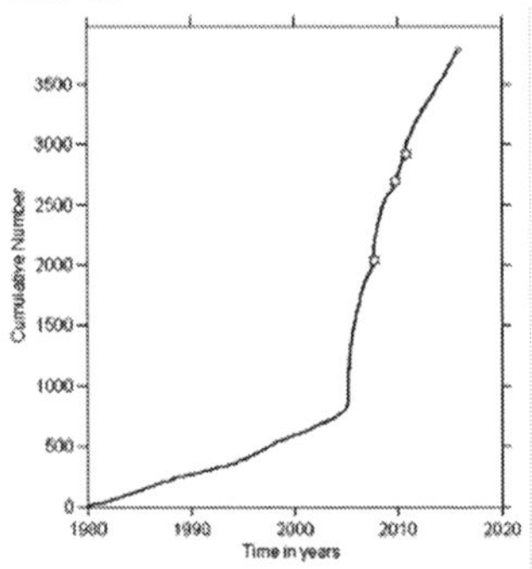

Figure 3. Cumulative Number Earthquake with Time Occured in Period from 1980 to 2015.

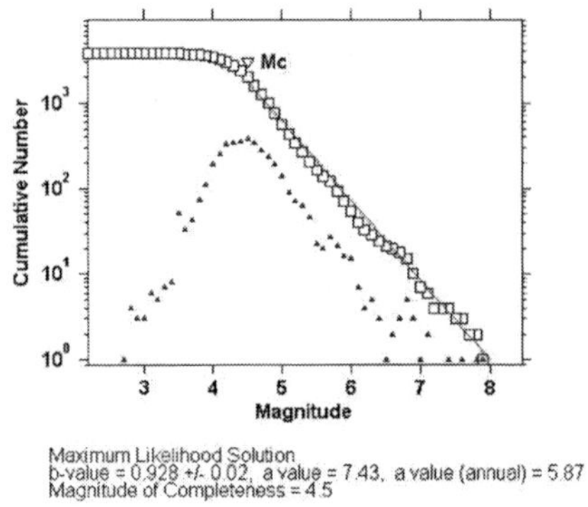

Figure 4. The Frequency - Magnitudo Distribution Distribution Curve in West Sumatera Region.

The results of calculating the values of a and b spatially can be seen in figures. 5a and 5b. Figure 5a shows the spatial distribution of b-values in West Sumatra. Based on the mapping there is a low-b anomaly value indicated by blue. This low b-value is near the Sumatra fault zone, subduction line, south of Nias island and in the Mentawai islands. These areas are dynamic seismic zones and potentially significant and destructive earthquakes. Areas that have a low-b value are also called asperities areas, an active seismic area that is still locked or called a locked zone. Locked zone may also cause a significant and destructive earthquake. Therefore, areas with low b values such as the Mentawai Islands, especially the southern island of Pagai and along the active faults of Sumatra should be suspected to experience significant and potentially damaging earthquakes in the future. While areas with a high b-value as seen in the north of Nias Island are referred to as creeping zones, are active fault zones that do not experience energy hoarding and do not experience slippage. This is in accordance with previous research that low b-values are associated with rock stress conditions in the area high, and vice versa [7].

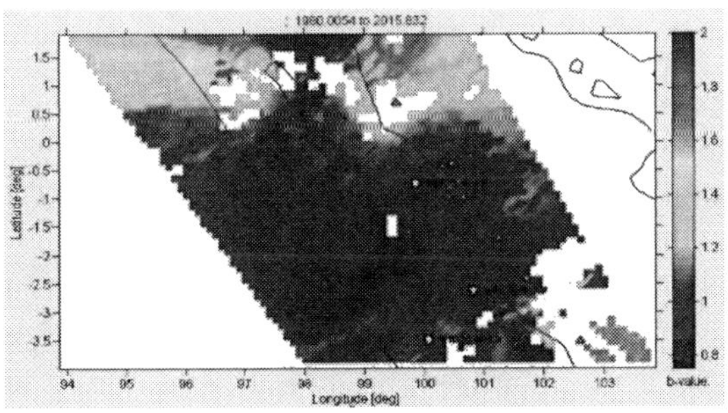

Figure 5a. Map of Distribution Spatial of b-value in West Sumatra Region

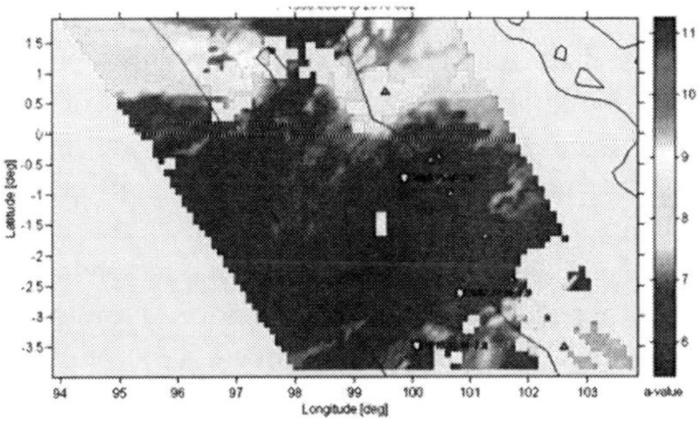

Figure 5b. Map of Distribution Spatial of a-value in West Sumatra Region

The spatial distribution of a-values is given by Fig. 5b which shows results almost equal to the spatial distribution of b-values. in Figure 6b it is seen that the low a-values, the blue color, are also seen in the area around the Sumatran fault zone, subduction path, south of Nias Island, and around the Mentawai Islands. Areas with low a-values are indicated to have relatively low seismicity. This low level of seismicity indicates the accumulation of energy (asperities) in the area. low seismicity levels associated with low spatial distribution of b-values. Areas with low seismic activity potentially experience significant earthquakes. In accordance with the previously mentioned that regions with low-b values also have great potential for significant earthquakes. While the high-a-values, as seen in the northern areas of Nias Island, indicate that the area has high seismicity, it is thought that the high seismicity in this area is due to frequent small earthquakes. Areas that often have small earthquakes are less likely to experience significant earthquakes in the future. This is inversely proportional to the rare areas of small earthquakes will potentially experience significant earthquakes later on. [13] This is in accordance with previous research results that low a-values indicate low seismic activity in the region. The lack of seismic activity means the accumulation of energy in this region.

The spatial distribution of z-values in West Sumatera at the beginning of 2015 is shown in figure 6. From this figure, found that there is a significant decrease in seismic activity or called seismic quiescence phenomenon. This phenomenon is seen around the Mentawai Islands. It is characterized by a positive z-value in the area indicated by a yellow to reddish color. According to experts in previous studies, a positive z-value indicates that the area has decreased seismicity, while negative z-values indicate an increase in seismicity.

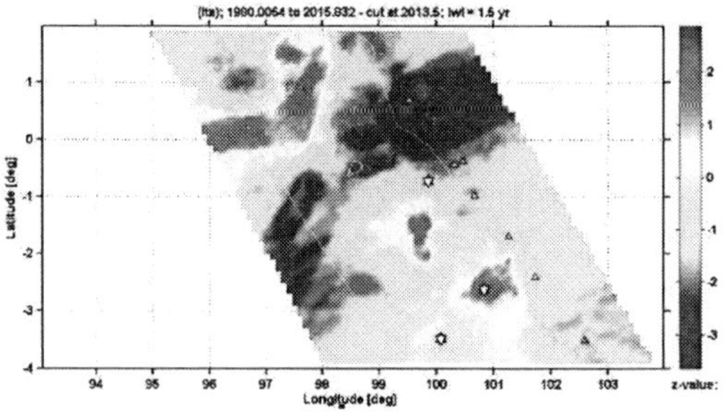

Figure 6. Map of Spatial Distribution of z-value in West Sumatera with Value of Tw was 1.5 year

The relatively low a-values and b-values in the area around the Sumatran fault zone, the subduction path, south of Nias Island, and around the Mentawai Islands indicate that these areas are likely to experience significant earthquakes in the future. This is due to the accumulation of energy in these areas. This accumulation of energy will lead to increased stress conditions in these areas. The decline in seismic activity and the accumulation of this energy will lead to the emergence of a phenomenon that usually always precedes a significant and destructive earthquake. The phenomenon in question is the seismic quiescence phenomenon that has been discussed in the previous paragraph, in the study of z-values.

Based on the observation and analysis that has been done that the phenomenon of seismic quiescence which become the early symptom of significant earthquake happened closely related to seismotectonic parameter (value-a and b-value).

4. Conclusion

Based on the analysis of the relationship between the two seismotectonic parameters and the seismic quiescence phenomenon in West Sumatera, it is concluded that the seismic quiescence phenomenon is related to seismotectonic parameters. This relationship can be seen from the result of the positive spatial distribution of z-values seen around the Mentawai Islands. Low z-values indicate that the area around the Mentawai Islands has decreased significantly in its seismic activity. This significant drop means that seismic activity in the area is reduced (a-low value). The rarity or loss of seismic activity means that in the area experiencing the accumulation of energy causing the level of local rock conditions in the area is also high (low-b value). These three things are closely related to each other and can be used as information to understand the potential of significant and damaging earthquakes in the future.

Acknowledgement

This work was supported by DIPA Universitas Negeri Padang year 2017 number. 1649/UN35.2/PG/2017. We also thank the operators of the BMKG and USGS-NIED for making their earthquake catalog.

References

[1] Natawidjaja D H and Triyoso W 2007 The Sumatran fault zone: from source to hazard *Journal of Earthquake and Tsunami* **1** 21-47

[2] Triyono, R 2015 *Ancaman Gempabumi di Sumatera Tidak Hanya Bersumber Dari Mentawai Megathrust.* Kepala Stasiun Geofisika Klas I Padang Panjang. *eoffice.bmkg.go.id/dokumen/artikel_Ancaman Gempabumi di Sumatera Tidak Hanya Bersumber dari Mentawai Megathrust*

[3] Putra R R and Kiyono J 2012 *Shaking Characteristic of Padang City, Indonesia.* Graduate School of Engineering, Kyoto University, Japan. Ono Y. Department of Urban Social System and Civil Engineering, Tottori University, Japan. 15 WCEE. LISBOA

[4] Wiemer S 2001 A software package to analyse seismicity: ZMAP *Seismological Research Letters* **72** 373–382

[5] Katsumata K 2011 A Long Term Seismic Quiescence Started 23 Years Before The 2011 of The Pacific Coast of Tohoku Earthquake (M=9.0) *Earth Planet Space* **63** 709-712

[6] Sunardi B, Sulastri and Nugraha J 2013 *Analisis Periodesitas dan Perubahan Tingkat Kegempaan Wilayah Jawa Barat Berbasis Statistik*. Conference Paper. BMKG Pusat 67 - 77

[7] Rohadi S, Grandis H and Ratag M A 2007 Variasi Spasial Seismisitas di Zona Subduksi Jawa *Jurnal Meteorologi dan Geofisika* **8** 42 – 47

[8] Scholz C H 1968 The frequency-magnitude relation of microfracturing in rock and its relation to earthquakes *Bull. Seis. Soc. Amer.* **58** 399- 415

[9] Wyss M, Wiemer S and Tsuboi S 1973 Mapping b-value anomalies in the subducting slabs beneath Japan, using the JUNEC data *Bull. Seism. Soc. Am. submitted*

[10] Gibowicz S J 1973 Variation of the frequency-magnitude relation during earthquake sequence in New Zealand, *Bull. Seimol. Soc. Am,* **63** 517 – 528

[11] Wiemer S and Wyss M 1994 Seismic quiescence before the 1993 M=7.5 Landers and M=6.5 Big Bear (California) earthquakes *Bull. Seis. Soc. Amer.* **84** 900-916

[12] Gutenberg B and Richter C F 1944 Frequency of earthquake in California *Bull. Seismol. Soc. Am.* **34** 185 – 188

[13] Ngadmanto D 2012 *Penentuan Potensi Gempabumi Merusak Berdasarkan Parameter Kegempaan di Wilayah Busur Banda. Jurnal Meteorologi dan Geofisika **8** 16 – 22*

IOP Conf. Series: Journal of Physics: Conf. Series **1040** (2018) 012051 doi:10.1088/1742-6596/1040/1/012051

Effect of sintering temperature on crystal structure and grain size of Manganese Ores from West Sumatera

A Fauzi, Ratnawulan and P J Putri

Department of Physics, Faculty of Mathematics and Natural Science, Universitas Negeri Padang,Indonesia

E-mail: afz_id@yahoo.com; ratnawulan@fmipa.unp.ac.id

Abstract. The effect of sintering temperature on crystal structure and grain size of manganese seed originating from West Sumatera has been investigated using X-ray Diffractometer (XRD). This investigation is important in relation to the emergence of new phases of economic value at a certain temperature. The sintering temperature variations used are 600^{0}C, 700^{0}C, 800^{0}C, 900^{0}C, and 1000^{0}C, respectively. The result of manganese ore investigation is known that sintering temperature variation, there is no change of structure from phases found in manganese ore which are equally tetragonal but there is change of grain size from the existing phases. For bixbyite and rhodonite phases, the grain size decreases with increasing sintering temperature. As for rhodonite and hausmannite phases, the grain size tends to increase with increasing sintering temperature.

1. Introduction

Manganese (Mn) is one of the most important raw materials for iron and steel production. Manganese is used in steel alloys to enhance beneficial characteristics such as strength, hardness and durability [1]. Manganese is an element that must always exist in steel in small quantities and as a deterrent to oxidation, thus every chemical process and metallurgical process can take place well. The addition of manganese elements in the alloy steel adds strength and heat resistance as well as a cleaner and shiny appearance.

Manganese ore contains high value oxide compounds such as pyrolusite, bixbyite, psilomelane, hausmannite, rhodokrosit and rhodonite [2] with varying concentrations in each region. The presence of manganese phase oxide content present in manganese ore causes manganese ore to have the potential to be developed as industrial materials. Currently, pyrolusite is the main ingredient for the dry battery industry. Bixbyite can also be used as a dry battery material that produces good performance and with long durability [3] [4]. In addition, hausmannite is widely used in industry as a catalyst, magnetism, electrochemistry, or for air decontamination [5].

Because of the magnitude of the benefits of manganese oxide in the form of pyrolusite, bixbyite, and hausmannite, then made various attempts to get it. One way is to sintering at a certain temperature so that there is a change of structure and phase tansformasi. The bixbyite and hausmannite phases can be obtained by heating from the pyrolusite phase. This phase change may occur due to the heating temperature. Pyrolusite phase can be bixbyite phase at sintering temperature above 400°C. the bixbyite phase can be a hausmannite phase at 400^{0}C sintering temperature. The difference in sintering temperature in the formation of pyrolusite to bixbyite, influenced by other content contained in pyrolusite. Just as it contains small amounts of Si, Al, Ca, P, K. Ba, Ag, and other elements. The same phases in each region have different compositions. The difference of oxide content content in manganese ore and phase formed due to geological order and phases process in every region.

From these reviews, it can be concluded that new phases of manganese oxide can be obtained by sintering at different temperatures and are influenced by the characteristics of the area in which the manganese ore is found. In this paper, we focus on the effect of the sintering temperature on crystal structure and crystallite size of the emerging phases. On the other hand, knowledge of crystal structure and crystallite size is helpful in understanding the surface and bulk characteristic of manganese oxide for the benefit of the application of manganese

2. Research Method

2.1. Preparation of manganese Powder

The sample of this research used manganese ore material obtained from West Sumatera, Indonesia. The cleansed manganese ore was crushed to obtain finer and homogeneous grains. The scour process is done for 3.5 hours. Furthermore, the crushed manganese ore sieved with an automatic sieve of 0.075 mm aims to have the same grain size. Samples were then disinterring by using furnaces at temperatures of 600^{0}C, 700^{0}C, 800^{0}C, 900^{0}C, and 1000^{0}C, with a duration of 3 hour detention.

2.2. Physical Measurements

Manganese ore samples after disinterring were then analyzed by crystal structure and grain size using XRD Cement (CuKα 1.5606 Å, 40 mA, 45 kV). Measurements were made from 10^0 to 90^0 with step size 0.010. The process of identifying phase and structure were obtained from search match using high score plus program. The grain size was calculated using the Scherrer formula, $D = 0.9\ \lambda/\ \beta \cos\theta$, where λ was the wavelength of X-ray radiation, β was the full width at half maximum (FWHM) of the peaks at the diffracting angle θ [6][7].

3. Results and discussion

3.1. Crystal Structure

Figure 1 shows the diffraction pattern of the sintered manganese sample at a temperature of 600^{0}C. The type of manganese oxide phase contained in the sintered sample at a temperature of 600° C is the bixbyite formed at the diffraction angle of 18.87^0, 38.28^0, and 65.95^0. Braunite is formed at the diffraction angle 23.25^0, 26.70^0, 33.01^0, 45.23^0, 49.56^0, 55.34^0, 85.16^0 and 91.14^0. In addition, the non-manganese oxide phase contained is the coesite formed at the diffraction angle 96.32^0.

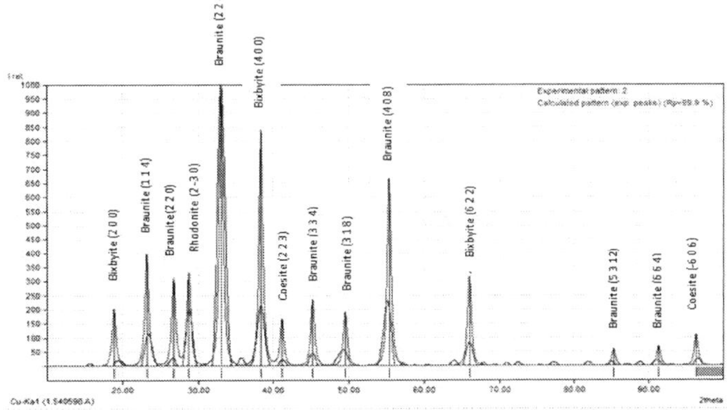

Figure 1. Spectra XRD of manganese powder with sintering at 600^{0}C

Figure 2 shows the diffraction pattern of the sintered manganese sample at a temperature of 700^{0}C. The diffraction results show in biical samples of bixbyite, braunite, hausmannite, rhodonite, braunite. The most dominant phase is braunite. The type of manganese oxide phase contained when sintered at a temperature of 700°C is Bixbyite which is seen at angles of 18.89^0, 38.28^0, 64.06^0, and 94.21^0. Braunite is visible at angles, 33.00^0, 45.29^0, 49.56^0, and 55.32^0. At diffraction angle 26.71^0, 28.56^0, mise oxide phase of rhodonite is formed

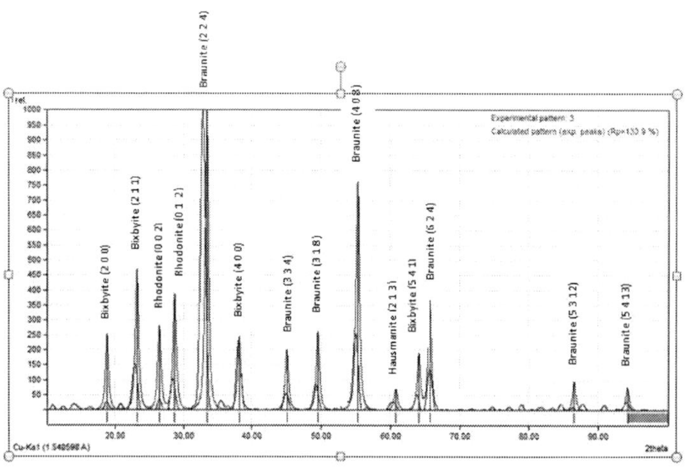

Figure 2. Spectra XRD of Manganese Powder with sintering at 700^{0}C

Figure 3 shows the diffraction pattern of the sintered manganese sample at a temperature of 800^{0}C. Results The diffraction pattern showed in samples containing braunite and bixbyite phases, hausmannite. The most dominant phases are braunite and bixbyite. The type of manganese oxide phase contained in the sample disinterring at a temperature of 800^{0}C is bixbyite formed at an angle 18.89^0, 38.28^0, 45.18^0 and 55.21^0. *Rhodonite* formed at an angle 26.70^0 and 28.66^0. *Braunite* formed at an angle 23.21^0, 49.46^0, 65.84^0, 69.48^0, 91.42^0, dan 94.05^0.

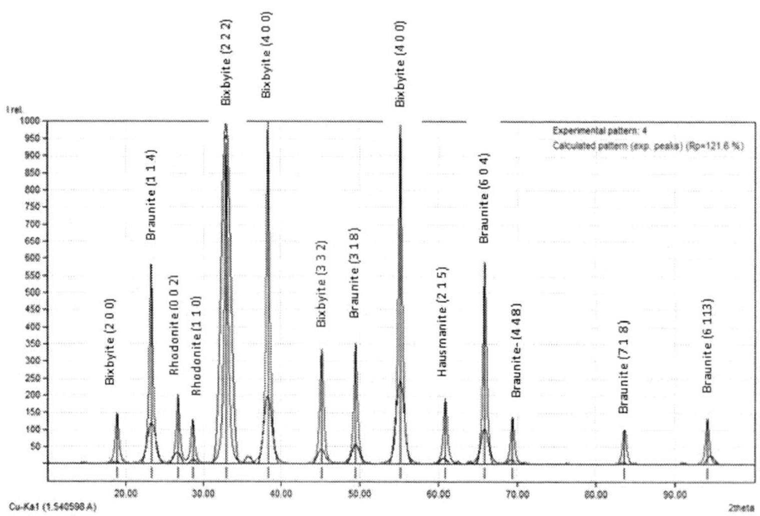

Figure 3. Spectra XRD of Manganese Powder with sintering at 800^{0}C

Figure 4 shows the diffraction pattern of the sintered manganese sample at a temperature of 900^{0}C. The type of manganese oxide phase present in the sample disinterring at a temperature of 900 ° C is a bixbyite formed at an angle 49.43^0, 55.25^0, *braunite* formed at an angle 23.18^0, 26.68^0, 33.00^0, 69.39^0, 74.12^0, 80.23^0, and 91.22^0. *Hausmannite* formed at an angle 18.07^0, 32.39^0, 36.17^0, 38.27^0, 45.41^0 and 58.7447^0.

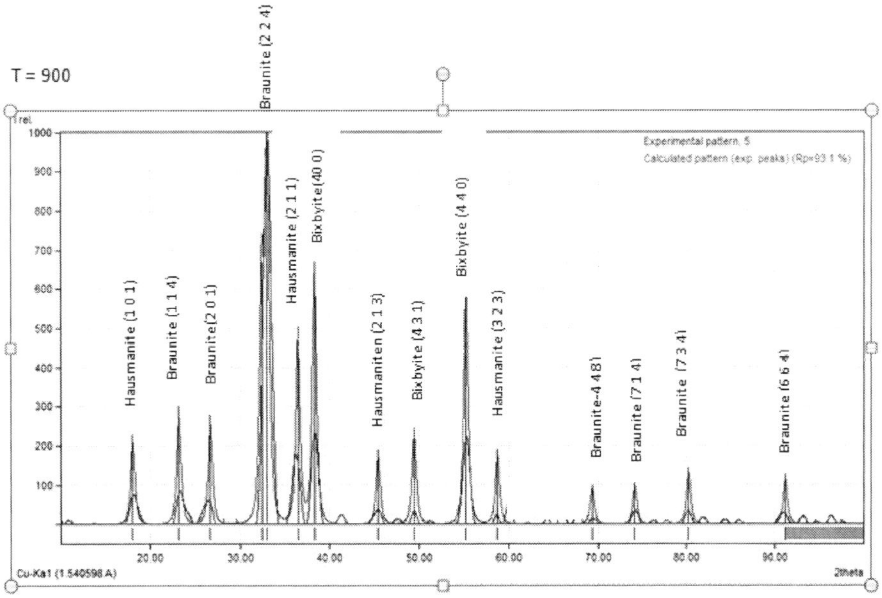

Figure 4. Spectra XRD of manganese powder with sintering at 900^{0}C

The type of manganese oxide phase present in the sample disinterring at 1000° C is shown in Figure 5. The phase type is a braunite formed at an angle 25.56^0, 26.64^0 and 36.12^0. *Hausmannite* formed at an angle 18.04^0, 28.94^0, 31.02^0, 32.39^0, 38.04^0, 44.48^0, 50.85^0, 55.31^0, 58.54^0, 59.91^0, 60.7981^0, 64.6520^0, 74.22^0, 80.03^0, and 89.59^0.

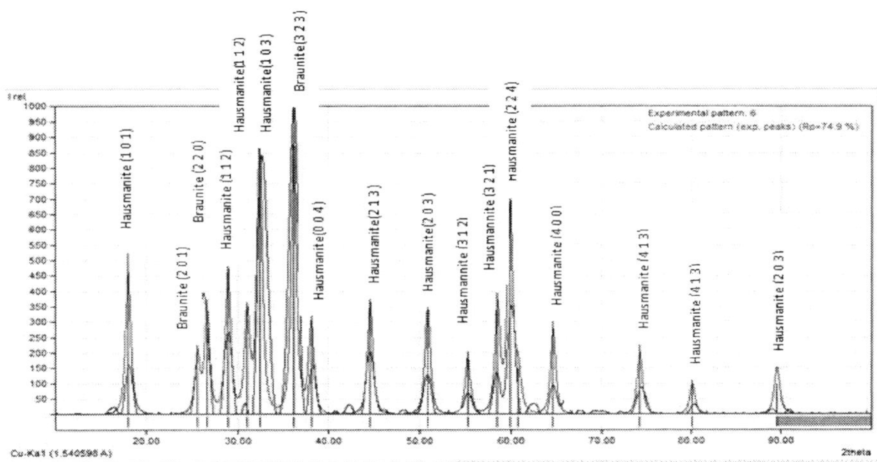

Figure 5. Spectra XRD of manganese powder with sintering at 1000^{0}C

Based on XRD data analysis for each sample it can be seen that sintering temperature variations influence the formation of manganese oxide phase. At a temperature of 1000^{0}C the other phases begin to disappear and the most dominant stay is hausmannite. The phase change starts at the sintering temperature of 600^{0}C, where the pyrolusite phase undergoes phase transformation to bixbyite. At 600^{0}C sintering temperatures all pyrolusite phases undergo phase transformation to bixbyite as a whole. This is evident from the absence of pyrolusite peaks on XRD patterns. At the sintering temperature of 700^{0}C-1000^{0}C there is also a phase transformation of bixbyite that joins the rhodonite phase to form braunite phase. At a sintering temperature of 1000^{0}C all bixbyite phases undergo transformation as a whole to be hausmannite, as can be seen from the increasing number of hausmannite peaks. The manganese oxide phase formed by the sintering temperature is shown in Table 1.

Table 1. The manganese oxide phase formed by the sintering temperature

Temperature	Phase					
	Pyrolusite	**Hausmannite**	**Rhodonite**	**Bixbyite**	**Braunite**	**Coesite**
26^{0}C	√					√
600^{0}C			√	√	√	√
700^{0}C		√	√	√	√	
800^{0}C		√	√	√	√	
900^{0}C		√		√	√	
1000^{0}C		√			√	

The occurrence of phase changes due to temperature changes is caused by the growth and alteration of one phase in manganese ore. Manganese oxide minerals are minerals formed from a combination of manganese elements with anion oxide (O) groups. Mineral manganese oxide is formed as a result of compounds between oxygen and manganese elements. The formation of these new phases is due to the pyrolusite oxidation process to bixbyite and hausmannite. While the braunite phase appears due to the bond between bixbyite and rhodonite [8]. The pyrolusite transformation reaction to bixbyite, begins in pyrolusite phase built on Mn3+ Ion arranged with tetragonal pattern and Mn2+ ions are arranged cubic. As a result of thermal energies causing atoms, in manganese ore especially pyrolusite vibrates. At that time there is a transfer of cations to the surface of the crystal and at the same time there is a cation vacuum. On the surface of the crystals the Mn2+ ion oxidizes into Mn3+ ions. The Mn3+ ion interacts with the binding of oxygen to form the edges of the bixbyite. Such a condition causes the presence of Mn3+ ions with tetragonal geometry and there is also a Mn3 + ion with cubic geometry paired with oxygen this is a feature of bixbyite. Based on the ion dissolution or dissolution reaction Mn2 + occurs due to the free electrons moving. The dissolution of the iom or the dissolution can be processed on Mn atoms present on the surface either in the oxidation or ion-exchange areas. When pyrolusite phase transformation occurs to bixbyite and then to hausmannite, there is also a change of structure of the manganese ore. The structural changes that occur are tetragonal structured pyrolusite phase with a 4.4000 Å and c = 2.8700Å axis cellular units to a cubic structured bixbyite phase with a cell unit a = b = c = 9.4091 Å and then to tetragonala = b = c = 5.7630 Å and tetragonala phase structures c = 9.4560Å. This structure change is caused by the continuous oxidation process in the sample. The oxidation process causes the emergence of Mn3 + ions, each pair of Mn3+ ions coupled with a cation vacuum. This thus affects the coordination of the atoms in the cell unit. The Mn2+ ion with tetragonal geometry and also the Mn3+ ion with cubic geometry paired with oxygen is characteristic of bixbyite. As the temperature is increased, the unit cell contains most of the Mn3+ ions this is a feature of hausmannite.

3.2. Grain size

The grain size of the sintered manganese samples at various temperature variations has been investigated simultaneously. Figure 6 shows that the grain size of the manganese oxide crystal changes with increasing temperature. The bixbyite and rhodonite phase increased grain size up to 600°C sintering temperature, then dropped and at 700°C and rose again at 800°C, then decreased and lost at 900°C. However, the grain size of the rhodonite phase drops at 800 °C and disappears at 900°C. The hausmannite phase has an increase in grain size from a temperature of 700° C to a sintering temperature of 1000°C.

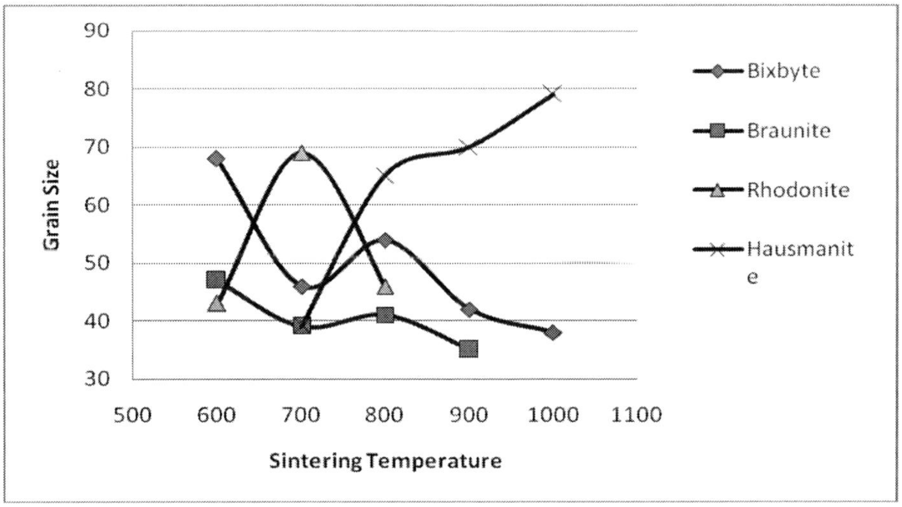

Figure 6. Grain size of manganese samples due to variation of sintering temperature 600°C, 700°C, 800°C, 900°C, dan 1000°C.

The size of coesite phase crystalline grains has increased. This increase occurs because at 600 ° C there is no phase transformation, so that the thermal energy provided by the manganese ore compounds to join each other and form a new bond. During the heating occurs there will be union between the particles. The increase in grain size is due to the transformation of the bixbyite phase to the hausmannite, so that the given thermal energy is used by bixbyite to join the other. The incorporation mechanism is preceded by pore-separated nuclei, due to continuous temperature, the pores.

4. Conclusion

Manganese powders were synthesized by dried method using natural manganese ores phase as the starting material. The crystal structure and grain size of manganese are depending on the sintering temperature. Giving variation of sintering temperature on manganese which influences change of structure of manganese oxide formed. The structural change that occurs is a tetragonal tetra-structured phase pyrolusite with a 4.4000 Å and c = 2.8700Å axis cell units to a cubic-structured bixbyite phase with a cell unit a = b = c = 9.4091 Å and then to a tetragonal a tethered phase a phenomenon = b = 5.7630 Å and c = 9.4560Å. In addition, the sintering temperature affects the size of the manganese oxide grains formed. For the hausmanite phase, the grain size increases with increasing sintering temperature

Acknowlegment
The authors thank to RISTEK DIKTI for financial support through Hibah MP3EI 2016-2017 for this work.

References

[1] Mohanty J K, Rao D S, Paul A K and Khaoash S 2009 *Journal of Geology and Mining Research* **1**(7) 149
[2] Talat J and Siddiqui F N 2014 *International Journal of Geosciences* **5** 710
[3] Wei J, Nagarajan N and Zhitomirsky I 2007 *J. Material Proc. Tech.* **186** 356
[4] Patrice R, Dupont L, Aldon L, Jumas J C, Wang E and Tarascon J M 2004 *J. Chem. Mater.* **16** 2772
[5] Zhanqiang L, Wen L W, Xianming L, Minchang W and Li Z Z 2004 *J. Solid State Chem.* **177** 1585
[6] Riazian M and Bahari A 2012 *Int. J.NanoDimens.* **3(2)** 127
[7] Ratnawulan, Fauzi A and Sukma H A E 2017 *Proc. The 4th Int.Conf. on Research, Implementation, and Education for Developing Scientific Attitude in Sciences and Mathematics* (Yogjakarta: AIP Conference Proceedings) **186** 060009
[8] Feng Q, Kanoh H and Ooi K 1999 *J. Mater. Chem.* 9 319

IOP Conf. Series: Journal of Physics: Conf. Series **1040** (2018) 012052 doi:10.1088/1742-6596/1040/1/012052

Development of Low Sidelobe Level Array Antenna for Synthetic Aperture Radar Sensor

Yohandri[1], Yulkifli[1] and J T S Sumantyo[2]

[1]Physics Department, Faculty of Mathematics and Natural Science, Universitas Negeri Padang, Jl. Prof. Hamka, Padang, Sumatera Barat, Indonesia 25131
[2]Microwave Remote Sensing Laboratory, Center for Environmental Remote Sensing, Chiba University, 1-33, Yayoi Cho, Inage Ku, Chiba Shi, Chiba, Japan 263-8522

E-mail: yohandri@fmipa.unp.ac.id

Abstract. The radiation pattern of the array antenna is an important characteristic in the synthetic aperture radar (SAR) sensor to maintain the quality of the images. In this work, an array antenna consists of five elements square microstrip antenna (SMA) having a low sidelobe level for SAR sensor is presented. The proposed antenna is designed to propagate in circular polarization with a corner-truncated technique. The low sidelobe level is obtained by implementing the Chebyshev polynomial to distribute the power to each element of the patch. For five elements square array antenna, the simulated axial ratio bandwidth (<3 dB) of about 11.2 MHz (0.88 %), which is consistent with the measured fabricated model of 10.5 MHz (0.83 %). The maximum sidelobe levels of the proposed antenna for measurement and simulation are about 20.0 dB and 21.5 dB, respectively. The both measured and simulated results are in line with the theoretical design made by using Chebyshev polynomial of 20 dB. The low side lobe level and reasonable axial ratio bandwidth indicate the antenna is satisfying the specifications of SAR sensor.

1. Introduction

A Synthetic Aperture Radar (SAR) sensor has played an important role in current remote-sensing technology. Various applications such as land mapping, land use, oceanography, and disaster monitoring are the implementing of this technology. Even further, the SAR sensor can be operated to measure the changing in the earth surface to millimeter accuracy [1]. As compare with the detection radar, the SAR sensor has a complex system due to the image generating as a result. In image construction, the SAR system receives the back scattering signal from the object or target area. The radiation pattern of the array antenna is an important characteristic in the synthetic aperture radar (SAR) sensor to maintain the quality to the images.

The low sidelobe level of the antenna is one of the critical parameters in radiation pattern to satisfy the SAR sensor requirements [2]. The high sidelobe level is potential to produce the error recorded data and resolution degradation of the SAR images. In mountainous areas or city with high buildings, the backscattering signal from sidelobe can be captured by the sensor causing errors data in recording the signal from the main lobe as illustrated in Figure 1. As can be seen in Figure 1, the backscattering signal from the sidelobe can be recorded by SAR sensor and disrupt the target signal from the mainlobe.

IOP Conf. Series: Journal of Physics: Conf. Series **1040** (2018) 012052 doi:10.1088/1742-6596/1040/1/012052

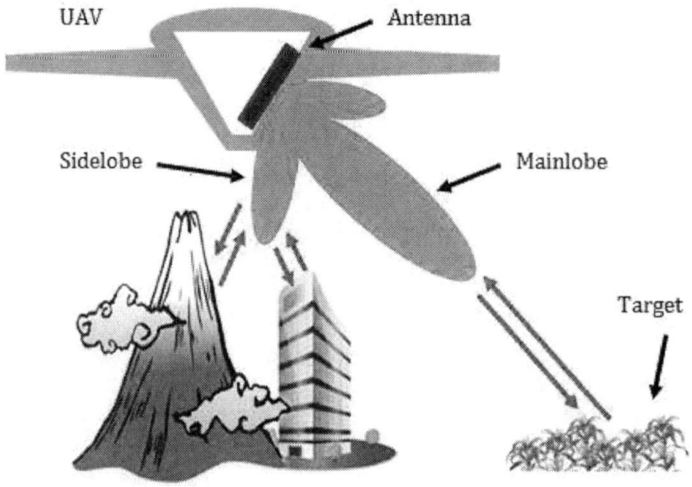

Figure 1. Illustration of the sidelobe effect on the data recording.

In previous work [3]–[6] several arrays antenna has been developed for the SAR sensor. The performance of the antenna, however, was not satisfactory in terms of the sidelobe level. In general, the high sidelobe is found in radiation pattern of the antenna with the sidelobe level of 12 dB. The sidelobe effect can be overcome by managing the difference between sidelobe and main lobe higher. Therefore, the compression of the sidelobe is obligatory for be done to avoid an error in recording data of the SAR sensor.

The purpose of this work is to present the development of low sidelobe level array antenna for SAR sensor. The low sidelobe level is obtained by arranging the power distribution of each element of the radiator. A Chebychev polynomial is adopted to distribute the power in the feed network of the proposed antenna. The coefficients of excitation are implemented in managing the weight of the power divider in the fed network in reduction the sidelobe level. The five elements Square Microstrip Antennas (SMA) are arranged in 1x5 array configuration and designed for of 20 dB sidelobe level. A corner truncated and proximity fed method are employed in generating the circular polarization. The circular polarization is beneficial since the future SAR requires a sensor that insensitive to Faraday rotation effect and produces more information from the target. The proposed antenna is operated in L-Band (1.27 GHz) and planned to be installed onboard an Unmanned Aerial Vehicle (UAV). In next section, the description of the SAR sensor will be presented. The design of proposed antenna and research result will be discussed in section three and four, respectively. Finally, the conclusion from this work will be presented in the last section.

2. SAR Sensor

In general, the SAR sensor consists of several parts, which is a transmitter (Tx), a receiver (Rx) and antennas as shown in Figure 1 [7]. The transmitter sub-system consists of a I-Q generator, an up-converter, a power amplifier (PA), switch, band-pass filters (BPF), High-power amplifier (HPA) and an isolator. On the other hand, the major component of the receiver subsystem consist of two band pass filters (BPF), Low-Noise Amplifier (LNA), I/Q demodulators and two Low-pass filter. The local oscillator is operated in both transmitter and receiver sub system.

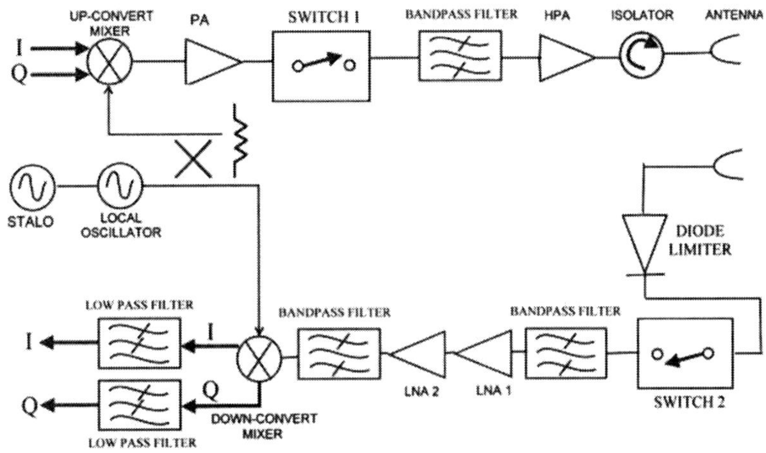

Figure 2. Block diagram of SAR sensor.

Based on the Figure 2, the antenna is operated as transmitter and receiver the electromagnetic wave. In the SAR system, the characteristics of the antenna are similar for both transmitter and receiver. In terms of power, the transmitter antenna is working with high power, while the receiver antenna with low power since receives the reflected signal from the target.

3. Antenna Design

3.1. Power Distribution

The power distribution of each element of the array antenna is calculated based on uniform space ($d = \lambda_0/2$) and sidelobe level of 20 dB. The excitation coefficient is used to arrange the power ratio in the feed network. The Array Factor (*AF*) for five elements array ($P = 5, M = 2$) can be derived as [8].

$$AF(u) = a_1 + a_2 \cos 2u + a_3 \cos 4u \quad (1)$$

In here, $u = 2\pi(d/\lambda)\cos\theta$, for uniform spaced ($d = \lambda/2$) then u can be simplified as $u = \pi \cos\theta$. The excitation coefficient of the array factor for five elements antenna can be obtained by substituting the Chebyshev polynomial and *AF* can be expressed as

$$AF(u) = 1 + 1.5022 \cos 2u + 0.4317 \cos 4u \quad (2)$$

3.2. Geometry Design

Five elements square microstrip antenna with corner truncated and proximity feed are employed in the proposed antenna. The power ratio in a corporate feed network is configured based on the array factor calculation. Figure 3 is shown the illustration of the power distribution to each element.

Five elements square microstrip antenna with corner truncated and proximity feed are employed in the proposed antenna. The power ratio in a corporate feed network is configured based on the array factor calculation. Figure 3 is shown the illustration of the power distribution to each element.

Figure 3. Power distribution for each patch.

The T junction power divider is implemented at the feed network as presented in Figure 4. In Figure 4, the input power P_1 is delivered to the intersection on a microstrip of width W_1 and impedance Z_1. The line then branches into two arms with power, width and impedance given by P_2, W_2, Z_2 and P_3, W_3, Z_3, respectively. The power on each arm can be obtained using the equations

$$P_2 = \frac{Z_1}{Z_2} P_1 \text{ and } P_3 = \frac{Z_1}{Z_3} P_1$$

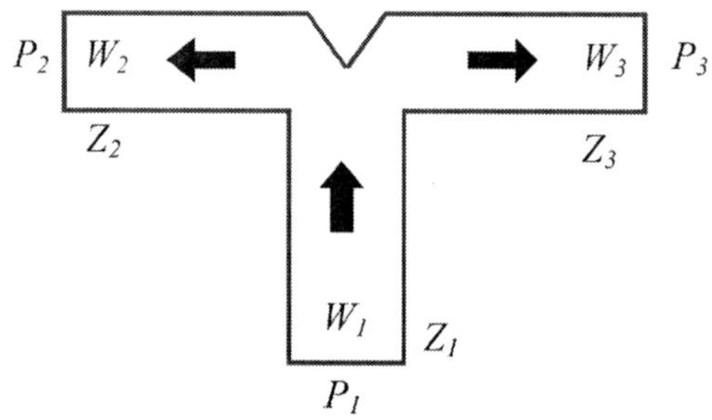

Figure 4. T Junction power divider.

The power distribution based on the excitation coefficient of the array factor can be realized in a corporate feed network as shown in Figure 5. In this Figure, the A, B and C T Junction are operated for dividing the source power as illustrated in Figure 4. The power dividing ratio each T junction is listed in Table 1.

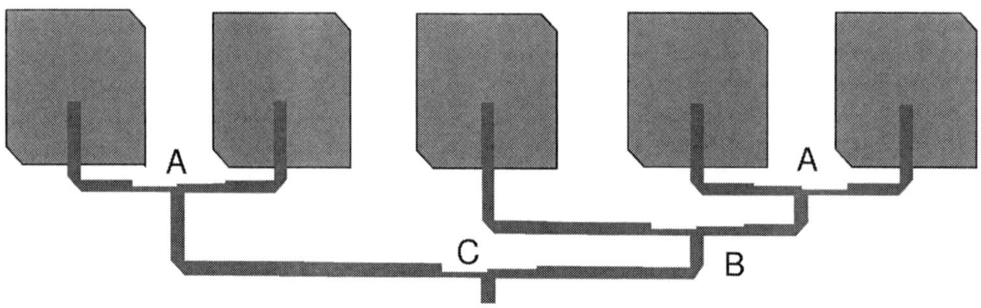

Figure 5. Geometry design of the Chebyshev array antenna.

Table 1. The power dividing ratio of the T Junction.

T Junction	P_2	P_3	Ratio (P_2/P_3)
C	1.5022+0.4317	1+1.5022+0.4317	0.66
B	1	1.5022+0.4317	0.52
A(right)	1.5022	0.4317	3.48
A(left)(mirror A right)	0.4317	1.5022	0.29

In this work, a Nippon Pillar substrate (NPC-H220A) is employed having a thickness 1.6 mm, a dielectric constant $\varepsilon_r = 2.17$ and a loss tangent $\delta = 0.0005$. The photograph of fabricated feed network and the patches are shown in Figure 6 and 7, respectively.

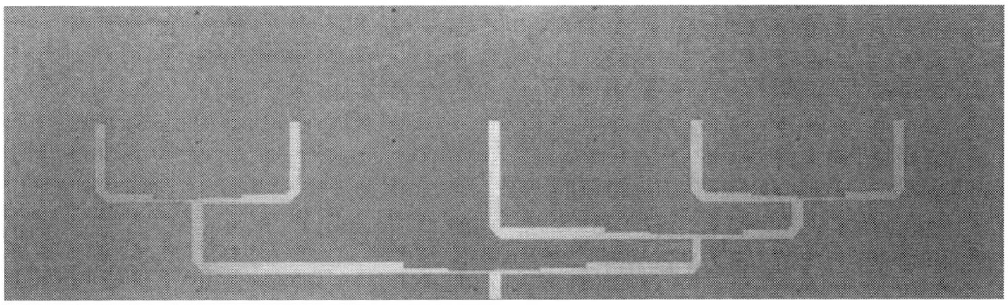

Figure 6. Power distribution on fabricated feed network.

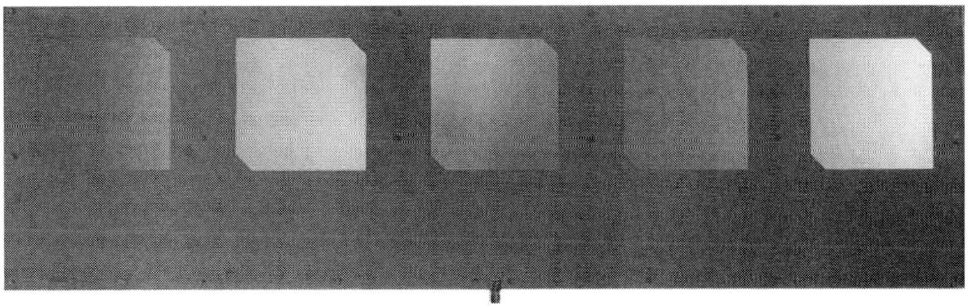

Figure 7. Photograph of fabricated 1x5 corner-truncated SMA.

4. Result and Discussion

Figures 8 to 11 shows the characteristics of the proposed antenna. In Figure 8, simulated and measured the reflection coefficient (S11) as a function of frequency is plotted. The return losses at 1.27 GHz are lower than 29.21 dB and 16.29 dB in the simulation and the measurement, respectively. Such degradation is probably caused by the transitions mismatch. The measured S11 characteristics show slightly broadened characteristics from 1.242 GHz to 1.312 GHz of 5.51%. Meanwhile, the simulated S11 shows slightly narrower than measured result ranging from 1.244 GHz to 1.300 GHz.

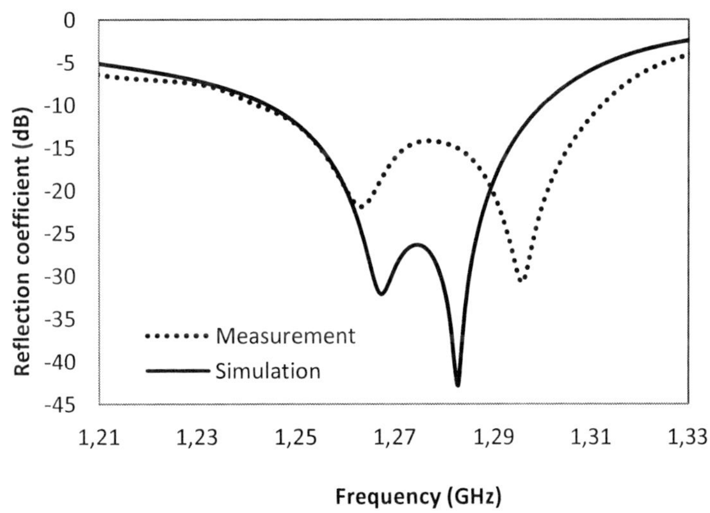

Figure 8. Reflection coefficient plotted as a function of frequency.

The polarization characteristic of the antenna is shown through the axial ratio (AR) as presented in Figure 9. The AR characteristics are slightly shifted to the higher frequency end, but the bandwidth remains almost the same in both simulation and measurement. A simulated axial ratio bandwidth of 0.9% is obtained from 1.264 GHz to 1.275 GHz where the measured results show a bandwidth of 0.83% from 1.268 GHz to 1.278 GHz.

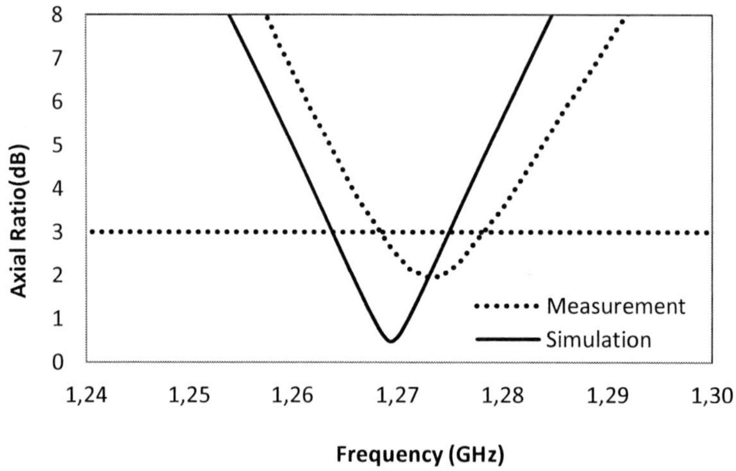

Figure 9. Axial ratio (*AR*) plotted as a function of frequency.

The performances of the antenna in terms of gain and sidelobe level are presented in Figure 10 and Figure 11. A maximum gain of 11.9 dBic in simulation and 10.4 dBic in measurement is obtained. Figure 11 shows the sidelobe level in radiation pattern of the antenna at the theta plane. The good agreements between calculation, simulation and measurement result are presented. The 3 dB beamwidth for calculation, measurement and simulation are 24, 23 and 22.25 degrees, respectively. The first side lobe appears at -36° and 32.4° with sidelobe level 20.8 dB and 19.5 dB for the simulation, for the measured are -58.7° and 36° with side lobe level 20.5 dB and 15.5 dB, respectively.

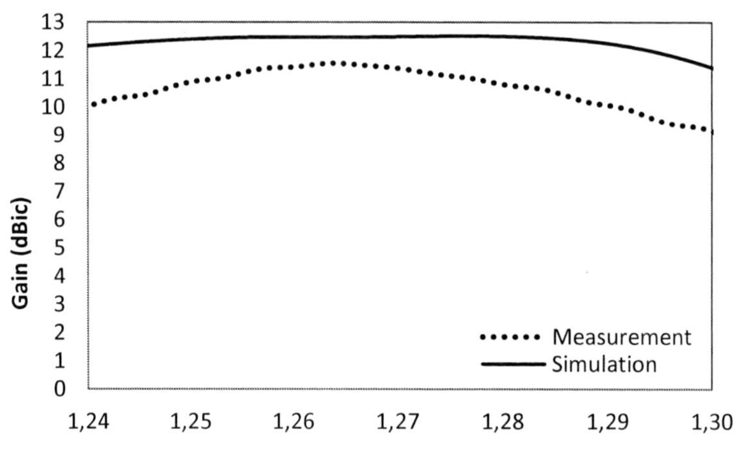

Figure 10. Relationship between antenna gain and frequency at θ angle of 0°.

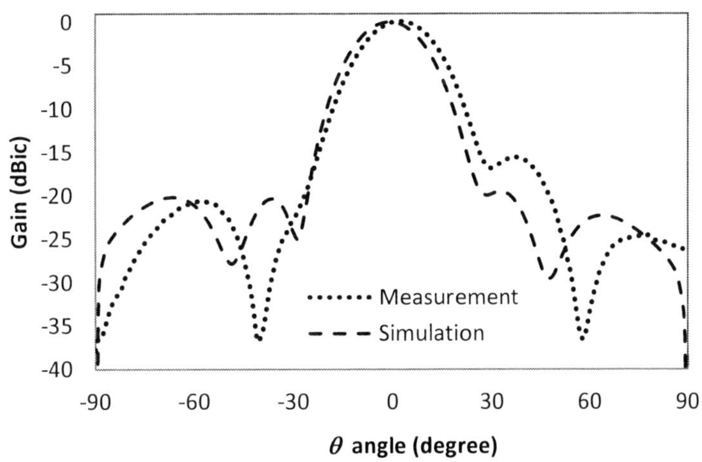

Figure 11. Normalized radiation pattern of the array antenna in the theta plane ($x-z$ plane) at f = 1.272 GHz.

5. Conclusion

In conclusion, the low sidelobe level array antenna for synthetic aperture radar sensor is designed and comprehensively characterized by simulation and experiment in 1.27 GHz centre frequency (L-band). Simulation results exhibit that proposed antenna can operate at low sidelobe level, which is in good agreement with the experiments. Moreover, the fabricated antenna has shown good performance in terms of return losses and axial ratio. This low sidelobe level and good properties of the array antenna is satisfying the requirements for a SAR sensor.

References

[1] Sri Sumantyo J T, Shimada M, Mathieu P P and Abidin H Z 2012 *IEEE Trans. Geosci. Remote Sens.* **50** (1) 259–270
[2] Baharuddin M Z, Izumi Y, Sumantyo J T S and Yohandri 2016 *IEICE Trans. Electron.* **E99.C** (10) 1174–1181
[3] Yohandri V, Wissan I, Firmansyah P, Rizki Akbar J T, Sri Sumantyo and Kuze H 2011 *Progress In Electromagnetics Research C* **19** 119-133 doi:10.2528/PIERC10121708
[4] Hussein M, Yohandri, Sri Sumantyo J T and Yahia A 2013 *Journal of Electromagnetic Waves and Applications* **27** (15) 1931-1941 https://doi.org/10.1080/09205071.2013.828577
[5] Yohandri, Asrizal and Sumantyo J T S 2016 Design of tilted beam circularly polarized antenna for CP-SAR sensor onboard UAV, in 2016 *International Symposium on Antennas and Propagation (ISAP)* pp. 658–659
[6] Yohandri, Sri Sumantyo J T and Kuze H 2011 Circularly polarized array antennas for synthetic aperture radar in Progress in Electromagnetics Research Symposium pp 1244–1247
[7] Koo V C, et al. 2012 *Progress In Electromagnetics Research* **122** 245-268
[8] Balanis C E 2005 *Antenna Theory: Analysis and Design, 3rd Edition* Constantine A. Balanis p. 1136